Agricultural Environmental Pollution, Risk Assessment, and Control

Agricultural Environmental Pollution, Risk Assessment, and Control

Guest Editors

Vinod Kumar
Pankaj Kumar

MDPI

Basel • Beijing • Wuhan • Barcelona • Belgrade • Novi Sad • Cluj • Manchester

Guest Editors

Vinod Kumar
Department of Zoology and Environmental Science
Gurukula Kangri (Deemed to Be University)
Haridwar
India

Pankaj Kumar
School of Environmental Studies
Maa Shakumbhari University
Saharanpur
India

Editorial Office

MDPI AG
Grosspeteranlage 5
4052 Basel, Switzerland

This is a reprint of the Special Issue, published open access by the journal *Agriculture* (ISSN 2077-0472), freely accessible at: https://www.mdpi.com/journal/agriculture/special_issues/0CXW868P40.

For citation purposes, cite each article independently as indicated on the article page online and as indicated below:

Lastname, A.A.; Lastname, B.B. Article Title. *Journal Name* **Year**, *Volume Number*, Page Range.

ISBN 978-3-7258-5485-1 (Hbk)
ISBN 978-3-7258-5486-8 (PDF)
https://doi.org/10.3390/books978-3-7258-5486-8

Cover image courtesy of Pankaj Kumar

Contents

About the Editors

Vinod Kumar

Vinod Kumar (M.Sc., Ph.D., F.A.N.S.F., F.A.E.S.A.) is a Senior Assistant Professor of Environmental Science in the Department of Zoology and Environmental Science at Gurukula Kangri (deemed-to-be-university), Haridwar, Uttarakhand, India. He has more than fourteen years of academic experience and over eighteen years of active research engagement. His primary research interests include agroecology, environmental pollution, bioremediation, and wastewater management. Dr. Kumar has published more than 148 research papers in reputed national and international journals and has played a leading role in advancing scientific knowledge in agriculture and environmental sciences. He is the Founder President of the Agriculture and Environmental Science Academy and serves as Editor-in-Chief of the Archives of Agriculture and Environmental Science. In addition, he contributes as an editorial board member and reviewer for more than fifteen well-recognized international journals. His research contributions are widely acknowledged, with 4,430 Google Scholar citations and an h-index of 38, alongside 1,956 Scopus citations with an h-index of 28. He has successfully supervised twenty-seven M.Sc. dissertations and six Ph.D. theses, with four doctoral students currently working under his guidance. Through his extensive teaching, research and editorial work, Dr. Kumar has made significant contributions to the advancement of environmental science, particularly in the areas of sustainable agriculture and pollution management.

Pankaj Kumar

Pankaj Kumar (M.Sc., Ph.D.) is an Assistant Professor in the School of Environmental Studies at Maa Shakumbhari University, Saharanpur, India, and affiliated with the Research and Development Division, Society for Agro-Environmental Sustainability, Dehradun. He has approximately eight years of research experience and three years of teaching experience. His research interests are biotechnology, bioremediation, wastewater management, bioenergy production, environmental modeling and related fields. Dr. Kumar has published over 110 research articles in high-impact journals and contributed to four books and twenty book chapters. His work has been cited approximately 3000 times on Google Scholar and 1800 times on Scopus with an h-index of 27. He has supervised five master's students and is actively involved in several national and international research projects as a collaborator. Through his publications, mentorship and active engagement in environmental research, Dr. Kumar continues to advance understanding and practice in sustainable environmental remediation and agricultural systems.

Preface

This reprint brings together selected contributions that examine the pressing issue of agricultural and environmental pollution in a rapidly industrializing and urbanizing world. The subject of this collection focuses on the interface of soil science, crop production, environmental health and sustainable resource management. Its scope extends from fundamental studies on pollutant dynamics to applied approaches for mitigation and remediation in agricultural systems.

The aim of this reprint is to provide an integrated scientific perspective on emerging pollutants, including pesticide residues, heavy metals, polyaromatic hydrocarbons, persistent organic compounds, microplastics, radionuclides and biological contaminants, that pose risks to soil fertility, crop quality and human health. Its purpose is not only to deepen scientific understanding but also to inform policymakers, practitioners and stakeholders about sustainable strategies for pollution control and agricultural resilience.

The motivation for compiling this reprint arises from the urgent need to address the environmental and socio-economic consequences of polluted agricultural soils and food chains. As editors, we recognize the importance of encouraging discussion between research disciplines and promoting practical solutions such as bioremediation, integrated organic farming and biological alternatives to chemical fertilizers.

This reprint is intended for researchers, agricultural professionals, environmental managers, graduate students and policy planners who seek to advance knowledge and practice in sustainable agriculture and pollution management. By presenting recent advances and diverse perspectives, we hope this collection will contribute to the development of safer and more resilient agricultural systems for the future.

Vinod Kumar and Pankaj Kumar
Guest Editors

agriculture

Editorial

Preface to the Special Issue "Agricultural Environmental Pollution, Risk Assessment, and Control"

Pankaj Kumar [1,2] and Vinod Kumar [1,*]

1 Agro-Ecology and Pollution Research Laboratory, Department of Zoology and Environmental Science, Gurukula Kangri (Deemed to Be University), Haridwar 249404, India; rs.pankajkumar@gkv.ac.in
2 Research and Development Division, Society for AgroEnvironmental Sustainability, Dehradun 248007, India
* Correspondence: drvksorwal@gkv.ac.in

Recent developments in agriculture have played a key role in meeting the food needs of 8.1 billion people worldwide. However, the sector is also responsible for 14% of total greenhouse gas emissions and 70% of irrigation water use. This makes agriculture one of the largest contributors to environmental pollution and global climate change. However, the excessive release of pollutants from urban, industrial, commercial, and other sectors has led to the contamination of agro-ecosystems, ultimately affecting the health of consumers. In particular, the excessive use of chemical fertilizers, pesticides, urban and industrial runoff, excessive tillage, machinery, etc., have contributed to the contamination and degradation of agro-ecosystems. The contamination of food crops leads to adverse effects on both plant and human health, while rapid changes in global climate have a significant impact on crop production, leading to food insecurity. Thus, there is an urgent need to accelerate the search and adoption of sustainable technologies for agricultural production.

This Special Issue on "Agricultural Environmental Pollution, Risk Assessment, and Control" published novel research that addressed current agricultural and environmental challenges affecting global food security and consumer health. This Special Issue contains a total of 18 published articles that focus on a wide range of agricultural and environmental pollution problems and provide cutting-edge solutions to address them. The first article in this Special Issue highlighted how the degradation and transport of Cd and kasugamycin (an aminoglycoside antibiotic) in cultivated spinach (*Spinacia oleracea* L.) could be mitigated by the use of wheat-straw-derived biochar [1]. In the second article, toxicological studies were conducted to investigate the effect of rockwool on earthworms (*Eisenia fetida*) when used as a nursery substrate [2]. Similarly, the third article dealt with the spatio-temporal and residual analysis of thirteen (13) antibiotics in livestock and poultry, i.e., pigs, chickens, cattle, and sheep, in China [3]. Similarly, the fourth article in this Special Issue addressed issues related to the emission of air pollutants from the overuse of tractors in Korea [4]. The fifth article focused on the use of curry leaf (*Murraya koenigii*) extract as a sustainable herbicide to suppress weeds while improving the growth and biochemical traits of two crops, wheat (*Triticum aestivum*) and chickpea (*Cicer arietinum*) [5]. The sixth article also investigated the uptake and bioaccumulation of Cu and Ni metals in cocksfoot grass (*Dactylis glomerata* L.) as influenced by zinc and organic fertilizer applications [6]. The seventh article highlighted that the combined use of biochar, halloysite, and alginate could improve the quality of soils contaminated with petroleum waste products [7]. The eighth paper described the feasibility of a tax scheme to balance the environmental and economic burdens of large-scale pig farming in China [8]. In the ninth article, AL-Huqail et al. [9] presented a novel remediation solution for agricultural soils contaminated with battery waste by using a weed grass species (*Cyperus rotundus* L.).

The tenth article emphasized an integrated ecological risk assessment system for monitoring agricultural lands, using chemical, ecotoxicological, and ecological indicators as key factors [10]. The eleventh article dealt with the evaluation of a dynamic tube method for estimating ammonia emissions from agricultural fields under manure application [11]. Similarly, the twelfth article demonstrated the development of a portable spectrometer for the efficient detection of chemical contaminants in irrigation water systems [12]. The thirteenth paper highlighted that livelihood capital played a significant role in determining the attitude of farmers in the Yilong Lake Basin, China, which ultimately affected the performance of agricultural pollution management [13]. Furthermore, in the fourteenth article, Signorelli et al. [14] applied a quality-specific environmental risk index to understand pollution-impacted hive locations. The fifteenth article highlighted that social network participation plays an important role in improving farmers' participation in irrigation water-saving system management in China [15]. The sixteenth article showed that irrigation with industrial wastewater could introduce toxic metalloids into the edible parts of rice crops grown in urban areas [16]. Similarly, the penultimate article highlighted that the use of biopreparations can help to reduce ammonia emissions from cattle farms, considering manure storage time and ventilation intensities as important factors [17]. The very last article in this Special Issue dealt with the biomonitoring of eight heavy metals (As, Cd, Cu, Cr, Hg, Ni, Pb, and Zn) in topsoil samples from cultivated fields in Laiyang County, East China [18].

Therefore, this Special Issue published articles that addressed various problems related to agricultural and environmental pollution and provided useful information for their management. Overall, this Special Issue serves as a valuable resource for policymakers, researchers, and practitioners to promote effective pollution control and sustainable agricultural practices.

Author Contributions: Conceptualization, P.K. and V.K.; data curation, P.K. and V.K.; resources, P.K. and V.K.; supervision, P.K. and V.K.; validation, P.K. and V.K.; visualization, P.K. and V.K.; writing—original draft, P.K.; writing—review and editing, P.K. and V.K. All authors have read and agreed to the published version of the manuscript.

Conflicts of Interest: The authors declare no conflicts of interest.

References

1. Cui, L.; Ma, J.; Quan, G.; Yan, J.; Ippolito, J.; Wang, H. Biochar and Cd Alter the Degradation and Transport of Kasugamycin in Soil and Spinach. *Agriculture* **2023**, *13*, 2172. [CrossRef]
2. Zheng, S.; Wang, C.; Ju, J.; Dai, Q.; Zhao, H.; Liu, P.; Wang, X. Is Rockwool Potentially Harmful to the Soil Environment as a Nursery Substrate? Taking *Eisenia fetida* as an Example for Toxicological Analysis. *Agriculture* **2023**, *13*, 1964. [CrossRef]
3. Li, Y.; Li, Q.; Ji, Z.; Andom, O.; Wang, X.; Guo, X.; Li, Z. Current Status and Spatiotemporal Evolution of Antibiotic Residues in Livestock and Poultry Manure in China. *Agriculture* **2023**, *13*, 1877. [CrossRef]
4. Kim, M.; Kim, S. Estimation of Air Pollutant Emissions by Tractor Utilization in Korea. *Agriculture* **2023**, *13*, 1811. [CrossRef]
5. Vashishth, D.; Bachheti, A.; Bachheti, R.; Alhag, S.; Al-Shuraym, L.; Kumar, P.; Husen, A. Reducing Herbicide Dependency: Impact of *Murraya koenigii* Leaf Extract on Weed Control and Growth of Wheat (*Triticum aestivum*) and Chickpea (*Cicer arietinum*). *Agriculture* **2023**, *13*, 1678. [CrossRef]
6. Kuziemska, B.; Wysokinski, A.; Klej, P. The Content, Uptake and Bioaccumulation Factor of Copper and Nickel in Grass Depending on Zinc Application and Organic Fertilization. *Agriculture* **2023**, *13*, 1676. [CrossRef]
7. Wyszkowska, J.; Borowik, A.; Zaborowska, M.; Kucharski, J. Biochar, Halloysite, and Alginite Improve the Quality of Soil Contaminated with Petroleum Products. *Agriculture* **2023**, *13*, 1669. [CrossRef]
8. Yan, T.; Zhang, T.; Zhu, Z. The Environmental Tax Scheme in China's Large-Scale Pig Farming: Balancing Economic Burden and Responsibility. *Agriculture* **2023**, *13*, 1576. [CrossRef]
9. AL-Huqail, A.; Taher, M.; Širić, I.; Goala, M.; Adelodun, B.; Choi, K.; Kumar, P.; Kumar, V.; Kumar, P.; Eid, E. Bioremediation of Battery Scrap Waste Contaminated Soils Using Coco Grass (*Cyperus rotundus* L.): A Prediction Modeling Study for Cadmium and Lead Phytoextraction. *Agriculture* **2023**, *13*, 1411. [CrossRef]
10. Klimkowicz-Pawlas, A.; Smreczak, B.; Maliszewska-Kordybach, B. Integrated Ecological Risk Assessment of the Agricultural Area under a High Anthropopressure Based on Chemical, Ecotoxicological and Ecological Indicators. *Agriculture* **2023**, *13*, 1353. [CrossRef]
11. ten Huf, M.; Olfs, H. Evaluation of the Dynamic Tube Method for Measuring Ammonia Emissions after Liquid Manure Application. *Agriculture* **2023**, *13*, 1217. [CrossRef]
12. Zainurin, S.; Wan Ismail, W.; Wan Azlan, W.; Zainul Ariffin, K.; Wan Ahmad Kamil, W. Developing a Portable Spectrometer to Detect Chemical Contaminants in Irrigation Water. *Agriculture* **2023**, *13*, 1202. [CrossRef]

13. Zhou, N.; Zhong, F.; Yin, Y. Does Livelihood Determine Attitude? The Impact of Farmers'; Livelihood Capital on the Performance of Agricultural Non-Point Source Pollution Management: An Empirical Investigation in Yilong Lake Basin, China. *Agriculture* **2023**, *13*, 1036. [CrossRef]
14. Signorelli, D.; D'Auria, L.; Di Stasio, A.; Gallo, A.; Siciliano, A.; Esposito, M.; De Felice, A.; Rofrano, G. Application of a Quality-Specific Environmental Risk Index for the Location of Hives in Areas with Different Pollution Impacts. *Agriculture* **2023**, *13*, 998. [CrossRef]
15. Wang, L.; Zhao, Y. Will Social Network Relationship Significantly Enhance Farmers' Participation in the Supply of Small Water-Saving Irrigation and Water Conservancy Facilities in China? *Agriculture* **2023**, *13*, 216. [CrossRef]
16. AL-Huqail, A.; Kumar, P.; Eid, E.; Adelodun, B.; Abou Fayssal, S.; Singh, J.; Arya, A.; Goala, M.; Kumar, V.; Širić, I. Risk Assessment of Heavy Metals Contamination in Soil and Two Rice (*Oryza sativa* L.) Varieties Irrigated with Paper Mill Effluent. *Agriculture* **2022**, *12*, 1864. [CrossRef]
17. Mažeikienė, R.; Bleizgys, R. Use of Bio-Preparations to Reduce Ammonia Emissions from Cattle Farming: Effects of Manure Storage Time and Ventilation Intensity. *Agriculture* **2022**, *12*, 1626. [CrossRef]
18. Yu, L.; Liu, H.; Liu, W.; Qin, P.; Yu, J.; Zhou, B.; Zhang, F.; Chen, Z.; Zhao, Y.; Shi, Z. Spatial Distribution, Pollution Characteristics, and Health Risk Assessment of Heavy Metals in Soils from a Typical Agricultural County, East China. *Agriculture* **2022**, *12*, 1565. [CrossRef]

agriculture

Article

Biochar and Cd Alter the Degradation and Transport of Kasugamycin in Soil and Spinach

Liqiang Cui [1,2,†], Jingwen Ma [1,2,†], Guixiang Quan [1,2], Jinlong Yan [1,2,*], James A. Ippolito [3] and Hui Wang [1,2]

1 School of Environmental Science and Engineering, Yancheng Institute of Technology, No. 211 Jianjun East Road, Yancheng 224051, China; lqcui@ycit.cn (L.C.); 18351278031@163.com (J.M.); qgx@ycit.cn (G.Q.); whsl@ycit.cn (H.W.)
2 Jiangsu Engineering Research Center of Biomass Waste Pyrolytic Carbonization & Application, Yancheng 224051, China
3 School of Environment and Natural Resources, The Ohio State University, 2021 Coffey Road, Columbus, OH 4321, USA; ippolito.38@osu.edu
* Correspondence: yjlyt@ycit.cn
† These authors contributed equally to this work.

Abstract: Biochar has been widely studied to reduce multiple contaminant sources in one matrix (e.g., several heavy metals in soils), yet less attention has been paid to accelerating pesticide degradation while in the presence of any heavy metals, such as when kasugamycin (KSM) and cadmium (Cd) are both present in soil. While KSM has low toxicity compared to other pesticides, it can be a potential health risk when applied to vegetable crops, especially when KSM is used or overapplied to achieve rapid reductions in insect and disease pressure. The degradation behavior of KSM (2 kg ha^{-1}) in the presence of Cd (20 mg kg^{-1}) and biochar (5% by wt.) when growing spinach (*Spinacia oleracea*) was studied. The biochar increased spinach shoot and root biomass by 51.0–54.8% and 24.4–39.0%, respectively, compared to the KSM treatment only. Compared to the treatments that did not receive biochar, the biochar application increased the KSM degradation in the soil by 8.4–68.4% and, subsequently, less KSM was absorbed by the spinach roots (18.0–48.4%) and shoots (33.0–33.2%). The KSM degradation rate, as a function of soil depth, was enhanced in the presence of Cd. The biochar also effectively decreased the KSM concentration with soil depth, reducing downward KSM migration. The KSM degradation, increased by the biochar, led to smaller organic moieties and some macromolecular organic phases. In soils that are contaminated with Cd and where vegetables are raised, biochar may be used as an environmentally friendly proponent for increasing KSM degradation, reducing KSM downward transport and, thus, protecting environmental and human health.

Keywords: biochar; deterioration; downward migration; pesticide; soil

1. Introduction

Pesticides are widely used to protect crops and, consequently, to allow producers to maintain or increase yields. Globally, pesticide use increased from 1990 to 2020, with approximately 20,000 commercial products used for plant protection worldwide [1]. In 2022, the predicted pesticide volumes were 1.76×10^6 Mg for China, followed by USA and Brazil, with 4.08×10^5 and 2.05×10^5 Mg, respectively [2]. However, more than 90% of applied pesticides are lost or decomposed in the environment, which has led to an obvious waste of resources, along with a multitude of environmental problems, such as pesticide or organic pollution accumulation in sensitive ecosystems [3]. Some of the more environmentally friendly pesticides are naturally occurring, such as the naturally occurring antibiotic, kasugamycin (KSM; $C_{14}H_{25}N_3O_9$).

Specifically, KSM is an aminoglycoside antibiotic produced by two independently isolated strains of *Streptomyces kasugaensis*. It is composed of three structurally distinct sub-components: a glycine imine, an unusual amino-sugar, kasugamine, and D-chiro-inositol, which is created by the fungus *Magnaporthe oryzae* [4]. Furthermore, KSM is an aminoglycoside antibiotic that exhibits considerable medical potential [5]. For example, KSM has been used to counteract some recalcitrant human diseases, such as by controlling *Mycobacterium tuberculosis*, inhibiting herpes simplex virus-2, and even showing promising activity reductions against COVID-19 [6]. However, KSM is sensitive to light and heat, and it is unstable under alkaline conditions. For example, the degradation half-life of KSM at pH 9.0 and 50 °C is only 7.28 h and, therefore, it has largely been limited in its practical application [7]. Fortunately, these extreme conditions typically do not occur in the environment. However, alkaline conditions do occur globally; thus, under these conditions KSM application is often widespread and excessive due to its known degradation rate.

Within agricultural systems, KSM has been used for over half a century to combat rice-sheath blight [8]. In addition, KSM has been widely used for pest and disease prevention in crops and vegetables, and for reducing blossom infections [9]. More specifically, Chen [10] reported that terminal residual levels of KSM in Chinese cabbage were not detected 14 days after application under field-test conditions, with degradation rates related to temperature and precipitation. Huang [11] showed that KSM inhibited the growth of some aquatic bacteria, but also stimulated the growth of other resistant bacteria. Zhang [12] found that KSM controlled plant diseases by disturbing the transport of secondary metabolites in plants. Although KSM has been proven to be helpful for pest and disease prevention, the presence of any residual KSM may be a threat to environmental and human health. Biochar might be a promising material to promote KSM degradation, thus reducing its fate and transport in the environment.

Biochar is a carbon-rich material that features low costs, a large specific surface area, and strong adsorption capacity. Biochars often contain abundant functional groups (e.g., lactone, phenolic, and carbonyl groups), negative charge due to the presence of Pi electrons, and pores ranging from the micro to the meso pore size, all of which can contribute to sorption and the physical entrapment of contaminants [13–15]. Thus, biochars have been applied widely to reduce soil inorganic and organic pollution [16,17]. Furthermore, biochar's physicochemical properties can improve soil quality and water relations [18,19] and contribute to carbon sequestration, and they have the proven potential to sorb pesticides, reducing their mobility and bioavailability and enhancing their degradation [20]. The land application of biochar, as an eco-friendly sorbent, has certainly emerged as a promising technology to minimize pesticide-associated health and environmental risks [21].

Specifically, Jones [22] found that modified biochar promoted 10% greater simazine degradation in soil compared to soil not receiving biochar. Yang and Sheng [23] showed that the addition of wheat-straw biochar (1%; *w:v*) enhanced diuron sorption in water compared to a control. Biochar has been shown to reduce the negative effects of pesticides on certain microorganisms, all leading to the decreased phytotoxicity of 2,4-D in soils and plants [24]. Abas [25] found that biochar was effective in removing soil chlorantraniliprole mass (by 90% to 99%), a promising use of biochar for removing chlorantraniliprole in agricultural runoff. Cui [26] found that biochar promoted 2,4,6-trichlorophenol degradation into a combination of benzene derivatives and low-molecular-weight organic compounds. Unfortunately, the above studies did not focus attention on organic compound degradation in the presence of heavy metals.

Research has focused attention on biochar's use for improving plant growth in the presence of heavy metals. In a study of the application of wheat-straw biochar (5% by wt.), green pepper and eggplant biomass were significantly increased (16.9% to 519.9%), while the Cd concentration in both plants was decreased by 6.8% to 11.5% and by 15.1% to 15.4%, respectively [27]. Li [28] showed that the application of 30 Mg ha^{-1} biochar to Cd-polluted soil increased eggplant growth and reduced Cd concentrations in roots, shoots, and fruits by 12.2%, 12.5%, and 18.5%, respectively, compared to a control. Pot experiments proved that a 0.5% to 1% (by wt.) application of an iron-based biochar inhibited Cd and As bioaccumulation in *Brassica chinensis* L. by 23.6–36.5% and 43.1–45.1%, respectively, compared to a control [29]. Unfortunately, the above studies did not focus attention on biochar's ability to reduce both Cd and organic contaminant bioavailabilities simultaneously. One particular study, by Gao [30], found that biochar effectively reduced Cd and atrazine bioavailability in soils where pak choi was raised. The authors noted decreased plant stress, along with an increased rate of photosynthesis and antioxidant production.

The utilization of biochar to improve vegetable production and reduce pollution in soils containing both heavy metals (e.g., Cd) and residual pesticides (e.g., KSM) is yet to be fully explored. Thus, the objectives of this study were to (i) investigate the effectiveness of biochar in reducing the bioavailability and plant uptake of KSM, either in the absence or in the presence of excessive soil Cd, and (ii) identify the degradation pathways of KSM, in soil, when in the presence of biochar.

2. Materials and Methods

2.1. Experimental Materials and Design

Soil (0 cm to 20 cm depth) was collected in the Tinghu District (33°24′47.70″ N, 120°17′43.85″ E), Yancheng City, Jiangsu Province. The local climate was humid subtropical, with a mean annual temperature and precipitation of 14 °C and 1000 mm, respectively. The soil was described as a mainly coastal, salinized fluvo-aquic soil, according to the China soil-classification system (an Aridisol, according to the USDA Soil Taxonomic System) [31]. The soil was air-dried, stones and litter were removed, and then the soil was passed through a 5-millimeter sieve. Biochar was prepared by pyrolysis of wheat straw at 450 °C, with a hold time of 4 h, using a continuous anoxic production method at the Sanli New Energy Company, Shangqiu, Henan Province, China. Basic properties of soil and biochar were determined according to methods recommended by Lu [32] and IBI [33], and are presented in Table 1. Briefly, soil and biochar organic carbon contents were quantified using the dichromate oxidation method, cation-exchange capacity (CEC) was determined with the sodium-acetate-exchange method, cations (Ca, Mg, K, Na) determined via atomic absorption spectrometry (AAS; TAS-986, Persee, Beijing, China) following the NaOAc method, and biochar ash content was measured using the muffle-furnace method, with weight remaining determined using an electronic balance (to 0.0001 g; XPR105DR/AC, Mettler Toledo, Zurich, Switzerland) [32]. The pH was measured using a glass electrode in deionized water with soil-to-solution rate of 1:2.5. Total nitrogen (total N) was determined using the semi-micro-Kjeldahl procedure with Kjeldahl determination apparatus (UDK139, VELP, Usmate, Italy). Total phosphorus (total P) was measured using the NaOH fusion colorimetric method with P concentration determined via a spectrophotometer (TU-1901, Persee, Beijing, China). Total potassium (total K) was determined via NaOH fusion, with K concentration determined using AAS.

Table 1. Basic properties of soil and biochar at time of collection of both materials.

	pH	CEC (cmol kg^{-1})	EC (dS m^{-1})	OC (g kg^{-1})	Total N (g kg^{-1})	Total P (g kg^{-1})	Total K (g kg^{-1})	Ash (%)
Soil	8.7	2.4	4.1	6.5	0.7	0.4	7.3	NA
Biochar	10.4	10.6	4.4	650.2	6.3	20.4	16.9	28.6

CEC: cation-exchange capacity; EC: electrical conductivity; OC: organic carbon; NA: not applicable.

Wooden boxes (length × width × height = 1 × 0.5 × 0.5 m) were used as individual experimental units. Four treatments, replicated three times, were used: (1) KSM ($C_{14}H_{25}N_3O_9$, ≥90%, Sinopharm, China) (PZ1); (2) Cd and KSM (PZ2); (3) KSM and biochar (PZ3); and 4) KSM, Cd and biochar (PZ4). Wooden boxes were filled with 30 cm (compacting and watering every 10 cm) of soil containing either 0% (PZ1, PZ2) or 5% biochar (w:w; PZ3, PZ4; the biochar was mixed with soil before filling the box.). The soil bulk density in all boxes was similar to field conditions (1.41 g cm^{-3}). The total amount of soil in each box was 60 kg. For the Cd treatments (PZ2, PZ4), 250 mL of 4.8 g L^{-1} Cd ($Cd(NO_3)_2 \cdot 4H_2O$, 99%, Sinopharm, China) was mixed with 500 g of soil. The soil was allowed to dry and was then spread evenly over the top of the soil in the wooden boxes; the application rate was equivalent to 20 mg Cd kg^{-1}. All soils were then incubated outdoors and irrigated with tap water once per day in the summer, weekly in the spring and autumn, and monthly in winter, for three years.

After three years, spinach seeds (*Spinacia oleracea* L. NO. 21; purchased from a local seed company) were planted in all boxes; the boxes remained outside. Initially, 2 g of seed was evenly placed in the soil in every box, and 50 plants were kept after plants reached the two-leaf stage. Boxes were usually irrigated weekly, based on natural precipitation events and visual inspection. Five months after sowing and one month before harvest, KSM (100 mg box^{-1}; equivalent to 2 kg ha^{-1}) was sprayed on the leaf surfaces. Spinach-leaf and root samples were collected at 28 days after KSM application.

Spinach leaves and roots were washed three times with tap water and then three times with deionized water. A portion of each plant part was then stored at −20 °C to reduce KSM degradation prior to analysis. Another portion of each plant part was placed in an envelope, dried for 30 min at 105 °C in an oven, removed, and then dried to constant weight in an oven at 60 °C. Dried samples were crushed, mixed to homogenize, and then stored in polyethylene bags prior to chemical analysis.

Soil from depths of 0–5 cm was collected at 1 h, 1, 2, 7, 14, and 28 days after KSM application. Soil from depths of 0–5, 5–10, 10–15, 15–20, and 20–25 cm were also collected from all boxes at 14 and 28 days after KSM treatment. A portion of all soil samples was stored at −20 °C to reduce KSM degradation prior to analysis. Another portion of all soil samples was air-dried, ground to pass through a 2-millimeter sieve, and stored for subsequent analyses. Some of the 2-millimeter-sieved, air-dried soil was further ground to pass through a 0.15-millimeter sieve and used for additional soil analyses. Soil pH and SOC were determined as described above. Total soil Cd content was determined by placing 0.5000 g soil in polyfluortetraethylene crucibles, adding 25 mL of a mixed solution of HF, HNO_3, and $HClO_4$ (10:2.5:2.5, V:V:V), and placing the crucibles on a hot plate. The mixtures were digested at 100 °C for 60 min, and then at 250 °C until the volume was reduced to approximately 2 mL. The solutions were then removed, cooled, brought to a final volume of 25 mL, and filtered through a 0.45-micrometer-membrane filter. The Cd concentration in these solutions was determined using AAS.

2.2. KSM Extraction and Chromatographic Analysis Methods

Extraction of KSM from plant samples: A 5.0000 g (accurate to 0.0001 g) spinach sample was placed into a 50 mL centrifuge tube, 20 mL of methanol (for HPLC, ≥99.9%, Sinopharm, Shanghai, China) was added, the mixture was homogenized at high speed (2500 rpm; Rotator, JOANLAB, Huzhou, China) for 1 min, and then the pH was adjusted to 5.7 with formic acid (0.5 mol L^{-1}). Samples were centrifuged for 5 min at 4000 rpm, passed through a 0.22 μm nylon membrane filter, and then stored at 4 °C prior to analysis.

Extraction of KSM from soil samples: A 5.0000 g soil sample (accurate to 0.0001 g) was placed into a 50 mL centrifuge tube, 10 mL of 30% methanol (*v:v*, adjusted to pH = 5) was added, and then the mixture was sonicated for 30 min. Samples were then centrifuged for 5 min at 4000 rpm, followed by transferring the supernatant to a 50 mL volumetric flask. These steps were repeated three times, combining the supernatant in the 50 mL volumetric flask. The combined supernatant was then rotary-evaporated until nearly dry. Next, 2 mL of 30% methanol was added to redissolve the solid phase, the liquid was passed through a 0.22 μm nylon membrane filter, and then it was stored at 4 °C prior to analysis. Both plant and soil KSM extractions and analyses followed those of Zhang et al. [34]. All samples were run in triplicate. The average KSM recovery from soil was 83.2–91.5%, with a standard deviation of 5.63–8.45%. The KSM recoveries were not performed on plant material.

All KSM filtrates were analyzed by high-performance liquid chromatography (HPLC, PerkinElmer Flexar-15) with a UV detector equipped with an Athena C18 column (4.0 mm × 250 mm, 5 μm) with an isocratic mobile phase of 10:90 ($MeOH:H_2O$, V:V,

10 mmol L^{-1} sodium heptanesulfonate buffer solution, pH = 3), column temperature of 25 °C, and flow velocity of 1 mL min^{-1}. All determinations were performed in triplicate.

Degradation products of KSM were detected by liquid chromatography–triple-quadrupole mass spectrometry (LC-TQMS, Waters XevoTQS-micro, Massachusetts, USA) with a UV detector (215 nm) equipped with a Waters HPLC BEH HILIC column (2.1 mm × 100 mm, 1.7 μm) with isocratic mobile phases consisting of a mixture of 0.2% formic acid (A) and an aqueous 0.2% formic acid acetonitrile solution (B). Initial chromatographic conditions were created using a linear ramp up over 6 min to 90% A and 10% B. A second linear ramp was carried out over 7 min to 30% A and 70% B, and these conditions were maintained for 1 min, before moving back to initial conditions in 2 min and equilibrating the column for 8 min. The flow rate was 0.3 mL min^{-1} and columns were kept at 40 °C.

Mass spectrometry was performed in a Waters XevoTQS-micro (Waters, Milford, MA, USA) triple-quadrupole mass spectrometer equipped with electrospray ionization (ESI) as the ionization source. Electrospray voltage was set at 3.0 kV and temperatures of the ion-transfer tube and vaporizer were both held at 600 °C.

2.3. Functional-Group Characterization of Biochar and Soil

Functional groups of raw biochar and soils were detected by Fourier-transform infrared spectroscopy (FTIR, NEXUS-670, NICOLET, Waltham, MA, USA).

2.4. Statistical Analysis

Data were expressed as means ± standard deviation. Significant differences between treatments were determined using analyses of variance (ANOVA) at a $p < 0.05$. When present, differences between means were determined using a Tukey post hoc test. All statistical analyses were carried out using SPSS version 22.0 (SPSS Institute, IBM, Armonk, NY, USA).

3. Results and Discussion

3.1. Effects of Biochar on pH and SOC

The biochar treatments increased the soil pH (Figure 1A,B; Figure S1A) and soil organic carbon (SOC; Figure 1C,D; Figure S1B) over time and depth more often than when no biochar was added (i.e., the PZ3 and PZ4 treatments compared to the PZ1 and PZ2 treatments). The biochar had a relatively high pH (10.4; Table 1), which is likely to have been due to the presence of oxides, hydroxides, and carbonate mineral phases [35]. The biochar also contained a significantly greater amount of C than the soil used in the study (expressed as OM, Table 1). Increases in soil OC (SOC) following biochar application have been shown by many others [36–38]. Increases in SOC via biochar application play an important role in soil oxidation/reduction and sorption/desorption processes and, in conjunction with changes in soil pH, they can affect soils' buffering capacity. This, in turn, affects pesticide sorption and desorption in soils [21]. Liang [3] showed that pesticide stability was related to pH, as in the case of benzoic imine bonds, which were relatively stable in neutral or alkaline solutions, but easily broken under acidic conditions.

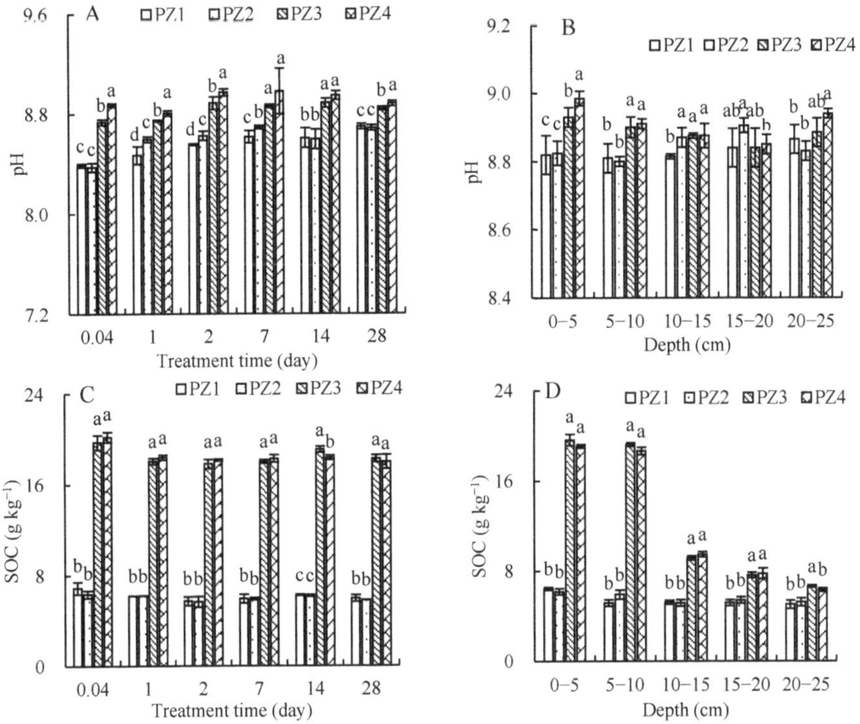

Figure 1. Changes in soil pH and soil organic carbon (SOC). (**A**) Soil pH (0–5 cm in depth) as a function of time; (**B**) soil pH as a function of soil depth, obtained at 14 days after KSM application; (**C**) SOC as a function of time; (**D**) SOC as a function of soil depth, obtained at 14 days after KSM application. Treatment PZ1 = 100 mg box^{-1} was sprayed on spinach leaves in unamended soil; soil containing PZ2 = 20 mg Cd kg^{-1} added to an unamended soil surface, and 100 mg box^{-1} sprayed on spinach leaves; soil amended with PZ3 = 5% biochar (*w:w*), and 100 mg box^{-1} sprayed on spinach leaves; and soil containing PZ4 = 20 mg Cd kg^{-1} added to the surface of a soil amended with 5% biochar (*w:w*), and 100 mg box^{-1} sprayed on spinach leaves. Different lower-case letters above error bars (error bars: the standard deviation of the mean; n = 3; $p < 0.05$, determined via a Tukey post hoc test) indicate statistically significant differences between treatments within a particular time (**A**,**C**) or depth (**B**,**D**).

3.2. Effects of Biochar on KSM Distribution in Soil and Spinach

The biochar's effects on the spinach-shoot and -root biomass are presented in Table 2. When biochar was applied (PZ3 and PZ4 treatments), both the shoot and the root biomass significantly increased compared to when biochar was not applied (PZ1 and PZ2 treatments). The addition of Cd, either with or without biochar, had little effect on the spinach-shoot or -root biomass (i.e., PZ2 vs. PZ1 and PZ4 vs. PZ3 treatments).

The KSM concentrations in the soils and in the spinach roots and shoots are shown in Figure 2. The soil KSM concentrations in the PZ3 and PZ4 treatments were decreased by 17.3–62.7% and 8.4–68.4% compared to the PZ1 and PZ2 treatments over the 28 d of the study (Figure 2A). Similarly, You [39] observed a decrease in soil metalaxyl (a pesticide) bioavailability in the presence of biochar and associated this reduction with the biochar's sorption capacity. When comparing the Cd treatments (e.g., PZ2 versus PZ1, or PZ4 versus PZ3), it appears that Cd also played a slight role in decreasing the presence of the soil KSM. However, based on the data in Figure 2A, it appears that the biochar application increased the KSM degradation to a greater degree than the Cd.

Table 2. Spinach-shoot and -root biomass. Treatment PZ1 = 100 mg box^{-1} (equivalent to (2 kg ha^{-1}) was sprayed on spinach leaves in unamended soil; soil containing PZ2 = 20 mg Cd kg^{-1} added to an unamended soil surface, and 100 mg box^{-1} sprayed on spinach leaves; soil amended with PZ3 = 5% biochar (*w:w*), and 100 mg box^{-1} sprayed on spinach leaves; and soil containing PZ4 = 20 mg Cd kg^{-1} added to the surface of a soil amended with 5% biochar (*w:w*), and 100 mg box^{-1} sprayed on spinach leaves. Different lower-case letters within a column indicate statistically significant differences between treatments, as determined by a Tukey post hoc test (p < 0.05. error = the standard deviation of the mean; n = 3).

Treatment	Shoot Biomass	Root Biomass
	(g box^{-1})	
PZ1	76.7 ± 1.2 b	11.8 ± 1.0 ab
PZ2	74.0 ±2.3 b	10.7 ± 0.8 b
PZ3	108.9 ± 4.2 a	14.7 ± 2.2 a
PZ4	117.3 ± 5.6 a	14.9 ± 3.4 a

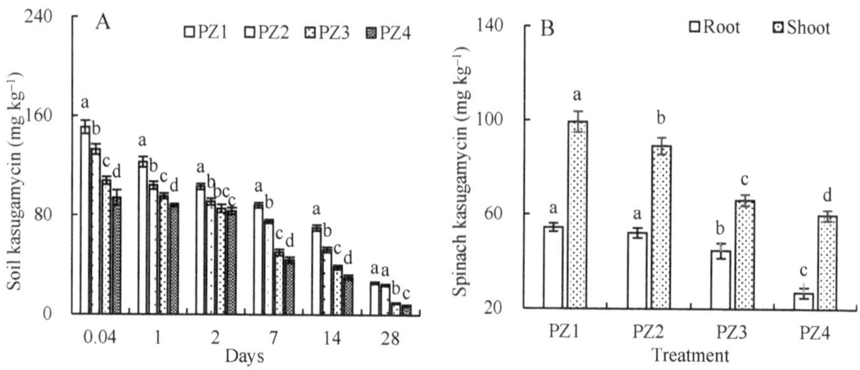

Figure 2. KSM concentrations in (**A**) soil (0–5 cm in depth) as a function of time and (**B**) spinach roots and shoots at the end of the study (28 days after KSM application). Treatment PZ1 = 100 mg box^{-1} (equivalent to (2 kg ha^{-1}) was sprayed on spinach leaves in unamended soil; soil containing PZ2 = 20 mg Cd kg^{-1} added to an unamended soil surface, and 100 mg box^{-1} sprayed on spinach leaves; soil amended with PZ3 = 5% biochar (*w:w*), and 100 mg box^{-1} sprayed on spinach leaves; and soil containing PZ4 = 20 mg Cd kg^{-1} added to the surface of a soil amended with 5% biochar (*w:w*), and 100 mg box^{-1} sprayed on spinach leaves. Different lower-case letters above error bars (error bars: the standard deviation of the mean; n = 3; p < 0.05, determined via a Tukey post hoc test) indicate statistically significant differences between treatments.

The KSM's degradation by the biochar, in turn, led to significant differences in KSM concentration within the spinach roots and shoots. The KSM concentrations in the roots and shoots of the PZ3 and PZ4 treatments were significantly decreased, by 18.0–48.4% and 33.0–33.2%, respectively, compared to the PZ1 and PZ2 treatments (Figure 2B). In the case of the Cd, however, KSM was present at significantly lower concentrations when the biochar and Cd were co-applied (PZ4) compared to the Cd alone (PZ2). In the soil treated with Cd–KSM–biochar, the Cd could have formed a bridge between the negative sites on the biochar and the negative sites on the KSM, helping to further hold the KSM in place for degradation to occur, a phenomenon proven by Awan [40]. You [39] observed a decrease in the uptake of metalaxyl by lettuce (*Lactuca sativa*) in the presence of wood-waste-derived biochar. These authors also found that the biochar amendment effectively decreased the shoot concentrations of metalaxyl and their metabolites by 57.7–86.3%, similar to the observations in the current study. The uptake of organic pollutants varies among different plants, and plants' capacity to absorb KSM also varies with biochar application [41].

The migration and distribution of the KSM concentrations at different soil depths are shown in Figure 3A. The KSM soil concentration decreased with increasing soil depth, indicating that soils do play a role in reducing the downward migration of KSM. However, when KSM was added to the soils that received biochar, the KSM concentration present at the soil surface was greatly reduced, and the KSM concentration changes with depth were less dramatic than with the absence of biochar. These findings suggest that biochar leads to either KSM degradation and/or strong KSM sorption. In either case, the KSM concentrations at different soil depths were reduced in the presence of biochar.

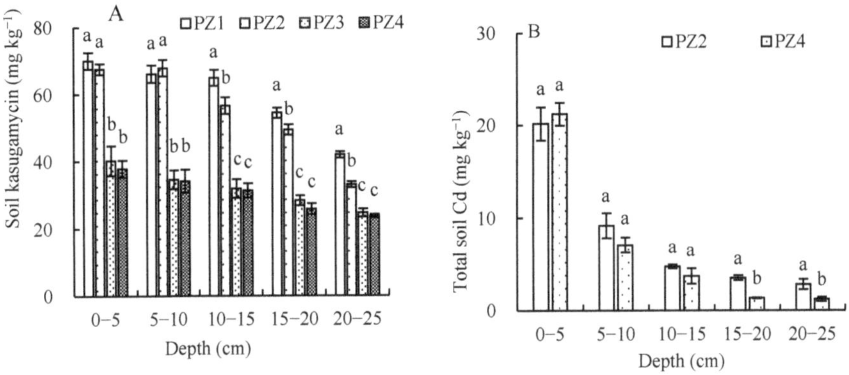

Figure 3. Soil KSM (**A**) and Cd (**B**) concentrations by soil depth at the end of the study (14 days after KSM application). Treatment PZ1 = 100 mg KSM box^{-1} (equivalent to (2 kg ha^{-1}) was sprayed on spinach leaves in unamended soil; soil containing PZ2 = 20 mg Cd kg^{-1} added to an unamended soil surface, and 100 mg box^{-1} sprayed on spinach leaves; soil amended with PZ3 = 5% biochar (*w:w*), and 100 mg box^{-1} sprayed on spinach leaves; and soil containing PZ4 = 20 mg Cd kg^{-1} added to the surface of a soil amended with 5% biochar (*w:w*), and 100 mg box^{-1} sprayed on spinach leaves. Different lower-case letters above error bars (error bars: the standard deviation of the mean; n = 3; $p < 0.05$, determined via a Tukey post hoc test) indicate statistically significant differences between treatments within a particular depth.

The KSM degradation rate, according to soil depth, tended to increase in the presence of Cd without (PZ1 versus PZ2) or with biochar (PZ2 versus PZ4) (Figure 3A). Specifically, the downward transport of KSM was reduced by the presence of Cd for the PZ2-versus-PZ1 treatments at the soil depths of 10–15, 15–20, and 20–25 cm. Compared to when no biochar was applied, the presence of the biochar reduced the spinach-leaf and -root Cd concentrations and reduced the soil's Cd bioavailability (Tables S1 and S2). Furthermore, the downward leaching of the Cd was reduced by the biochar application (Figure 3B). These results support the contention that the Cd may have formed a bridge between the negative sites on the biochar and KSM, to hold the KSM in place and to enhance the KSM degradation.

Similarly, Gao [42] showed that swine biochar could co-sorb Cd and glyphosate. Other works have shown that a modified biochar (polyacrylic acid, Fe and Co) could quickly (4 h) sorb 73% of Pb^{2+} while concomitantly degrading 81% of the quinclorac present in contaminated water [43]. Yu [44] found that the application of 1% biochar to soil decreased the total plant contents of chlorpyrifos and carbofuran by 10% and 25% of a control, respectively. The authors noted that the biochar was particularly effective in reducing the phytoavailability of both pesticides due to its high affinity for and ability to sequester pesticide residues.

3.3. KSM and Degradation Products

The concentrations of pesticides in soil tend to diminish with biochar application, both of which affect pesticide sorption, mineralization, and degradation [45]. The current study was not a biochar-rate study, but the degradation-product results should still provide evidence pointing towards biochar acting as a sorbent or a degrader of KSM. The soil (0–5 cm) KSM and degradation products' chromatogram full-scan mass spectra, after 1 day following the PZ4 treatment, are shown in Figure S2. The mass spectra show a partly protonated molecule $[M+4H]^+$ (m/z 384) and the loss of O from a protonated molecule due to in-source collision-induced dissociation (m/z 368). It appears that the KSM was quickly degraded in the soil and produced an abundance of low-molecular-weight organic moieties or larger molecular molecules (Table 3, Figure S4). Slack [46] showed that apple-blossom blight-control efficacy was significant with KSM usage, but stated that KSM is vulnerable to photodegradation. If photodegradation had been dominant in the current study, differences in the presence of KSM would not have been observed with or without biochar (or Cd), yet differences existed. In a chili-field study, Lu [7] showed that the degradation rate of KSM was relatively fast and that, upon degradation, smaller molecular substances were created. In the current study, and based on the degradation-product findings, the results suggest that the biochar did, in fact, lead to a more rapid KSM degradation. Others have shown similarly rapid degradation results when using biochar to degrade other pesticides (e.g., atrazine; Jablonowski [47]; Qiu [48]).

Table 3. KSM-soil-degradation products 1 day after application to spinach leaves (PZ4).

$C_{14}H_{25}O_9N_3$	$C_{14}H_{29}O_8N_3$	$C_{14}H_{27}O_7N_3$	$C_{12}H_{24}O_8N_2$	$C_{13}H_{21}O_7N$	$C_8H_{17}O_8N$
$C_8H_{13}O_4N_3$	$C_7H_{13}ON_3$	$C_7H_{14}ON_2$	$C_6H_{14}O_2N_2$	$C_6H_{10}O$	C_3H_9N

3.4. Functional-Group Characteristics of Raw Biochar and Soil

Soil and raw biochar functional groups were detected using FTIR (Figure 4) and XPS (Figure S3), with the functional groups for the band assignments provided in Table S3. A broad-spectrum band at 3419 cm^{-1} was the result of -OH stretching vibration ($\nu_{O\text{-}H}$), which could have been due to soil and biochar organic carbon [49]. The stretching of $-CH_2-$ or $-CH_3$ in aliphatics or alicyclics ($\nu_{C\text{-}H}$) in peaks at 2950–2860 cm^{-1} was mainly present in the soil, with these peaks slightly increasing with the biochar application. A peak at 1632 cm^{-1} represented the stretching of C=O groups associated with aromatic rings, or -COOH ($\nu_{C=O}$), which were abundant in the soil regardless of the biochar application. A red shift (i.e., the wavenumber was shifted in the soil compared to the raw biochar) of the stretching appeared at 1020 cm^{-1} in the biochar and 1087 cm^{-1} in the soil with biochar, which is likely to have been due to the presence of aromatic C-O ($\nu_{C\text{-}O}$). The aliphatic CH_2 deformation (773 cm^{-1}) was decreased after the biochar was applied to the soil. The biochar contained Si-O functional groups, as shown at 466 cm^{-1}. The XPS data also show functional groups of biochar C1s, i.e., C-C (284.8 eV), C-O (286.1 eV), and C=O (287.0 eV), and four peaks for the O1s were C=O at 532.5 eV, O-H at 530.3 eV, O-C=O at 532.8 eV, and C-OO at 533.8 eV. The above data support the fact that biochar contains abundant oxygen-containing functional groups, as shown by our research group in the past [50]. Sanchez-Hernandez [51] showed that biochar can inactivate metabolites of organophosphorus pesticides, which they mainly attributed to the abundant functional groups in biochar. Zhang [52] proved that organophosphorus pesticide-degradation mechanisms can be ascribed to biochar's high surface area and porosity, as well as to strong π–π interaction and H-bonding between biochar surfaces and organophosphorus pesticides. Our findings, in conjunction with those observed by others, suggest that biochar has the potential to sorb, sequester, and degrade pesticides.

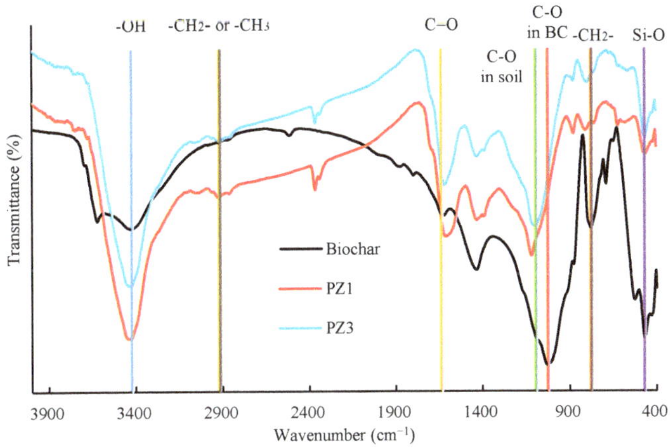

Figure 4. FTIR of raw biochar and soil treatment 1 day following KSM application to spinach leaves. Note: PZ1, 100 mg box^{-1} (equivalent to (2 kg ha^{-1}) was sprayed on spinach leaves in unamended soil; soil amended with PZ3, 5% biochar (*w*:*w*), and 100 mg box^{-1} sprayed on spinach leaves.

3.5. The Relationship between KSM, Biochar, Soil Cd, and Soil Properties

The biochar application improved the soil properties, decreased the soil KSM concentration and increased degradation, and reduced the plants' KSM uptake and transfer, with the direct or indirect pathways illustrated via PCA and correlation analysis (Figure 5, Table S4). The PCA results showed that the KSM concentration in the soil and plants was correlated with the soil properties (e.g., SOC, pH, soil Cd) in PC1 (81.7%). The factors of PC2 mainly included SOC and soil pH, which were mainly related to the biochar-application rates. The factors of PC1 and PC2 explained 95.5% of the variability for all the factors. Positive relationships between soil Cd, SOC (i.e., in this case, biochar application), soil KSM concentration with depth, and plant-root KSM content were observed (Table S4). In particular, and importantly for the current study, the biochar and soil heavy-metal concentrations were found to be key factors in reducing KSM concentrations in soils and plants (Figure S5). This is similar to the findings of others. For example, Mazarji [53], using a PCA analysis, found that the level of biochar amendment affected the degree of total PAH degradation in the presence or absence of heavy metals; the biochars also sorbed and prevented heavy-metal transport, leading to moderate PAH degradation. Kong [54] showed that biochar and heavy metals played an important role in mitigating organic toxicity with respect to seed germination and plant growth.

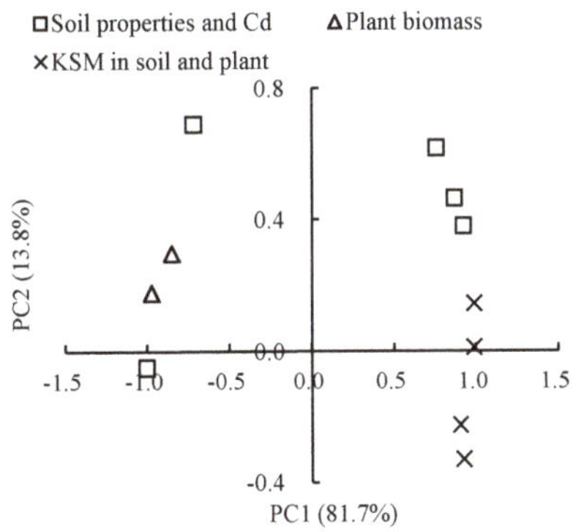

Figure 5. Principal component analysis as a function of soil properties and Cd presence, plant biomass, and KSM in both soils and plants.

4. Conclusions

Kasugamycin (KSM) is often used as an agricultural fungicide to control blight and blossom diseases; residual KSM in soils or plants can pose a risk to the environment and humans. In order to reduce the environmental/human health risk of KSM, accelerating its degradation post-application is important. Biochar was shown to effectively decrease the bioavailable KSM concentration, accelerate KSM degradation, and lessen KSM's downward transport in soil. The KSM degradation was further enhanced in the presence of both biochar and Cd, which is possibly attributable to the Cd forming a bridge between the negative sites on both the biochar and the KSM and, thus, further stimulating KSM sorption and degradation. In an equally important finding, the biochar increased the spinach biomass while decreasing the KSM concentration in the leaves and roots. It appears that biochar can act as both a sorbent and a degrader of KSM, reduce KSM uptake by plants, and in the future, potentially lessen pesticides' effects on the environment.

Supplementary Materials: The following supporting information can be downloaded at: https://www.mdpi.com/article/10.3390/agriculture13112172/s1. Figure S1: Changes in soil (A) pH and (B) soil organic carbon (SOC) as a function of depth at the end of the study (28 days after KSM application); Figure S2: Chromatogram of soil KSM and degradation products after 1 day following treatment (soil containing PZ4; 20 mg Cd kg^{-1} added to the surface of a soil amended with 5% biochar (wt:wt), and 100 mg box^{-1} sprayed on spinach leaves) full-scan mass spectra; Figure S3: XPS peak fits of C1s (A), O1s and (B) and maps (C) from biochar; Figure S4: Possible KSM degradation products; Figure S5: Possible KSM degradation routes; Table S1: Total Cd concentration in spinach leaves and roots (mg kg^{-1}). Treatment PZ2 did not feature biochar, while treatment PZ4 featured the application of 5% biochar to soil; Table S2: Soil bioavailable Cd concentrations by depth (DTPA, mg kg^{-1}). Treatment PZ2 did not feature biochar, while treatment PZ4 featured the application of 5% biochar to soil; Table S3: Band assignments in FTIR spectra of biochar; Table S4. Band assignments in.

Author Contributions: Conceptualization, J.Y. and L.C.; methodology, L.C. and H.W.; software, J.M.; validation, L.C., G.Q. and J.M.; formal analysis, L.C.; investigation, J.M.; resources, L.C. and J.M.; data curation, J.M.; writing—original draft preparation, J.M. and H.W.; writing—review and editing, L.C., G.Q., J.M. and H.W.; visualization, H.W.; supervision, J.Y. and J.A.I.; project administration, J.Y.; funding acquisition, L.C., H.W. and J.Y. All authors have read and agreed to the published version of the manuscript.

Funding: This research was funded by Natural Science Foundation of Jiangsu Province, China, grant number SBK2022020100, and National Natural Science Foundation of China, grant numbers 21677119, 22006127, and 41501339.

Institutional Review Board Statement: Not applicable.

Data Availability Statement: All data generated for this study are presented within the manuscript and the Supplementary Files.

Acknowledgments: The FTIR was performed in Analytical and Testing Center of Yancheng Institute of Technology.

Conflicts of Interest: The authors declare no conflict of interest.

References

1. Maggi, F.; Tang, F.H.M.; Black, A.J.; Marks, G.B.; McBratney, A. The pesticide health risk index—An application to the world's countries. *Sci. Total Environ.* **2021**, *801*, 149731. [CrossRef] [PubMed]
2. Food and Agriculture Organization of the United Nations. FAOSTAT. In *Database Collection of the Food and Agriculture Organization of the United Nations 2022*; Food and Agriculture Organization of the United Nations: Rome, Italy, 2022. Available online: https://www.fao.org/3/cb6034en/cb6034en.pdf (accessed on 12 December 2022).
3. Liang, Y.; Duan, Y.; Fan, C.; Dong, H.; Yang, J.; Tang, J.; Tang, G.; Wang, W.; Jiang, N.; Cao, Y. Preparation of kasugamycin conjugation based on ZnO quantum dots for improving its effective utilization. *Chem. Eng. J.* **2019**, *361*, 671–679. [CrossRef]
4. Kasuga, K.; Sasaki, A.; Matsuo, T.; Yamamoto, C.; Minato, Y.; Kuwahara, N.; Fujii, C.; Kobayashi, M.; Agematu, H.; Tamura, T.; et al. Heterologous production of kasugamycin, an aminoglycoside antibiotic from *Streptomyces kasugaensis*, in Streptomyces lividans and *Rhodococcus erythropolis* L-88 by constitutive expression of the biosynthetic gene cluster. *Appl. Microbiol. Biotechnol.* **2017**, *101*, 4259–4268. [CrossRef]

5. Zhang, Y.; Aleksashin, N.A.; Klepacki, D.; Anderson, C.; Vázquez-Laslop, N.; Gross, C.A.; Mankin, A.S. The context of the ribosome binding site in mRNAs defines specificity of action of kasugamycin, an inhibitor of translation initiation. *Proc. Natl. Acad. Sci. USA* **2022**, *119*, e2118553119. [CrossRef] [PubMed]
6. Rattinam, R.; Basha, R.S.; Wang, Y.L.; Wang, Z.C.; Hsu, N.S.; Lin, K.H.; Zadeh, S.M.; Adhikari, K.; Lin, J.P.; Li, T.L. KasQ an Epimerase Primes the Biosynthesis of Aminoglycoside Antibiotic Kasugamycin and KasF/H Acetyltransferases Inactivate Its Activity. *Biomedicines* **2022**, *10*, 212. [CrossRef]
7. Lu, L.; Zhao, S.; Deng, L.; Chen, Y.; Liu, X.; Li, D. Residues and Dynamics of Kasugamycin in Chilli and Soil. *Bull. Environ. Contam. Toxicol.* **2012**, *89*, 649–653. [CrossRef] [PubMed]
8. Alechaga, E.; Moyano, E.; Teresa Galceran, M. Simultaneous analysis of kasugamycin and streptomycin in vegetables by liquid chromatography-tandem mass spectrometry. *Anal. Methods* **2015**, *7*, 3600–3607. [CrossRef]
9. Jurgens, A.G.; Babadoost, M. Sensitivity of *Erwinia amylovora* in Illinois Apple Orchards to Streptomycin, Oxytetracyline, Kasugamycin, and Copper. *Plant Dis.* **2013**, *97*, 1484–1490. [CrossRef]
10. Chen, G.; Qiao, Y.; Liu, F.; Zhang, X.; Liao, H.; Zhang, R.; Dong, J. Dissipation and dietary risk assessment of kasugamycin and saisentong in Chinese cabbage. *Environ. Sci. Pollut. Res.* **2020**, *27*, 35228–35238. [CrossRef]
11. Huang, C.-Y.; Ho, C.-H.; Lin, C.-J.; Lo, C.-C. Exposure effect of fungicide kasugamycin on bacterial community in natural river sediment. *J. Environ. Sci. Health Part B-Pestic. Food Contam. Agric. Wastes* **2010**, *45*, 485–491. [CrossRef]
12. Zhang, H.; Zhang, C.; Xiang, X.; Zhang, Q.; Zhao, W.; Wei, G.; Hu, A. Uptake and transport of antibiotic kasugamycin in castor bean (*Ricinus communis* L.) seedlings. *Front. Microbiol.* **2022**, *13*, 948171. [CrossRef] [PubMed]
13. Su, Y.; Wen, Y.; Yang, W.; Zhang, X.; Xia, M.; Zhou, N.; Xiong, Y.; Zhou, Z. The mechanism transformation of ramie biochar's cadmium adsorption by aging. *Bioresour. Technol.* **2021**, *330*, 124947. [CrossRef] [PubMed]
14. Guo, J.; Xiao, H.; Zhang, J.B.; Dai, C.; Li, T.; Gao, M.-t.; Hu, J.; Li, J. Characterization of highly stable biochar and its application for removal of phenol. *Biomass Convers. Biorefinery* **2022**. [CrossRef]
15. Klüpfel, L.; Keiluweit, M.; Kleber, M.; Sander, M. Redox Properties of Plant Biomass-Derived Black Carbon (Biochar). *Environ. Sci. Technol.* **2014**, *48*, 5601–5611. [CrossRef]
16. Pan, X.; Gu, Z.; Chen, W.; Li, Q. Preparation of biochar and biochar composites and their application in a Fenton-like process for wastewater decontamination: A review. *Sci. Total Environ.* **2021**, *754*, 142104. [CrossRef] [PubMed]
17. Natasha, N.; Shahid, M.; Khalid, S.; Bibi, I.; Naeem, M.A.; Niazi, N.K.; Tack, F.M.G.; Ippolito, J.A.; Rinklebe, J. Influence of biochar on trace element uptake, toxicity and detoxification in plants and associated health risks: A critical review. *Crit. Rev. Environ. Sci. Technol.* **2022**, *52*, 2803–2843. [CrossRef]
18. Rogovska, N.; Laird, D.A.; Rathke, S.J.; Karlen, D.L. Biochar impact on Midwestern Mollisols and maize nutrient availability. *Geoderma* **2014**, *230*, 340–347. [CrossRef]
19. Varjani, S.; Kumar, G.; Rene, E.R. Developments in biochar application for pesticide remediation: Current knowledge and future research directions. *J. Environ. Manag.* **2019**, *232*, 505–513. [CrossRef]
20. Ogura, A.P.; Lima, J.Z.; Marques, J.P.; Massaro Sousa, L.; Rodrigues, V.G.S.; Espíndola, E.L.G. A review of pesticides sorption in biochar from maize, rice, and wheat residues: Current status and challenges for soil application. *J. Environ. Manag.* **2021**, *300*, 113753. [CrossRef]
21. Khalid, S.; Shahid, M.; Murtaza, B.; Bibi, I.; Natasha; Asif Naeem, M.; Niazi, N.K. A critical review of different factors governing the fate of pesticides in soil under biochar application. *Sci. Total Environ.* **2020**, *711*, 134645. [CrossRef]
22. Jones, D.L.; Edwards-Jones, G.; Murphy, D.V. Biochar mediated alterations in herbicide breakdown and leaching in soil. *Soil Biol. Biochem.* **2011**, *43*, 804–813. [CrossRef]
23. Yang, Y.; Sheng, G. Enhanced pesticide sorption by soils containing particulate matter from crop residue burns. *Environ. Sci. Technol.* **2003**, *37*, 3635–3639. [CrossRef] [PubMed]
24. Oleszczuk, P.; Jośko, I.; Futa, B.; Pasieczna-Patkowska, S.; Pałys, E.; Kraska, P. Effect of pesticides on microorganisms, enzymatic activity and plant in biochar-amended soil. *Geoderma* **2014**, *214–215*, 10–18. [CrossRef]
25. Abas, K.; Brisson, J.; Amyot, M.; Brodeur, J.; Storck, V.; Montiel-León, J.M.; Duy, S.V.; Sauvé, S.; Kõiv-Vainik, M. Effects of plants and biochar on the performance of treatment wetlands for removal of the pesticide chlorantraniliprole from agricultural runoff. *Ecol. Eng.* **2022**, *175*, 106477. [CrossRef]
26. Cui, L.; Yin, C.; Chen, T.; Quan, G.; Ippolito, J.A.; Liu, B.; Yan, J.; Hussain, Q. Biochar Immobilizes and Degrades 2,4,6-Trichlorophenol in Soils. *Environ. Toxicol. Chem.* **2019**, *38*, 1364–1371. [CrossRef]
27. Sun, J.; Fan, Q.; Ma, J.; Cui, L.; Quan, G.; Yan, J.; Wu, L.; Hina, K.; Abdul, B.; Wang, H. Effects of biochar on cadmium (Cd) uptake in vegetables and its natural downward movement in saline-alkali soil. *Environ. Pollut. Bioavailab.* **2020**, *32*, 36–46. [CrossRef]
28. Li, Z.; Qi, X.; Fan, X.; Du, Z.; Hu, C.; Zhao, Z.; Isa, Y.; Liu, Y. Amending the seedling bed of eggplant with biochar can further immobilize Cd in contaminated soils. *Sci. Total Environ.* **2016**, *572*, 626–633. [CrossRef]
29. Yang, D.; Yang, S.; Wang, L.; Xu, J.; Liu, X. Performance of biochar-supported nanoscale zero-valent iron for cadmium and arsenic co-contaminated soil remediation: Insights on availability, bioaccumulation and health risk. *Environ. Pollut.* **2021**, *290*, 118054. [CrossRef]
30. Gao, Z.; Liu, Y.; Huang, Q.; Li, B.; Ma, T.; Qin, X.; Zhao, L.; Sun, Y.; Xu, Y. Effects of sepiolite and biochar on the photosynthetic and antioxidant systems of pakchoi under Cd and atrazine stress. *J. Environ. Sci. Health Part B-Pestic. Food Contam. Agric. Wastes* **2022**, *57*, 897–904. [CrossRef]

31. Gong, Z.T.; Lei, W.J.; Chen, Z.C.; Gao, Y.X.; Zeng, S.G.; Zhang, G.L.; Xiao, D.N.; Li, S.G. *Chinese Soil Taxonomy*; Science Press: Beijing, China, 2001.
32. Lu, R. Methods of inorganic pollutants analysis. In *Soil and Agro-Chemical Analysis Methods*; China Agriculture Science and Technique Press: Beijing, China, 2000; pp. 205–266.
33. IBI. Standardized product definition and product testing guidelines for biochar that is used in soil. In *IBI Biochar Standards*; IBI: Toronto, ON, Canada, 2012.
34. Zhang, Q.; Zhou, Y.; Tang, L.; Zhang, N.; Zhang, Z. Residue and decline study of kasugamycin in paddy water and paddy soil. *Chin. J. Pestic. Sci.* **2012**, *14*, 533–538.
35. Novak, J.M.; Busscher, W.J.; Laird, D.L.; Ahmedna, M.; Watts, D.W.; Niandou, M.A.S. Impact of Biochar Amendment on Fertility of a Southeastern Coastal Plain Soil. *Soil Sci.* **2009**, *174*, 105–112. [CrossRef]
36. Xing, Y.; Wang, J.X.; Kinder, C.E.S.; Yang, X.; Slany, M.; Wang, B.; Song, H.; Shaheen, S.M.; Leinweber, P.; Rinklebe, J. Rice hull biochar enhances the mobilization and methylation of mercury in a soil under changing redox conditions: Implication for Hg risks management in paddy fields. *Environ. Int.* **2022**, *168*, 107484. [CrossRef] [PubMed]
37. Tang, C.Y.; Liu, B.L.; Cheng, K.; Antonietti, M.; Yang, F. Moderating carbon dynamics in black soil by combined application of biochar and an artificial humic substance. *Land Degrad. Dev.* **2023**, *34*, 1352–1362. [CrossRef]
38. Cui, L.; Liu, Y.; Yan, J.; Hina, K.; Hussain, Q.; Qiu, T.; Zhu, J. Revitalizing coastal saline-alkali soil with biochar application for improved crop growth. *Ecol. Eng.* **2022**, *179*, 106594. [CrossRef]
39. You, X.; Suo, F.; Yin, S.; Wang, X.; Zheng, H.; Fang, S.; Zhang, C.; Li, F.; Li, Y. Biochar decreased enantioselective uptake of chiral pesticide metalaxyl by lettuce and shifted bacterial community in agricultural soil. *J. Hazard. Mater.* **2021**, *417*, 126047. [CrossRef]
40. Awan, S.; Ippolito, J.A.; Ullman, J.L.; Ansari, K.; Cui, L.; Siyal, A.A. Biochars reduce irrigation water sodium adsorption ratio. *Biochar* **2021**, *3*, 77–87. [CrossRef]
41. Alfonso, R.-V. Changes on the Phytoavailability of Nutrients in a Mine Soil Reclaimed with Compost and Biochar. *Water Air Soil Pollut.* **2016**, *227*, 453. [CrossRef]
42. Gao, W.K.; Yang, Y.Y.; Zong, H.Y.; Liu, J.; Song, N.N.; Wang, F.L. Simultaneously Sorptive Reduction in Cadmium and Glyphosate Diffuse loss by Biochar-Amended Soil. *Fresenius Environ. Bull.* **2020**, *29*, 4545–4555.
43. Li, R.; Wen, Y.; Liu, M.; Su, L.; Wang, Y.; Li, S.; Zhong, M.-e.; Zhou, Z.; Zhou, N. Simultaneous removal of organic inorganic composite contaminants by in situ double modified biochar: Performance and mechanisms. *J. Taiwan Inst. Chem. Eng.* **2022**, *139*, 104523. [CrossRef]
44. Yu, X.-Y.; Ying, G.-G.; Kookana, R.S. Reduced plant uptake of pesticides with biochar additions to soil. *Chemosphere* **2009**, *76*, 665–671. [CrossRef]
45. Cara, I.G.; Țopa, D.; Puiu, I.; Jităreanu, G. Biochar a Promising Strategy for Pesticide-Contaminated Soils. *Agriculture* **2022**, *12*, 1579. [CrossRef]
46. Slack, S.M.; Walters, K.J.; Outwater, C.A.; Sundin, G.W. Effect of Kasugamycin, Oxytetracycline, and Streptomycin on In-orchard Population Dynamics of Erwinia amylovora on Apple Flower Stigmas. *Plant Dis.* **2021**, *105*, 1843–1850. [CrossRef]
47. Jablonowski, N.D.; Borchard, N.; Zajkoska, P.; Fernández-Bayo, J.D.; Martinazzo, R.; Berns, A.E.; Burauel, P. Biochar-Mediated [14C]Atrazine Mineralization in Atrazine-Adapted Soils from Belgium and Brazil. *J. Agric. Food Chem.* **2013**, *61*, 512–516. [CrossRef] [PubMed]
48. Qiu, Y.; Pang, H.; Zhou, Z.; Zhang, P.; Feng, Y.; Sheng, G.D. Competitive biodegradation of dichlobenil and atrazine coexisting in soil amended with a char and citrate. *Environ. Pollut.* **2009**, *157*, 2964–2969. [CrossRef] [PubMed]
49. Cui, L.; Fan, Q.; Sun, J.; Quan, G.; Yan, J.; Hina, K.; Wang, H.; Zhang, Z.; Hussain, Q. Changes in surface characteristics and adsorption properties of 2,4,6-trichlorophenol following Fenton-like aging of biochar. *Sci. Rep.* **2021**, *11*, 4293. [CrossRef]
50. Cui, L.; Chen, T.; Quan, G.; Xiao, B.; Ma, Y.; Pan, M.; Liu, Y.; Liu, B.; Yin, C.; Yan, J.; et al. Renewable Material-derived Biochars for the Efficient Removal of 2,4-Dichlorophen from Aqueous Solution: Adsorption/Desorption Mechanisms. *Bioresources* **2017**, *12*, 4912–4925. [CrossRef]
51. Sanchez-Hernandez, J.C.; Cares, X.A.; Pérez, M.A.; del Pino, J.N. Biochar increases pesticide-detoxifying carboxylesterases along earthworm burrows. *Sci. Total Environ.* **2019**, *667*, 761–768. [CrossRef]
52. Zhang, S.; Hua, Z.; Yao, W.; Lü, T.; Chen, Y.; Fang, Z.; Zhao, H. Use of corn straw-derived biochar for magnetic solid-phase microextraction of organophosphorus pesticides from environmental samples. *J. Chromatogr. A* **2021**, *1660*, 462673. [CrossRef]
53. Mazarji, M.; Minkina, T.; Sushkova, S.; Mandzhieva, S.; Barakhov, A.; Barbashev, A.; Dudnikova, T.; Lobzenko, I.; Giannakis, S. Decrypting the synergistic action of the Fenton process and biochar addition for sustainable remediation of real technogenic soil from PAHs and heavy metals. *Environ. Pollut.* **2022**, *303*, 119096. [CrossRef]
54. Kong, L.; Liu, J.; Zhou, Q.; Sun, Z.; Ma, Z. Sewage sludge derived biochars provoke negative effects on wheat growth related to the PTEs. *Biochem. Eng. J.* **2019**, *152*, 107386. [CrossRef]

agriculture

MDPI

Article

Is Rockwool Potentially Harmful to the Soil Environment as a Nursery Substrate? Taking *Eisenia fetida* as an Example for Toxicological Analysis

Shengyang Zheng [1], Chenzhe Wang [1], Jing Ju [1,*], Qigen Dai [2], Haitao Zhao [1], Ping Liu [1] and Xin Wang [1]

1 College of Environmental Science and Engineering, Yangzhou University, Yangzhou 225127, China; syzheng@yzu.edu.cn (S.Z.); mx120210702@stu.yzu.edu.cn (C.W.); htzhao@yzu.edu.cn (H.Z.); liuping329098@163.com (P.L.); mz120211231@stu.yzu.edu.cn (X.W.)

2 Joint International Research Laboratory Agricultural & Agricultural Product Safety, Ministry Education China, Yangzhou University, Yangzhou 225009, China; qgdai@yzu.edu.cn

* Correspondence: jujing@yzu.edu.cn

Abstract: We studied the effect of rockwool matrix on the conventional physical and chemical properties of soil and analyzed its toxicological effect on *Eisenia fetida*. The physical and chemical properties of rockwool were studied with characterization tests. By measuring earthworm enzymes and earthworm intestinal microorganisms, the effects of different rockwool particle sizes and additive amount on *Eisenia fetida* were analyzed. The results indicate that a low concentration of rockwool (<30 g/kg) had little effect on the soil physicochemical properties and the activity of *Eisenia fetida*, and played a positive role in improving the soil porosity. A high concentration of rockwool (>100 g/kg) reduced the exchangeable Ca and Mg content in the soil, and had a significant impact on the enzyme activity of *Eisenia fetida*. Mechanism studies have shown that high concentrations of rockwool (>200 g/kg) can have a significant impact on the nervous system of earthworm tissue. In addition, small particle size and low concentration of rockwool is conducive to the increase in intestinal microbial species of *Eisenia fetida*. This study clarifies the effects of emerging rockwool substrates on soil and soil organisms and provides theoretical support for the safe and reliable application of rockwool substrates in agricultural production.

Keywords: rockwool; substrate; earthworm; soil; toxicology

1. Introduction

Cultivation substrates are commonly used in modern greenhouse agriculture. The physicochemical properties of the substrate determine the water and nutrient absorption performance and air content of the substrate, which not only affects the crop root system's ability to absorb and transport water and nutrients, but also affects the growth and development of the plant root system. The substrate materials used should have the following characteristics: (1) excellent permeability and water-holding capacity; (2) low impact on irrigation water and no effect on the recycling of drainage fluid; (3) no pollution; and (4) low cost [1]. Commonly used inorganic substrates include rockwool, vermiculite, sand, and perlite, and commonly used organic substrates include peat, rice husk charcoal, and bark [2–4]. One of the most used and successful soilless substrates is rockwool. The main reason for its success is that it is perfectly matched to the irrigation systems of modern greenhouses. For example, 95% of tomatoes in European regions are grown in greenhouses, and the key to this relies on rockwool slabs as a substrate as well as excellent water management and nutrient management [5]. In the Nordic regions, such as the Netherlands, where greenhouse cultivation is the main agricultural production, rockwool substrate is one of the most commonly used substrates [6], and rockwool, as an inorganic substrate, prevents the growth of some pathogens [7]. Since peat is a non-renewable resource, it is no longer used as a substrate in numerous countries [8], and it has been shown that crushed rockwool can replace peat and mixtures with alternative components in peat substrates with an optimal replacement ratio of 35% [9].

Rockwool was introduced in China in the 1990s and was popular with horticultural growers at the time, but is still mostly grown for vegetables. In the early 21st century, China used rockwool as a rice seedling substrate, but due to the immaturity of the technology and residual problems, rockwool seedling entered a relatively awkward position. In recent years, due to the improvement of the rockwool production process and the maturity of the substrate seedling technology, rockwool seedling has gradually developed.

Rockwool seedlings are soilless seedlings cultivated using rockwool as a substrate. Rockwool is an inert substrate, making it possible to control the nutrient availability and the root environment. Rockwool is weakly alkaline (pH = 7.5) and almost ion-free (electric conductivity = 50–100 μS/cm) [10]. The production process is conducted at a high temperature. Rockwool is sterile and non-toxic and is made of basalt, dolomite, and other raw materials via melting, high-speed centrifugation, blowing gas into flocculent fibers, and the addition of phenolic resin and other adhesives via pendulum, wrinkling, drying, and other technical processes [11]. Rockwool is light in texture, with a porosity of 90–96%. The water absorption rate of rockwool can reach 90%, and rockwool exhibits strong permeability, uniform pore spaces, and better permeability than other substrates, such as mudstone [12]. Moreover, the application of rockwool saves labor and effort in seedling cultivation; rockwool also inhibits weed growth and can be sown, mulched, and fertilized at the same time to improve efficiency. The residual amount of rockwool was also reduced to 0.8 g/m^2 when mechanical planting was performed. Since the main components of rockwool are basalt, dolomite, and other soil components and are not easily degraded, rockwool can be collected and treated together with the roots of rice.

The current research on rockwool is mostly focused on its application as a substrate for seedlings and partly on its effect on the microbial community. For example, Gundersen V studied the comparative effect of rockwool as a substrate on the concentration of major elements and for elements in tomatoes [6]; G. Carmassi used rockwool as a cultivation substrate to establish a comprehensive model of water requirements for closed-loop soilless cultivation of room-temperature tomatoes [13], studying the comparison of rockwool and coir fiber on the growth of greenhouse cucumbers [14]; and Prommart K studied the quantitative characteristics of four different substrates, including rockwool, on the population density and population dynamics of indigenous microorganisms [15]. However, due to the non-degradable nature of rockwool, researchers have not specifically investigated the physicochemical properties of rockwool on soil and its toxicological effects on earthworms. Therefore, basic data on the effects of rockwool on soil physicochemical properties and on earthworm toxicology need to be obtained.

Hence, the aim of this study was to investigate the effects of rockwool on the physicochemical properties of soil and the toxicological effects on earthworms. To address the abovementioned issues, this work selected the common earthworm *Eisenia fetida* as an earthworm representative and used an integrated approach of determining the earthworm biomass and enzyme activity, as well as conducting 16SrRNA sequencing and macrogenetic analysis [16–18], to assess the risk of rockwool in regard to soil biotoxicity. The effects of rockwool addition on earthworm activity in the presence of earthworms and on soil physicochemical properties were evaluated. This study provides the necessary data to guide the rational application of rockwool as a seedling substrate, and the results obtained will help the safe and reliable application of rockwool substrate in agricultural production.

2. Materials and Methods

2.1. Test Materials

The test materials were soil, rockwool, and earthworms.

Soil: The soil was collected from the practical training base of Yangzhou University, Jiangsu Province, China. The soil used in the experiment was Anthrosols-Plaggic. The soil type was chalky loam with low nutrient content. The basic physicochemical properties were pH 7.42, an electrical conductivity of 139.5 μS/cm, 12.85 g/kg organic matter, 0.67 g/kg total nitrogen, and 12.77 mg/kg effective phosphorus. The indicators are based on soil dry weight. The soil was air-dried, sieved through a 2 mm sieve, and prepared for earthworm cultivation.

Rockwool: Rockwool was obtained from Huaian Hand Agricultural Technology Co. (Huai'an City, China).

Earthworms: Earthworms were purchased from Wang Jun Earthworm Farm in Jurong City, Jiangsu Province, China. The earthworms were placed in a plastic box lined with moist filter paper, sealed with a perforated plastic lid, and placed in a constant temperature and humidity incubator (temperature 20 °C, humidity 60–70%) for 24 h. The filter paper was changed during the test, and the earthworms were removed and washed after the test.

2.2. Experimental Design

The experimental design was divided into 11 treatments, and each treatment was added to the corresponding rockwool, as shown in Table 1. Three parallel trials were set up for each treatment. A black opaque plastic box of 40 cm × 20 cm × 30 cm was used as the culture container for the experiment, and each pot was filled with 1 kg of air-dried soil. Deionized water was added to the soil to maintain the soil moisture content at 60% of the maximum water-holding capacity, and the soil was fully stirred. Forty washed earthworms were taken after 24 h, weighed, and placed on the soil surface. Soil without rockwool was used as the control group. During the experiment, the boxes were placed in an incubator at 20 ± 2 °C and covered with a perforated plastic cover to ensure adequate ventilation. Soil moisture was weighed every other day and water was added if needed. Because rockwool is an inert and hard-to-degrade substrate, this study evaluated the response of earthworms to rockwool exposure in the med-term (28 days sub-chronic phase) [19].

Table 1. Earthworm treatments.

No.	Treatment	Explanation
1	CK	1 kg of soil without rockwool
2	T10M1	Add 10 g of 10-mesh sieved rockwool to 1 kg of soil
3	T10M3	Add 30 g of 10-mesh sieved rockwool to 1 kg of soil
4	T10M5	Add 50 g of 10-mesh sieved rockwool to 1 kg of soil
5	T10M10	Add 100 g of 10-mesh sieved rockwool to 1 kg of soil
6	T10M20	Add 200 g of 10-mesh sieved rockwool to 1 kg of soil
7	T20M1	Add 10 g of 20-mesh sieved rockwool to 1 kg of soil
8	T20M3	Add 30 g of 20-mesh sieved rockwool to 1 kg of soil
9	T20M5	Add 50 g of 20-mesh sieved rockwool to 1 kg of soil
10	T20M10	Add 100 g of 20-mesh sieved rockwool to 1 kg of soil
11	T20M20	Add 200 g of 20-mesh sieved rockwool to 1 kg of soil

2.3. Characterization and Determination of Rockwool Morphology

The morphology of the rockwool was observed and analyzed using a field emission scanning electron microscope (Hitachi S-4800, Chiyoda City, Japan) equipped with energy-dispersive X-ray energy spectroscopy (EDS). The specific surface area and pore size distribution of the rockwool were obtained by measuring the N_2 adsorption–desorption curves using a specific surface analyzer (BET, Autosorb-iQ, Shanghai City, China). The water retention capacity of the rockwool at high temperatures was simulated with an oven.

2.4. Determination of Physical and Chemical Properties of Soils

The organic matter content of the soil was determined based on the national standard "Determination of soil organic matter" (NY/T 85-1988) [20]. The Ca^{2+} and Mg^{2+} contents of the soil were determined based on the national standard "Determination of cation exchange and exchangeable salt group in neutral soil" (NY/T 295-1995) [21]. The SiO_2 content in the soil was determined based on the standard "Determination of effective silicon in soil" (NY/T 1121.15-2006) [22].

2.5. Determination of Earthworm Avoidance Rate

The earthworm avoidance rate behavior test was performed using the two-chamber method of the International Organization for Standardization ISO 2007. A clear plastic box of 40 cm × 20 cm × 30 cm was selected, a baffle was inserted in the middle, and 500 g of test soil was placed on one side and clean uncontaminated soil on the other side. The baffle was removed and 10 earthworms were placed in the middle line and sealed with gauze to ensure earthworm movement. After 48 h of incubation, the baffles were inserted back into place and the number of earthworms in the soil on each side was counted based on the position of the earthworm's head pointing. Each group was set up in three parallels. The formula for calculating the earthworm avoidance rate is shown below:

$$NR = \frac{C - T}{N} \times 100\% \quad (1)$$

where NR is the net earthworm avoidance rate; C is the number of earthworms in the control soil; T is the number of earthworms in the test group; and N is the total number of earthworms added.

Earthworms in the fasted state were removed and washed the day before the test, placed in a plastic box lined with moistened filter paper, sealed with a perforated plastic lid, placed in a constant temperature and humidity incubator (25 °C, 60–70% humidity), and purged for 24 h. The filter paper was changed during the period, and the earthworms were removed and washed after purging and placed in the test box. Earthworms in a satiated state did not need to be fasted and were cleaned and placed directly into the test box.

2.6. Determination of Enzyme Activity of Earthworms

The determination of earthworm enzymatic activity is uniformly performed by the kit-microdosing method. The test was performed by Suzhou Kemin Bio on behalf of Suzhou Kemin Biological Company (Suzhou, China). In order to reflect the effect of rockwool on earthworms, the highest and lowest concentrations of 10-mesh rockwool and 20-mesh rockwool were selected for the assay of enzyme activity, i.e., five treatments of CK, T10M1, T10M20, T20M1, and T20M20.

Earthworm soluble protein content: under alkaline conditions, cysteine, cystine, tryptophan, tyrosine, and peptide bonds in the protein can reduce Cu^{2+} to Cu^{+}; two molecules of bicinchoninic acid combine with Cu^{+} to produce a purple complex with absorption peaks at 540–595 nm, with the strongest absorption peak at 562 nm.

GST: the spectrophotometer/enzymatic standard was preheated for 30 min, and the wavelength was adjusted to 340 nm and zeroed with distilled water. Reagent III was kept at 25 °C (general species) or 37 °C (mammalian) to keep it warm. GST catalyzes the binding of glutathione to 2,4-dinitrochlorobenzene (CDNB), and the light absorption peak of its binding product is at 340 nm; the GST activity can be calculated by measuring the rate of absorbance rise at 340 nm.

SOD: the superoxide anion (O^{2-}) is produced by xanthine and the xanthine oxidase reaction system. O^{2-} can react with WST-8 to produce the water-soluble dye methyl wax, which shows absorption at 450 nm, and the SOD activity can be calculated by measuring the absorbance at 450 nm.

CAT: H_2O_2 has a characteristic absorption peak at 240 m. CAT is able to decompose H_2O_2 so that the absorbance of the reaction solution at 240 nm decreases with reaction time, and the CAT activity can be calculated from the rate of change in absorbance.

T-AOC: the total antioxidant capacity is reflected by the ability of antioxidant substances to reduce Fe^{3+}-tripyridyltriacridine (Fe^{3+}-TPTZ) to produce blue Fe^{2+}-TPTZ in an acidic environment.

MDA: MDA was condensed with thiobarbituric acid (TBA) to produce a red product with a maximum absorption peak at 532 nm, which can be used to estimate the content of lipid peroxide in the sample after colorimetric analysis; in addition, the absorbance at 600 nm was measured, and the difference between the absorbance at 532 and 600 nm was used to calculate the MDA content.

CL: the anthrone colorimetric method was used to determine the content of reducing sugars produced by the CL-catalyzed degradation of sodium carboxymethyl cellulose.

AchE: AchE catalyzes the hydrolysis of Ach to produce choline, which interacts with dithiobenzoic acid (DTNB) to produce 5-mercapto-nitrobenzoic acid (TNB), which has an absorption peak at 412 nm. The AchE activity was calculated by measuring the rate of increase in absorbance at 412 nm.

2.7. Determination of Earthworm Macrogenes

Fresh tissue samples were taken, washed with phosphate-buffered saline to remove surface microorganisms, and wiped with anhydrous ethanol. The tissues were then blocked, snap-frozen using liquid nitrogen, and stored at −80 °C for subsequent sequencing. Macrogenome sequencing was conducted based on Illumina high-throughput sequencing, using the paired-end sequencing method to construct small fragment libraries for sequencing, and was performed by Beijing BioMarker Technologies Co., Ltd. (Beijing, China). After the sample genomic DNA was tested and qualified, the DNA was broken and the fragmented DNA was subjected to end repair. The 3′-end addition of A, ligation of sequencing connectors, ligation of product purification sieves, library amplification, and product purification were performed to construct a sequencing library, which was sequenced using the Illumina sequencer after qualified library quality control. The raw reads obtained from sequencing were quality-controlled and filtered to obtain clean reads for subsequent bioinformatics analysis. The clean reads were assembled by splicing, predicting coding genes, constructing non-redundant gene sets, and performing functional annotation and taxonomic analysis on the non-redundant gene sets in both the general and specialized databases, counting the sample species composition and abundance information, and completing the analysis closure report based on BMKCloud.

Treatments for the assay: in order to represent the effect of rockwool on earthworms, the highest and lowest concentrations of 10-mesh rockwool and 20-mesh rockwool were selected for the macrogenome assay, i.e., five treatments of CK, T10M1, T10M20, T20M1, and T20M20.

2.8. Statistical Analysis

The data were statistically analyzed using Excel software and SPSS 23.0 software, and using the one-way analysis of variance model test. Differences between treatments were shown to be significant using the new complex polarization method (Duncan analysis), $p < 0.05$. OriginPro 9.1 software was used for plotting. The macrogenetic sequencing data were analyzed on the BMKCloud platform.

3. Results and Discussion

3.1. Material Characterization

3.1.1. Morphological and Elemental Analysis of Rockwool

According to the scanning electron microscopy images, elemental mapping, and EDS spectra of rockwool shown in Figures 1 and 2, it can be seen that rockwool has a fibrous structure and contains irregular spindles inside, which are mainly composed of C^{4+}, O^{2-}, Si^{4+}, Ca^{2+}, and Mg^{2+}, among which the O^{2-}, Si^{4+}, and Ca^{2+} contents are higher and the C^{4+} and Mg^{2+} contents are lower.

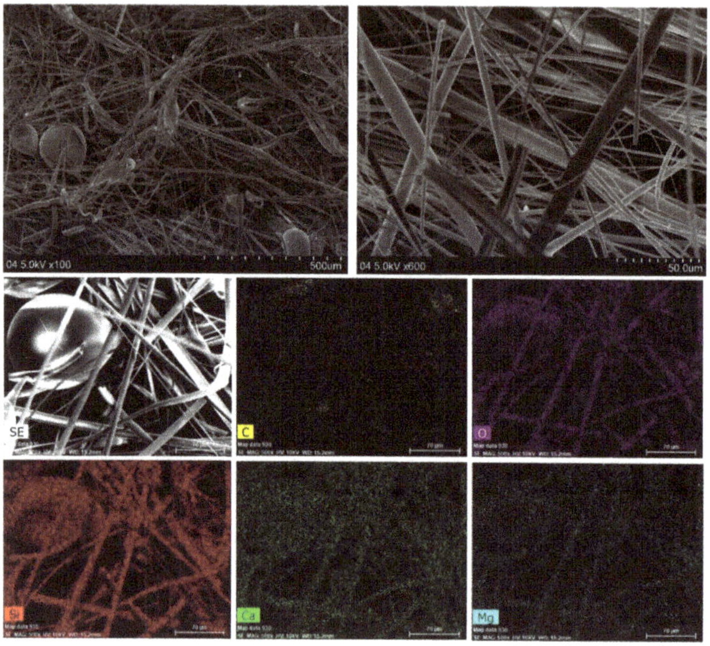

Figure 1. SEM image and element mapping of rockwool.

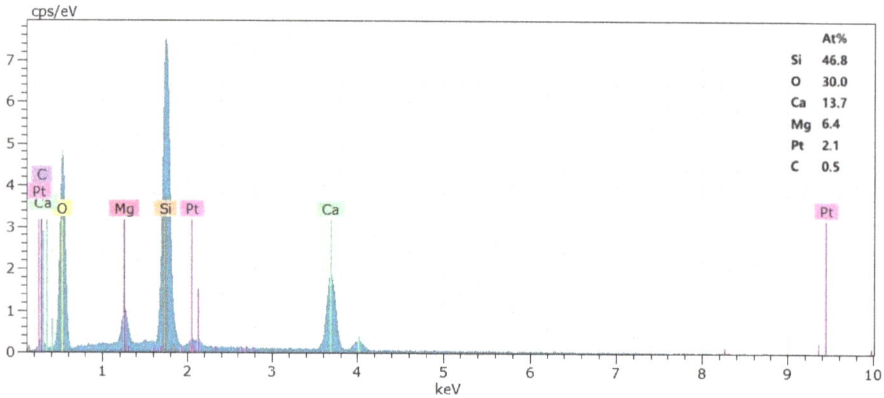

Figure 2. EDS profile of rockwool.

3.1.2. BET Analysis

N_2 isothermal adsorption experiments were conducted to further investigate the porosity and specific surface area of the materials. As shown in Table 2, the specific surface area of rockwool was 2.74 $m^2 \cdot g^{-1}$, the average pore volume was 0.01 $cm^3 \cdot g^{-1}$, and the average pore size was 4.31 nm.

Table 2. BET surface area and pore characteristics of rockwool.

Materials	Specific Surface Area ($m^2 \cdot g^{-1}$)	Pore Volume ($cm^3 \cdot g^{-1}$)	Average Pore Size (nm)
Rockwool	2.74	0.01	4.31

3.1.3. Water Retention Capacity of Rockwool

The same thickness of material and the same mass of pure water were placed into Petri dishes, and a high-temperature evaporation experiment was conducted to simulate the water retention capacity of various materials. As shown in Figure 3, at 30 °C, the rockwool contained 19.41 g of water after 8 h of evaporation, the vermiculite substrate with a water retention agent added had 18.92 g of water remaining, and the amounts of water remaining in the Petri dishes of pure vermiculite substrate, pure soil, and pure water were 17.66, 17.86, and 17.87 g, respectively. At 40 °C, evaporation was faster in the first 2 h because of the high surface moisture of each material, but after that the decline of the residual moisture was linear up to 8 h. At 8 h, rockwool was the best water retention material with 13.01 g of water remaining, followed by vermiculite matrix with a water retention agent added with 12.13 g of water remaining, pure vermiculite matrix with 11.07 g of water remaining, pure soil with 14.50 g of water remaining, and pure water with 9.35 g of water remaining. This experiment demonstrated that rockwool had the best water retention capacity, followed by vermiculite substrate with a water retention agent, and pure soil had the worst water retention capacity. The higher the temperature, the more significant the water retention capacity of rockwool.

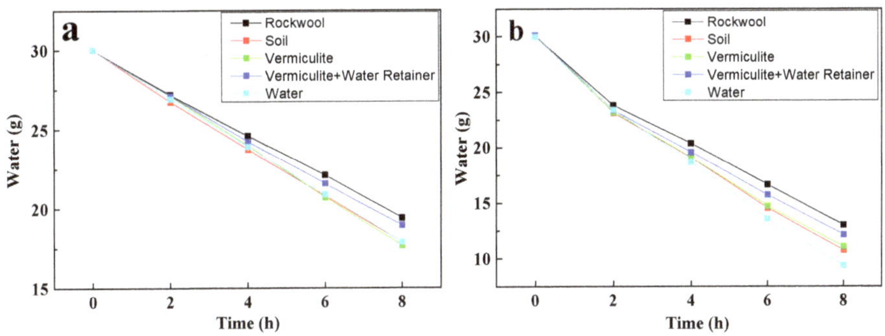

Figure 3. Water retention capacity of each material at (**a**) 30 °C and (**b**) 40 °C.

3.1.4. Summary

The above characterization indicates that rockwool has a fibrous structure, high porosity, and good air permeability in addition to an outstanding water retention capacity. According to the elemental analysis, its elemental composition is basically the same as that of the soil, and it contains Ca^{2+}, Mg^{2+}, and SiO_2, which are beneficial to plant growth and meet the requirements for use as a substrate [1]. In addition, silicon fertilizer and calcium–magnesium phosphate fertilizer are used for rice in the form of SiO_2, CaO, and MgO, which are the main components of rockwool. It has been found that the application of silica fertilizer can thicken rice stalks, improve the mechanical strength of stalks, and effectively enhance the resistance of rice to collapse [23]. Therefore, the use of rockwool as a seedling substrate may promote the growth and development of seedlings.

3.2. Analysis of Soil Physical and Chemical Properties

3.2.1. Changes in Organic Matter Content in the Soil

Soil organic matter (SOM) alters the physicochemical reactions that affect the active composition of micronutrients [24,25]. As shown in Figure 4, the SOM content tended to decrease as the rockwool content in the soil increased, but the particle size of rockwool had no significant effect on the SOM content. This suggests that the increase in rockwool content has an effect on the normal activities of earthworms, thereby decreasing the degree of soil fecalization and thus the SOM content. However, the SOM content was slightly increased at low concentrations of 10% rockwool, indicating that rockwool at that concentration could be used as an additive to change the soil interstices without affecting the activities of earthworms.

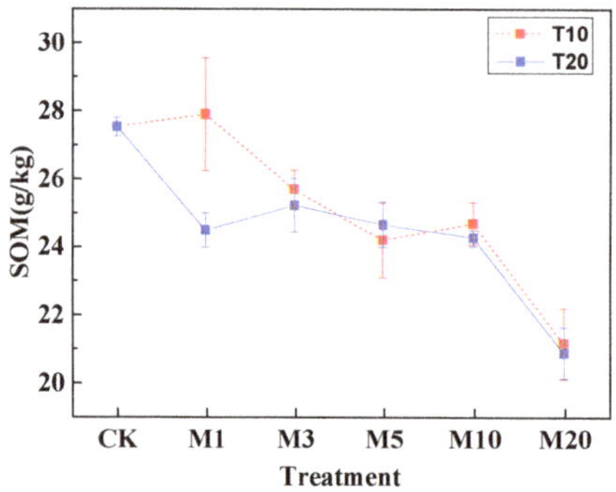

Figure 4. Changes in soil organic matter content.

3.2.2. Changes of Ca^{2+}, Mg^{2+}, and SiO_2 in the Soil

As shown in Figure 5a,b, as the content of rockwool in soil increased, the calcium and magnesium content in soil exhibited a decreasing trend, but the change in calcium and magnesium content only differed significantly at the highest rockwool concentration. The effect of 20-mesh rockwool on the calcium–magnesium–silica content of soil was smaller than that of 10-mesh rockwool. This indicates that the increase in rockwool content has little effect on the calcium and magnesium content in soil. As shown in Figure 5c, the concentration content of rockwool in the soil has a large effect on the effective silicon content in the soil. After the addition of rockwool, the effective silica content in the soil was showed a decreasing trend compared to the blank control. The decrease of 10-mesh rockwool was 7.25~17.10%, and the decrease of 10-mesh rockwool was 15.22~23.42%.

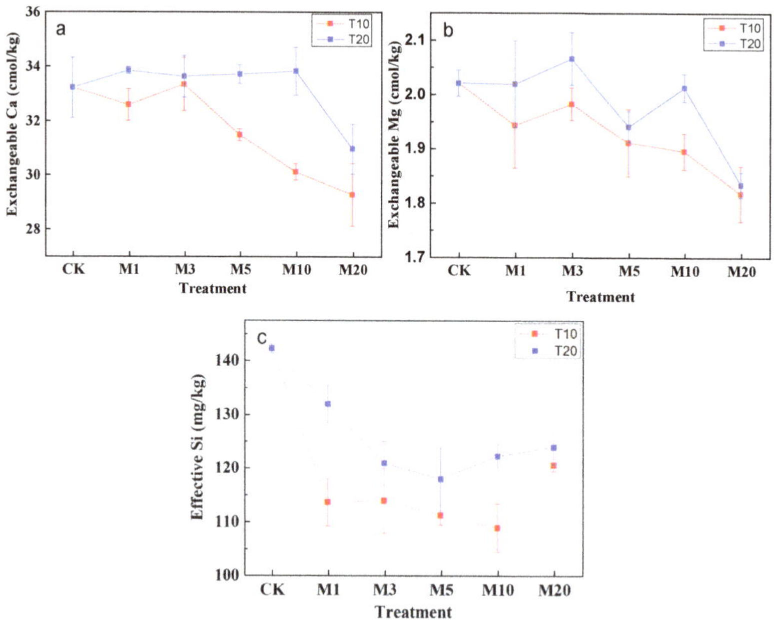

Figure 5. Exchangeable calcium content (**a**), exchangeable magnesium content (**b**), and effective silicon content (**c**) in the soil.

From the three figures, it is clear that the high concentration of rockwool addition has the highest effect on the effective state of Ca^{2+}, Mg^{2+}, and SiO_2 in the soil. The reason for this is that the high concentration of rockwool reduces the activity of earthworms, thus affecting the Ca^{2+}-Mg^{2+}- SiO_2 content in the soil.

3.3. Effect of Rockwool on the Survival of Earthworms

3.3.1. Survival of Earthworms in Rockwool-Amended Soil

From Figure 6, it can be seen that the survival rate of earthworms decreased after the addition of 10% rockwool, but there was no significant change in the survival rate of earthworms with the increase in the amount of rockwool. After the addition of 20% rockwool, the survival rate of earthworms decreased gradually with the increase in the amount of rockwool, and the lowest survival rate was 59.17% with the addition of 20% rockwool. However, the lowest earthworm survival rate after the addition of powdered rockwool was still higher than that after the addition of most granular rockwool, indicating that the size of rockwool had a significant effect on the survival rate of earthworms, and the larger the granules, the lower the survival rate of earthworms. The possible reason for this is that the granular rockwool (10-mesh) is more irritating to earthworms.

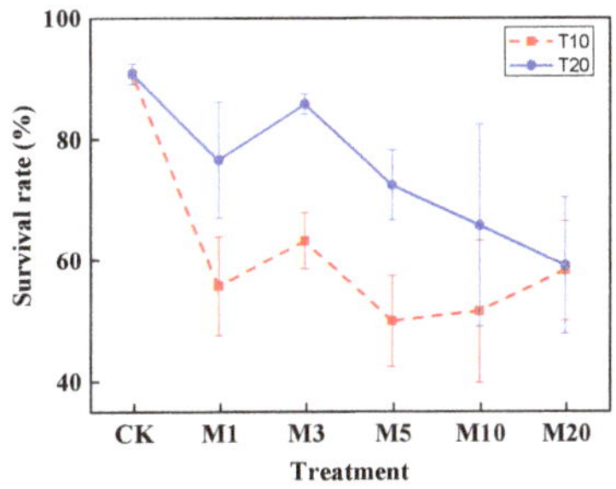

Figure 6. Change in survival rate of earthworms in added soil.

3.3.2. Earthworm Avoidance Rate Test

Earthworms exhibit rejection behaviors for unfavorable soil, and different behavior patterns when they are satiated and starved. As can be seen from Figure 7, when the rockwool consisted of large particles and the concentration was 10 g/kg and below, i.e., under low-concentration conditions, the earthworm avoidance rate was less than 20% and the avoidance behavior was not obvious, and the pollutant probably had no obvious toxic effect on the earthworms and did not constitute pollution stress. With the addition of rockwool, the avoidance behavior of earthworms gradually became obvious, indicating that the coercive effect of rockwool on earthworms gradually increased with an increasing amount of rockwool. There was no significant difference between the avoidance rate of earthworms in the satiated and starved states.

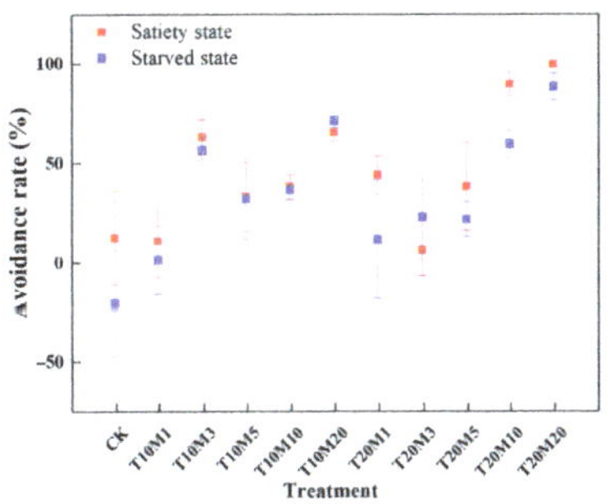

Figure 7. Evasion rate of earthworms in the satiated and starved states.

3.3.3. Summary

Earthworm avoidance behavior is considered to be convincing proof for soil quality assessment [26]. Due to the incorporation of rockwool, the survival and avoidance results shown by earthworms in soils containing high concentrations of rockwool were consistent with a greater risk to earthworms. However, there was little variability between satiated and starved earthworms, suggesting that high concentrations of rockwool are for the most part a valid hazard for earthworms [27,28]. In contrast, low concentrations of rockwool can be beneficial to earthworms in specific situations [29].

3.4. *Effect of Rockwool on the Enzymatic Activity of Earthworms*

3.4.1. Changes in the SP Content of Earthworm Tissues

As shown in Figure 8, the content of earthworm tissue SP gradually increased with increasing rockwool concentrations at 10-mesh levels, and the highest tissue SP content was observed at the highest rockwool concentration. This indicated that as the concentration of rockwool increased, the earthworm tissues accelerated the breakdown of metabolic proteins to obtain energy to resist the stress caused by the addition of rockwool. Similarly, Yang [30] found that SP levels in gophers were increased when they were subjected to external stimuli. With the addition of 20-mesh rockwool, the SP content of earthworm tissues increased and then decreased, and was highest at the lowest concentration, but was still lower than the highest value under the 10-mesh treatment, indicating a strong adaptation of earthworms to the powdered rockwool. This may have occurred because the 20-mesh-treated rockwool was less stimulating to earthworms than the 10-mesh-treated rockwool.

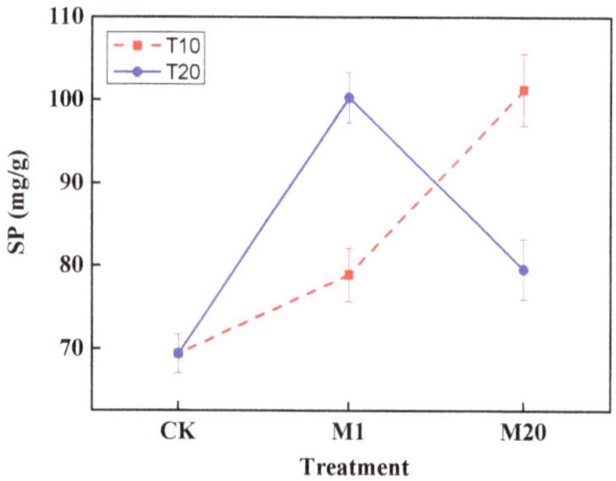

Figure 8. Variation of SP content of earthworm tissue in different soil additions.

3.4.2. Changes in the Antioxidant Capacity of Earthworm Tissues

Using earthworms as test organisms to assess the toxicity of pollutants to soil, biomarkers are often used as diagnostic and predictive early warning indicators of soil environmental pollution stress. Antioxidant activities, such as SOD and CAT activities, scavenge excess reactive oxygen species (ROS), thereby protecting earthworm cells from the adverse effects of ROS [31]. Normally, SOD and CAT activities can be kept in dynamic balance to meet the needs of the organism and eliminate ROS; however, this balance can be easily disrupted if the organism is subjected to environmental stress [32,33]. GST is a powerful detoxifying enzyme in biological systems that promotes the binding of GSH to electrophilic substrates in toxic components to produce thiol uric acid and excrete it from the body [34]. As shown in Figure 9, the level of enzyme activities of earthworm tissues differed among treatments, and the activities of T-AOC and SOD did not change significantly among the five treatments. In the 10-mesh rockwool environment, earthworm tissue GST activity gradually decreased with the increasing rockwool concentration, and under the influence of the high rockwool concentration there was a significant decrease in GST activity of 37.5% compared with the control (CK). Meanwhile, CAT activity was directly decreased and all earthworm tissues in the 10-mesh treatment exhibited significant differences in CAT activity compared with CK, with decreases of 34.3% and 35.6%, respectively, in the M1 and M20 treatments. In the 20-mesh rockwool environment, the GST activity of earthworm tissues was the lowest in the M1 treatment, but was still higher than the lowest value in the 10-mesh treatment, while the GST content of earthworm tissues increased instead under the influence of high rockwool concentration, although the GST content was not significantly different from that in CK. By contrast, the CAT activity gradually decreased with the increasing rockwool concentration, but the lowest value in the 20-mesh treatment was still higher than the CAT activity in the 10-mesh treatment [35,36].

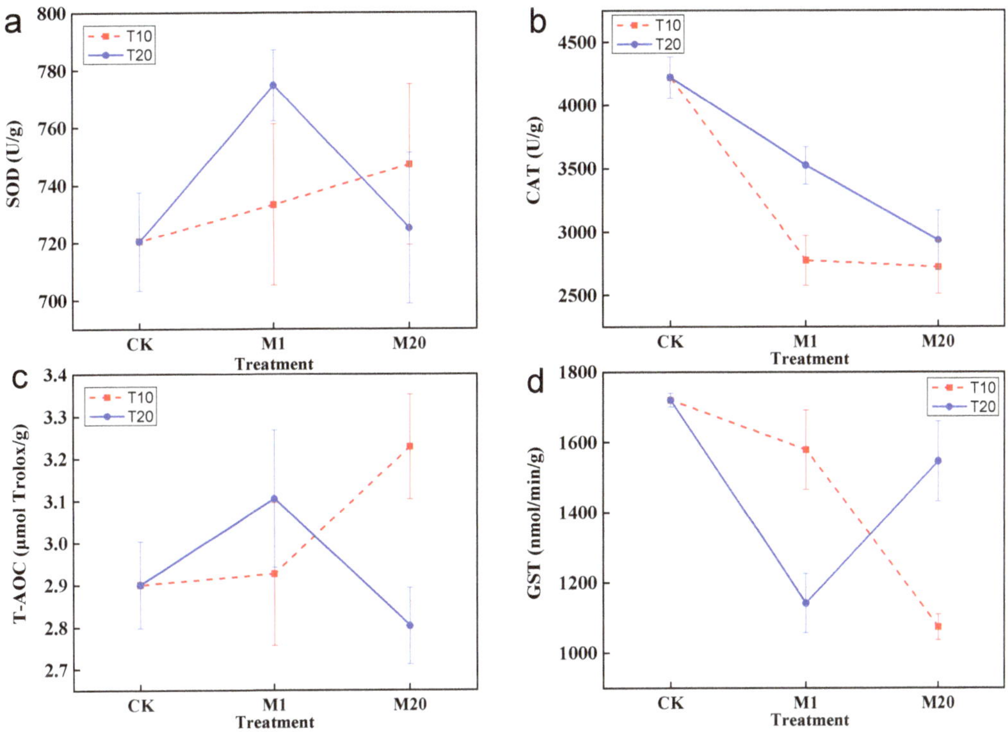

Figure 9. Changes in the activities of SOD (**a**), CAT (**b**), T-AOC (**c**), and GST (**d**) of earthworm tissues in different treatments.

In summary, the variation of the antioxidant capacity of earthworm tissues showed that the effect of a high concentration of 10-mesh rockwool on earthworms was higher than that of 20-mesh rockwool, but the effect of a low concentration of 10-mesh rockwool on earthworms was less than that of 20-mesh rockwool. The large particles of 10-mesh rockwool, compared with 20-mesh powdered rockwool, were more likely to affect the normal activities of earthworms in the soil and disrupt the internal balance of earthworms, while the high concentration of 20-mesh rockwool excluding CAT showed a tendency to return to CK treatment. The reason for this may be that the high concentration of 20-mesh rockwool is in powder form and when mixed with soil, it allows earthworms to adapt to the mixed soil faster than the 1% concentration treatment due to the 20% concentration share, therefore returning the earthworm's antioxidant capacity to near-normal values.

3.4.3. Changes in Oxidative Damage Indicators in Earthworm Tissues

The MDA content reflects the degree of peroxidation of body lipids in earthworms and briefly reflects the degree of cellular damage, where the higher the content, the greater the degree of damage [31,33,37]. As shown in Figure 10 (left), the MDA activity of earthworm tissues increased significantly with the addition of rockwool, especially at the 1% concentration, from 10.73 to 12.81 nmol/g, but there was no significant difference in the MDA content of all rockwool addition treatments [36]. This finding showed that the addition of rockwool caused cell damage in the earthworms, but that the effect was not dependent on the amount of rockwool added.

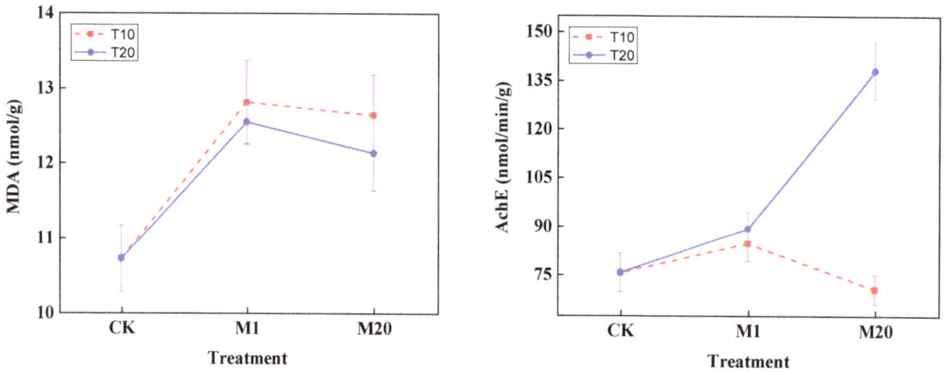

Figure 10. Changes in MDA and AchE contents of earthworm tissues in different soil additions.

AchE is a key enzyme in biological nerve conduction. Between cholinergic synapses, the enzyme degrades acetylcholine, terminating the excitatory effect of the neurotransmitter on the postsynaptic membrane and ensuring the normal transmission of nerve signals in biological tissues [38]. As shown in Figure 10 (right), only the highest concentration of 20-mesh powdered rockwool affected the nervous system of earthworms, with AchE increasing from 75.87 to 138.29 nmol/min/g, with no significant change in the remaining treatments, similar to the findings of LUO Y-R [39]. This finding showed that the addition of a high concentration of 20-mesh rockwool caused significant changes in AchE content, indicating significant effects on the nervous system of earthworms, but the addition of 10-mesh rockwool had no significant effect on the AchE content of earthworm bodies.

3.4.4. Changes in CL Activity of Earthworm Tissues

CL is an important digestive enzyme that directly affects the ability of earthworms to break down plant litter and other cellulosic material and is therefore widely used as a biomarker to detect the effects of pollutants on earthworms [39,40]. As can be seen from Figure 11, compared with CK, there was a significant increase in CL content in all treatments except for under the low concentration of 20-mesh powdered rockwool, indicating that the earthworm bodies in these groups produced higher CL content in their tissues to obtain more energy to survive to obtain sufficient nutrients. KIM S study found that earthworms had upregulated cellulase activity after attack by LPS, Gram-negative, or Gram-positive bacteria [41]. This finding showed that the 10-mesh granular rockwool and the high concentration of 20-mesh powdered rockwool had an effect on the survival of earthworms.

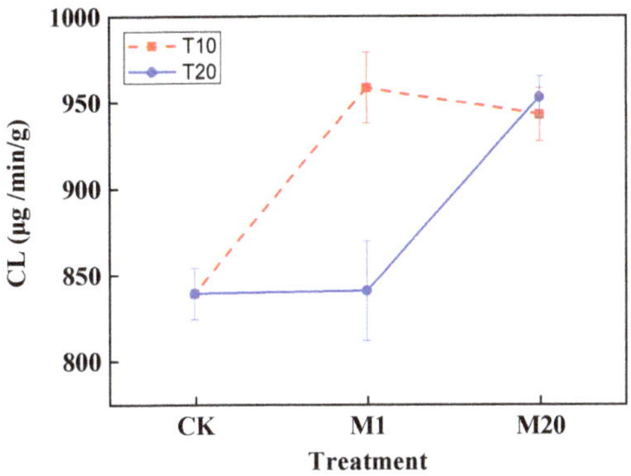

Figure 11. Variation of earthworm tissue CL content in different soil additions.

3.4.5. Summary

Measurement of the enzyme activity of earthworms showed that 10-mesh rockwool had a greater effect on earthworms than 20-mesh rockwool. The results are consistent with the survival and avoidance rates of earthworms described above. In the experiment, the external stimulation of earthworms by rockwool caused some damage to their nervous system [30]. In contrast, the addition of 20-mesh high concentration of rockwool reduced some of the enzyme activities of the earthworms to control levels. It was shown that earthworms were more adapted to the contaminated soil than to the soil with 10-mesh rockwool added during the 28-day test period [42]. Therefore, the optimal rockwool addition concentration is 10 g/kg of 20-mesh rockwool, taking into account the survival and evasion rates.

3.5. Determination of Macrogenes in the Earthworm Intestine

3.5.1. Analysis of Species Composition

The microbial composition of the earthworm gut varied among treatments at the species level (Figure 12), with 1016 microorganisms common to each treatment and 43, 25, 11, 52, and 15 microorganisms unique to the six treatments CK, T10M1, T10M20, T20M1, and T20M20, respectively. This indicated that the addition of rockwool changed the microbial composition of the earthworm gut. In addition, 43 unique microbial communities were observed in CK, indicating that these microorganisms were sensitive to rockwool addition. From Figure 13, it can be seen that the more rockwool added, the fewer the microbial species, and the smaller the rockwool particle size, the more microbial species. This demonstrated that a small amount of small-particle-size rockwool had an elevated effect on the earthworm gut microorganisms. This may have been due to the fact that a small amount of small-particle-size rockwool had less of an effect on earthworms, but had a greater effect on the gut microorganisms due to the lower concentration and no decline in microbial species, but an increase in species for self-protection.

In terms of beta diversity analysis (Figure 14), non-metric multidimensional scaling suggested the separation of T10M20 and the remaining four groups of treated specimens, indicating that the addition of large particle sizes and a high concentration of rockwool resulted in significant changes in the structure of the earthworm gut microbial community. This finding was consistent with the previous results of enzyme activity analysis. According to the hierarchical clustering tree, the clusters of T10M20 were significantly separated from the clusters of the remaining four groups of treatments. The clusters of CK, T10M1, T20M1, and T20M20 were grouped together.

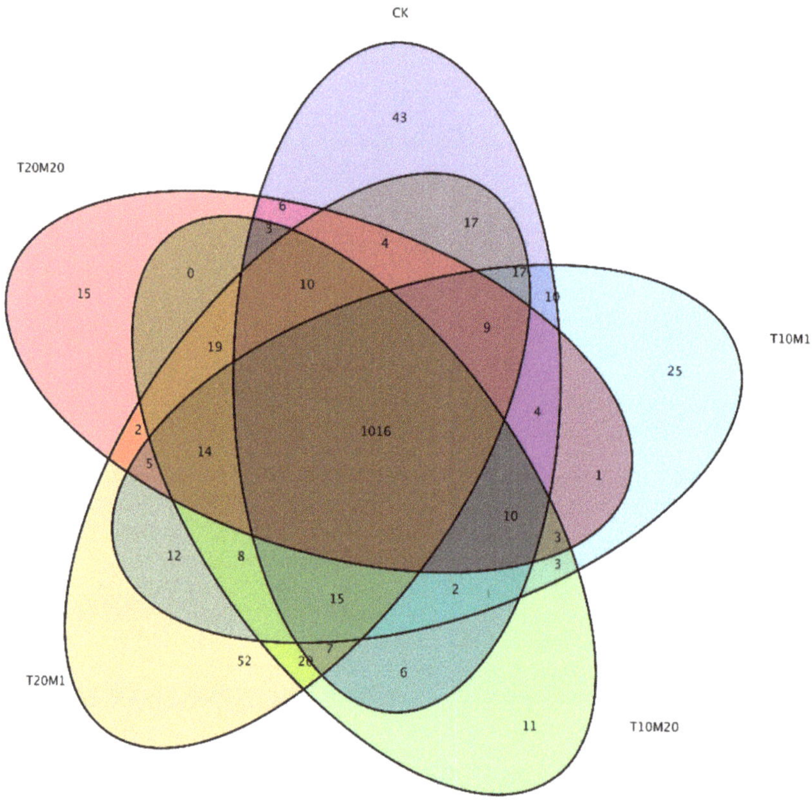

Figure 12. Venn diagram of the earthworm gut microbial community at the genus level.

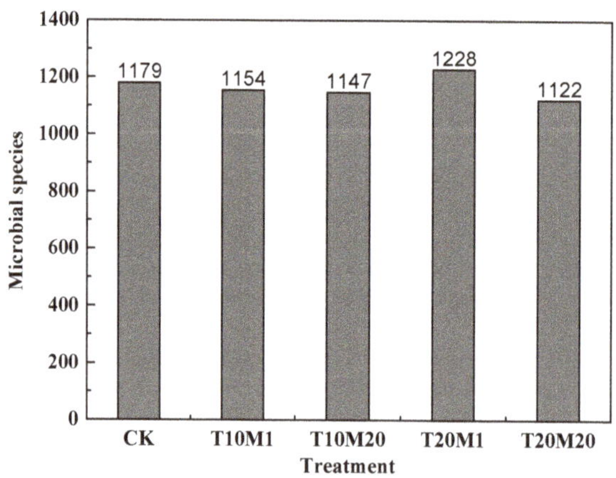

Figure 13. Number of earthworm gut microbial species at the species level.

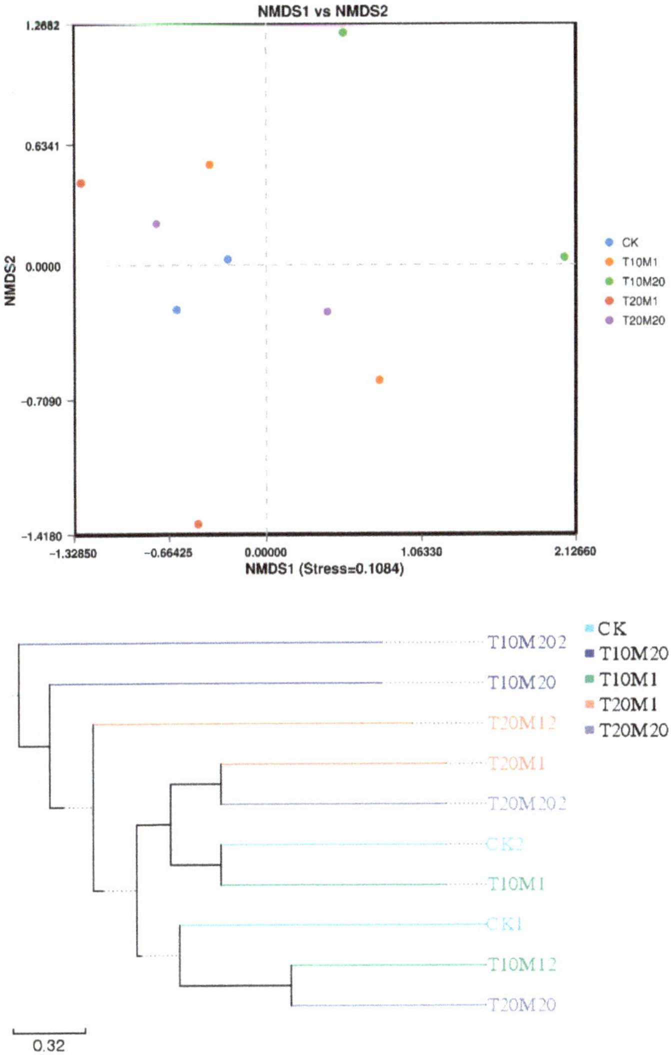

Figure 14. Beta diversity (change in community structure) of soil microbial communities among the five taxa. Top: non-metric multidimensional scaling (NMDS). Bottom: hierarchical cluster analysis (UPGMA).

3.5.2. Characterization of Functional Genes in Earthworms

The highest abundance of ko01100, ko01110, ko00190, and ko01130 in the Kyoto Encyclopedia of Genes and Genomes (KEGG) metabolic pathway is shown in Figure 15, where ko01100, ko01110, and ko01130 belong to the global and overview maps of the secondary metabolic pathway and ko00190 belongs to the energy metabolism secondary metabolic pathway. The abundances of most metabolism-related pathways were higher in the T20M1-treated soils than in the remaining treatments, while there was no significant difference between the remaining treatments.

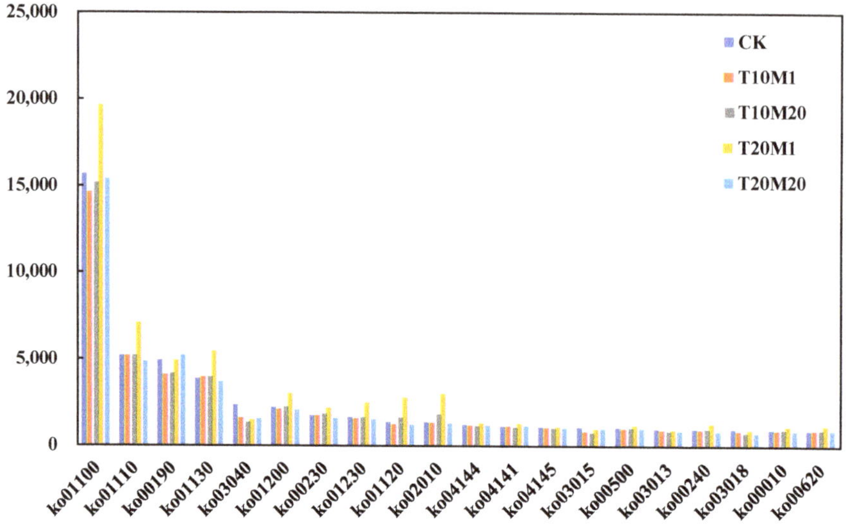

Figure 15. Abundance map of the top 20 dominant pathways.

In addition, as shown by the abundance of the top 20 dominant pathways listed in Table 3, many of the highly abundant pathways fell into the metabolism category, including global and overview maps (ko01100, ko01110, ko01130, ko01200, ko01230, and ko01120), energy metabolism (ko00190), nucleotide metabolism (ko00230 and ko00240), sugar metabolism (ko00500, ko00010, and ko00620), cellular transport and catabolism (ko04144 and ko04145), translation (ko03015 and ko03013), transcription (ko03040), folding, sorting, and degradation (ko04141 and ko03018), and membrane transport (ko02010).

Table 3. Differences in abundance of the 20 dominant pathways at the second level ("a" indicates the variability between the data).

KEGG Pathway Level 1	KEGG Pathway Level 2	Abundance				
		CK	T10M1	T10M20	T20M1	T20M20
Cellular Processes	Cell growth and death	283.75 ± 23.75 a	299.25 ± 34.75 a	266.00 ± 25.50 a	324.00 ± 46.50 a	330.25 ± 32.25 a
	Transport and catabolism	1150.5 ± 60.5 a	1120 ± 69.5 a	1085.5 ± 59.5 a	1206.25 ± 113.25 a	1110.75 ± 76.25 a
Genetic Information Processing	Translation	1033 ± 49 a	873 ± 38 a	811 ± 45.5 a	962 ± 29.5 a	923.5 ± 80 a
	Transcription	2354	1612.5	1356	1491	1554.5
	Folding, sorting, and degradation	1051.25 ± 89.75 a	993.75 ± 138.75 a	916.5 ± 154 a	1109 ± 178 a	952.5 ± 205 a
Metabolism	Global and overview maps	4996.67 ± 2221.16 a	4784.83 ± 2064.17 a	4981.75 ± 2121.26 a	6747 ± 2685.31 a	4798.25 ± 2196.11 a
	Energy metabolism	4947	4095.5	4158	4937	5239
	Nucleotide metabolism	1355 ± 387.5 a	1336.25 ± 404.75 a	1408.75 ± 434.75 a	1727 ± 450.5 a	1206.75 ± 379.25 a
	Carbohydrate metabolism	959.17 ± 52.1 a	933.83 ± 32.66 a	969.33 ± 36.28 a	1173.33 ± 19.69 a	924.17 ± 40.33 a
Environmental Information Processing	Membrane transport	1369	1335.5	1801.5	3000	1312.5

In the experiment with five treatments, there was no significant difference between treatments at KEGG pathway level 2 due to variations in the amount of rockwool added. However, one functional gene, ko01100, was more abundant in the treatment with 10 g of 20-mesh rockwool than in the remaining treatments. This functional gene belonged to the metabolic pathways, which indicated a significant effect on the metabolic pathways of earthworms with the addition of low concentrations of 20-mesh rockwool. However, based on the previous analysis, although low concentration of 20-mesh rockwool changed the abundance of earthworm metabolic pathways, it had a relatively small impact on earthworm survival activity. The size of rockwool particles is still the main reason for affecting earthworm activity.

4. Conclusions

As an inorganic substrate prepared from rock, rockwool has a high degree of elemental overlap with soil, and its porous nature ensures water retention when it is used as a substrate. The addition of a low concentration of rockwool (10 g/kg) to the soil had a minimal effect on the soil and *Eisenia fetida* and increased the soil porosity; however, the addition of a high concentration of rockwool (200 g/kg) significantly affected the SOM and calcium–magnesium–silica content of the soil and had a greater effect on the enzyme activity of *Eisenia fetida*, making its toxic effect significantly higher than that of the blank test group. The findings showed that the residual particle size of the rockwool was too large, even at low levels, to irritate and thus affect the activity of the earthworm *Eisenia fetida*, but had less of an effect on the earthworm gut microorganisms, indicating that it was the surface of the earthworm that was affected rather than the body. In summary, rockwool, as an unbiodegradable inorganic substrate, affects the physicochemical properties of soil, and the addition of a high concentration of rockwool particles has a toxicological effect on earthworms. In agricultural production, rockwool as a seedling substrate can be used safely with regular removal.

Author Contributions: All authors contributed to the study conception and design. S.Z.: Data curation, Formal analysis, Investigation, Writing—original draft, Methodology. C.W.: Formal analysis, Methodology, Experimental operation. J.J.: Methodology, Conceptualization, Formal analysis, Supervision, Validation, Writing—review and editing. Q.D.: Funding acquisition, Resources, Writing—review and editing. H.Z.: Writing—review and editing. P.L.: Writing—review and editing. X.W.: Experimental operation. All authors have read and agreed to the published version of the manuscript.

Funding: This research was supported by the Key Research and Development Plan of Jiangsu Province (BE2019343) and a Project Funded by the Priority Academic Program Development of Jiangsu Higher Education Institutions (PAPD). The views and ideas expressed herein are solely the authors' and do not represent the ideas of the funding agencies in any form.

Data Availability Statement: The data presented in this study are available on request from the corresponding author. Data could not be made public due to subsequent research.

Conflicts of Interest: The authors declare no conflict of interest.

References

1. Di Gioia, F.; De Bellis, P.; Mininni, C.; Santamaria, P.; Serio, F. Physicochemical, agronomical and microbiological evaluation of alternative growing media for the production of rapini (*Brassica rapa* L.) microgreens. *J. Sci. Food Agric.* **2016**, *97*, 1212–1219. [CrossRef] [PubMed]
2. Clematis, F.; Minuto, A.; Gullino, M.L.; Garibaldi, A. Suppressiveness to *Fusarium oxysporum* f. sp. radicis lycopersici in re-used perlite and perlite–peat substrates in soilless tomatoes. *Biol. Control* **2009**, *48*, 108–114. [CrossRef]
3. Acuña, R.A.; Bonachela, S.; Magán, J.J.; Marfà, O.; Hernández, J.H.; Cáceres, R. Reuse of rockwool slabs and perlite grow-bags in a low-cost greenhouse: Substrates' physical properties and crop production. *Sci. Hortic.* **2013**, *160*, 139–147. [CrossRef]
4. Wisdom, B.; Nyembezi, M.; Agathar, K. Effect of different vermiculite and pine bark media substrates mixtures on physical properties and spiral rooting of radish (*Raphanus sativus* L.) in float tray system. *Rhizosphere* **2017**, *3*, 67–74. [CrossRef]
5. Meot-Duros, L.; Le Floch, G.; Meot, B.; Letousey, P.; Jacob, B.; Barbier, G. Eelgrass Slabs, a Soilless Culture Substrate That Inhibits Adhesion of Fungi and Oomycetes and Enhances Antioxidant Activity in Tomato. *J. Agric. Food Chem.* **2011**, *59*, 10913–10918. [CrossRef] [PubMed]
6. Gundersen, V.; McCall, D.; Bechmann, I.E. Comparison of Major and Trace Element Concentrations in Danish Greenhouse Tomatoes (*Lycopersicon esculentum* Cv. Aromata F1) Cultivated in Different Substrates. *J. Agric. Food Chem.* **2001**, *49*, 3808–3815. [CrossRef] [PubMed]
7. van der Gaag, D.J.; Wever, G. Conduciveness of different soilless growing media to Pythium root and crown rot of cucumber under near-commercial conditions. *Eur. J. Plant Pathol.* **2005**, *112*, 31–41. [CrossRef]
8. Xiong, J.; Tian, Y.Q.; Wang, J.G.; Liu, W.; Chen, Q. Comparison of Coconut Coir, Rockwool, and Peat Cultivations for Tomato Production: Nutrient Balance, Plant Growth and Fruit Quality. *Front. Plant Sci.* **2017**, *8*, 1327. [CrossRef]
9. Dubsky, M.; Sramek, F. The effect of rockwool on physical properties of growing substrates for perennials. *Hortic. Sci.* **2009**, *36*, 38–43. [CrossRef]
10. Rijck, G.D.; Schrevens, E. Distribution of nutrients and water in rockwool slabs. *Sci. Hortic.* **1998**, *72*, 277–285. [CrossRef]

11. Medeiros, M.G.; Nadaleti, W.C.; Rocha, J.C.; Cheriaf, M.; Gleise, P.J.P.; de Castilhos, A.B. A cleaner material production by the incorporation of the rockwool waste into portland cement matrices. *J. Clean. Prod.* **2021**, *293*, 126059. [CrossRef]
12. Jørgensen, E. "Grodan" Stone Wool as Medium for Propagation and Culture. *Acta Hortic.* **1975**, *54*, 137–142. [CrossRef]
13. Carmassi, G.; Incrocci, L.; Maggini, R.; Malorgio, F.; Tognoni, F.; Pardossi, A. An aggregated model for water requirements of greenhouse tomato grown in closed rockwool culture with saline water. *Agric. Water Manag.* **2007**, *88*, 73–82. [CrossRef]
14. He, L.; Ding, X.; Jin, H.; Zhang, H.; Cui, J.; Chu, J.; Li, R.; Zhou, Q.; Yu, J. Comparison of rockwool and coir for greenhouse cucumber production: Chemical element, plant growth, and fruit quality. *Heliyon* **2022**, *8*, e10930. [CrossRef] [PubMed]
15. Koohakan, P.; Ikeda, H.; Jeanaksorn, T.; Tojo, M.; Kusakari, S.-I.; Okada, K.; Sato, S. Evaluation of the indigenous microorganisms in soilless culture: Occurrence and quantitative characteristics in the different growing systems. *Sci. Hortic.* **2004**, *101*, 179–188. [CrossRef]
16. Cycoń, M.; Markowicz, A.; Borymski, S.; Wójcik, M.; Piotrowska-Seget, Z. Imidacloprid induces changes in the structure, genetic diversity and catabolic activity of soil microbial communities. *J. Environ. Manag.* **2013**, *131*, 55–65. [CrossRef]
17. Newman, M.M.; Hoilett, N.; Lorenz, N.; Dick, R.P.; Liles, M.R.; Ramsier, C.; Kloepper, J.W. Glyphosate effects on soil rhizosphere-associated bacterial communities. *Sci. Total Environ.* **2016**, *543*, 155–160. [CrossRef]
18. Wang, C.; Wang, F.; Zhang, Q.; Liang, W. Individual and combined effects of tebuconazole and carbendazim on soil microbial activity. *Eur. J. Soil Biol.* **2016**, *72*, 6–13. [CrossRef]
19. OECD. *Earthworm Acute Toxicity Tests*; Organization for Economic Cooperation and Development: Paris, France, 1984; Volume 207, pp. 1–9.
20. Ministry of Agriculture of the People's Republic of China. *Soil Testing. Part 15: Method for Determination of Soil Avalible Silicon*. NY/T 1121.15-2006.2006-07-10. Available online: http://www.moa.gov.cn/govpublic/SCYJJXXS/201006/t20100606_1532967.htm (accessed on 1 October 2023).
21. Ministry of Agriculture of the People's Republic of China. *Soil Testing. Part 13: Method for Determination of Soil Exchangeable Calcium and Magnesium*. NY/T 1121.13-2006. 2006-07-10. Available online: http://www.moa.gov.cn/govpublic/SCYJJXXS/201006/t20100606_1532967.htm (accessed on 1 October 2023).
22. Ministry of Agriculture of the People's Republic of China. *Soil Testing Part 6: Method for Determination of Soil Organic Matter*. NY/T 1121.6-2006. 2006-07-10. Available online: http://www.moa.gov.cn/govpublic/SCYJJXXS/201006/t20100606_1532967.htm (accessed on 1 October 2023).
23. Mehrabanjoubani, P.; Abdolzadeh, A.; Sadeghipour, H.R.; Aghdasi, M.; Bagherieh-Najjar, M.B.; Barzegargolchini, B. Silicon increases cell wall thickening and lignification in rice (*Oryza sativa*) root tip under excess Fe nutrition. *Plant Physiol. Biochem.* **2019**, *144*, 264–273. [CrossRef]
24. Rengel, Z. Availability of Mn, Zn and Fe in the rhizosphere. *J. Soil Sci. Plant Nutr.* **2015**, *15*, 397–409. [CrossRef]
25. Agrawal, R.; Kumar, B.; Priyanka, K.; Narayan, C.; Shukla, K.; Sarkar, J.; Anshumali, A. Micronutrient Fractionation in Coal Mine-Affected Agricultural Soils, India. *Bull. Environ. Contam. Toxicol.* **2016**, *96*, 449–457. [CrossRef] [PubMed]
26. Hund-Rinke, K.; Achazi, R.; Rmbke, J.; Warnecke, D. Avoidance test withEisenia fetida as indicator for the habitat function of soils: Results of a laboratory comparison test. *J. Soils Sediments* **2003**, *3*, 7–12. [CrossRef]
27. Slimak, K.M. Avoidance responses as a sublethal effect of pesticides on *Lumbricus terrestris* (Oligochaeta). *Soil Biol. Biochem.* **1997**, *29*, 713–715. [CrossRef]
28. Pereira, J.L.; Antunes, S.C.; Ferreira, A.C.; Goncalves, F.; Pereira, R. Avoidance behavior of earthworms under exposure to pesticides: Is it always chemosensorial? *J. Environ. Sci. Health Part B-Pestic. Food Contam. Agric. Wastes* **2010**, *45*, 229–232. [CrossRef] [PubMed]
29. Ečimović, S.; Velki, M.; Mikuška, A.; Bažon, J.; Kovačić, L.S.; Kristek, S.; Jović, J.; Nemet, F.; Perić, K.; Lončarić, Z. How the Composition of Substrates for Seedling Production Affects Earthworm Behavior. *Agriculture* **2022**, *12*, 2128. [CrossRef]
30. Yang, Y.C.; Kim, J.Y.; Park, I.K. Neurotoxin 6-aminonicotinamide affects levels of soluble proteins and enzyme activities in various tissues of golden hamsters. *Int. J. Biochem. Cell Biol.* **2000**, *32*, 549–556. [CrossRef]
31. Maity, S.; Roy, S.; Chaudhury, S.; Bhattacharya, S. Antioxidant responses of the earthworm Lampito mauritii exposed to Pb and Zn contaminated soil. *Environ. Pollut.* **2008**, *151*, 1–7. [CrossRef]
32. Liu, S.; Zhou, Q.; Wang, Y. Ecotoxicological responses of the earthworm Eisenia fetida exposed to soil contaminated with HHCB. *Chemosphere* **2011**, *83*, 1080–1086. [CrossRef]
33. Hu, S.; Zhang, W.; Li, J.; Lin, K.; Ji, R. Antioxidant and gene expression responses of Eisenia fetida following repeated exposure to BDE209 and Pb in a soil-earthworm system. *Sci. Total Environ.* **2016**, *556*, 163–168. [CrossRef]
34. Maity, S.; Banerjee, R.; Goswami, P.; Chakrabarti, M.; Mukherjee, A. Oxidative stress responses of two different ecophysiological species of earthworms (*Eutyphoeus waltoni* and *Eisenia fetida*) exposed to Cd-contaminated soil. *Chemosphere* **2018**, *203*, 307–317. [CrossRef]
35. Shi, L.; Zhang, P.; Xu, J.; Wu, X.; Pan, X.; He, L.; Dong, F.; Zheng, Y. Systematic assessment of cyflumetofen toxicity in soil-earthworm (*Eisenia fetida*) microcosms. *J. Hazard. Mater.* **2023**, *452*, 131300. [CrossRef] [PubMed]
36. Xue, Y.; Li, Z.; Liu, C.; Liu, D.; Wang, J.; Liu, C.; Xia, X. Effect of different exposure times and doses of cyantraniliprole on oxidative stress and genotoxicity in earthworms (*Eisenia fetida*). *Chemosphere* **2023**, *319*, 138023. [CrossRef] [PubMed]
37. Zhang, Q.; Zhang, G.; Yin, P.; Lv, Y.; Yuan, S.; Chen, J.; Wei, B.; Wang, C. Toxicological effects of soil contaminated with spirotetramat to the earthworm *Eisenia fetida*. *Chemosphere* **2015**, *139*, 138–145. [CrossRef] [PubMed]

38. Bednarska, A.J.; Choczyński, M.; Laskowski, R.; Walczak, M. Combined effects of chlorpyriphos, copper and temperature on acetylcholinesterase activity and toxicokinetics of the chemicals in the earthworm *Eisenia fetida*. *Environ. Pollut.* **2017**, *220*, 567–576. [CrossRef]
39. Luo, Y.-R.; Wang, S.-H.; Yun, M.-X.; Li, X.-Y.; Wang, J.-J.; Sun, Z.-J. The toxic effects of ionic liquids on the activities of acetylcholinesterase and cellulase in earthworms. *Chemosphere* **2009**, *77*, 313–318. [CrossRef]
40. Tejada, M.; Gómez, I.; Hernández, T.; García, C. Response of Eisenia fetida to the application of different organic wastes in an aluminium-contaminated soil. *Ecotoxicol. Environ. Saf.* **2010**, *73*, 1944–1949. [CrossRef]
41. Kim, S.; Jeon, D.; Lee, J.-Y.; Cho, S.-J.; Lim, Y.; Eyun, S.-i.; Park, S.C.; Seo, Y.-J. Upregulation of cellulase activity and mRNA levels by bacterial challenge in the earthworm *Eisenia andrei*, supporting the involvement of cellulases in innate immunity. *Biochem. Biophys. Res. Commun.* **2020**, *521*, 15–18. [CrossRef]
42. Wang, Y.; Xing, M.Y.; Yang, J. Earthworm (*Eisenia fetida*) Eco physiological Characteristics in Vermifiltration System for Wastewater Treatment Through Analyzing Differential Proteins. *Water Air Soil Pollut.* **2017**, *228*, 60. [CrossRef]

agriculture

MDPI

Article

Current Status and Spatiotemporal Evolution of Antibiotic Residues in Livestock and Poultry Manure in China

Yanli Li, Qingjie Li, Zhengyu Ji, Okbagaber Andom, Xiaoxing Wang, Xueqi Guo and Zhaojun Li *

Key Laboratory of Plant Nutrition and Fertilizer, Ministry of Agriculture, China-New Zealand Joint Laboratory for Soil Molecular Ecology, Institute of Agricultural Resources and Regional Planning, Chinese Academy of Agricultural Sciences, Beijing 100081, China; liyanli02@caas.cn (Y.L.); 82101221185@caas.cn (Q.L.); 82101191167@caas.cn (Z.J.); 2020y90100055@caas.cn (O.A.); 82101201133@caas.cn (X.W.); 82101211162@caas.cn (X.G.)

* Correspondence: lizhaojun@caas.cn

Abstract: The use of antibiotics in the livestock and poultry industries has raised significant concern about environmental and health problems. In light of this, accurate knowledge of antibiotic residues in livestock and poultry manure is important for pollution management and strategic decision-making at the national level. This study aims to provide a comprehensive report on antibiotic residues in livestock and poultry manure in China using the published data of 3751 livestock and poultry feces in 29 provincial-level units over the past 20 years. In this study, the overall status of antibiotic residues in livestock and poultry feces was analyzed by mathematical statistics. Moreover, the spatio-temporal variation characteristics were analyzed by spatial statistics, and the differences among livestock and poultry species were evaluated by subgroup analysis. The finding indicated that tetracyclines (TCs), quinolones (QLs), sulfonamides (SAs), and macrolides (MLs) were the most abundant residues in livestock and poultry manure. The spatial and temporal variation revealed that the overall trend of antibiotic residues decreased gradually, and the spatial distribution was primarily concentrated in the southeast of Hu Line, exhibiting a "northeast-southwest" distribution. The distribution range also decreased slightly, with the residues of tetracyclines (TCs), quinolones (QLs), sulfonamides (SAs), and pleuromutilins (PMs) showing a significant spatial hot spot. The center of gravity of antibiotic residue shifted to the southwest between 2003 and 2021. In comparison to cow and sheep manure, the tetracyclines (TCs), sulfonamides (SAs), and macrolides (MLs) in pig and chicken manure were higher. The results can serve as a reference for the control and reduction of antibiotic pollution in livestock and poultry manure, as well as the wise utilization of those resources and achieving goals for clean agriculture.

Keywords: antibiotic; manure of livestock and poultry; spatial distribution; livestock and poultry species

1. Introduction

The environmental and health problems caused by antibiotics have received widespread attention. The three main sources of antibiotics in the environment are livestock farms, wastewater treatment plants, and chemical manufacturing plants [1]. Antibiotics used in livestock and poultry are not fully absorbed by animals. Approximately 60–90% of the antibiotics remain in animal manure in their original form or as metabolites, and these antibiotics eventually make their way into agricultural soils as organic fertilizers [2]. The introduction of antibiotics into the environment may affect soil microorganisms, thereby affecting soil nutrient cycling, enriching and poisoning plant tissues, and leading to the development of drug-resistant bacteria and antibiotic resistance genes (ARGs) in the environment [3,4]. According to One Health's concept of human-animal-environmental integration, antibiotics left in animal feces could end up posing potential health risks to humans [5].

At present, there is no threshold limit for antibiotic residues applied in animal manure as organic fertilizer in China. In order to improve the measures of returning livestock and poultry manure to farmland and to prevent antibiotic pollution of the farming environment from the source, it is helpful to understand the antibiotic residues in livestock and poultry manure. However, large-scale investigations of antibiotic residues in livestock and poultry manure face many challenges. On the one hand, antibiotic detection requires high-performance analytical techniques (e.g., detection limit, limit of quantitation) [6]. On the other hand, it is currently difficult to advance a general method for veterinary antibiotic determination due to the abundance of species and complex residues [7]. Because of these challenges, large-scale antibiotic testing investigations are costly and difficult to achieve. Subject to the constraints of the aforementioned factors, the main problem to be addressed in this study is how to express the spatio-temporal characteristics of antibiotic residues in livestock and poultry manure in China.

Meta-analysis is an efficient method for extracting empirical data through statistical analysis using published literature in order to achieve a comprehensive study of the entire scale system [8]. Currently, the method has been extensively used in agriculture, ecology, and other fields. In recent years, meta-analysis has been used to characterize antibiotic residues and pollution in the environment [9,10], proving its feasibility in the study of antibiotic residues in livestock and poultry manure. However, the object of relevant research is mainly single livestock and poultry manure, and the residual environment is concentrated primarily in soil, surface waters, and vegetation. Overall, antibiotic residues in livestock and poultry manure in China are not well analyzed and researched.

Therefore, in this study, we focus on the spatio-temporal pattern and driving factors of antibiotic residues in livestock and poultry manure in China. A total of 151 articles published between 2003 and 2021 were retrieved from the study cases. In addition to characterizing the antibiotic residues in manure from different livestock and poultry species, spatial statistics were used to characterize the spatial distribution and migratory patterns of antibiotics. Moreover, subgroup analysis was conducted in addition to the analysis of key driving factors. The results of this study will contribute to the management of the resource use of livestock and poultry manure in China and the formulation of relevant policies.

2. Materials and Methods

2.1. Data Sources and Extraction

In this study, all case data were collected from major research articles on antibiotic residues in livestock and poultry manure in mainland China published between 1 January 2003 and 31 October 2022. Using "livestock and poultry" or "pigs, chickens, cattle, sheep", "antibiotics" or single antibiotics (tetracyclines, quinolones, sulfonamides, macrolides, pleuromutilins, β-lactam, aminoglycosides, lincosamides, chloramphenicol, cephalosporins, polypeptides, nitroimidazoles, nitrofurans) and "feces" or "manure" as the key words, the search was performed on the China National Knowledge Network (CNKI) database and Web of Science, using a search of literature.

Selection criteria for the study cases are: (1) the study site must be within mainland China; (2) it must be field measured data, excluding artificial summary data; (3) the data must be the study of antibiotic levels in livestock and poultry manure, excluding urine, sewage, etc.; (4) the source of antibiotics in fecal matter must be routine farmed excretion, excluding artificial addition and feeding control.

2.2. Data Extraction and Preprocessing

2.2.1. Data Extraction

The data extracted from each paper includes the following: (1) first author, title, and year of publication; (2) the sampling location, which should be subject to the provincial unit; (3) sampling time (year); (4) type of antibiotic/mean value/number of samples (with each farm as a sample); (5) species of livestock and poultry, including pigs, chickens, cattle, and sheep.

2.2.2. Tests and Sensitivity Analyses of the Data

Following data extraction, the database was separated into six files based on the types of antibiotics used and then classified into studies based on region and livestock species. Then, data extremes and publication bias were tested. According to distribution statistics, there are extreme values. However, since this study is observational, extreme values must be included in the analysis as objective conditions. This study suffers from a significant publication bias, as researchers and readers tend to pay more attention to livestock and poultry with a high level of antibiotic residues in their feces. On the other hand, it is preferable to publish papers in the area of specific livestock and poultry (livestock and poultry that account for a large proportion of the breeding industry). Thus, the subgroup analysis of cattle and poultry species in the present study may well compensate for publication bias.

2.3. Geographic Centroid and Trajectory of Migration Calculations

2.3.1. Geographic Centroid Calculations

In this study, the centroid model was used to compute the geographical centroid of antibiotic residues. Using this method, the location (longitude and latitude) can be estimated from the centroid of antibiotic residues over the time period under study. The coordinates of the geographic centroid of antibiotic residues were calculated using the following equation:

$$X_t = \frac{\sum_{i=1}^{n}(A_{i,t} \times X_i)}{\sum_{i=1}^{n} A_{i,t}}$$

$$Y_t = \frac{\sum_{i=1}^{n}(A_{i,t} \times Y_i)}{\sum_{i=1}^{n} A_{i,t}}$$

where X_t and Y_t are the coordinates of the geographic centroid of antibiotic residues in the t year of province i, X_i and Y_i are the coordinates of targeted province i, n represents the number of provinces in which antibiotic residues were detected in livestock and poultry manure, and $A_{i,t}$ denotes mean value of antibiotic residues in the t year of province i (μg/kg). The values of t are four sections: 2003–2007, 2008–2012, 2013–2017, and 2018–2021.

2.3.2. Trajectory of Migration of the Geographic Centroid

The angle (θ) and distance (d) of the centroid moving were calculated using the following equations:

$$\theta = \arctan\left[\frac{Y_{k+m} - Y_k}{X_{k+m} - X_k}\right]$$

The distance of the centroid moving as follows:

$$d = \sqrt[2]{\left[(X_{k+m} - X_k)^2 + (Y_{k+m} - Y_k)^2\right]}$$

θ is the angle of centroid moving; d is the distance of centroid moving (km); G_k (X_k, Y_k) and G_{k+m} (X_{k+m}, Y_{k+m}) denote coordinates of the geographic centroid of antibiotic residues in the k year and $k + m$ year, respectively; m represents the interval between the k year and the $k + m$ year.

2.4. Standard Deviational Ellipse

The standard deviation ellipse (SDE) is a spatial statistical method that can accurately depict the spatial distribution characteristics of geographical elements [11]. The center of the SDE is the geographic centroid of the antibiotic in the manure. The direction of the major axis of the SDE (the angle between the major axis and true north and clockwise rotation) reflects the trend direction of antibiotic residue distribution in livestock and poultry manure, while eccentricity (the ratio of the major axis to the minor axis) reflects the spatial distribution of elements. The surface of the SDE reflects the dispersion of the elements. The direction of SDE is determined using the formula below:

$$\alpha = \frac{\left[\left(\sum_{i=1}^{n} \tilde{x}_i^2 - \sum_{i=1}^{n} \tilde{y}_i^2\right) + \sqrt{\left(\sum_{i=1}^{n} \tilde{x}_i - \sum_{i=1}^{n} \tilde{y}_i\right)^2 + 4\left(\sum_{i=1}^{n} \tilde{x}_i\tilde{y}_i\right)^2}\right]}{2\sum_{i=1}^{n} \tilde{x}_i\tilde{y}_i}$$

The standard deviation of the XY axis in SDE is expressed as follows:

$$d_x = \sqrt{2}\sqrt{\frac{\sum_{i=1}^{n}\left(\tilde{x}_i\cos\alpha - \tilde{y}_i\sin\alpha\right)^2}{n}}$$

$$d_y = \sqrt{2}\sqrt{\frac{\sum_{i=1}^{n}\left(\tilde{x}_i\sin\alpha + \tilde{y}_i\cos\alpha\right)^2}{n}}$$

where, $\tilde{x}_i$ and $\tilde{y}_i$ denote the coordinates of each study object to the SDE center; α is the angle of elliptic rotation (i.e., the angle between the major axis and true north and clockwise rotation); d_x and d_y represent the standard deviation along the X-axis and Y-axis, respectively.

2.5. Hotspot Analysis (Getis-Ord Gi)*

Hotspot analysis is a spatial statistical method to study spatial clustering. Using hotspot analysis, it is possible to obtain the location of clusterings of high- or low-value elements in space. Its principle is to look at each factor in the adjacent factor environment. The factor should have a high value and be surrounded by other factors with a high value; that is, the factor should become a hot spot with significant statistical needs.

2.6. Data Analysis

Excel 2021 was used for data preprocessing and statistical analysis. SPSS 26 was used for one-way analysis of variance. ArcMap 10.8 and Origin 2022 were used for spatial statistical analysis and mapping, respectively.

3. Results

3.1. Overall Status of Antibiotic Residues in Livestock and Poultry Manure

The overall status of antibiotic residues in livestock and poultry manure in China is presented in Table 1 below. Antibiotic residues were found in the feces of 3751 livestock and poultry farms in 29 provinces. As illustrated in Table 1, monitoring studies mainly focus on tetracyclines (TCs), quinolones (QLs), sulfonamides (SAs), macrolides (MLs). and pleuromutilins (PMs), indicating that they have higher attention and data quality.

Table 1. Statistics on antibiotic residues in livestock manure in China.

Element	Tetracyclines	Quinolones	Sulfonamides	Macrolides	Pleuromutilins	Others	Total
Provinces	29	26	24	19	16	19	29
Farms	3402	2456	2613	1701	583	1658	3751
Mean [a]	18,779.52	1689.09	1929.71	1414.92	336.45	1457	7845.99
Standard deviation	80,722.66	21,724.75	5551.21	8269.28	753.80	5278.50	50,370.85
Species	pig, chicken, cattle, sheep	pig, chicken, cattle, sheep	pig, chicken, cattle, sheep	pig, chicken, cattle, sheep	pig, cattle	pig, chicken, cattle	pig, chicken, cattle, sheep

[a] Unit: μg/kg.

The mean of antibiotic residues could reflect the overall situation of antibiotic residues in livestock and poultry manure, such that TCs > SAs > QLs > others > MLs > PMs. The degree of scatter in the data was reflected in the standard deviation. Results indicated that the contents of all kinds of antibiotics showed a high degree of dispersion, i.e., the residual amount of antibiotics in livestock and poultry manure was highly variable across the country. Therefore, it was crucial to analyze the space and time dimensions and subgroups.

3.2. Spatial Distribution of Antibiotic Residues in Livestock and Poultry Manure

According to the natural break point (Jenks) classification method, the antibiotic residues in TCs, QLs, SAs, MLs, PMs, and others were categorized into six levels, representing the average antibiotic residues in livestock and poultry manure in each province (Figure 1). Dark-to-light colors represent antibiotic residues from high to low concentrations. Using the mean of each case as the weight, the SDE was calculated for each type of antibiotic residue. SDE covers 63% of the antibiotic residues in China, indicating the trend and rule of sample distribution with antibiotic residues as the weight.

Generally, the spatial distribution of the antibiotics showed that areas with high antibiotic residue were primarily concentrated in the southeastern part of the Hu Line (the eastern and southern provinces of China). Hu Line, which is a contrast line for dividing the population density in China, has also become the dividing line of urbanization level. The spatial distribution of antibiotic residues in China indicated significant variations, and the overall distribution of the residual antibiotic concentrations was in the "Northeast-Southwest" ward. The distribution of residual antibiotic concentrations in TCs and others was noticeably in the direction of "Northeast-Southwest", with the rotation angles of SDE 40.34° and 58.52°, respectively. The SDE rotation angle of antibiotic residue QLs, on the other hand, was 11.28°, and the direction of the distribution was closer to the "South-North" distribution. The SDE rotation angles of SAs, MLs, and PMs were 86.39°, 84.84°, and 77.51°, respectively, and the distribution direction was closer to the "East-West" distribution.

The area under the SDE curve reflected the spatial dispersion of antibiotic residues in livestock and poultry manure. According to the SDE area, QLs, PMs, and others showed a higher spatial aggregation effect than TCs, SAs, and MLs. The spatial accumulation of QLs was concentrated primarily in most of the regions of the provinces of Hebei, Shandong, Henan, Anhui, Jiangsu, Zhejiang, and Hubei. Spatially aggerated PMs are concentrated specifically in the provinces of Sichuan, Hunan, Jiangxi, Yunnan, and Guangdong. Others were mainly concentrated in Guangxi, Hunan, Hubei, Fujian, and Guangdong provinces.

Further revealing the spatial accumulation characteristics of antibiotic residues in cattle and poultry manure would aid in a more complete analysis of the spatial correlation of antibiotic residues. In this study, hot spot analysis (Getis-Ord Gi*) (95% confidence interval, CI) was performed for six classes of antibiotics to determine the local correlation types in each region from the perspective of statistics.

Results indicated that the six classes of antibiotics were statistically different in space, and there was a significant spatial correlation (95% confidence interval, CI) among the antibiotic residues of TCs, QLs, SAs, and PMs in livestock and poultry, and all of them were hot spots (Figure 2). The hotspot regions and their neighboring provinces had significant levels of antibiotic residues in the feces of livestock and poultry. Thus, the high-residue area of TCs could be judged to be Guangdong and its neighboring provinces. On the other hand, Shandong Province, Jiangsu Province, and neighboring provinces were regions with high residual QLs. Yunnan Province, Guangdong Province, and neighboring provinces were regions with high residual SAs. High residual PM was observed in Guizhou Province and its neighboring provinces.

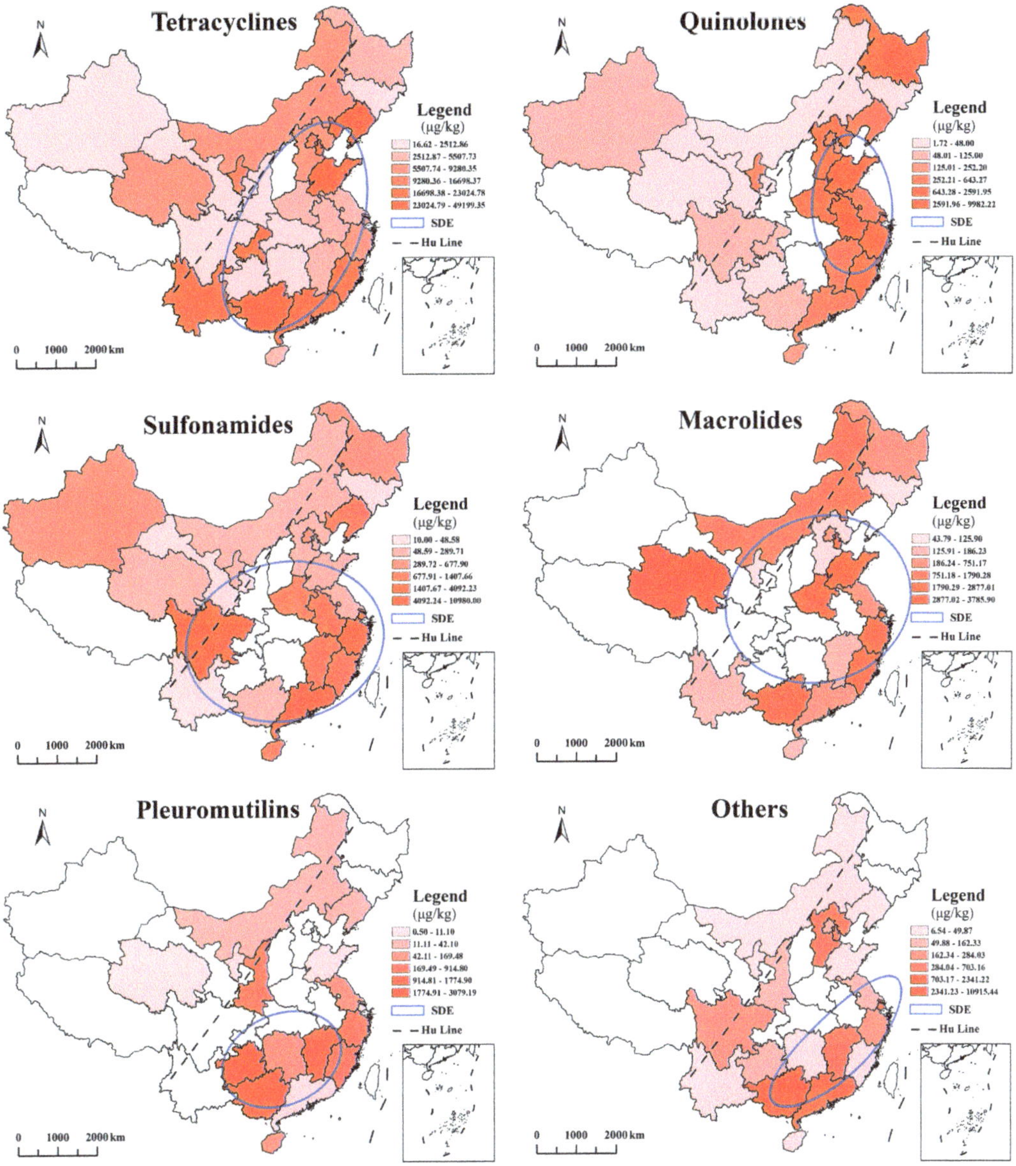

Figure 1. Classification maps of antibiotic residues in livestock and poultry manure (the standard map with the approval number GS (2022) 1873 downloaded from the Standard Map Service Network of the Ministry of Natural Resources, People's Republic of China, is made without modification of the base map).

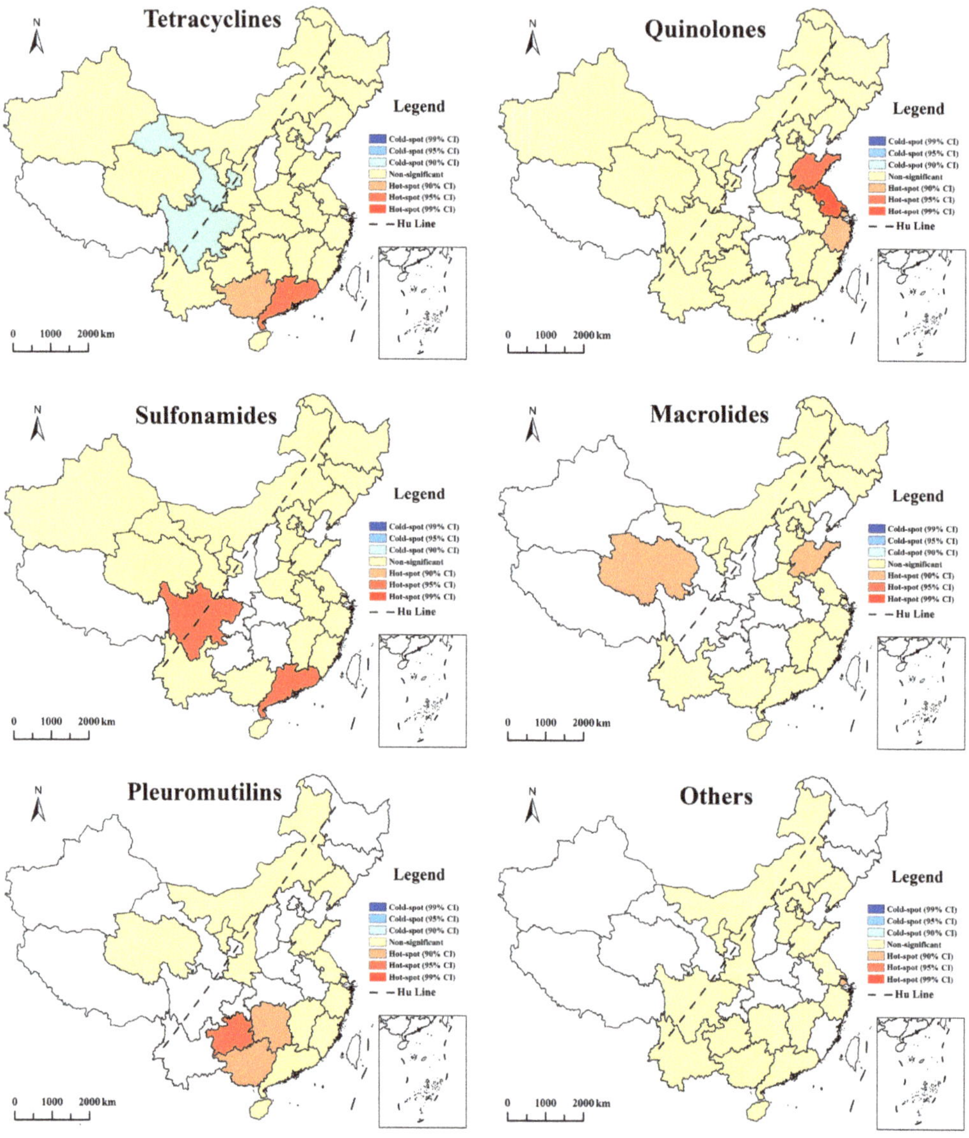

Figure 2. Cold and hot spot distribution maps of antibiotic residues in livestock and poultry manure (the standard map with the approval number GS (2022) 1873 downloaded from the Standard Map Service Network of the Ministry of Natural Resources, People's Republic of China, is made without modification of the base map).

3.3. Temporal Variation of Antibiotic Residues in Livestock and Poultry Manure

In this study, the residual amount of antibiotics in livestock and poultry manure in China was linearly fitted, and the overall trend showed a slow decline from 2003 to 2021, $R^2 = 0.58$ (Figure 3A). The results indicated that the antibiotic residues during this period decreased from 7457.56 μg/kg to 5342.86 μg/kg and fell by 28.36% with a mean annual rate of decline of 111.30 μg/kg.

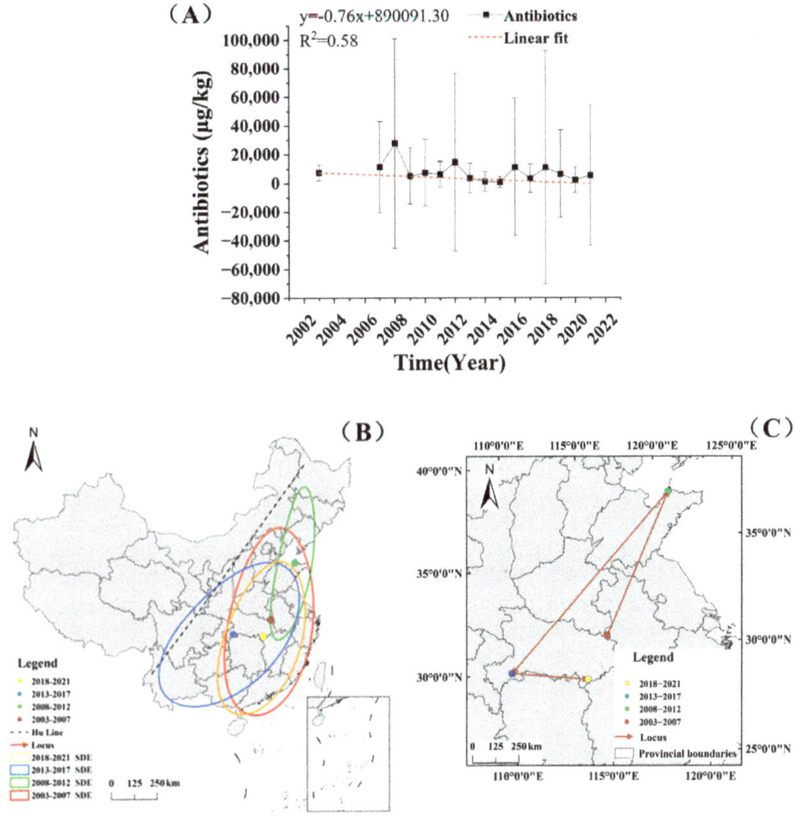

Figure 3. The spatial and temporal patterns of antibiotic residues in livestock and poultry manure in China. (**A**) Interannual trends in antibiotics. (**B**) Distribution of antibiotics in different periods (**C**) Migration trajectory of the antibiotic center of gravity. (The standard map with the approval number GS (2022) 1873 downloaded from the Standard Map Service Network of the Ministry of Natural Resources, People's Republic of China, is made without modification of the base map).

Due to the small number of sampling points in some years from 2003 to 2021, it was impossible to represent the whole year. Therefore, this study considered only five years as a stage and divided it into four periods for analysis. As it was clearly indicated, the X standard distance of the SDE exhibits a "northeast-southwest" trend throughout the entire time period (Figure 3C, Table 2). The X standard distance of remnant antibiotics over the 19 years was 9.08 km. The Y standard distance and the area of the ellipse were decreased by 228.71 km and 41.25 $\times$ 10^4 km^2, respectively. This study showed that the accumulation effect of antibiotic residues in cattle and poultry manure was slightly serious in the "northeast-southwest" direction, and the overall range of distribution was reduced. The clustering effect increased substantially between 2008 and 2012. At 120.54° E and 38.06° N, the ellipse with 258.65 km and 1096.30 km as the X and Y standard distances, respectively, might contain 63% of the antibiotic residues in the manure of livestock and poultry in China, with the district covering most of the provinces of Jilin, Liaoning, Hebei, Shandong, Jiangsu, and Anhui. This result may be related to the changes in the main livestock species and common livestock diseases in different regions.

Table 2. Migration and change of geographical center of gravity of antibiotic residues in livestock and poultry manure in China from 2003 to 2021.

Time	Center X (°)	Center Y (°)	X Stddist (km)	Y Stddist (km)	Area (10^4 km^2)	Direction (°)
2003–2007	115.45	31.77	531.87	1299.48	217.10	9.78
2008–2012	120.54	38.06	258.65	1096.30	89.05	15.28
2013–2017	109.87	30.34	675.38	1178.73	250.08	42.22
2018–2021	114.11	29.66	522.79	1070.77	175.85	19.50

The geographic centroid of antibiotic residues in livestock and poultry manure showed noticeable spatial migration between 2003 and 2021. It migrated 252.66 km to the southwest by 9.03°, and the geographic centroid of antibiotic residues migrated from the west of Anhui to the southeast of Hubei. During each time period, the mean migration rate was 853.82 km/5a. The geographic centroid rate of antibiotic residue during 2008–2012 and 2013–2017 was the highest, and it migrated 1310.69 km in the southwest direction at 28.4°.

3.4. *The Variation of Antibiotic Residue Contents among Different Livestock and Poultry Species*

Since livestock and poultry species could also affect antibiotic residues in manure, the variation of antibiotic residues among different livestock and poultry species was presented in this study. A subgroup analysis of antibiotic residues in the manure of different livestock and poultry species was conducted (Figure 4). According to the weighted mean values, there were three distinct groupings of variation. TCs, SAs, and PMs, which have larger residues in pig manure, were part of the first group. In contrast, the second group (MLs) had the highest content of chicken manure. The third group (QLs, Others) had higher residues in cow manure.

TCs, SAs, and MLs residues showed significant differences in all kinds of livestock and poultry manure ($p < 0.05$). TCs in pig and chicken manure were significantly higher than those in cattle and sheep manure. On the other hand, the residues in cattle manure were significantly higher than those in sheep. The residual amount of SAs in pig manure was significantly higher than that of other animals' manure understudy. Moreover, the residual amount of SAs in chicken and cattle manure was significantly higher than that of sheep. Moreover, the residue of MLs in pig and chicken manure was significantly higher than that of sheep. Based on the forementioned expression, considering all animal manure understudy, it can be summarized that the highest and lowest antibiotic residues were investigated in pig and sheep manure, respectively, in this study. The differences between the residues of feces of QLs, PMs, and others in various types of livestock and poultry were not significant; thus, we shall refrain from delving into specifics at this point.

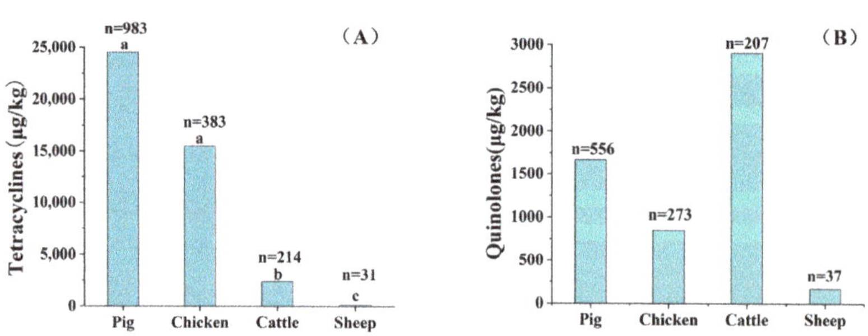

Figure 4. *Cont.*

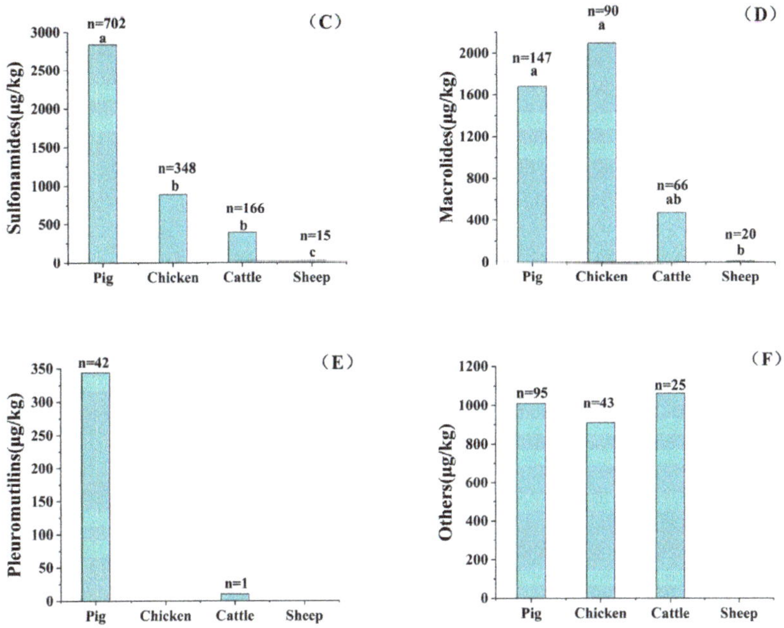

Figure 4. Antibiotic residues in the feces of different livestock and poultry species ("*n*" represents the number of samples in each group. The letter "abc" indicates the significance of the difference between the treatments, and two treatments with the same letter are not significant).

4. Discussion

4.1. Analysis of the Causes of Antibiotic Residues in Livestock and Poultry Manure in China

Antibiotics used in animal farms in China can be divided into feed medication, drug administration by drinking water, parenteral administration, and other routes of administration. According to the Official Veterinary bulletin (Table 3), 24,391.82 tons of antibiotics were injected into the bodies of livestock and poultry through feed medication and drinking water administration during 2018–2020, accounting for 78.42% of the total amount of antibiotics administered throughout the year, among which 60–90% of antibiotics would be eliminated from the body with urine and manure [12–14]. Therefore, antibiotics excreted directly with drug prototypes were the main source of antibiotic residues in manure, which were absorbed into the bodies of livestock and poultry through feed medication and drug administration by drinking water.

Table 3. Ways of antibiotic administration for livestock and poultry in China.

Time	Feed Medication (%)		Drug Administration by Drinking Water (%)		Parenteral Administration (%)		Other Routes of Administration (%)	
	Dosage (t)	Ratio (%)	Dosage (t)	Ratio (%)	Dosage (t)	Ratio (%)	Dosage (t)	Ratio (%)
2018	18,525.66	62.22	5702.44	19.15	3566.02	11.98	1979.98	6.65
2019	19,102.79	61.81	5451.56	17.64	3492.65	11.30	2856.66	9.24
2020	13,184.73	40.23	11,208.27	34.20	3572.53	10.90	4810.76	14.68
Mean	16,937.73	54.75	7454.09	23.66	3543.73	11.39	3215.80	10.19

Based on the purpose of use, antibiotics for animals in China could be divided into two categories: growth promotion and therapeutic (before the Announcements 194 and 246 issued by the Ministry of Agriculture and Rural Affairs of the People's Republic of China). Growth-promoting antibiotics were mainly used to prevent disease and promote the growth of livestock and poultry. Research has shown that such antibiotics could promote the weight increase of livestock and poultry by 5–6% and increase the feed conversion rate by 3–4% [15,16]. There were several hypotheses about its growth-promoting mechanism, which mainly focused on intestinal microbial regulation and immune system regulation [17,18]. Therapeutic antibiotics were used to treat diseases in livestock and poultry. In China, from 2018 to 2020, antibiotics that have been used for growth promotion accounted for 53.2%, 48.12%, and 28.69% of total dosing, respectively. Despite the proportion gradually decreasing, dosing was still high at 15,412.68 t, 14,871.05 t, and 9403.21 t, respectively. In addition, most growth-promoting antibiotics were feed medication and drinking water administration by means of feed addition, and the main types were TCs (chlorotetracycline and oxytetracycline), accounting for about 22% of the total dosage, which was consistent with the highest antibiotic residues of TCs in livestock and poultry manure. Therefore, antibiotics used for growth promotion are one of the main causes of antibiotic residues in livestock and poultry manure.

In addition, the mean of antibiotic residues was that TCs > SAs > QLs > others > MLs > PMs in China. The most major reason maybe growth-promoting antibiotics were feed medication and drinking water administration by means of feed addition, and the main types were TCs (aureomycin, oxytetracycline), accounting for about 22% of the total dosage, which was consistent with the highest antibiotic residues of TCs in livestock and poultry manure. The residual amount of SAs ranked second, mainly because there are many varieties of SAs, which are mainly used to prevent and treat bacterial infections such as diarrhea. At present, it is also one of the most used veterinary antibiotics; the adsorbability and degradability of SAs in animal intestines are poor, and a large proportion of drug prototypes will be excluded in vitro. In addition, veterinary antibiotic half-lives in animal manure are also important factors affecting residual quantity. TCs, QLs, and SAs have a long half-life of approximately 100 days, whereas MLs and β-Lactam have a very short half-life.

One of the major reasons for antibiotic residues in the manure of different livestock and poultry species was the imbalanced development of the livestock and poultry industries. According to the China Animal Husbandry and Veterinary Yearbook, the number of pigs and poultry raised in China in 2020 will account for 35.27% and 41.68%, respectively, while the total number of cattle and sheep raised will account for 12.9%. Rapid farming development was the main reason for antibiotic residues in pig and poultry manure. Moreover, the physiological structure of livestock and poultry may also cause the difference in antibiotic residues in manure. For example, the intestine of poultry was shorter than that of vertebrates, which may result in shorter gut residence times, less absorption, and greater excretion of antibiotics administered in mixed diets [19].

4.2. The Cause of Spatial and Temporal Variation in Antibiotic Residues

4.2.1. The Cause of Spatial Variation

The spatial distribution of antibiotic residue content in China showed significant variations. One of the reasons was that the distribution characteristics of antibiotic residues in livestock and poultry manure were directly affected by the development of the industry. An important index for the development of the breeding industry is the number of livestock and poultry raised. According to the China Animal Husbandry and Veterinary Yearbook (Figure 5A,B), the quantity of livestock and poultry breeding in China showed an overall increase of 9.38% from 2008 to 2021. Based on the spatial distribution, the breeding quantity in southeast and south-central regions, which were the centers of domestic industry, maintained rapid growth, and the distribution characteristics of antibiotic residues in livestock and poultry manure were largely consistent.

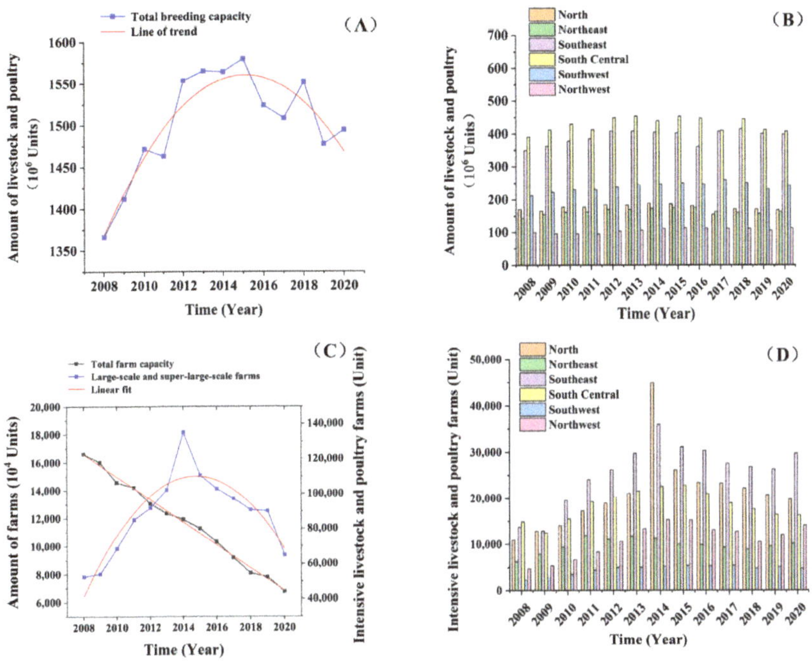

Figure 5. Development of the industry of livestock and poultry in China. (**A**). Amount of livestock and poultry which is converted to pig equivalent; (**B**) Amount of Livestock and poultry in six region of China; (**C**) Total number of farms and Intensive livestock and poultry farms; (**D**) Amount of intensive livestock and poultry farms in six region of China.

4.2.2. The Cause of Temporal Variation

Overall antibiotic residues in livestock and poultry manure decreased from 2003 to 2020 as a result of productivity, methods of breeding, policies, and regulations.

Intensive and large-scale farming could reduce antibiotic overuse in terms of environmental control and management, resulting in reduced antibiotic residues in livestock and poultry manure. According to the China Animal Husbandry and Veterinary Yearbook (Figure 5C,D), the total number of livestock and poultry farms decreased by 63.70% from 2008 to 2020. On the other hand, the number of large-scale and super large-scale livestock operations increased from 46,891 to 94,558 in 2008, with an overall increase of 101.65%. This clearly indicates that China's livestock and poultry breeding trends were shifting toward intensive large-scale breeding, while the number of small and medium aquaculture farms was gradually declining. Intensive and large-scale breeding can improve the efficiency of biosecurity prevention and control, help in the prevention and control of animal diseases, standardize production management, and effectively reduce antibiotic overuse, which is also one of the key factors for the decreasing trend of antibiotic residues in livestock and poultry manure in China. As the abuse of antibiotics in livestock and poultry farming has received increasing attention, the National Action Plan for Reducing the Use of Veterinary Antimicrobial Drugs (2021–2025) and the Veterinary Medicine Production Quality Management Practice (2020 revision) have been issued, which prohibit the production and sale of antibacterial feed additives for growth promotion by producers and sellers of veterinary drugs. Antibiotic abuse was fundamentally prohibited by policy and regulation. At the same time, antibiotic substitutes such as enzyme preparations, microbial preparations, traditional Chinese medicine preparations, and antimicrobial peptides have been developed and produced successively, realizing the reduction of antibiotics in a more environmentally friendly and sustainable way [20–22].

4.2.3. The Cause of Temporal Variation

The geographic centroid of antibiotic residues in livestock and poultry manure migrated visibly between 2003 and 2021, which was related to multiple factors such as industrial development, policy, and blight.

Since the 1990s, it can be broken down into two stages, with 2014 serving as the dividing point. The first stage was the development stage of improving quality and efficiency (1997–2014), and the second stage was the comprehensive transformation and upgrading stage focusing on environmental protection (2015–present) [23]. In the first stage, China promulgated a series of policies and measures to promote the development of animal husbandry, such as the Opinions of The State Council on Promoting the Sustainable and Healthy Development of Animal Husbandry, the Opinions of The State Council on Promoting the Sustainable and Healthy Development of the Dairy Industry and the National Development Plan for Beef and Mutton Production (2013–2020), and so on [24]. In this stage, China's animal husbandry was developing rapidly, and the breeding quantity in northeast China increased by 19.5%, higher than that in other parts of the country (Figure 5B). Its rapid growth may be accompanied by the abuse of antibiotics, which may be one of the important reasons for the migration of the geographic centroid of antibiotic residues to the northeast of China. In the second phase, in the context of vigorously promoting green agricultural development in China, livestock and poultry farming is an important part of Chinese agriculture, and the livestock and poultry pollution control problem has become a political issue. The Regulation on Pollution Prevention and Control of Large-scale Livestock and Poultry Farming is China's first administrative regulation on pollution prevention and control of agriculture and rural areas at the national perspective. Since then, pollution prevention and control in livestock and poultry farming have reached a turning point, and the policy of no-breeding areas has been implemented nationwide. The Action Plan for Soil Pollution Prevention, Control and the Action Plan for Water Pollution Prevention and Control and other environmental protection policies have been promulgated successively, resulting in "production restraint benefits" for livestock and poultry breeding. Moreover, the implementation of the above policies in the eastern region was better than that in the central and western regions, which directly led to the migration of livestock and poultry breeding areas to the west [25].

From 2003 to 2021, a number of large-scale livestock and poultry outbreaks occurred in China, including H7N9 bird flu in 2012 and 2014, Peste des Petits Ruminants (PPR) in 2014, and African swine fever in 2018. On one hand, the epidemic affected the quantity of livestock and poultry produced, and on the other hand, the densely populated eastern region and developed logistics increased the risk of epidemic inflow. These factors may also have contributed to the westward migration of the breeding industry and the relocation of the geographic centroid of antibiotic residues in the region [26,27].

4.3. Influence on Environmental Safety and Human Health

Farmland is the primary source of livestock and poultry manure. On agricultural land, manure (containing residual antibiotics) is often used as organic fertilizer. The introduction of antibiotics into the environment, such as farmland fertilization, will lead to environmental pollution [28]. Currently, there has been a great deal of research into the environmental pollution of antibiotics, which manifests itself primarily in three aspects: impact on soil micro-organisms, toxicity to plants and induction of resistance genes.

Antibiotics are not easy to degrade once applied to soil with manure [29], as they can be built up in the soil over a long time after biotransformation, leaching, adsorption/resolution, and other processes [30]. It is also possible that the mixed accumulation of multiple antibiotics may produce combined toxicity, which may affect the microbial species and community structure in the soil [31]. It has been reported that antibiotics can interfere with the community structure by changing the relative abundance of some microorganisms through selective action (inhibiting protein/cell wall/DNA synthesis), thus affecting the conversion of substances and nutrients in the soil. Moreover, the effects of antibiotics on the composition of the microbial flora can cause changes in the soil microbial population and genetic diversity [32].

Similar to the above-stated side effects, the toxic effects of antibiotics on plants are manifested in their effects on the rate of germination, root system, hypocotyl, and leaf growth, which can directly affect the yield of the crop [33–35]. Plant uptake of antibiotics is primarily active and complemented by passive uptake. Moreover, antibiotics present in soil are bioconcentrated (especially in vegetables) and can accumulate in plant tissues [36], posing a threat to the health of humans.

Residues of antibiotics in livestock and poultry manure directly lead to a pool of ARGs, which can cause pollution of ARGs in the environment [37–39]. ARGs have become a global public health problem [40]. The antibiotic screening pressure results in the production of antibiotic-resistant bacteria and ARGs. ARGs can be transferred between generations and between different bacteria, which may cause non-pathogenic bacteria to pass ARGs on to pathogenic bacteria, and the spread of ARGs will cause potential threats to the environment and human health [7,41]. It has been estimated that ARGs cause 25,000 deaths per year in the European Union alone and 700,000 deaths globally [42].

5. Conclusions

This paper reviewed monitoring-related studies about residues of antibiotics (i.e., TCs, QLs, SAs, MLs, PMs, etc.) in livestock and poultry manure from 2003 to 2021 in China. The average residues of TCs, QLs, SAs, and MLs in livestock and poultry waste in China are higher than others. The spatial and temporal changes showed that the overall residue showed a slow decline trend, and the spatial distribution was mainly concentrated in the southeast of the Hu Huanyong line, and the overall distribution was "northeastern-southwest", in which the residues of TCs, QLs, SAs, and truncated PMs showed significant spatial hot spots. From 2003 to 2021, the center of gravity of antibiotic residues shifted to the southwest. The TCs, SAs, and MLs in pig manure and chicken manure were higher than those in cow manure and sheep manure. Finally, we believe that the results of this study provide data support for understanding the status of antibiotic pollution in livestock and poultry manure in China and serve as a reference for the control and reduction of antibiotic pollution in livestock and poultry manure, which ultimately contributes to the implementation of wise utilization of those resources and achieving goals for clean agriculture in China.

Author Contributions: Conceptualization, Z.L.; methodology, Z.J. and Q.L.; formal analysis, O.A.; data curation, X.W. and X.G.; writing—original draft preparation, Y.L.; All authors have read and agreed to the published version of the manuscript.

Funding: This research was supported by the National Natural Science Foundation of China (52370141), the National Natural Science Foundation of China (31972943), the earmarked fund for CARS (China Agriculture Research System, CARS-29-ZP-10), and the China State Construction Engineering Corporation Technology Research and Development Project (CSCEC-2023-Z-21).

Data Availability Statement: Not applicable.

Conflicts of Interest: The authors declare no conflict of interest.

References

1. Danner, M.C.; Robertson, A.; Behrends, V.; Reiss, J. Antibiotic pollution in surface fresh waters: Occurrence and effects. *Sci. Total Environ.* **2019**, *664*, 793–804. [CrossRef]
2. Sarmah, A.K.; Meyer, M.T.; Boxall, A.B.A. A global perspective on the use, sales, exposure pathways, occurrence, fate and effects of veterinary antibiotics (VAs) in the environment. *Chemosphere* **2006**, *65*, 725–759. [CrossRef] [PubMed]
3. Wu, J.; Wang, J.Y.; Li, Z.T.; Guo, S.M.; Li, K.J.; Xu, P.S.; Ok, Y.S.; Jones, D.L.; Zou, J.W. Antibiotics and antibiotic resistance genes in agricultural soils: A systematic analysis. *Crit. Rev. Environ. Sci. Technol.* **2022**, *53*, 847–864. [CrossRef]
4. Liu, Y.W.; Feng, M.M.; Johansen, A.; Cheng, D.M.; Xue, J.M.; Feng, Y.; Fan, S.H.; Li, Z.J. Composting reduces the risks of antibiotic resistance genes in maize seeds posed by gentamicin fermentation waste. *Sci. Total Environ.* **2023**, *870*, 161785. [CrossRef] [PubMed]

5. McEwen, S.A.; Collignon, P.J. Antimicrobial resistance: A one health perspective. *Microbiol. Spectr.* **2018**, *6*, 521–547. [CrossRef] [PubMed]
6. Feng, Y.; Wei, C.J.; Zhang, W.J.; Liu, Y.W.; Li, Z.J.; Hu, H.Y.; Xue, J.M.; Davis, M. A simple and economic method for simultaneous determination of 11 antibiotics in manure by solid-phase extraction and high-performance liquid chromatography. *J. Soils Sediments* **2016**, *16*, 2242–2251. [CrossRef]
7. Lu, J.; Wang, Y.; Jin, M.; Yuan, Z.; Bond, P.; Guo, J. Both silver ions and silver nanoparticles facilitate the horizontal transfer of plasmid-mediated antibiotic resistance genes. *Water Res.* **2020**, *169*, 115229. [CrossRef] [PubMed]
8. Moher, D.; Liberati, A.; Tetzlaff, J.; Altman, D.G. The PRISMA Group, Preferred reporting items for systematic reviews and meta-analyses: The PRISMA statement. *Ann. Intern. Med.* **2009**, *151*, 264–269. [CrossRef]
9. Frey, L.; Tanunchai, B.; Glaser, B. Antibiotics residues in pig slurry and manure and its environmental contamination potential. A meta-analysis. *Agron. Sustain. Dev.* **2022**, *42*, 31. [CrossRef]
10. Lu, G.Y.; Chen, Q.; Li, Y.P.; Liu, Y.T.; Zhang, Y.Y.; Zhu, L. Status of antibiotic residues and detection techniques used in Chinese milk: A systematic review based on cross-sectional surveillance data. *Food Res. Int.* **2021**, *147*, 110450. [CrossRef]
11. Lefever, D.W. Measuring geographic concentration by means of the standard deviational ellipse. *Am. J. Sociol.* **1926**, *32*, 88–94. [CrossRef]
12. Martínez, J.L. Antibiotics and antibiotic resistance genes in natural environments. *Science* **2008**, *321*, 365–367. [CrossRef] [PubMed]
13. Li, X.W.; Xie, Y.F.; Wang, J.F.; Christakos, G.; Si, J.L.; Zhao, H.N.; Ding, Y.Q.; Li, J. Influence of planting patterns on fluoroquinolone residues in the soil of an intensive vegetable cultivation area in northern China. *Sci. Total Environ.* **2013**, *458*, 63–69. [CrossRef]
14. Haller, M.Y.; Müller, S.R.; McArdell, C.S.; Alder, A.C.; Suter, M.J.F. Quantification of veterinary antibiotics (sulfonamides and trimethoprim) in animal manure by liquid chromatography–mass spectrometry. *J. Chromatogr.* **2002**, *952*, 111–120. [CrossRef] [PubMed]
15. Shao, Y.T.; Wang, Y.P.; Yuan, Y.W.; Xie, Y.J. A systematic review on antibiotics misuse in livestock and aquaculture and regulation implications in China. *Sci. Total Environ.* **2021**, *798*, 149205. [CrossRef]
16. Stutz, M.W.; Lawton, G.C. Effects of diet and antimicrobials on growth, feed efficiency, intestinal Clostridium perfringens, and ileal weight of broiler chicks. *Poult. Sci.* **1984**, *63*, 2036–2042. [CrossRef]
17. Rahman, M.; Fliss, I.; Biron, E. Insights in the development and uses of alternatives to antibiotic growth promoters in poultry and swine production. *Antibiotics* **2022**, *11*, 766. [CrossRef] [PubMed]
18. Looft, T.; Allen, H.K.; Cantarel, B.L.; Levine, U.Y.; Bayles, D.O.; Alt, D.P.; Henrissat, B.; Stanton, T.B. Bacteria, phages and pigs: The effects of in-feed antibiotics on the microbiome at different gut locations. *ISME J.* **2014**, *8*, 1566–1576. [CrossRef] [PubMed]
19. Wei, S.; Morrison, M.; Yu, Z. Bacterial census of poultry intestinal microbiome. *Poult. Sci.* **2013**, *92*, 671–683. [CrossRef] [PubMed]
20. Dong, J.P.; Liu, L.Z. The impact of environmental regulations on the reduction of veterinary drugs by meat duck farmers: Based on a survey data from 10 provinces. *Res. Agric. Mod.* **2022**, *43*, 616–626. (In Chinese with English Abstract)
21. Zhang, J.; Yu, H.; Zhang, H.; Zhao, Q.; Si, W.; Qin, Y.; Zhang, J. Dietary Epimedium extract supplementation improves intestinal functions and alters gut microbiota in broilers. *J. Anim. Sci. Biotechnol.* **2023**, *14*, 14. [CrossRef] [PubMed]
22. Zamojska, D.; Nowak, A.; Nowak, I.; Macierzyńska-Piotrowska, E. Probiotics and postbiotics as substitutes of antibiotics in farm animals: A review. *Animals* **2021**, *11*, 3431. [CrossRef] [PubMed]
23. Wang, M.L. China's livestock industry development: Achievements, experiences and future trends. *Issues Agric. Econ.* **2018**, *8*, 60–70.
24. Xu, Y.N.; Luan, W.X.; Yang, Y.J.; Wang, H. Evolutionary game for the stakeholders in livestock pollution control based on circular economy. *J. Clean. Prod.* **2021**, *282*, 125403.
25. Xiong, B.; Wang, R.M. Structural adjustment and environmental effect of livestock and poultry breeding under policy of prohibiting livestock and poultry production. *J. China Agric. Univ.* **2022**, *27*, 291–304. (In Chinese with English Abstract)
26. Guo, Y.; Li, R.; Sun, X.; Zhang, Z.H.; Zheng, H.C.; Han, L.X.; Cui, Y.N.; Zhang, D.X.; Liu, M.C. In vitro Antibiotic Susceptibility, Virulence Genes Profiles and Integrons of Streptococcus suis Isolates from Pig Herds in Liaoning Province of China. *Pak. Vet. J.* **2022**, *42*, 117–121.
27. Liu, J.J.; Wang, X.; Bi, C.L.; Ali, F.; Saleem, M.U.; Qin, J.H.; Ashfaq, H.; Han, Z.Q.; Abdullah, F. Epidemiological investigation of Staphylococcus aureus Infection in Dairy Cattle in Anhui, China. *Pak. Vet. J.* **2022**, *42*, 580–583. [CrossRef]
28. Schmitt, H.; Stoob, K.; Hamscher, G.; Smit, E.; Seinen, W. Tetracyclines and tetracycline resistance in agricultural soils: Microcosm and field studies. *Microb. Ecol.* **2006**, *51*, 267–276. [CrossRef] [PubMed]
29. Kim, K.R.; Owens, G.; Kwon, S.I.; So, K.H.; Lee, D.B.; Ok, Y.S. Occurrence and environmental fate of veterinary antibiotics in the terrestrial environment. *Water Air Soil Pollut.* **2011**, *214*, 163–174. [CrossRef]
30. Jechalke, S.; Heuer, H.; Siemens, J.; Amelung, W.; Smalla, K. Fate and effects of veterinary antibiotics in soil. *Trends Microbiol.* **2014**, *22*, 536–545. [CrossRef] [PubMed]
31. Backhaus, T.; Porsbring, T.; Arrhenius, A.; Brosche, S.; Johansson, P.; Blanck, H. Single-substance and mixture toxicity of five pharmaceuticals and personal care products to marine periphyton communities. *Environ. Toxicol. Chem.* **2011**, *30*, 2030–2040. [CrossRef]
32. Cycoń, M.; Mrozik, A.; Piotrowska-Seget, Z. Antibiotics in the soil environment—Degradation and their impact on microbial activity and diversity. *Front. Microbiol.* **2019**, *10*, 338. [CrossRef]

33. Xiong, J.Q.; Zhao, C.Y.; Qin, J.Y.; Cui, P.; Zhong, Q.L.; Ru, S. Metabolic perturbations of *Lolium perenne* L. by enrofloxacin: Bioaccumulation and multistage defense system. *J. Hazard. Mater.* **2022**, *427*, 127893. [CrossRef] [PubMed]
34. Opriş, O.; Copaciu, F.; Soran, M.L.; Ristoiu, D.; Niinemets, Ü.; Copolovici, L. Influence of nine antibiotics on key secondary metabolites and physiological characteristics in Triticum aestivum: Leaf volatiles as a promising new tool to assess toxicity. *Ecotoxicol. Environ. Saf.* **2013**, *87*, 70–79. [CrossRef] [PubMed]
35. Zahid, N.; Ahmed, M.j.; Tahir, M.M.; Maqbool, M.; Shah, S.Z.A.; Hussain, S.J.; Khaliq, A.; Rehmani, M.I.A. Integrated effect of urea and poultry manure on growth, yield and postharvest quality of cucumber (*Cucumis sativus* L.). *Asian J. Agric. Biol.* **2021**, *1*, 1–9.
36. Bassil, R.J.; Bashour, I.I.; Sleiman, F.T.; Abou-Jawdeh, Y.A. Antibiotic uptake by plants from manure-amended soils. *J. Environ. Sci. Health Part B* **2013**, *48*, 570–574. [CrossRef] [PubMed]
37. Do, K.H.; Byun, J.W.; Lee, W.K. Antimicrobial Resistance, Adhesin and Toxin Genes of Porcine Pathogenic *Escherichia coli* Following the Ban on Antibiotics as the Growth Promoters in Feed. *Pak. Vet. J.* **2021**, *41*, 519–523.
38. Zarei-Baygi, A.; Smith, A.L. Intracellular versus extracellular antibiotic resistance genes in the environment: Prevalence, horizontal transfer, and mitigation strategies. *Bioresour. Technol.* **2021**, *319*, 124181. [CrossRef] [PubMed]
39. Malik, F.; Nawaz, M.; Anjum, A.A.; Firyal, S.; Shahid, M.A.; Irfan, S.; Ahmed, F.; Bhatti, A.A. Molecular Characterization of Antibiotic Resistance in Poultry Gut Origin Enterococci and Horizontal Gene Transfer of Antibiotic Resistance to *Staphylococcus aureus*. *Pak. Vet. J.* **2022**, *42*, 383–389.
40. Rossolini, G.M.; Arena, F.; Pecile, P.; Pollini, S. Update on the antibiotic resistance crisis. *Curr. Opin. Pharmacol.* **2014**, *18*, 56–60. [CrossRef]
41. Lima, T.; Domingues, S.; Da Silva, G.J. Manure as a potential hotspot for antibiotic resistance dissemination by horizontal gene transfer events. *Vet. Sci.* **2020**, *7*, 110. [CrossRef] [PubMed]
42. Gajdács, M.; Urbán, E.; Stájer, A.; Baráth, Z. Antimicrobial resistance in the context of the sustainable development goals: A brief review. *Eur. J. Investig. Health Psychol. Educ.* **2021**, *11*, 71–82.

agriculture

Article

Biochar, Halloysite, and Alginite Improve the Quality of Soil Contaminated with Petroleum Products

Jadwiga Wyszkowska *, Agata Borowik, Magdalena Zaborowska and Jan Kucharski

Department of Soil Science and Microbiology, Faculty of Agriculture and Forestry, University of Warmia and Mazury in Olsztyn, Plac Łódzki 3, 10-727 Olsztyn, Poland; agata.borowik@uwm.edu.pl (A.B.); m.zaborowska@uwm.edu.pl (M.Z.); jan.kucharski@uwm.edu.pl (J.K.)

* Correspondence: jadwiga.wyszkowska@uwm.edu.pl

Abstract: Investigations into the effective, fast, and economically viable remediation of soils polluted with petroleum-derived products are still relevant. The vegetative pot experiment was conducted at the Didactic-Experimental Garden greenhouse (NE, Poland, 53.759° N, 20.452° E) on loamy sand (LS) and sandy loam (SL) soils. Its main research objective was to assess the effectiveness of biochar (B), halloysite (H) and alginite (A) in the biological regeneration of contaminated soil diesel oil (DO) and petrol (P). The assessment was conducted by determining the magnitude of the adverse impact of these xenobiotics on the growth and development of *Zea mays*, as well as the activity of seven soil enzymes. The impact of the tested contaminants and sorbents was assessed based on the impact factors (IF) of DO and P, as well as B, H, and A on *Zea mays* biomass and enzymatic activity of the soil. Soil contamination with petroleum-derived products disrupted the growth and development of *Zea mays*. DO had a stronger inhibitory effect on plant growth compared to P. *Zea mays* cultivated in LS, which was less resistant to the effects of these contaminants compared to that cultivated in SL. The impact of DO and P on enzyme activity depended on the soil texture. DO stimulated enzyme activity in LS and SL, while P only did so in LS. All remediation substances, and biochar in particular, led to an increase in plant biomass in the DO-contaminated soils. Both biochar, halloysite and alginite also improved the biochemical quality index (BA) of SL and LS. Despite the unquestionable remediation potential of the analyzed sorbents, their highest efficacy can only be achieved by their application on soils with physicochemical properties corresponding to their characteristics, which is a valuable guideline for further research.

Keywords: sorbents; soil; diesel oil; petrol; soil enzymes; biochemical soil quality index

1. Introduction

Viable techniques for the separation of petroleum-derived substances have long been a challenge in the oil industry [1–4]. Issues encountered during production, transportation, and storage have been regulated for years by the Seveso III Directive (2012/18/EU) [5], which sets provisions regarding hazard control. It obliges for entrepreneurs to obtain necessary permits, conduct audits, and monitoring. The Seveso III Directive (2012/18/EU) mandates that industrial facilities develop and implement a risk management system and take preventive actions to minimize the risk of major industrial accidents. Enterprises are also obliged to prepare emergency plans. It is important to note that obligations and regulations may vary among individual Member States of the European Union. Therefore, it is crucial for enterprises to comply with the regulations and guidelines stipulated by the relevant regulatory authorities in their respective countries [5].

Preventive actions are highly justified, due to the fact that soils contaminated with petroleum-derived products contain numerous toxic compounds such as benzene, toluene, xylene (BTEX) or polycyclic aromatic hydrocarbons (PAHs) [6]. They raise multiple concerns due to their hydrophobic nature, low biodegradability, and consequently high persistence in the soil environment [7]. This translates into disturbances in soil properties, such as mechanical properties, permeability, shear strength, texture, and porosity [8,9], fertility characteristics (N, P, and K contents) and its chemical properties (soil conductivity, pH, and organic matter content) [10]. Soil homeostasis disruption also reflects the plant's response to the pressure of these pollutants. It involves changes in plant physiology and biochemistry, including reductions in plant biomass, chlorophyll, and root vitality [11,12]. These findings are consistent with the observed highly toxic impact of diesel oil on the growth and development of *Avena sativa* L. [13], *Lolium perenne* L. × *hybridum*, *Poa pratensis*, *Festuca rubra*, *F. arundinacea*, *Phleum pratense*, and *Dactylis glomerata*, with plants showing significantly greater sensitivity to petrol. Moreover, the plant species also influence the extent of degradation of individual PAH groups [14].

One of the techniques used to separate petroleum-derived substances from soil involves the use of minerals [15–17]. Iron- and calcium-rich minerals can react with pollutants in the soil binding and stabilize them, thus causing their immobilization [18,19]. Therefore, phytostabilization is a stage that can contribute to reducing the toxicity and availability of pollutants to living organisms by adsorbing pollutants on mineral surfaces [20–22]. This process leads to the formation of less toxic compounds. The coupling of phytostabilization and phytoextraction, which utilizes plants capable of accumulating contaminants, can yield even better results [23–25]. Hence, novel techniques are required for the effective protection of soil, water in the soil, and essential nutrients crucial for plants and microorganisms, ultimately leading to the enhancement of environmental quality.

Biochar, also known as plant charcoal or pyrolytic carbon, is a product obtained through the pyrolysis of biomass, involving its decomposition under limited oxygen conditions. It has been found capable of effectively mitigating the emission of toxic compounds [26]. Biochar exhibits immense potential in regulating soil properties and fertility, promoting microbial activity, and enhancing plant growth [26–30]. As described by Hamidzadeh et al. [26], biochar's carboxylic acids and phenolic compounds, derived from biomass lignin, augment the specific surface area of biochar, enabling the elimination of pollutants and proper absorption and distribution of nutrients in the soil. Furthermore, by applying current environmental modifications to pyrolysis, soil pH and the efficacy of aromatic compounds in biochar can be improved [26,31]. Selvarajoo et al. [32] demonstrated that, apart from enhancing soil properties such as pH, biochar exerted positive effects on nutrient cycling, water retention, and exchangeable cations. It was also shown to enhance water-holding capacity, reduce soil density, nitrogen content, leaching of pesticide residues, and essential mineral loss [26,28,29,31]. Additionally, biochar can also act as a carrier for microorganisms, influencing various soil biological processes, such as organic matter decomposition and mineral transformation [30,33]. This sorbent increases the mobility of anionic toxic elements [34]. Due to its porous structure, biochar can adsorb and retain organic pollutants [35,36], heavy metals [37–39], and other harmful compounds [15].

Halloysite is a mineral that mainly consists of silicon dioxide and aluminium oxides [40]. Clay minerals, such as kaolin, bentonite, and halloysite, belong to the kaolinite-serpentine subgroup and are described by the general chemical formula $Al_2Si_2O_5(OH)_4$. These clay silicate minerals are utilized as fuel additives for the thermal conversion of biomass and waste [41]. They are currently used in the energy sector to mitigate the risk of ash deposition and high-temperature corrosion in low-quality fuels such as biomass and waste. Maj and Matus [41] demonstrated that the use of clay silicates minimized combustion-related issues by immobilizing alkalis in highly volatile compounds. In turn, Masoudniaragh et al. [42] noted that the application of halloysite, particularly in agricultural fields, could lead to increased production. Additionally, according to these authors, halloysite can act as a carrier for biostimulants or bioactive components, which may contribute to their improved stability. In the study conducted by Masoudniaragh et al. [42], halloysite proved effective as a carrier of pyroligneous acid in mitigating the effects of soil salinity.

The application of alginite, similar to biochar, influences an increase in water retention capacity [43,44]. Consequently, soil amended with alginite becomes more drought-resistant, and plants have access to water for a longer period. Alginite, as a natural rock, contains essential nutrients such as macro- and micronutrients [45,46]. By adding alginite to the soil, these nutrients can be supplied to plants, leading to their healthier growth, increased productivity and condition [47–49]. Zhao et al. [43,50] observed that alginite significantly reduces the activity of stress enzymes, catalase, and peroxidase, indicating potential damage to the plant's defense system. This natural, ecological, and versatile polymer is commonly used as a carrier for microbial cultures, mainly *Pseudomonas* and *Azotobacter* [51,52]. Alginite has the ability to regulate soil pH. It acts as a natural buffer, helping to maintain the optimal pH for plant growth. If the soil is too acidic or alkaline, the addition of alginite can assist in balancing and maintaining the optimal pH [53].

The described properties of the sorbents have prompted authors to formulate a research hypothesis that biochar, halloysite, and alginite are effective in the bioremediation of soil contaminated with petroleum-derived products. Therefore, a study was conducted with the aim to assess the applicability of these environmentally friendly materials in mitigating the adverse effects of diesel oil and petrol on the growth and development of *Zea mays*, as well as the activity of soil enzymes belonging to the oxidoreductase and hydrolase classes. For the phytoremediation of soils contaminated with petroleum-derived products, it was decided to utilize a popular crop that is distributed worldwide [19], with its cultivation area steadily increasing [54,55] characterized by a short life cycle [56,57]. The implementation of this defined objective allowed evaluating of the effectiveness of biochar, halloysite, and alginite in enhancing the quality of soils contaminated with petroleum-derived products.

2. Materials and Methods

2.1. Material Characteristics

2.1.1. Soil

Two soils with different textures were utilized, loamy sand (LS) and sandy loam (SL), in the vegetative pot experiment conducted in the greenhouse. They were selected for the study as sandy soils occupy approximately 50% of the agricultural land in Poland [58]. Based on the classification by the International Union of Soil Sciences [59], they were classified as Eutric Cambisols. LS and SL were collected from the top layer (0–0.20 m) of a field located in northeastern Poland (NE, Poland, 53,713° N, 20,432° E), where winter oilseed rape was cultivated. The soils were then transported to the greenhouse and air-dried. The soil used for the vegetative experiment was sieved through a 5 mm mesh sieve, while the soil used for the basic physico-chemical, chemical, and biochemical analyses was sieved through a 2 mm mesh sieve. A detailed characterization of the soils is presented in Table 1.

Table 1. Some properties of the soil used in the experiment.

Abbreviation	Properties	Unit	Soil	
			Loamy Sand (LS)	Sandy Loam (SL)
		Grain-size Composition		
Sand	0.05–2.0 mm	%	74.30	70.38
Silt	0.02–0.05 mm		23.69	27.19
Clay	<0.002 mm		2.01	2.43
		Chemical and Physicochemical Properties		
N_{tot}	Total Nitrogen	$g\ kg^{-1}$ dm	0.98	1.01
C_{org}	Organic Carbon		11.20	11.50
$P_{available}$	Phosphorus Available	$mg\ kg^{-1}$ dm	164.05	172.73
$K_{available}$	Potassium Available		53.95	78.85
$Mg_{available}$	Magnesium Available		46.00	38.00
pH	Soil pH_{KCl} Reaction	1 mol KCl dm^{-3}	6.98	7.13

Table 1. *Cont.*

Abbreviation	Properties	Unit	Soil	
			Loamy Sand (LS)	Sandy Loam (SL)
EBC	Sum of Exchangeable Base Cations	mM (+) kg^{-1} dm	84.20	181.80
HAC	Hydrolytic Acidity		8.00	5.70
CEC	Cation Exchange Capacity	%	92.20	187.50
ACS	Alkaline Cation Saturation		91.32	96.96
Enzymatic Activity per 1 kg dm h^{-1}				
Deh	Dehydrogenases	µM TFF	5.426	8.007
Cat	Catalase	M O_2	0.211	0.369
Ure	Urease	mM N-NH_4	0.153	1.114
Pac	Acid Phosphatase	mM PN	2.291	0.954
Pal	Alkaline Phosphatase		2.037	1.572
Glu	β-glucosidase		0.542	0.914
Aryl	Arylsulphatase		0.386	0.721

2.1.2. Plant

Zea mays PR39H32 (variety registered in the European Union) easily adapts to unfavorable environmental conditions. An additional advantage for the utilization of *Zea mays* in the study was the fact that its biomass can be used in alternative energy production [60–62].

2.1.3. Petroleum-Derived Products

The latest generation fuels meeting European requirements for sulfur-free fuels were used as the contaminating substances: VERVA diesel oil (DO) and VERVA 98 petrol (P). The density of diesel oil ranges from 820–845 kg m^{-3}, whereas that of petrol ranges from 720–775 kg m^{-3}. Both products were purchased from a PKN Orlen fuel station (Poland), and their detailed characteristics can be found on the website http://www.orlen.pl/ (accessed on 20 May 2021).

2.1.4. Sorbents

The vegetative experiment, aiming to reduce the mobility of petroleum-derived products in soil and neutralize their impact on the growth and development of *Zea mays*, as well as the biochemical properties of the soil, was conducted with three substances with sorptive properties: biochar (Fluid, Sędziszów, Poland), halloysite (Halosorb minerals sorbents, Intermark, Gliwice, Poland), and alginite (Vázsony-Szövetkezeti Kft., Nagyvázsony Hungary). Biochar is produced through the pyrolysis of various raw materials under limited oxygen conditions [47,63,64]. The biochar used in the study was produced from the biomass waste of willow and miscanthus. Halloysite ($Al_2Si_2O_5(OH)_4$) is a clay mineral from the kaolinite group characterized by high porosity (pore diameter of 10–20 nm). The halloysite used in the experiment was extracted from the Dunino mine, which is the only deposit of this mineral in Europe and one of three worldwide [47,65]. Alginite is a naturally occurring mineral from the liptinite group, formed from fossilized algal biomass and weathered tuffs. Its pH ranges from 7.15 to 7.78. A comprehensive characterization of biochar and halloysite was presented in our previous publications [38,39,47], and alginite [17,47] was also previously described.

2.2. Research Design

The pot experiment was conducted in the greenhouse of the University of Warmia and Mazury in Olsztyn (NE, Olsztyn, Poland, 53.759° N, 20.452° E) with 5 replications. The procedure for setting up and conducting the experiment was as follows:

(1) Preparation of 2.5 kg of LS or SL soils whose characteristics are presented in Table 1.
(2) Soil amendment with:

- N, P, K and Mg in the following amounts (mg kg^{-1} dm of soil): 112, 39, 112 and 15, respectively. Nitrogen was applied as N_2H_4CO, phosphorus as—KH_2PO_4, potassium as—KH_2PO_4 and KCl, and magnesium as—$MgSO_4 \times 7H_2O$;
- petroleum-derived products: diesel oil (DO) and petrol (P) at a rate of 0 and 7 cm^3 kg^{-1} dm of soil;
- sorbents: biochar (B), halloysite (H) or alginite (A) at a rate of 0 and 10 g kg^{-1} dm of soil.

(3) Mixing the soils was conducted with mineral fertilizers and, in appropriate treatments, with petroleum-derived products and sorbents. The soil was packed into polyethylene pots with a capacity of 3.0 dm^3 (upper base diameter: 18.5 cm; lower base diameter: 14 cm; height: 15 cm) and the soil moisture was adjusted to 60% of the maximum water-holding capacity. This moisture level was maintained throughout the entire plant growth period (60 days).
(4) Sowing 8 *Zea mays* seeds and leaving 5 plants in each pot after germination.
(5) Harvesting the aerial and root parts of *Zea mays* at the Biologische Bundesanstalt, Bundessortenamt and Chemical (BBCH) stage 59. On the same day as the plant harvest, soil samples were collected for biochemical analyses. These samples were sieved through a 2.0 mm mesh sieve.

2.3. Methodology for Physico-Chemical and Chemical Determinations

Prior to the experiment setup, using the methods described in our previous studies [66,67], the soil texture was determined using the aerometric method, and pH was measured in a 1 mol KCl dm^{-3} solution using a potentiometric method with an HI 2221 m (Hanna Instruments, Washington, UK) in three replicates of air-dried soils. Additionally, the sum of exchangeable base cations (EBC) and hydrolytic acidity (HAC) were determined, based on which the cation exchange capacity (CEC) and alkaline cation saturation (ACS) were calculated. The chemical properties of the soil were assessed by determining the contents of total nitrogen (N_{tot}), organic carbon (C_{org}), and available P, K, and Mg. The Buchi B-324 distillation unit (Buchi, Flawil, Switzerland), Jenway 6705 UV/VIS spectrophotometer (Jenway Ltd., Staffordshire, UK), Jenway PFP 7 flame photometer (Jenway Ltd., Staffordshire, UK), and GBC 932AA atomic absorption spectrophotometer (GBC Scientific Equipment, Braeside, Australia) were used for these analyses.

2.4. Methodology for Determining Soil Enzyme Activities

Before the initiation of the vegetation experiment and after the harvest of *Zea mays*, the activities of dehydrogenases (Deh), catalase (Cat), and β-glucosidase (Glu) were used as indicators of the carbon cycle, urease (Ure) as an indicator of the nitrogen cycle, acid phosphatase (Pac) and alkaline phosphatase (Pal) as indicators of the phosphorus cycle, and arylsulfatase (Aryl) as an indicator of the sulphur cycle, and were determined using specific enzyme assay procedures (buffers, incubation times, termination of reactions, and temperature) described in our previous studies [37,67], in soils with a moisture content of approximately 60% of the maximum water-holding capacity (three replicates). The activities of Deh, Ure, Pac, Pal, Glu, and Aryl were measured using a Perkin-Elmer Lambda 25 spectrophotometer (Waltham, MA, USA).

2.5. Calculations

By considering the biomass of *Zea mays'* aerial and root parts, as well as the activity of seven enzymes involved in C, N, P, and S metabolism, in both uncontaminated and petroleum-contaminated soils supplemented and non-supplemented with of biochar (B), halloysite (H), and alginite (A), the Impact Index (IF) of DO, P, and Ad on these parameters was determined using Formulas (1)–(4).

$$IF_{DO} = \frac{DO}{C2} - 1 \text{ or } IF_P = \frac{P}{C2} - 1 \quad (1)$$

where:

IF—impact index of DO or P,
DO—biomass of *Zea mays* or the activity value of the tested enzyme in DO-contaminated soil or P-contaminated soil without Ad,
C2—biomass of *Zea mays* or the activity value of the tested enzyme in the control soil uncontaminated by DO or P without Ad.

$$IF_{Ad} = \frac{Ad\_C2}{C2} - 1 \quad (2)$$

where:

IF—impact index of sorbents: B, H or A,
Ad_C2—biomass of *Zea mays* or the activity value of the tested enzyme in the control soil uncontaminated by DO or P with the addition of Ad (B, H, A),
C2—the explanation is provided in formula number 1.

$$IF_{Ad\_DO} = \frac{Ad\_DO}{DO} - 1 \quad (3)$$

where:

IF_{Ad_DO}—impact index of sorbents: B, H lub A in soil contaminated DO,
Ad_DO—biomass of *Zea mays* or the activity value of the tested enzyme in DO-contaminated soil with the addition of Ad (B, H, or A),
DO—biomass of *Zea mays* or the activity value of the tested enzyme in DO-contaminated soil without Ad.

$$IF_{Ad\_P} = \frac{Ad\_P}{P} - 1 \quad (4)$$

where:

IF_{Ad_P}—impact index of sorbents: B, H lub A in soil contaminated P,
Ad_P—biomass of *Zea mays* or the activity value of the tested enzyme in P-contaminated soil with the addition of Ad (B, H, or A),
P—biomass of *Zea mays* or the activity value of the tested enzyme in P-contaminated soil without Ad.

The above indexes can range from −1 to 1. An index of 1.00 indicates a 100% increase in the value of the measured parameter, whereas −1.00 represents a 100% decrease in its value. The indexes were presented on heat maps using RStudio 2023.06.0 [68] software with R 4.2.2 [69] and the gplots library [70].

Based on the activities of the analyzed enzymes, the soil quality index (BA) was calculated using formula number 5:

$$BA = Deh + Cat + Ure + Pac + Pal + Aryl + Glu \quad (5)$$

The BA index of soil quality assessment has been recognized as one of the most reliable, composite indicators of soil fertility from the pool of 21 indicators determining the condition of the soil. This is justified by the fact that it takes into account the activity of as many as seven soil enzymes presented in the formula, recognized not only as very sensitive indicators of early processes indicating the disturbance of soil equilibrium under the pressure of xenobiotics, but also indicators of the intensity of biological processes related to the physicochemical properties of soils [71].

2.6. Statistical Analyses

η2 coefficients were calculated and presented in a pie chart created using the Circos 0.68 package [72]. Principal Component Analysis (PCA) was performed to highlight the interdependencies between the data. The Shapiro–Wilk test was used to check for data normality. Statistical analysis, conducted separately for each tested soil, was carried out using the Tukey test with a significance level of $p < 0.05$. Standard deviations (SD) were also calculated for the yield of aerial parts and roots of *Zea mays* and for soil enzyme activity [73].

3. Results

3.1. The Biomass Yield of *Zea mays* under the Influence of Diesel Oil, Petrol, Biochar, Halloysite, and Alginite

Soil contamination with both DO and P had an adverse effect on the growth and development of *Zea mays*. However, plant response varied depending on the type of petroleum product and the soil texture (Table 2).

Table 2. The effect of soil contamination with diesel oil (DO) and petrol (P) and sorbents on *Zea mays* biomass.

Type of Sorbent	Objects	Loamy Sand (LS)			Sandy Loam (SL)		
		Aerial Parts (Ap)	Roots (r)	Ap/r	Aerial Parts (Ap)	Roots (r)	Ap/r
		g dm of pot^{-1}			g dm of pot^{-1}		
Control (C1)	C2	40.712 ^{b} ±1.721	8.818 ^{bc} ±0.192	4.617	46.708 ^{a} ±1.031	8.533 ^{ab} ±0.293	5.474
	DO	4.903 ^{e} ±0.690	1.656 ^{f} ±0.238	2.960	28.122 ^{d} ±2.060	4.545 ^{e} ±0.816	6.188
	P	33.378 ^{d} ±3.270	6.826 ^{d} ±0.131	4.890	42.422 ^{b} ±1.594	5.723 ^{de} ±0.468	7.412
Biochar (B)	C2	41.930 ^{b} ±0.408	11.022 ^{a} ±1.495	3.804	46.733 ^{a} ± 0.726	9.501 ^{a} ±0.206	4.919
	DO	7.214 ^{e} ±0.148	4.953 ^{e} ±0.488	1.457	31.745 ^{c} ±1.011	5.302 ^{de} ±1.462	5.987
	P	35.805 ^{d} ±1.940	8.486 ^{cd} ±0.523	4.219	43.042 ^{b} ±0.489	6.588 ^{cd} ±0.363	6.534
Halloysite (H)	C2	47.858 ^{a} ±1.433	10.186 ^{ab} ±0.623	4.698	48.907 ^{a} ±0.893	9.212 ^{a} ± 0.629	5.309
	DO	6.025 ^{e} ±0.022	3.885 ^{e} ±0.188	1.551	31.124 ^{cd} ±1.421	8.385 ^{ab} ±0.855	3.712
	P	32.799 ^{d} ± 1.070	8.492 ^{cd} ± 0.603	3.862	40.200 ^{b} ±1.270	8.181 ^{abc} ±0.819	4.914
Alginite (A)	C2	40.097 ^{bc} ± 1.850	9.234 ^{bc} ± 0.835	4.342	47.796 ^{a} ±2.553	8.404 ^{ab} ±0.418	5.687
	DO	7.304 ^{e} ± 0.768	3.876 ^{e} ± 0.012	1.884	33.092 ^{c} ±0.699	7.031 ^{bcd} ±0.922	4.707
	P	36.641 ^{cd} ± 2.069	8.058 ^{cd} ± 1.098	4.547	46.807 ^{a} ±1.064	5.682 ^{de} ±0.367	8.237

C1—soil without sorbent, C2—soil uncontaminated with DO or P; DO—diesel oil VERVA; P—VERVA 98 petrol. Homogeneous groups denoted with letters (a–f) were calculated separately for each part of the plant and each kind of soil and each kind of soil.

DO had a stronger inhibitory impact on the growth of the aerial parts and roots of *Zea mays* compared to P (Figure 1a,b). This effect was observed in both loamy sand (LS) and sandy loam (SL). Under DO pressure, the biomass of aerial parts and roots of *Zea mays* in LS was reduced by 88.0% and 81.2%, respectively, whereas in SL it was reduced by 39.8% and 46.7%, respectively. As a result of the different effects of DO and P in LS and SL, higher values of the aerial parts-to-root biomass ratio were determined in *Zea mays* grown in SL compared to LS. When P was added to both LS and SL, it had a significantly smaller disruptive effect on the growth of *Zea mays*, as indicated by the higher IF_P values compared to IF_{DO} (Figure 1a,b). Specifically, the IF_B values of the aerial parts of *Zea mays* in LS and SL were higher by 0.700 and 0.306, respectively, whereas the values for roots were higher by 0.586 and 0.138, respectively.

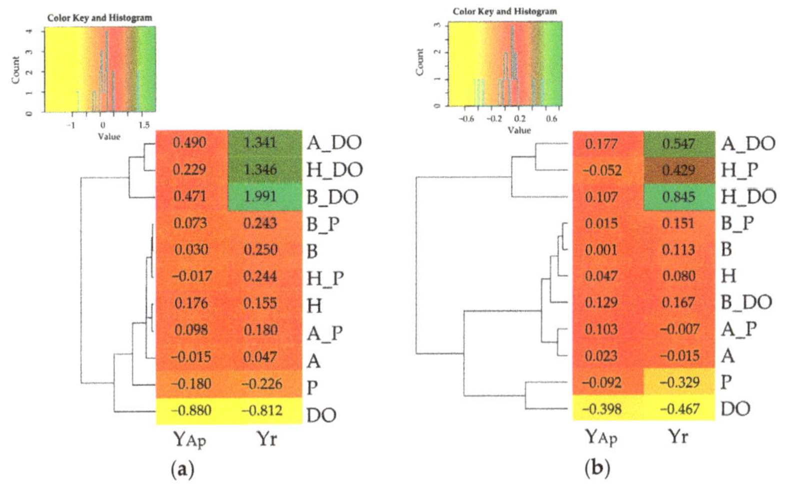

Figure 1. Index of the impact of DO and P and sorbents on the yield (Y) of *Zea mays* cultivated on loamy sand (**a**) and sandy loam (**b**). The abbreviations are explained in the Table 2.

In both soils not exposed to petroleum products, two out of the three sorbents (biochar and halloysite) among biochar, halloysite, and alginite induced the growth of the aerial parts and roots of *Zea mays* (Table 2, Figure 1a,b).

The greatest increase in the aboveground biomass (17.6%) was observed with the application of halloysite in LS, whereas for roots, the highest increase (25.0%) was observed with biochar, also in LS. Generally, the effectiveness of these sorbents was more pronounced in LS compared to SL.

The assessment of biochar, halloysite, and alginite for effectiveness in mitigating the toxic impact of DO and P revealed their potential in alleviating the adverse impact of petroleum products on the yield of *Zea mays*. However, this potential varied depending on contaminant type and soil texture. In LS, the effects of all sorbents on the yield of aerial parts and roots were significantly higher in the case of DO-contaminated soil compared to the P-contaminated soil, whereas in SL, such effects were only observed for the aboveground biomass (Table 2). These findings were supported by the influence indices of individual sorbents on the aboveground and roots biomass of *Zea mays*, which consistently showed positive values in DO contaminated LS and SL. Positive values were observed only in LS and SL supplemented with biochar, as well as LS supplemented with alginite in P-contaminated soils (Figure 1a,b).

3.2. *Changes in Soil Enzyme Activity under the Influence of Diesel Oil, Petrol, Biochar, Halloysite, and Alginite*

The contamination of LS and SL soils with petroleum products resulted in a significant destabilization of the enzymatic properties of the soils, as shown in Tables 3 and 4 and Figures 2 and 3. The response of the enzymes to these pollutants varied depending on soil type. In LS, both DO and P stimulated the activity of Deh, Cat, Ure, Pac, and inhibited Glu and Aryl (Table 3), whereas in SL, they stimulated Pac and Glu activity (Table 4). The response of the other enzymes differed depending on contaminant type. In LS, DO significantly stimulated the activity of Pal, whereas in SL, it enhanced the activity of Deh, Cat, Ure, Pal, and Aryl. On the other hand, P decreased the activity of these enzymes. The response of soil enzymes to the application of sorbents to the soil not exposed to petroleum products was determined by multiple independent variables (Tables 3 and 4,

Figures 2 and 3). Biochar, halloysite, and alginite significantly increased Deh activity (IF values were 0.538, 0.735, 0.385, respectively), Ure activity (IF—1.631, 3.269, 2.719), Pac activity (IF—0.245, 0.113, 0.098), and Aryl activity (IF—0.105, 0.067, 0.374) in LS. In SL, they significantly enhanced Pac activity (IF—0.522, 0.361, 0.397) and inhibited Cat activity (IF −0.379, −0.427, −0.410). Additionally, in LS, biochar induced Cat activity (IF 0.084) and Glu activity (IF 0.075), and halloysite induced Glu activity (IF 0.151), while in SL, Ure activity was stimulated by alginite (IF 0.227) and inhibited by biochar (IF −0.366).

Sorbents exhibited different effects on enzyme activity in the heavier soil contaminated with DO and P (Figure 3). In the DO series, Cat activity was significantly suppressed by all sorbents, while Ure and Aryl activities were reduced only by biochar. In the soil amended with P, the inhibitory effect of biochar was noted for Deh (IF −0.182) and Pac (IF −0.037), that of halloysite on Pal (IF −0.093), and that of alginite on Cat (IF −0.251). Conversely, in the soil contaminated with DO, biochar exhibited a stimulating effect on Pac (IF 0.175) and Pal (IF 0.078) activity, while P stimulated Ure (IF 0.142) and Glu (IF 0.079) activity. Halloysite increased the activity of Deh (IF 0.084) and Ure (IF 0.300) in the soil samples with DO and that of Ure (IF 0.142), Pac (IF 0.345), and Glu (IF 0.358) in the objects with P. Alginite enhanced Ure (IF 0.245–0.187) and Pac (IF 0.134–0.581) activity in the soil with DO and P, as well as Deh (IF 0.385) and Pal (IF 0.179) activity in the SL with DO, and Glu (IF 0.181) activity in P-treated soil.

Table 3. The impact of diesel oil (DO) and petrol (P) contamination, as well as sorbents, on enzyme activity in 1 kg dm of soil × h^{-1} of loamy sand (LS).

Type of Sorbent	Object	Deh	Cat	Ure	Pac	Pal	Glu	Aryl
C1	C2	5.901 ^{h} ±0.090	0.227 ^{g} ±00.008	0.160 ^{g} ±0.125	2.473 ^{g} ±0.004	2.086 ^{de} ±0.001	0.571 ^{cd} ±0.004	0.401 ^{d} ±0.002
	DO	6.824 ^{g} ±0.087	0.339 ^{c} ±0.000	0.264 ^{f} ±0.027	3.551 ^{bc} ±0.012	2.606 ^{b} ±0.010	0.476 ^{fg} ±0.003	0.385 ^{de} ±0.003
	P	7.388 ^{ef} ±0.219	0.386 ^{a} ±0.011	0.266 ^{f} ±0.045	3.369 ^{cd} ±0.094	1.986 ^{e} ±0.005	0.446 ^{h} ±0.007	0.256 ^{g} ±0.007
B	C2	9.075 ^{c} ±0.001	0.246 ^{f} ±0.004	0.421 ^{d} ±0.025	3.079 ^{e} ±0.024	2.131 ^{cd} ±0.060	0.614 ^{b} ±0.006	0.443 ^{c} ±0.009
	DO	7.614 ^{e} ±0.183	0.347 ^{bc} ±0.008	0.311 ^{e} ±0.027	3.722 ^{ab} ±0.190	2.619 ^{b} ±0.017	0.515 ^{e} ±0.004	0.009
	P	7.105 ^{fg} ±0.242	0.310 ^{d} ±0.008	0.334 ^{e} ±0.025	3.829 ^{a} ±0.113	2.123 ^{cd} ±0.026	0.494 ^{ef} ±0.024	0.275 ^{g} ±0.003
H	C2	10.240 ^{b} ±0.069	0.227 ^{g} ±0.008	0.683 ^{b} ±0.025	2.752 ^{f} ±0.010	2.204 ^{c} ±0.009	0.657 ^{a} ±0.002	0.428 ^{c} ±0.009
	DO	7.707 ^{e} ±0.173	0.384 ^{a} ±0.004	0.287 ^{e} ±0.013	3.289 ^{de} ±0.020	2.738 ^{a} ±0.006	0.493 ^{ef} ±0.001	0.476 ^{b} ±0.007
	P	6.686 ^{g} ±0.090	0.287 ^{e} ±.0.000	0.421 ^{d} ±0.025	3.180 ^{de} ±0.002	2.110 ^{cd} ±0.010	0.501 ^{ef} ±0.004	0.298 ^{f} ±0.003
A	C2	8.173 ^{d} ±0.007	0.234 ^{fg} ±0.005	0.595 ^{c} ±0.025	2.715 ^{f} ±0.008	2.148 ^{cd} ±0.009	0.589 ^{bc} ±0.005	0.551 ^{a} ±0.005
	DO	11.594 ^{a} ±0.256	0.364 ^{b} ±0.008	0.893 ^{a} ±0.013	3.591 ^{b} ±0.055	2.754 ^{a} ±0.026	0.558 ^{d} ±0.017	0.009
	P	5.732 ^{h} ±0.135	0.227 ^{g} ±0.000	0.465 ^{d} ±0.025	3.314 ^{d} ±0.015	2.046 ^{de} ±0.006	0.453 ^{gh} ±0.003	0.375 ^{e} ±0.003

C1—soil without sorbent, C2—soil uncontaminated with DO or P; B—biochar, H—halloysite, A—alginite, DO—VERVA diesel oil; P—VERVA 98 petrol; Deh—dehydrogenases; Cat—catalase; Ure—urease; Pac—acid phosphatase; Pal—alkaline phosphatase; Glu—β-glucosidase; Aryl—arylsulfatase. Homogeneous groups denoted with letters (a–h) were calculated separately for each enzyme.

Table 4. The impact of diesel oil (DO) and petrol (P) contamination, as well as sorbents, on enzyme activity in 1 dm of soil × h^{-1} of sandy loam (SL).

Type of Sorbent	Object	Deh	Cat	Ure	Pac	Pal	Glu	Aryl
C1	C2	8.550^{d} ±0.327	0.393^{b} ±0.004	1.238^{de} ±0.066	1.011^{h} ±0.014	1.619^{de} ±0.013	0.986^{e} ±0.004	0.765^{de} ±0.014
	DO	11.395^{c} ±0.116	0.437^{a} ±0.012	1.664^{b} ±0.069	1.640^{d} ±0.011	2.318^{c} ±0.018	1.092^{cd} ±0.020	0.934^{a} ±0.021
	P	7.588^{e} ±0.125	0.303^{ef} ±0.004	0.979^{g} ±0.020	1.454^{f} ±0.031	1.553^{de} ±0.044	1.044^{cde} ±0.041	0.677^{f} ±0.034
B	C2	8.746^{d} ±0.490	0.244^{g} ±0.004	0.785^{h} ±0.012	1.539^{e} ±0.025	1.679^{d} ±0.011	0.998^{e} ±0.009	0.757^{e} ±0.004
	DO	11.168^{c} ±0.162	0.338^{d} ±0.008	1.074^{fg} ±0.069	1.927^{b} ±0.008	2.499^{b} ±0.128	1.100^{c} ±0.020	0.858^{bc} ±0.018
	P	6.206^{f} ±0.276	0.310^{e} ±0.008	1.118^{ef} ±0.025	1.400^{g} ±0.006	1.515^{ef} ±0.008	1.126^{c} ±0.004	0.681^{f} ±0.017
H	C2	8.769^{d} ±0.207	0.225^{h} ±0.007	1.282^{d} ±0.025	1.376^{g} ± 0.006	1.628^{de} ±0.011	1.004^{de} ±0.009	0.815^{cde} ±0.035
	DO	12.351^{b} ±0.300	0.374^{c} ±0.004	2.163^{a} ±0.026	1.644^{d} ±0.003	2.531^{b} ±0.019	1.127^{c} ±0.012	0.914^{ab} ±0.017
	P	7.651^{e} ±0.139	0.287^{f} ±0.004	1.118^{ef} ±0.025	1.956^{b} ±0.012	1.408^{f} ±0.064	1.418^{a} ±0.086	0.680^{f} ±0.021
A	C2	9.210^{d} ±0.149	0.232^{gh} ±0.004	1.519^{c} ±0.045	1.412^{fg} ±0.012	1.621^{de} ±0.011	1.004^{de} ±0.005	0.817^{cd} ±0.008
	DO	15.779^{a} ±0.466	0.354^{d} ±0.008	2.072^{a} ±0.026	1.859^{c} ±0.036	2.733^{a} ±0.021	1.109^{c} ±0.026	0.908^{ab} ±0.008
	P	6.884^{ef} ±0.344	0.227^{gh} ±0.004	1.162^{def} ±0.025	2.299^{a} ±0.015	1.568^{de} ±0.009	1.233^{b} ±0.020	0.654^{f} ±0.013

Abbreviations are explained under the Table 3. Homogeneous groups denoted with letters (a–h) were calculated separately for each enzyme.

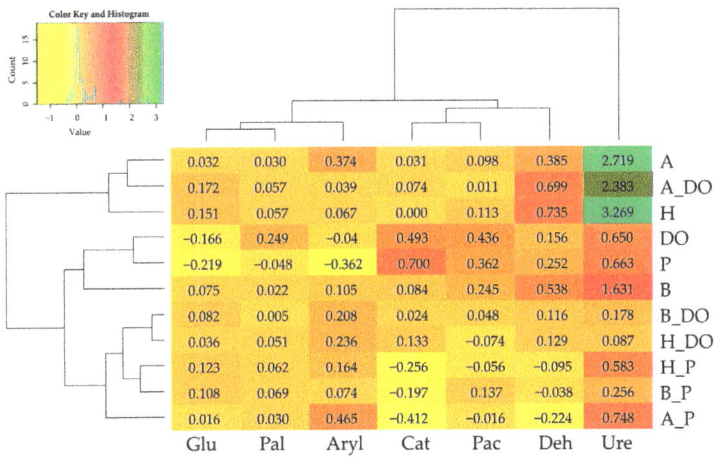

Figure 2. The index of the impact of DO, P, and sorbents on enzyme activity in loamy sand. On the *X*-axis are abbreviations of enzyme names, and on the *Y*-axis are abbreviations of research objects. Abbreviations are explained under the Table 3.

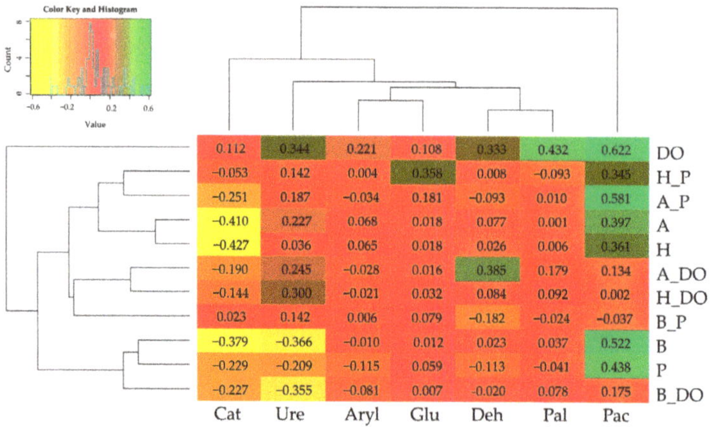

Figure 3. The index of the impact of DO, P, and sorbents on enzyme activity in sandy loam. On the *X*-axis are abbreviations of enzyme names, and on the *Y*-axis are abbreviations of research objects. Abbreviations are explained under the Table 3.

The changes in soil enzyme activity observed under the influence of petroleum products and sorbents were reflected in the biochemical soil quality index (BA), which was 11.819 in LS and 14.563 in SL (Figure 4).

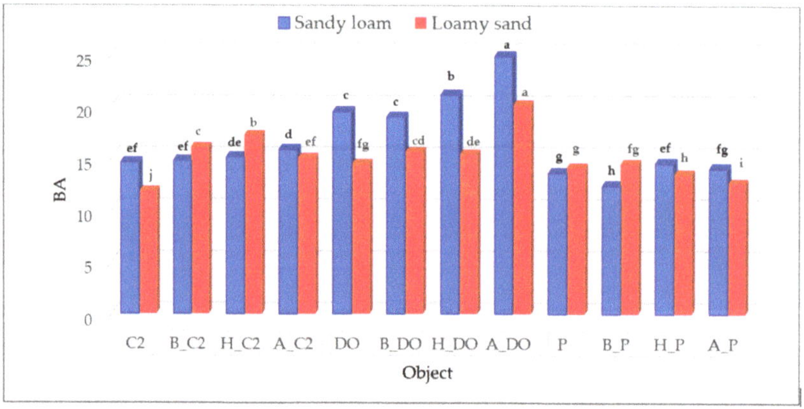

Figure 4. The effect of DO and P on the biochemical soil quality Index (BA). Abbreviations are explained under the Table 3. Homogeneous groups denoted with letters (a–j) were calculated separately for each soil.

In the LS, both DO and P increased the BA value, while in SL, only DO exerted this effect. In the LS, all sorbents applied to the uncontaminated soil significantly increased the BA value, whereas only alginite exerted this effect in the SL. The greatest changes in the BA index in soil contaminated with DO were observed under the influence of halloysite and alginite, which increased the BA value by 6.4% to 39.5% in LS and by 8.3% to 27.4% in the SL. The sorbents tested for their effect on P caused significantly smaller fluctuations in the BA index in the experimental series.

3.3. Correlations between Zea mays Biomass and Soil Enzyme Activity—Independent Variable Analysis and PCA

The analysis of independent variables (η2) indicates that, in both soils, enzyme activity was mainly determined by soil contamination with petroleum products (Figure 5). An exception was the activity of Ure in LS, which was more dependent on soil amendment with sorbents than its contamination with DO and P. In LS, DO and P predominantly determined the activity of Pal (94.98% and 95.25% respectively), while sorbents influenced Ure activity (59.27% in LS) and Cat activity (35.00% in SL) to a greater extent. The biomass of the aerial parts of *Zea mays* in LS was primarily influenced by DO and P to a significant extent (97.51%), followed by root biomass (84.73%), whereas in SL, the influence was 91.28% and 47.08% respectively. Sorbents, on the other hand, influenced the above-ground biomass by 0.37% and 9.57% in LS and 3.01% and 24.25% in SL.

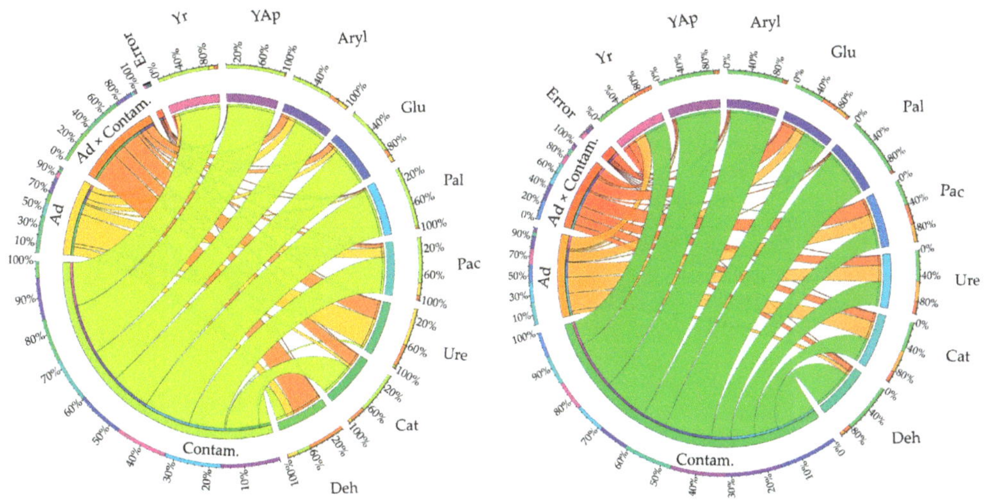

Figure 5. The share of independent variables (η2) in shaping the dependent variables in loamy sand (**a**) and sandy loam (**b**). Abbreviations are explained under Tables 2 and 3.

PCA analysis, considering all sorbents (B, H, and A) as well as petroleum products (DO and P), was performed separately for LS (Figure 6a) and SL (Figure 6b). In LS, positive correlations were observed between the activities of Pac and Pal with Cat, as well as Deh with Ure and Glu. In SL, Pal was correlated with Deh; Cat, and Ure; Aryl with Deh; Ure, and Pal, Deh with Ure, and Glu with Pac (Figure 7). Significant negative correlations were found between Cat, Pac, and Pal activities with the biomass of the aerial parts and roots of *Zea mays* in LS, as well as between Deh, Cat, Ure, Pal, and Aryl activities with the biomass of the aerial parts in SL. In both soils, DO had a greater destabilizing effect on *Zea mays* growth compared to P, resulting in a lower biomass yield of this plant (Figure 6a,b). The independent variables formed three groups: one in the soil not exposed to the pressure of petroleum products, the second in the soil with DO, and the third in the soil with P.

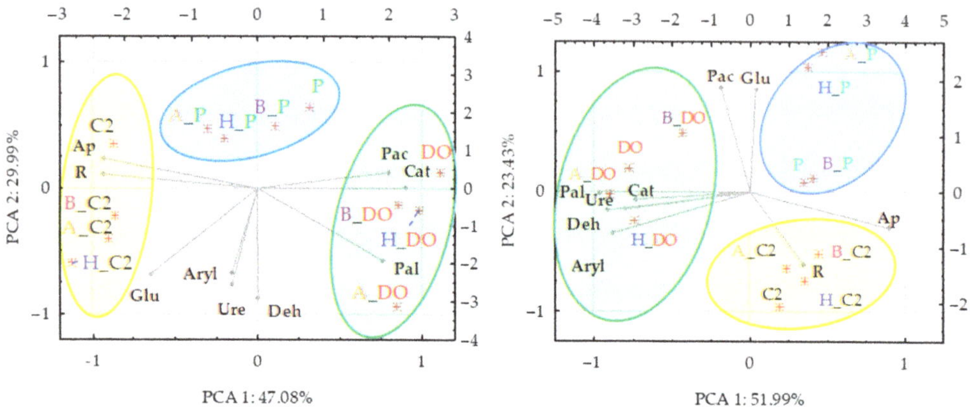

(**a**) (**b**)

Figure 6. The share of independent variables (η2) in shaping the dependent variables in loamy sand (**a**) and sandy loam (**b**). Abbreviations are explained under Tables 2 and 3.

Soil	Variable	YAp	Yr	Deh	Cat	Ure	Pac	Pal	Glu	Aryl	Soil
Loamy sand	YAp	1.000	0.534	−0.705 *	−0.760 *	−0.578 *	−0.335	−0.845 *	−0.234	−0.677 *	Sandy loam
	Yr	0.949 *	1.000	−0.075	−0.398	−0.013	−0.434	−0.300	−0.209	−0.042	
	Deh	−0.110	−0.010	1.000	0.510	0.779 *	0.161	0.928 *	−0.141	0.873 *	
	Cat	−0.784 *	−0.774 *	0.146	1.000	0.483	−0.181	0.602 *	−0.041	0.556	
	Ure	0.052	0.105	0.810 *	−0.144	1.000	0.073	0.713 *	0.005	0.747 *	
	Pac	−0.630 *	−0.583 *	0.031	0.698 *	−0.031	1.000	0.263	0.704 *	−0.067	
	Pal	−0.909 *	−0.814 *	0.401	0.614 *	0.195	0.442	1.000	−0.096	0.857 *	
	Glu	0.431	0.534	0.614 *	−0.559	0.484	−0.615 *	−0.052	1.000	−0.330	
	Aryl	−0.137	−0.008	0.295	−0.262	0.255	−0.393	0.393	0.545	1.000	

Figure 7. Pearson's simple correlation coefficients. * significant at $p = 0.05$, $n = 12$. —Sandy loam, —Loamy sand. Abbreviations are explained under the Tables 2 and 3.

4. Discussion

4.1. Response of Zea mays and Soil Enzymes to Soil Contamination with Petroleum-Derived Products

The toxicity of petroleum-derived substances determines the quality of soils as well as the health of plants, animals, and even humans [20,22,74,75]. Therefore, the remediation of soil contaminated with organic compounds poses a challenge for scientists and practitioners [76–78], considering that there are approximately 2.8 million sites contaminated with diesel fuel, petrol, mineral oils, and heating oil in Europe alone [79]. Bica [80] estimates that the number of these sites will continue to increase. Therefore, the classification of contaminated areas into risk classes [80,81] and the selection of appropriate remediation measures are crucial for the restoration of these soils [82–84]. When choosing a remediation method, the type and extent of contamination, soil texture, size of the contaminated area, and its economic purpose use should be taken into account [85,86]. Therefore, the plants can play an invaluable role in mitigating potential hazards associated with the impact of petroleum-derived products [14,86,87]. They have the ability to eliminate organic pollutants through the acceleration of their degradation by microorganisms inhabiting the rhizosphere [14,88,89]. For successful phytoremediation, plants, especially those from the *Poaceae* family, should adapt well to the contaminated environment, be easy to cultivate, and produce a high biomass [89,90]. Worldwide research on phytoremediation potential [14,86,91] demonstrates that leguminous plants belonging to the *Fabaceae* family [85] and plants belonging to the *Poaceae* family [14,92–95] are most effective in degrading petroleum hydrocarbons in contaminated soils. All these plants have well-developed root systems capable of reaching deep soil layers where petroleum pollutants can accumulate [96–98]. The success of these plants in phytoremediation relies on the synergistic cooperation between plant roots and microorganisms inhabiting the rhizosphere [99], which participate in the breakdown and degradation of petroleum compounds [14,23]. Bacteria, fungi, and protozoa are the main producers of enzymes in the soil [100]. They secrete enzymes involved in the breakdown and degradation of organic compounds into simpler forms, releasing nutrients.

In the present research, we utilized *Zea mays*, belonging to the *Poaceae* family, whose biomass obtained from areas contaminated with organic compounds can be used for bioenergy production [62,101]. In our studies, both diesel oil and petrol disrupted the growth and development of *Zea mays*, although to varying degrees. The use of plants, particularly *Poaceae*, for the remediation of soils contaminated with petroleum substances offer a viable remediation strategy [85,102,103]. Diesel fuel and petrol containing aromatic hydrocarbons, benzene, toluene, and other toxic compounds, can exhibit strong phytotoxic effects [74]. The adverse impact of petroleum-derived products on plants, as observed in the present study, could be attributed to the reduction in nutrient levels and available water for plants, as well as disturbances in nutrient uptake [85,104], leading to oxidative stress in plants resulting from the overproduction of reactive oxygen species (ROS) impairing the synthesis of amino acids, enzymes, and proteins [105,106]. Dian et al. [104] and Meudec et al. [107] highlight the negative influence of hydrocarbons and heavy metals contained in petroleum-derived products on the reduction of chlorophyll levels and inhibition of photosynthesis. The extent of the impact of the tested petroleum-derived products in our studies, which are mixtures of organic compounds primarily composed of hydrogen and carbon atoms, as well as smaller amounts of nitrogen, oxygen, sulfur, and trace amounts of metallic components [108,109] on the growth and development of *Zea mays*, was determined by contaminant type and soil texture. Diesel oil exhibited more toxic effects on plants compared to petrol.

According to Al-Rubaye et al. [75], the toxicological effects of hydrocarbons depend on their molecular weight. Toxic substances present in diesel fuel can also increase the concentration of heavy metal salts [102,110,111], negatively affecting populations and activity of soil microorganisms, which play a significant role in the nutrient cycling and biological processes in the soil [4,67,74,103,112,113]. These changes can have an adverse impact on root development and the absorption of nutrients by plants. Compounds present in diesel fuel can therefore cause metabolic dysfunctions in plants, leading to their growth inhibition [85]. Since hydrocarbons persist in ecosystems for a long time, they can accumulate in animal and plant tissues, transferring from one trophic level to another in the food chain, causing death or genetic mutations. The interaction between petroleum-derived products and plants, as well as the soil enzyme activity, depends on the soil particle size distribution [67,87,112]. The present study results demonstrated that diesel fuel and petrol had significantly less disruptive effects on the growth of *Zea mays* cultivated on sandy loam compared to loamy sand. In loamy sand, both products increased the value of the biochemical soil quality index, which reflects the activity of seven enzymes, whereas in sandy loam, only diesel fuel did so. The differences in the effects of petroleum-derived products on soils with different agronomic categories arise from changes in the bioavailability of these products in individual soils [114,115].

4.2. The Role of Sorbents in Improving the Quality of Soil Contaminated with Petroleum-Derived Products

In recent years, there has been an intensive development of methods supporting the phytoremediation of soils contaminated with petroleum-derived compounds [90]. A modern approach to soil restoration under the pressure of organic compounds not only focuses on the use of biostimulation, bioaugmentation, and the application of biosurfactants [116], but also necessitates the search for newer remediation compounds. In our research, biochar, halloysite, and alginite were applied to enhance the phytoremediation potential. The higher impact indices noted for biochar, halloysite, and alginite in the soil exposed to DO pressure than P are a due to the stronger negative influence of DO on *Zea mays* biomass in both soils compared to P, as well as the unique properties of sorbents such as large surface area, porosity, carbon content, and increased water-retention capacity [43,44,90].

These properties of biochar, halloysite, and alginite induced an increase in the activity of dehydrogenases, catalase, urease, alkaline phosphatase, β-glucosidase, and arylsulfatase in loamy sand, as well as acidic and alkaline phosphatases in sandy loam. As a result of our research, a negative correlation was observed between *Zea mays* yield and soil enzymatic activity, most likely due to the negative, toxic interaction of petroleum-derived pollutants with plants. This negative correlation explains the positive response of *Zea mays* cultivated on soils contaminated with petroleum products to the applied sorbents (biochar, halloysite, and alginite). The sorbents used may have contributed to the improvement of nutrient availability for plants. Disturbances in the development of *Zea mays* and soil enzymatic activity were mitigated by the applied sorbents, although their effects varied depending on the tested soil. All remediation substances resulted in increased plant biomass in soils contaminated with DO. This is likely due to the adsorption of nutrients on the surface or within the sorbents, thereby protecting them from leaching or oxidation in the soil. The successive release of nutrients in a controlled and plant-adapted manner contributed to improved nutrient availability for plants, thus supporting the degradation of pollutants. This concept is supported by studies conducted by Fahimizadeh et al. [117], Guopeng et al. [118], and Wang et al. [119]. The use of sorbents for the remediation of soil contaminated with petroleum-derived products in combination with plants can be highly valuable. Based on our own research, it was found that biochar, halloysite, and alginite can be recommended for remediating soils contaminated with petroleum products, especially in coarser-textured soils. Differences in plant and enzyme response to soil contamination with petroleum products are attributable to the variability in colloidal clay content, which plays a similar role in the soil as the tested sorbents. This explains the underlying mechanism of the positive influence of sorbents on the development of *Zea mays* and enzyme activity in loamy sand. According to Zhang et al. [120], phosphorus-enriched biochar has great potential for use as a soil fertilizer. Additionally, Khosravi et al. [121] states that the use of biochar-based fertilizers can prevent nutrient leaching. The safe application of this sorbent is also supported by its certification based on two main credentials: the certification program of the Swiss-based European Biochar Certificate (EBC) [122] and the US-based International Biochar Initiative (IBI) [123].

5. Conclusions

The study demonstrates a negative correlation between the yield of *Zea mays* and soil enzymatic activity, which was mainly caused by the toxic effect of diesel oil and petrol on this crop. The magnitude of disturbances in the growth and development of *Zea mays* was determined by the type of contaminant and the texture of soils. Diesel oil exerted a more toxic effect on plants than petrol. Both pollutants caused greater disturbances in the development of aerial parts and roots of plants grown in loamy sand soils compared to sandy loam soils. This finding explains the positive response of *Zea mays*, cultivated in soils contaminated with petroleum-derived products, to the applied sorbents (biochar, halloysite, and alginite). All of these sorbents can be recommended for the remediation of soils contaminated with petroleum-derived products, especially in coarser-textured soils, as in the case of loamy sand, where they can improve enzymatic activity, as indicated by the magnitude of the biochemical soil quality index. The present study results demonstrate that combining the determination of biochemical indicators with the evaluation of plant responses to diesel oil and petrol pressure is a viable approach in soil quality assessment. The conducted research indicates that effective soil remediation depends on a thoughtful selection of a sorbent, based on the knowledge of its characteristics combined with well-defined physicochemical properties of the soil and the type of petroleum-derived substances. In the case of selecting the appropriate sorbent, one should consider its sorption capacity, ability to immobilize pollutants, biodegradability, and environmental impact. Simultaneously, we suggest that further research is necessary to better understand the successful phytoremediation processes, particularly regarding changes in soil microbiome quality, plants cultivated in contaminated areas, and the interactions between soil, plants, pollutants, and soil enzyme activity.

Author Contributions: Conceptualization, J.W., A.B., M.Z. and J.K.; Methodology, J.W., A.B., M.Z. and J.K.; Investigation, J.W. and A.B.; Formal & Statistical Analysis, J.W. and A.B.; Writing—Original Draft Preparation J.W. and A.B., Writing—Review & Editing, J.W., A.B. and M.Z.; Supervision, J.K., Funding Acquisition, J.W. All authors have read and agreed to the published version of the manuscript.

Funding: This research was funded by the University of Warmia and Mazury in Olsztyn, Faculty of Agriculture and Forestry, Department of Soil Science and Microbiology (grant No. 30.610.006-110).

Institutional Review Board Statement: Not applicable.

Data Availability Statement: Data are available by contacting the authors.

Conflicts of Interest: The authors declare no conflict of interest.

Abbreviations

DO—VERVA diesel oil, P—VERVA 98 petrol, B—biochar, H—halloysite, A—alginite, SL—sandy loam; LS—loamy sand; Deh—dehydrogenases; Cat—catalase; Ure—urease; Pac—acid phosphatase; Pal—alkaline phosphatase; Aryl—arylsulfatase, Glu—β-glucosidase; BA—biochemical soil quality index; C_{org}—total organic carbon; N_{total}—total nitrogen; HAC—hydrolytic acidity, CEC—cation exchange capacity, ACS—alkaline cation saturation.

References

1. Ambaye, T.G.; Chebbi, A.; Formicola, F.; Prasad, S.; Gomez, F.H.; Franzetti, A.; Vaccari, M. Remediation of Soil Polluted with Petroleum Hydrocarbons and its Reuse for Agriculture: Recent Progress, Challenges, and Perspectives. *Chemosphere* **2022**, *293*, 133572. [CrossRef] [PubMed]
2. Fu, L.; Wei, M.; Liao, K.; Qianli, M.; Shao, M.; Gu, F.; Fan, Y.; Longjie, L.; Yanfeng, H. Application of Environmentally Stimuli-Responsive Materials in the Development of Oil and Gas Field. *J. Pet. Sci. Eng.* **2022**, *219*, 111088. [CrossRef]
3. Ossai, I.C.; Ahmed, A.; Hassan, A.; Hamid, F.S. Remediation of Soil and Water Contaminated with Petroleum Hydrocarbon: A Review. *Environ. Technol. Innov.* **2020**, *17*, 100526. [CrossRef]
4. Borowik, A.; Wyszkowska, J.; Kucharski, M.; Kucharski, J. Implications of Soil Pollution with Diesel Oil and BP Petroleum with ACTIVE Technology for Soil Health. *Int. J. Environ. Res. Public. Health* **2019**, *16*, 2474. [CrossRef]
5. Directive 2012/18/EU of the European Parliament and of the Council of 4 July 2012 on the Control of Major-Accident Hazards Involving Dangerous Substances, Amending and Subsequently Repealing Council Directive 96/82/EC Text with EEA Relevance. Available online: https://eur-lex.europa.eu/eli/dir/2012/18/oj (accessed on 13 July 2023).
6. Khalilova, H.K. The impact of oil contamination on soil ecosystem. *Biol. Chem. Res.* **2015**, *2015*, 133–139.
7. Mohan, S.V.; Kisa, T.; Ohkuma, T.; Kanaly, R.A.; Shimizu, Y. Bioremediation technologies for treatment of PAH-contaminated soil and strategies to enhance process efficiency. *Rev. Environ. Sci. Biotechnol.* **2006**, *5*, 347–374. [CrossRef]
8. Nasehi, S.A.; Uromeihy, A.; Nikudel, M.R.; Morsali, A. Influence of gas oil contamination on geotechnical properties of fine and coarse-grained soils. *Geotech. Geol. Eng.* **2016**, *34*, 333–345. [CrossRef]
9. Kucharski, J.; Jastrzębska, E. Effects of heating oil on the count of microorganisms and physico-chemical properties of soil. *Pol. J. Environ. Stud.* **2005**, *14*, 189–198.
10. Bolan, N.; Sarkar, B.; Yan, Y.; Li, Q.; Wijesekara, H.; Kannan, K.; Tsang, D.C.W.; Schauerte, M.; Bosch, J.; Noll, H.; et al. Remediation of poly- and perfluoroalkyl substances (PFAS) contaminated soils—To mobilize or to immobilize or to degrade? *J. Hazard. Mater.* **2021**, *401*, 123892. [CrossRef]
11. Haghollahi, A.; Fazaelipoor, M.H.; Schaffie, M. The effect of soil type on the bioremediation of petroleum contaminated soils. *J. Environ. Manag.* **2016**, *180*, 197–201. [CrossRef]
12. Yuan, L.; Wu, Y.; Fan, Q.; Li, P.; Liang, J.; Liu, Y.; Ma, R.; Li, R.; Shi, L. Remediating petroleum hydrocarbons in highly saline–alkali soils using three native plant species. *J. Environ. Manag.* **2023**, *339*, 117928. [CrossRef] [PubMed]
13. Borowik, A.; Wyszkowska, J. Response of *Avena sativa* L. and the soil microbiota to the contamination of soil with Shell diesel oil. *Plant Soil. Environ.* **2018**, *64*, 102–107. [CrossRef]

14. Wyszkowska, J.; Borowik, A.; Kucharski, J. The Resistance of *Lolium Perenne* L. × Hybridum, *Poa Pratensis, Festuca Rubra, F. Arundinacea, Phleum Pratense* and *Dactylis Glomerata* to Soil Pollution by Diesel Oil and Petroleum. *Plant Soil. Environ.* **2019**, *65*, 307–312. [CrossRef]
15. Yu, J.; Feng, H.; Tang, L.; Pang, Y.; Wang, J.; Zou, J.; Xie, Q.; Liu, Y.; Feng, C.; Wang, J. Insight into the Key Factors in Fast Adsorption of Organic Pollutants by Hierarchical Porous Biochar. *J. Hazard. Mater.* **2021**, *403*, 123610. [CrossRef] [PubMed]
16. Wyszkowska, J.; Borowik, A.; Zaborowska, M.; Kucharski, J. The Usability of Sorbents in Restoring Enzymatic Activity in Soils Polluted with Petroleum-Derived Products. *Materials* **2023**, *16*, 3738. [CrossRef]
17. da Silva, L.J.; Alves, F.C.; de França, F.P. A Review of the Technological Solutions for the Treatment of Oily Sludges from Petroleum Refineries. *Waste Manag. Res.* **2012**, *30*, 1016–1030. [CrossRef]
18. Bandura, L.; Woszuk, A.; Kołodyńska, D.; Franus, W. Application of Mineral Sorbents for Removal of Petroleum Substances: A Review. *Minerals* **2017**, *7*, 37. [CrossRef]
19. Padhye, L.P.; Srivastava, P.; Jasemizad, T.; Bolan, S.; Hou, D.; Shaheen, S.M.; Rinklebe, J.; O'Connor, D.; Lamb, D.; Wang, H. Contaminant Containment for Sustainable Remediation of Persistent Contaminants in Soil and Groundwater. *J. Hazard. Mater.* **2023**, *455*, 131575. [CrossRef]
20. Wyszkowski, M.; Kordala, N. Role of Different Material Amendments in Shaping the Content of Heavy Metals in Maize (*Zea mays* L.) on Soil Polluted with Petrol. *Materials* **2022**, *15*, 2623. [CrossRef]
21. Valdiviezo Gonzales, L.G.; Castañeda-Olivera, C.A.; Cabello-Torres, R.J.; García Ávila, F.F.; Cerrón, R.V.M.; Alfaro Paredes, E.A. Scientometric Study of Treatment Technologies of Soil Pollution: Present and Future Challenges. *Appl. Soil. Ecol.* **2023**, *182*, 104695. [CrossRef]
22. Wyszkowski, M.; Wyszkowska, J.; Kordala, N.; Borowik, A. Effects of Coal and Sewage Sludge Ashes on Macronutrient Content in Maize (*Zea mays* L.) Grown on Soil Contaminated with Eco-Diesel Oil. *Materials* **2022**, *15*, 525. [CrossRef] [PubMed]
23. Borowik, A.; Wyszkowska, J. Remediation of Soil Contaminated with Diesel Oil. *J. Elementol.* **2018**, *23*, 767–788. [CrossRef]
24. Stepanova, A.Y.; Gladkov, E.A.; Osipova, E.S.; Gladkova, O.V.; Tereshonok, D.V. Bioremediation of Soil from Petroleum Contamination. *Processes* **2022**, *10*, 1224. [CrossRef]
25. Kisić, I.; Hrenović, J.; Zgorelec, Ž.; Durn, G.; Brkić, V.; Delač, D. Bioremediation of Agriculture Soil Contaminated by Organic Pollutants. *Energies* **2022**, *15*, 1561. [CrossRef]
26. Hamidzadeh, Z.; Ghorbannezhad, P.; Ketabchi, M.R.; Yeganeh, B. Biomass-Derived Biochar and Its Application in Agriculture. *Fuel* **2023**, *341*, 127701. [CrossRef]
27. Jabborova, D.; Annapurna, K.; Paul, S.; Kumar, S.; Saad, H.A.; Desouky, S.; Ibrahim, M.F.M.; Elkelish, A. Beneficial Features of Biochar and Arbuscular Mycorrhiza for Improving Spinach Plant Growth, Root Morphological Traits, Physiological Properties, and Soil Enzymatic Activities. *J. Fungi* **2021**, *7*, 571. [CrossRef]
28. Song, X.; Li, H.; Song, J.; Chen, W.; Shi, L. Biochar/Vermicompost Promotes Hybrid Pennisetum Plant Growth and Soil Enzyme Activity in Saline Soils. *Plant Physiol. Biochem.* **2022**, *183*, 96–110. [CrossRef]
29. Wang, Y.; Ma, Z.; Wang, X.; Sun, Q.; Dong, H.; Wang, G.; Chen, X.; Yin, C.; Han, Z.; Mao, Z. Effects of Biochar on the Growth of Apple Seedlings, Soil Enzyme Activities and Fungal Communities in Replant Disease Soil. *Sci. Hortic.* **2019**, *256*, 108641. [CrossRef]
30. Ajeng, A.A.; Abdullah, R.; Ling, T.C.; Ismail, S.; Lau, B.F.; Ong, H.C.; Chew, K.W.; Show, P.L.; Chang, J.-S. Bioformulation of Biochar as a Potential Inoculant Carrier for Sustainable Agriculture. *Environ. Technol. Innov.* **2020**, *20*, 101168. [CrossRef]
31. Li, C.; Ahmed, W.; Li, D.; Yu, L.; Xu, L.; Xu, T.; Zhao, Z. Biochar Suppresses Bacterial Wilt Disease of Flue-Cured Tobacco by Improving Soil Health and Functional Diversity of Rhizosphere Microorganisms. *Appl. Soil. Ecol.* **2022**, *171*, 104314. [CrossRef]
32. Selvarajoo, A.; Wong, Y.L.; Khoo, K.S.; Chen, W.-H.; Show, P.L. Biochar Production via Pyrolysis of Citrus Peel Fruit Waste as a Potential Usage as Solid Biofuel. *Chemosphere* **2022**, *294*, 133671. [CrossRef]
33. Ren, H.-Y.; Wei, Z.-J.; Wang, Y.; Deng, Y.-P.; Li, M.-Y.; Wang, B. Effects of Biochar Properties on the Bioremediation of the Petroleum-Contaminated Soil from a Shale-Gas Field. *Environ. Sci. Pollut. Res.* **2020**, *27*, 36427–36438. [CrossRef]
34. Guo, M.; Song, W.; Tian, J. Biochar-Facilitated Soil Remediation: Mechanisms and Efficacy Variations. *Front. Environ. Sci.* **2020**, *8*, 521512. [CrossRef]
35. Fedeli, R.; Alexandrov, D.; Celletti, S.; Nafikova, E.; Loppi, S. Biochar Improves the Performance of Avena Sativa L. Grown in Gasoline-Polluted Soils. *Environ. Sci. Pollut. Res.* **2023**, *30*, 28791–28802. [CrossRef] [PubMed]
36. Wei, Z.; Wei, Y.; Liu, Y.; Niu, S.; Xu, Y.; Park, J.-H.; Wang, J.J. Biochar-Based Materials as Remediation Strategy in Petroleum Hydrocarbon-Contaminated Soil and Water: Performances, Mechanisms, and Environmental Impact. *J. Environ. Sci.* **2024**, *138*, 350–372. [CrossRef]
37. Gao, J.; Han, H.; Gao, C.; Wang, Y.; Dong, B.; Xu, Z. Organic Amendments for in Situ Immobilization of Heavy Metals in Soil: A Review. *Chemosphere* **2023**, *335*, 139088. [CrossRef]
38. Wyszkowska, J.; Borowik, A.; Zaborowska, M.; Kucharski, J. Evaluation of the Usefulness of Sorbents in the Remediation of Soil Exposed to the Pressure of Cadmium and Cobalt. *Materials* **2022**, *15*, 5738. [CrossRef] [PubMed]
39. Wyszkowska, J.; Borowik, A.; Zaborowska, M.; Kucharski, J. Mitigation of the Adverse Impact of Copper, Nickel, and Zinc on Soil Microorganisms and Enzymes by Mineral Sorbents. *Materials* **2022**, *15*, 5198. [CrossRef]
40. Tamsu Selli, N.; Basaran, N. Controlling the Hardness and Wear Resistance of Opaque White Glaze by Addition of Halloysite Clay in the Composition. *Bol. Soc. Esp. Ceram. Vidr.* **2023**, *62*, 243–256. [CrossRef]

41. Maj, I.; Matus, K. Aluminosilicate Clay Minerals: Kaolin, Bentonite, and Halloysite as Fuel Additives for Thermal Conversion of Biomass and Waste. *Energies* **2023**, *16*, 4359. [CrossRef]
42. Masoudniaragh, A.; Oraei, M.; Gohari, G.; Akbari, A.; Faramarzi, A. Using Halloysite Nanotubes as Carrier for Proline to Alleviate Salt Stress Effects in Sweet Basil (*Ocimum Basilicum* L.). *Sci. Hortic.* **2021**, *285*, 110202. [CrossRef]
43. Gömöryová, E.; Vass, D.; Pichler, V.; Gömöry, D. Effect of Alginite Amendment on Microbial Activity and Soil Water Content in Forest Soils. *Biologia* **2009**, *64*, 585–588. [CrossRef]
44. Mukherjee, A.; Zimmerman, A.R.; Harris, W. Surface Chemistry Variations among a Series of Laboratory-Produced Biochars. *Geoderma* **2011**, *163*, 247–255. [CrossRef]
45. Hippmann, S.; Ahmed, S.S.; Fröhlich, P.; Bertau, M. Demulsification of Water/Crude Oil Emulsion Using Natural Rock Alginite. *Colloids Surf. A Physicochem. Eng. Asp.* **2018**, *553*, 71–79. [CrossRef]
46. Wang, B.; Gao, B.; Zimmerman, A.R.; Zheng, Y.; Lyu, H. Novel Biochar-Impregnated Calcium Alginate Beads with Improved Water Holding and Nutrient Retention Properties. *J. Environ. Manag.* **2018**, *209*, 105–111. [CrossRef]
47. Strachel, R.; Wyszkowska, J.; Baćmaga, M. An Evaluation of the Effectiveness of Sorbents in the Remediation of Soil Contaminated with Zinc. *Water Air Soil. Pollut.* **2018**, *229*, 235. [CrossRef]
48. Matvieieva, N.; Duplij, V.; Vozár, L'.; Kovár, P.; Hric, P. Stimulation Effect of Alginite on *Rhodiola Rosea* L. in Vitro Growth. *Agrobiodivers Improv.* **2023**, *7*, 7–16.
49. Pour, M.M.; Riseh, R.S.; Ranjbar-Karimi, R.; Hassanisaadi, M.; Rahdar, A.; Baino, F. Microencapsulation of *Bacillus Velezensis* Using Alginate-Gum Polymers Enriched with TiO_2 and SiO_2 Nanoparticles. *Micromachines* **2022**, *13*, 1423. [CrossRef]
50. Zhao, L.; Angel Hernandez-Viezcas, J.; Peralta-Videa, J.R.; Bandyopadhyay, S.; Peng, B.; Munoz, B.; Keller, A.A.; Gardea-Torresdey, J.L. ZnO Nanoparticle Fate in Soil and Zinc Bioaccumulation in Corn Plants (*Zea mays*) Influenced by Alginate. *Environ. Sci. Process. Impacts* **2013**, *15*, 260–266. [CrossRef]
51. Szopa, D.; Mielczarek, M.; Skrzypczak, D.; Izydorczyk, G.; Mikula, K.; Chojnacka, K.; Witek-Krowiak, A. Encapsulation Efficiency and Survival of Plant Growth-Promoting Microorganisms in an Alginate-Based Matrix—A Systematic Review and Protocol for a Practical Approach. *Ind. Crops Prod.* **2022**, *181*, 114846. [CrossRef]
52. Saberi Riseh, R.; Hassanisaadi, M.; Vatankhah, M.; Soroush, F.; Varma, R.S. Nano/Microencapsulation of Plant Biocontrol Agents by Chitosan, Alginate, and Other Important Biopolymers as a Novel Strategy for Alleviating Plant Biotic Stresses. *Int. J. Biol. Macromol.* **2022**, *222*, 1589–1604. [CrossRef]
53. Song, Y.; Bechtel, A.; Sachsenhofer, R.F.; Groß, D.; Liu, Z.; Meng, Q. Depositional Environment of the Lower Cretaceous Muling Formation of the Laoheishan Basin (NE China): Implications from Geochemical and Petrological Analyses. *Org. Geochem.* **2017**, *104*, 19–34. [CrossRef]
54. OECD-FAO. *OECD-FAO Agricultural Outlook 2021–2030*; OECD Publishing: Paris, France, 2021; ISBN 978-92-64-43607-7.
55. Erenstein, O.; Chamberlin, J.; Sonder, K. Estimating the Global Number and Distribution of Maize and Wheat Farms. *Glob. Food Secur.* **2021**, *30*, 100558. [CrossRef]
56. Yi, L.; Shenjiao, Y.; Shiqing, L.; Xinping, C.; Fang, C. Growth and Development of Maize (*Zea mays* L.) in Response to Different Field Water Management Practices: Resource Capture and Use Efficiency. *Agric. Meteorol.* **2010**, *150*, 606–613. [CrossRef]
57. Aghili, S.; Golzary, A. Greening the Earth, Healing the Soil: A Comprehensive Life Cycle Assessment of Phytoremediation for Heavy Metal Contamination. *Environ. Technol. Innov.* **2023**, *32*, 103241. [CrossRef]
58. Usowicz, B.; Lipiec, J. Spatial Variability of Soil Properties and Cereal Yield in a Cultivated Field on Sandy Soil. *Soil Tillage Res.* **2017**, *174*, 241–250. [CrossRef]
59. IUSS Working Group WRB. *World Reference Base for Soil Resources 2014, Update 2015. International Soil Classification System for Naming Soils and Creating Legends for Soil Maps*; World Soil Resources Reports No. 106.; FAO: Rome, Italy, 2015.
60. Morales-Máximo, C.N.; López-Sosa, L.B.; Rutiaga-Quiñones, J.G.; Corral-Huacuz, J.C.; Aguilera-Mandujano, A.; Pintor-Ibarra, L.F.; López-Miranda, A.; Delgado-Domínguez, S.N.; Rodríguez-Magallón, M.D.C.; Morales-Máximo, M. Characterization of Agricultural Residues of *Zea mays* for Their Application as Solid Biofuel: Case Study in San Francisco Pichátaro, Michoacán, Mexico. *Energies* **2022**, *15*, 6870. [CrossRef]
61. Wyszkowska, J.; Borowik, A.; Zaborowska, M.; Kucharski, J. Calorific Value of *Zea mays* Biomass Derived from Soil Contaminated with Chromium (VI) Disrupting the Soil's Biochemical Properties. *Energies* **2023**, *16*, 3788. [CrossRef]
62. Liao, C.; Xu, W.; Lu, G.; Liang, X.; Guo, C.; Yang, C.; Dang, Z. Accumulation of Hydrocarbons by Maize (*Zea mays* L.) in Remediation of Soils Contaminated with Crude Oil. *Int. J. Phytoremediation* **2015**, *17*, 693–700. [CrossRef]
63. Murtaza, G.; Ahmed, Z.; Eldin, S.M.; Ali, B.; Bawazeer, S.; Usman, M.; Iqbal, R.; Neupane, D.; Ullah, A.; Khan, A.; et al. Biochar-Soil-Plant Interactions: A Cross Talk for Sustainable Agriculture under Changing Climate. *Front. Environ. Sci.* **2023**, *11*, 1059449. [CrossRef]
64. Ayaz, M.; Feizienė, D.; Tilvikienė, V.; Akhtar, K.; Stulpinaitė, U.; Iqbal, R. Biochar Role in the Sustainability of Agriculture and Environment. *Sustainability* **2021**, *13*, 1330. [CrossRef]
65. Radziemska, M.; Gusiatin, Z.M.; Kumar, V.; Brtnicky, M. Co-Application of Nanosized Halloysite and Biochar as Soil Amendments in Aided Phytostabilization of Metal(-Oid)s-Contaminated Soil under Different Temperature Conditions. *Chemosphere* **2022**, *288*, 132452. [CrossRef] [PubMed]
66. Zaborowska, M.; Wyszkowska, J.; Borowik, A.; Kucharski, J. Effect of Separate and Combined Toxicity of Bisphenol A and Zinc on the Soil Microbiome. *Int. J. Mol. Sci.* **2022**, *23*, 5937. [CrossRef] [PubMed]

67. Borowik, A.; Wyszkowska, J.; Wyszkowski, M. Resistance of Aerobic Microorganisms and Soil Enzyme Response to Soil Contamination with Ekodiesel Ultra Fuel. *Environ. Sci. Pollut. Res.* **2017**, *24*, 24346–24363. [CrossRef]
68. RStudio Team RStudio Team. RStudio: Integrated Development for R. RStudio Version 2023.06.0, PBC, Boston, MA, USA. 2020. Available online: http://www.rstudio.com/ (accessed on 10 July 2023).
69. R Core Team R Core Team. R: A Language and Environment for Statistical Computing; R Foundation for Statistical Computing, Version 4.2.2. Vienna, Austria. 2022. Available online: https://www.r-project.org/ (accessed on 10 July 2023).
70. Warnes, G.R.; Bolker, B.; Bonebakker, L.; Gentleman, R.; Huber, W.; Liaw, A.; Lumley, T.; Maechler, M.; Magnusson, M.; Moeller, S.; et al. Gplots: Various R Programming Tools for Plotting Data. Gplots 3.1.3. 2022. Available online: https://cran.r-project.org/web/packages/gplots/ (accessed on 10 July 2023).
71. Wyszkowska, J.; Borowik, A.; Kucharski, M.; Borowik, A. Applicability of biochemical indices to quality assessment of soil polluted with heavy metal. *J. Elementol.* **2013**, *18*, 733–756. [CrossRef]
72. Krzywinski, M.; Schein, J.; Birol, İ.; Connors, J.; Gascoyne, R.; Horsman, D.; Jones, S.J.; Marra, M.A. Circos: An Information Aesthetic for Comparative Genomics. *Genome Res.* **2009**, *19*, 1639–1645. [CrossRef]
73. TIBCO Software Inc Statistica, Version 13; Data Analysis Software System; Tibco Software Inc.: Palo Alto, CA, USA. 2021. Available online: http://statistica.io (accessed on 10 July 2023).
74. Thacharodi, A.; Hassan, S.; Singh, T.; Mandal, R.; Chinnadurai, J.; Khan, H.A.; Hussain, M.A.; Brindhadevi, K.; Pugazhendhi, A. Bioremediation of Polycyclic Aromatic Hydrocarbons: An Updated Microbiological Review. *Chemosphere* **2023**, *328*, 138498. [CrossRef]
75. Al-Rubaye, A.H.; Jasim, D.J.; Ameen, H.F.M.; Al-Robai, H.A.; Al-Assal, J.R. The Impacts of Petroleum on Environment. *IOP Conf. Ser. Earth Environ. Sci.* **2023**, *1158*, 032014. [CrossRef]
76. Panchenko, L.; Muratova, A.; Dubrovskaya, E.; Golubev, S.; Turkovskaya, O. Natural and Technical Phytoremediation of Oil-Contaminated Soil. *Life* **2023**, *13*, 177. [CrossRef]
77. Borowik, A.; Wyszkowska, J. Bioaugmentation of Soil Contaminated with Diesel Oil. *J. Elementol.* **2018**, *23*, 1161–1178. [CrossRef]
78. Wang, W.; Gao, Y.; Du, J.; Zheng, L.; Kong, X.; Wang, H.; Yang, X.; Duan, L.; Zhao, Q.; Liu, Y.; et al. Dose–Effect of Nitrogen Regulation on the Bioremediation of Diesel Contaminated Soil. *Environ. Technol. Innov.* **2023**, *32*, 103245. [CrossRef]
79. Pérez, A.P.; Eugenio, N.R. Status of Local Soil Contamination in Europe: Revision of the Indicator 'Progress in the Management Contaminated Sites in EUROPE'. JRC. 2018. Available online: https://data.europa.eu/doi/10.2760/093804 (accessed on 15 July 2023).
80. Bica, I. Contaminated Sites Investigation. Objectives and Methods. *E3S Web Conf.* **2020**, *169*, 02002. [CrossRef]
81. Brombal, D.; Wang, H.; Pizzol, L.; Critto, A.; Giubilato, E.; Guo, G. Soil Environmental Management Systems for Contaminated Sites in China and the EU. Common Challenges and Perspectives for Lesson Drawing. *Land Use Policy* **2015**, *48*, 286–298. [CrossRef]
82. Wanner, P.; Freis, M.; Peternell, M.; Kelm, V. Risk Classification of Contaminated Sites—Comparison of the Swedish and the German Method. *J. Environ. Manag.* **2023**, *327*, 116825. [CrossRef]
83. Gao, J.; Faheem, M.; Yu, X. Global Research on Contaminated Soil Remediation: A Bibliometric Network Analysis. *Land* **2022**, *11*, 1581. [CrossRef]
84. Wyszkowski, M.; Wyszkowska, J.; Kordala, N.; Borowik, A. Applicability of Ash Wastes for Reducing Trace Element Content in *Zea mays* L. Grown in Eco-Diesel Contaminated Soil. *Molecules* **2022**, *27*, 897. [CrossRef]
85. Rusin, M.; Gospodarek, J.; Barczyk, G.; Nadgórska-Socha, A. Antioxidant Responses of Triticum Aestivum Plants to Petroleum-Derived Substances. *Ecotoxicology* **2018**, *27*, 1353–1367. [CrossRef]
86. Mishra, P.; Kiran, N.S.; Romanholo Ferreira, L.F.; Yadav, K.K.; Mulla, S.I. New Insights into the Bioremediation of Petroleum Contaminants: A Systematic Review. *Chemosphere* **2023**, *326*, 138391. [CrossRef]
87. Borowik, A.; Wyszkowska, J.; Kucharski, M.; Kucharski, J. The Role of *Dactylis Glomerata* and Diesel Oil in the Formation of Microbiome and Soil Enzyme Activity. *Sensors* **2020**, *20*, 3362. [CrossRef]
88. Ptaszek, N.; Pacwa-Płociniczak, M.; Noszczyńska, M.; Płociniczak, T. Comparative Study on Multiway Enhanced Bio- and Phytoremediation of Aged Petroleum-Contaminated Soil. *Agronomy* **2020**, *10*, 947. [CrossRef]
89. Mitter, E.K.; Kataoka, R.; de Freitas, J.R.; Germida, J.J. Potential Use of Endophytic Root Bacteria and Host Plants to Degrade Hydrocarbons. *Int. J. Phytoremediation* **2019**, *21*, 928–938. [CrossRef] [PubMed]
90. Kafle, A.; Timilsina, A.; Gautam, A.; Adhikari, K.; Bhattarai, A.; Aryal, N. Phytoremediation: Mechanisms, Plant Selection and Enhancement by Natural and Synthetic Agents. *Environ. Adv.* **2022**, *8*, 100203. [CrossRef]
91. Meištininkas, R.; Vaškevičienė, I.; Dikšaitytė, A.; Pedišius, N.; Žaltauskaitė, J. Potential of Eight Species of Legumes for Heavy Fuel Oil-Contaminated Soil Phytoremediation. *Sustainability* **2023**, *15*, 4281. [CrossRef]
92. Gawryluk, A.; Stępniowska, A.; Lipińska, H. Effect of Soil Contamination with Polycyclic Aromatic Hydrocarbons from Drilling Waste on Germination and Growth of Lawn Grasses. *Ecotoxicol. Environ. Saf.* **2022**, *236*, 113492. [CrossRef]
93. Hou, J.; Wang, Q.; Liu, W.; Zhong, D.; Ge, Y.; Christie, P.; Luo, Y. Soil Microbial Community and Association Network Shift Induced by Several Tall Fescue Cultivars during the Phytoremediation of a Petroleum Hydrocarbon-Contaminated Soil. *Sci. Total Environ.* **2021**, *792*, 148411. [CrossRef]
94. He, M.; Li, Z.; Chen, C.; Mei, P. Impact of Soil Types and Root Exudates on Cadmium and Petroleum Hydrocarbon Phytoremediation by *Sorghum Sudanense*, *Festuca Arundinace*, and *Lolium Perenne*. *Front. Ecol. Evol.* **2022**, *10*, 1036765. [CrossRef]

95. Lin, M.-S.; Huang, C.-Y.; Lin, Y.-C.; Lin, S.-L.; Hsiao, Y.-H.; Tu, P.-C.; Cheng, P.-C.; Cheng, S.-F. Green Remediation Technology for Total Petroleum Hydrocarbon-Contaminated Soil. *Agronomy* **2022**, *12*, 2759. [CrossRef]
96. Yavari, S.; Malakahmad, A.; Sapari, N.B. A Review on Phytoremediation of Crude Oil Spills. *Water Air Soil. Pollut.* **2015**, *226*, 279. [CrossRef]
97. Hussein, Z.S.; Hamido, N.; Hegazy, A.K.; El-Dessouky, M.A.; Mohamed, N.H.; Safwat, G. Phytoremediation of Crude Petroleum Oil Pollution: A Review. *EJBO* **2022**, *62*, 611–640. [CrossRef]
98. Franchi, E.; Cardaci, A.; Pietrini, I.; Fusini, D.; Conte, A.; De Folly D'Auris, A.; Grifoni, M.; Pedron, F.; Barbafieri, M.; Petruzzelli, G.; et al. Nature-Based Solutions for Restoring an Agricultural Area Contaminated by an Oil Spill. *Plants* **2022**, *11*, 2250. [CrossRef]
99. Cook, R.L.; Hesterberg, D. Comparison of Trees and Grasses for Rhizoremediation of Petroleum Hydrocarbons. *Int. J. Phytoremediation* **2013**, *15*, 844–860. [CrossRef] [PubMed]
100. Aluvihara, S.; Weerawardena, T.E.; Gunaratne, H.D.A.K.; Deniyapahala, D.V.R.I.; Dissanayake, S.S.; Gunawardana, S.I. Advanced Review on Soil Microbiology and Fertility of Soils. *Mech. Agric. Conserv. Resour.* **2023**, *67*, 27–31.
101. Singh, H.; Pant, G. Phytoremediation: Low Input-Based Ecological Approach for Sustainable Environment. *Appl. Water Sci.* **2023**, *13*, 85. [CrossRef]
102. Tan, H.W.; Pang, Y.L.; Lim, S.; Chong, W.C. A State-of-the-Art of Phytoremediation Approach for Sustainable Management of Heavy Metals Recovery. *Environ. Technol. Innov.* **2023**, *30*, 103043. [CrossRef]
103. Borowik, A.; Wyszkowska, J.; Gałązka, A.; Kucharski, J. Role of *Festuca Rubra* and *Festuca Arundinacea* in Determinig the Functional and Genetic Diversity of Microorganisms and of the Enzymatic Activity in the Soil Polluted with Diesel Oil. *Environ. Sci. Pollut. Res.* **2019**, *26*, 27738–27751. [CrossRef] [PubMed]
104. Duan, Q.; Han, G.; Cui, B.; Rennenberg, H. The Effects of Petroleum Contaminated Soils on the Growth, Gas Exchange and Antioxidative Level of Sea-Buckthorn. *Plant Soil.* **2023**, *486*, 535–550. [CrossRef]
105. Houshani, M.; Salehi-Lisar, S.Y. Agronomic Crop Responses and Tolerance to Polycyclic Aromatic Hydrocarbon Toxicity. In *Agronomic Crops: Volume 3: Stress Responses and Tolerance*; Hasanuzzaman, M., Ed.; Springer: Singapore, 2020; pp. 265–283. ISBN 9789811500251.
106. da Silva Correa, H.; Blum, C.T.; Galvão, F.; Maranho, L.T. Effects of Oil Contamination on Plant Growth and Development: A Review. *Environ. Sci. Pollut. Res.* **2022**, *29*, 43501–43515. [CrossRef]
107. Meudec, A.; Poupart, N.; Dussauze, J.; Deslandes, E. Relationship between Heavy Fuel Oil Phytotoxicity and Polycyclic Aromatic Hydrocarbon Contamination in Salicornia Fragilis. *Sci. Total Environ.* **2007**, *381*, 146–156. [CrossRef]
108. Yemashova, N.; Murygina, V.; Zhukov, D.; Zakharyantz, A.; Gladchenko, M.; Appanna, V.; Kalyuzhnyi, S. Biodeterioration of Crude Oil and Oil Derived Products: A Review. *Rev. Environ. Sci. Biotechnol.* **2007**, *6*, 315–337. [CrossRef]
109. Koshlaf, E.; Ball, A.S. Soil Bioremediation Approaches for Petroleum Hydrocarbon Polluted Environments. *AIMS Microbiol.* **2017**, *3*, 25–49. [CrossRef]
110. Wyszkowski, M.; Kordala, N. Applicability of Compost and Mineral Materials for Reducing the Effect of Diesel Oil on Trace Element Content in Soil. *Materials* **2023**, *16*, 3655. [CrossRef]
111. Wyszkowski, M.; Kordala, N. Trace Element Contents in Petrol-Contaminated Soil Following the Application of Compost and Mineral Materials. *Materials* **2022**, *15*, 5233. [CrossRef]
112. Borowik, A.; Wyszkowska, J.; Kucharski, J. Microbiological Study in Petrol-Spiked Soil. *Molecules* **2021**, *26*, 2664. [CrossRef]
113. Wyszkowska, J.; Borowik, A.; Kucharski, J. Response of *Avena Sativa*, Microorganisms and Enzymes to Contamination of Soil with Diesel Oil. *Plant Soil. Environ.* **2015**, *61*, 483–488. [CrossRef]
114. Dindar, E.; Şağban, F.O.T.; Başkaya, H.S. Variations of Soil Enzyme Activities in Petroleum-Hydrocarbon Contaminated Soil. *Int. Biodeterior. Biodegrad.* **2015**, *105*, 268–275. [CrossRef]
115. Wu, B.; Lan, T.; Lu, D.; Liu, Z. Ecological and Enzymatic Responses to Petroleum Contamination. *Environ. Sci. Process. Impacts* **2014**, *16*, 1501–1509. [CrossRef] [PubMed]
116. Song, Y.; Li, R.; Chen, G.; Yan, B.; Zhong, L.; Wang, Y.; Li, Y.; Li, J.; Zhang, Y. Bibliometric Analysis of Current Status on Bioremediation of Petroleum Contaminated Soils during 2000–2019. *Int. J. Environ. Res. Public Health* **2021**, *18*, 8859. [CrossRef]
117. Fahimizadeh, M.; Pasbakhsh, P.; Mae, L.S.; Tan, J.B.L.; Raman, R.K.S. Sustained-Release of Nutrients by Yeast Extract-Loaded Halloysite Nanotubes Supports Bacterial Growth. *Appl. Clay Sci.* **2023**, *240*, 106979. [CrossRef]
118. Teng, G.; Chen, C.; Jing, N.; Chen, C.; Duan, Y.; Zhang, L.; Wu, Z.; Zhang, J. Halloysite Nanotubes-Based Composite Material with Acid/Alkali Dual PH Response and Foliar Adhesion for Smart Delivery of Hydrophobic Pesticide. *Chem. Eng. J.* **2023**, *451*, 139052. [CrossRef]
119. Wang, C.; Gu, L.; Ge, S.; Liu, X.; Zhang, X.; Chen, X. Remediation Potential of Immobilized Bacterial Consortium with Biochar as Carrier in Pyrene-Cr(VI) Co-Contaminated Soil. *Environ. Technol.* **2019**, *40*, 2345–2353. [CrossRef]
120. Zhang, H.; Yang, H.; Shao, J.; Chen, Y.; Zhang, S.; Chen, H. Multifunctional Carboxymethyl Cellulose Sodium Encapsulated Phosphorus-Enriched Biochar Composites: Multistage Adsorption of Heavy Metals and Controllable Release of Soil Fertilization. *J. Chem. Eng.* **2023**, *453*, 139809. [CrossRef]
121. Khosravi, A.; Zheng, H.; Liu, Q.; Hashemi, M.; Tang, Y.; Xing, B. Production and Characterization of Hydrochars and Their Application in Soil Improvement and Environmental Remediation. *Chem. Eng. J.* **2022**, *430*, 133142. [CrossRef]

122. European-BioChar. The European Biochar Certificate. Available online: http://www.european-biochar.org/en (accessed on 10 August 2023).
123. International Biochar Initiative. About the IBI Biochar Certification Program. Available online: http://www.biochar-international.org/certification (accessed on 10 August 2023).

agriculture

MDPI

Article

Reducing Herbicide Dependency: Impact of *Murraya koenigii* Leaf Extract on Weed Control and Growth of Wheat (*Triticum aestivum*) and Chickpea (*Cicer arietinum*)

Deepti Singh Vashishth [1], Archana Bachheti [1,*], Rakesh Kumar Bachheti [2,3,4], Sadeq K. Alhag [5], Laila A. Al-Shuraym [6], Pankaj Kumar [7,8] and Azamal Husen [9]

1 Department of Environment Science, Graphic Era (Deemed to be University), Dehradun 248002, India
2 Bio-Process and Biotechnology Centre of Excellence, College of Applied Sciences, Addis Ababa Science and Technology University, Addis Ababa P.O. Box 16417, Ethiopia
3 Department of Industrial Chemistry, College of Applied Sciences, Addis Ababa Science and Technology University, Addis Ababa P.O. Box 16417, Ethiopia
4 Department of Allied Sciences, Graphic Era Hill University, Dehradun 248002, India
5 Biology Department, College of Science and Arts, King Khalid University, Muhayl Asser 61913, Saudi Arabia
6 Department of Biology, College of Science, Princess Nourah bint Abdulrahman University, Riyadh 11671, Saudi Arabia
7 Agro-Ecology and Pollution Research Laboratory, Department of Zoology and Environmental Science, Gurukula Kangri (Deemed to Be University), Haridwar 249404, India
8 Research and Development Division, Society for AgroEnvironmental Sustainability, Dehradun 248007, India
9 School of Public Health, Wolaita Sodo University, Wolaita P.O. Box 138, Ethiopia
* Correspondence: archanabachheti@geu.ac.in

Abstract: The objective of this study was to determine the effects of *Murraya koenigii* (curry leaf) aqueous extract on the seed germination and growth of two crops and their associated weeds (*Anagallis arvensis*, *Poa annua*, *Lepidium didymum*, and *Vicia sativa*). Wheat (*Triticum aestivum*) and Chickpea (*Cicer arietinum*) seeds were soaked in solutions containing 2, 4, 6, 8, and 10% concentrations of aqueous extracts of *M. koenigii* leaves, while distilled water was used as a control. Higher concentrations of the extract significantly increased the percentage of germination inhibition and decreased the radicle length and plumule length of both crops compared to the control. It was interesting to find that the pot-based experiment had no influence on the shoot length, plant dry mass, chlorophyll, carotenoid, proline, or protein content, stomatal conductance, or photosynthetic and transpiration rate in either crop. Similarly, in comparison to the control, the activity of antioxidant enzymes (peroxidase (PER), superoxide dismutase (SOD), and catalase (CAT)) had no significant influence. Soil physiochemical parameters and weed shoot growth were also investigated. In a Petri plate experiment, 6, 8, and 10% concentrations inhibited germination of both crops. In comparison, in the pot experiment, the growth of both crops was significant at lower concentrations, and 1% aqueous extract of *M. koenigii* was reported to be lethal to weeds without affecting the soil physiochemical properties. Based on the results of this study, it can be inferred that the reduced germination rate of the tested species after treatment with the extract could be attributed to damage caused by the extracts to the membrane system of the seed. This study concluded that an aqueous leaf extract of *M. koenigii* can inhibit the growth of weeds without affecting the germination, growth, photosynthetic, and biochemical properties of *C. arietinum* and *T. aestivum* or the physiochemical properties of soil; thus, it may be a potential candidate for the development of new bioherbicides.

Keywords: allelopathy; biochemical; bioherbicides; chlorophyll; germination; pesticides overuse; weeds

1. Introduction

The overuse of hazardous herbicides for weed removal is a major global obstacle to achieving optimal crop productivity. Weeds compete with crops for essential resources such as space, water, soil nutrients, light, and others, leading to yield reductions of up to 34% in major crops [1]. Weed control is an essential factor in crop production, and various weed management strategies are used on arable land around the world. Conventional methods of managing weeds involve manual weeding, mechanical tilling, and the application of chemical herbicides. These approaches help to minimize weed density while increasing agricultural production. Despite the benefits of these approaches in increasing agricultural output, they have some drawbacks. The biggest obstacles in hand weeding are diminishing availability and rising labor costs [2].

Due to increased soil turnover, mechanical weed management disturbs the soil structure and depletes soil fertility. Similarly, the usage of chemical herbicides for weed management has negative effects on soil quality and fertility, the environment, and human and animal health. Because of the difficulties involved with current weed control approaches, it is critical to create sustainable weed management strategies. Throughout history, natural methods have been utilized in traditional agricultural systems to control weeds. Crop rotation and intercropping are two common and effective techniques for managing weeds. In agroecosystems, the production and release of allelochemicals, which are phytotoxic compounds, by plants can influence coexisting species and have a significant impact on weed management. Rain percolating from upper parts can release these allelochemicals into the environment, affecting the growth of other plants [3]. The intercropping of plants or grasses not only covers the agricultural space but also feeds on the nutrients required by crop plants; using aqueous extracts will prove to be the most simple and non-dependent method in the coming era. The effects of allelochemicals include impacts on cell division, plant hormone production, membrane permeability, pollen grain germination, mineral uptake, the opening and closing of stomata, plant pigment and protein synthesis, photosynthesis, photorespiration, nitrogen fixation, and specific enzyme activities [4].

To defend against stress, plants have several Reactive Oxygen Species (ROS) scavenging enzymes, such as superoxide dismutase (SOD), catalase (CAT), and peroxidase (PER), which are activated to strengthen plant resistance to abiotic and biotic stresses [5]. Under stressful conditions such as suboptimal temperature, excessive light, salinity, or pathogen infection, plants produce large quantities of ROS [6]. It is widely acknowledged that stress can impact various aspects of plant biology, including their morphology, physio-chemical processes, and biomolecular processes. As a result, stress can lead to growth inhibition, stomata closure, reduced transpiration, decreased chlorophyll content and photosynthesis inhibition, protein changes, and different responses of antioxidative enzymes [7]. Photosynthesis is necessary for the survival of all life on Earth, as it contributes to oxygen production and the food chain. Transpiration aids in nutrient uptake, water balance, turgor pressure, and plant survival under stress conditions, whereas stomatal conductance is required for gas exchange [8]. However, the use of chemical herbicides affects the nutrient status of soil in the long run, which can cause disruptions in various biochemical processes, such as litter formation, nitrogen fixation [9], or nitrogen cycling, which affects the decomposition and microorganism ecology of soil [10].

M. koenigii is a well-known plant used as a food flavor enhancer all over the world. It is a valued plant due to its distinctive perfume and therapeutic usefulness. Various chemical components have been recovered from every section of this plant. The primary phytochemicals responsible for the distinct and potent aroma are P-gurjunene, P-caryophyllene, P-elemene, and O-phellandrene; additionally, carbazole alkaloids are abundant in the plant [11]. *M. koenigii* has been used for generations in Indian cooking and plays various roles in traditional medicine. *M. koenigii* contains a range of chemical compounds that interact in a complex manner to elicit their pharmacodynamic effects.

The objective of this study is to systematically evaluate the efficacy of an aqueous extract of curry leaves in controlling several common weeds in wheat and chickpea crops, i.e., *A. arvensis*, *L. didymum*, and *P. annua*, *V. sativa*, and to analyze soil physiochemical properties to check the impact of the applied extract on the nutrient status of soil. Moreover, the growth performance and physiological and biochemical parameters of *T. aestivum* and *C. arietinum* crop plants are also examined.

2. Materials and Methods

2.1. Experimental Site

The current study was conducted in Graphic Era Deemed to be University, Clement Town, Dehradun, Uttarakhand (India). The coordinates of the city are 30.318° N and 78.029° E.

2.2. Plant Materials

M. koenigii leaves were used to make an aqueous extract for the study. *M. koenigii* is an aromatic plant found growing wild in forest and residential areas in Uttarakhand. The fresh young green leaves of mature plants were collected from a residential colony of Clement Town, Dehradun. The Systematic Botany Division of Forest Research Institute, Dehradun, identified and authenticated the plants.

2.3. Preparation of Extract

Firstly, the leaves of *M. koenigii* were rinsed with regular water to eliminate dust particles from the surface of the leaves. The leaves were then air dried in the lab before being shredded into small pieces to increase the surface area so that they could be in contact with the solvent, i.e., water. The leaves were then immersed in water in a closed glass vessel to begin the extract preparation. During this 24 h process, the leaves were heated five times every 4 h The boiling procedure was periodic, i.e., the leaves were allowed to boil for 10 min and then left to cool for 4 h. To obtain crude extracts, the extract was filtered using a double layer of muslin cloth and then centrifuged for 20 min at 1000 rpm to eliminate particulate matter. The resulting supernatant was collected and stored at 4 °C for further use [12].

2.4. Petri Plate Experiment and Observation of Seed Germination

The seeds of *C. arietinum* and *T. aestivum* were sterilized by immersing them in a 0.2% sodium hypochlorite solution for 10 min and then rinsed three times with distilled water to prevent any pathogen contamination. Dual layers of filter paper were fixed in Petri plates (radius = 45 mm); then, ten seeds of the recipient species were placed in each Petri plate and treated with 10 mL of the various dilutions (2%, 4%, 6%, 8%, and 10% concentrations) of *M. koenigii* aqueous extract (MKAE). Distilled water was used as a control, and each treatment group had three replicates. The plates were kept at an ambient temperature with a 12 h light/12 h dark cycle. Next, 2 mL of the corresponding extract was added every 48 h to maintain the moisture of the filter paper. The experiment was completed in one week, and the number of germinated seeds was counted from the second to the last day based to determine the germination percentage (Equation (1)) and germination inhibition percentage (Equation (2)). Furthermore, the lengths of the plumule and radicle of the seeds were also recorded, as given in Figure S1 [13].

$$\text{Germination } (\%) = \frac{\text{No. of germinated seeds after given treatment}}{\text{No. of germinated seeds in control}} \quad (1)$$

$$\text{Germination Inhibition } (\%) = 100 - \text{Germination percentage} \quad (2)$$

2.5. Pot Experiments, Growth Conditions, and Experimental Design

Pot experiments were set up in December 2021 and carried out until February 2022 in loamy soil. The overall layout of the experiments is given in Figure 1. The pots were kept under standard weather conditions (10–26 °C temperature; 50–60% relative humidity; 14 h light/10 h dark) that are comparable to normal field conditions in India's chickpea and wheat growing regions. The extract treatments were designed according to the observations recorded during the Petri plate experiment. A total of 3 kg of agricultural soil was filled, and four plants were potted in each pot; the first extract dose was given in soil 2 days before planting plants. The extract treatment was given in five different concentrations, i.e., 1%, 2%, 3%, 4%, and 5% (weight/volume). Distilled water was used as a control; each treatment group had three replicates. A total of two doses of the corresponding extract were given. The number of recipient species plants (*C. arietinum*, *T. aestivum*, *A. arvensis*, *P. annua*, *L. didymum*, and *V. sativa*) that died was counted, as depicted in Figures S2–S4 [13].

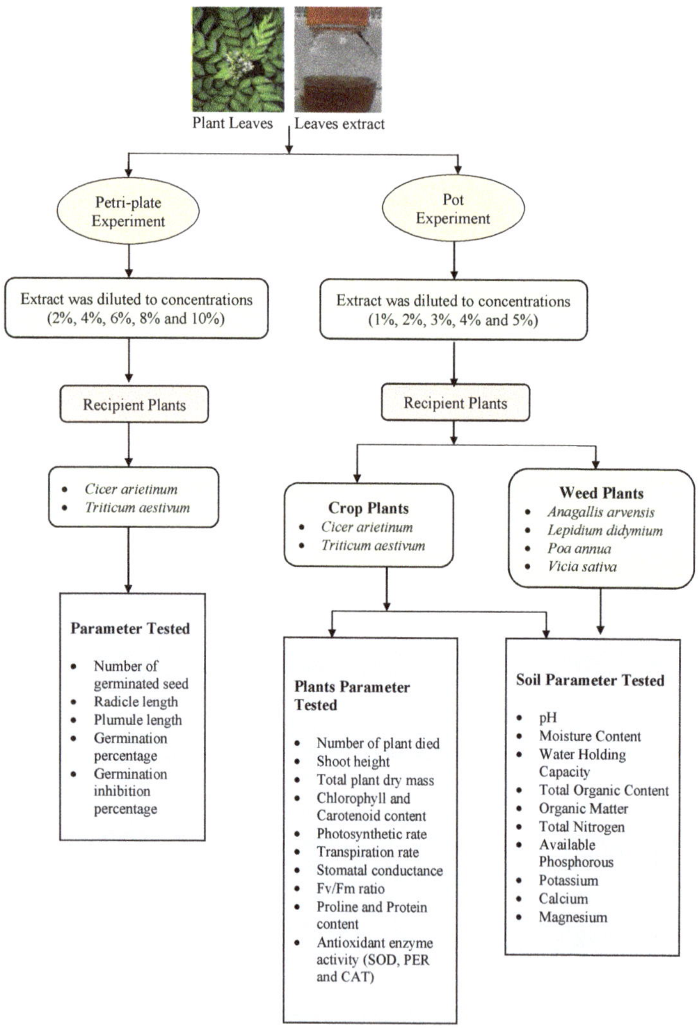

Figure 1. The layout of the experimental design adopted in this study.

2.6. Photosynthetic Parameter Measurements

To quantify photosynthetic pigments, the method of Arnon [14] was used. The chlorophyll content, fresh and dry weight of the root, and shoots of crop recipient species were calculated. Shoots were separated from the roots at the cotyledonary node, then dried at 65 °C for 48 h, and weighed. Next, 250 mg of leaves was grated in 10 mL of 80% acetone using a pestle and mortar. The paste was centrifuged at 3000 RPM for 5 min, then transferred to a 25 mL volumetric flask, and the volume was made up using acetone. The samples were analyzed on a spectrophotometer (Labtronics instrument LT-2800, Labtronics Panchkula, India) at 663 nm, at 645 nm for chlorophyll A (Equation (3)) and B (Equation (4)), at 652 nm for total chlorophyll (Equation (5)), and at 470 nm for carotenoids (Equation (6)) [14].

$$\text{Chlorophyll A (mg/g)} = 12.7\ (\text{A at } 663) - 2.69\ (\text{A at } 645) \times \frac{\text{V}}{1000 \times \text{W}} \quad (3)$$

$$\text{Chlorophyll B (mg/g)} = 22.9\ (\text{A at } 645) - 4.69\ (\text{A at } 663) \times \frac{\text{V}}{1000 \times \text{W}} \quad (4)$$

$$\text{Total Chlorophyll (mg/g)} = \frac{\text{A at } 652}{34.5} \times \frac{1000 \times \text{V}}{1000 \times \text{W}} \quad (5)$$

$$\text{Carotenoid (mg/g)} = \frac{1000\ (\text{A at } 470) - 1.8 \times \text{Chl. A} \ - 85.02 \times \text{Chl. B})}{198} \quad (6)$$

where A is the absorbance on the spectrophotometer, V is the final volume of supernatant, and W is the fresh weight of the sample.

The Li 6400 photosynthetic system (Li-Cor, Inc., Lincoln, NE, USA) was used to measure the net photosynthetic rate (Pn), stomatal conductance (gs), and transpiration rate (E). Fully grown and developed leaves from each replication were examined between 10 a.m.–12 p.m. under a photosynthetic flux density of 1000 μmol m^{-2} s^{-1}, provided by a LED light source. The leaf chamber had an air humidity of about 60%, a CO_2 concentration of 370–380 μmol mol^{-1}, an ambient air temperature of 26–28 °C, and a flow rate of 500 μmol s^{-1} [15]. Chlorophyll fluorescence measurements were taken in the morning between 10–11 a.m. using a portable Hansatech Plant Efficiency Analyzer (Hansatech, King's Lynn, UK). Before the measurements, the leaves were darkened for 20 min using leaf clips. Fluorescence was stimulated by red (actinic) radiation with a peak wavelength of 650 nm, emitted by light-emitting diodes. The irradiance was set to 2000 μmol m^{-2} s^{-1}. The maximum quantum yield of photosystem II was estimated by the ratio Fv/Fm = (Fm − Fo)/Fm [15].

2.7. Determination of Biochemical Parameters

2.7.1. Proline

The Bates method [16] was used for the proline assay. To begin, 0.2 g of leaves was homogenized with 4 mL of 3% sulphosalicylic acid solution; then, 2 mL of the resulting solution was combined with reactive ninhydrin (2 mL) in acetic acid (2 mL) and incubated in boiling water for 1 h. After completion of the reaction, the samples were immediately placed in an ice box. Once the samples had reached ambient temperature, they were allowed to settle. The samples were then mixed with 4 mL of toluene, and the resulting solution was vigorously shaken. An ultraviolet-visible spectrophotometer (Perkin Elmer Lambda 25) was used to calculate absorbance at 520 nm using the colored complex. Finally, the proline content (Equation (7)) was estimated in mol/gram of fresh weight (FW) using the given equation [17].

$$\text{Proline} = \frac{\text{Absorbance of sample} \times \text{ K value } \times \text{ dilution factor}}{\text{Weight of sample } \times 100} \quad (7)$$

2.7.2. Soluble Protein

The content of protein was measured from leaves using the method of Lowry [18]. This experiment is based on the hydrolysis of proteins, which release amino acids, resulting in the formation of colored complexes with the Fulvin reagent. The concentration of protein in each sample was determined using a standard curve of bovine serum albumin, and the protein content was calculated in mg/g DW (dry weight) [17]. Finally, the absorption of the samples at 660 nm was estimated using a spectrophotometer (Perkin Elmer Lambda 25).

2.7.3. Antioxidant Enzyme Assay

Leaf material from each sample (500 mg) was homogenized in an extraction buffer containing 0.5% Triton X-100 [$C_{14}H_{22}O(C_2H_4O)_n$] and 1% polyvinylpyrrolidone [$(C_6H_9NO)_n$] in 100 mM potassium phosphate buffer (pH 7.0). The crude extraction was prepared utilizing a chilled mortar and pestle. The supernatant was used for enzymatic assays after centrifuging the homogenate at 15,000 rpm for 20 min at 4 °C. A cold-temperature enzyme extract for superoxide dismutase (SOD) was then prepared. Firstly, the measured amount of leaf sample (1 g) was chilled in liquid nitrogen to prevent proteolytic activity. The chilled sample was ground with 10 mL of extraction buffer (0.1 M phosphate buffer, pH 7.5, including 0.5 mM EDTA and 1 mM ascorbic acid). The mixture was passed through the multilayer cheesecloth, and the resulting filtrate was centrifuged for 20 min at 15,000 rpm at 4 °C. The supernatant obtained after centrifugation was used as an enzyme [19]. The activity of SOD was assessed by measuring the reduction of optical density of nitroblue tetrazolium dye by the enzyme [20]. The reaction was started by adding 2 mM of riboflavin to the tubes and then placing them under two 15 W fluorescent lamps for 15 min. The reaction mixture used in this experiment consisted of 13 mM methionine, 25 mM nitroblue tetrazolium chloride, 0.1 mM EDTA, 50 mM phosphate buffer (pH 7.8), 50 mM sodium carbonate, and 0.1 mL of an enzyme extract. As a control, a complete reaction mixture without enzymes but producing the maximal color was used. The reaction was halted by turning off the light and leaving the tubes in the dark. The optical density was measured at 560 nm using a Perkin Elmer UV/VIS spectrophotometer Lamda 25, and one unit of enzyme activity was defined as the amount of enzyme that caused a 50% decrease in absorbance compared to tubes that did not contain an enzyme [15]. A non-irradiated complete reaction mixture was used as a control. PER activity was calculated with guaiacol as the substrate, as Husen [21] illustrated. The assay mixture contained 0.1 M phosphate buffer with a pH of 6.1, 4 mM guaiacol, 3 mM H_2O_2, and 0.4 mL crude enzyme extract, with a total volume of 1.2 mL. Enzyme activity was measured by recording the optical density (OD) at 420 nm using a spectrophotometer (Perkin Elmer Lambda 25), and the results were expressed as µmoles H_2O_2 destroyed per minute per milligram of protein [15]. Chandlee and Scandalios' [22] method was used to measure the catalase enzyme activity (CAT). In an ice box, a mixture of 2.5 mL of 50 mmol L^{-1} potassium phosphate and 0.3 mL of 3% H_2O_2 was prepared; then, 2.0 mL of enzyme extract was immediately added to the mixture. The absorbance variation curve at 240 nm was determined for 3–4 min. The change in absorbance was used to calculate the enzyme activity, which was expressed as mg/protein/min [17].

2.8. *HPTLC Analysis of Allelopathic Compounds*

HPTLC was used to obtain more detailed information on the chemical components present in the extract of *Murraya koenigii*. The extract was qualitatively analyzed for various phytoconstituents using TLC (thin layer chromatography) G-60 F254 silica gel (Merck, Rahway, NJ, USA) plates as stationary phase and a mobile phase consisting of a mixture of toluene, ethyl acetate, methanol, and formic acid. A Biostep AS-30 applicator and a filling syringe (AS-30) were used for application. HPTLC plates precoated with G-60 F254 silica gel were used for fingerprinting to obtain better separation of the compounds present (quercetin, kaempferol, and benzoic acid) in the extract. The plates were washed with methanol before use to remove any impurities that may have been picked up during laboratory storage. The fingerprint of each extract was obtained by scanning the chromatographic plate at 254 nm. Next, 1 mL of *M. koenigii* dried extract was precisely measured and dissolved in 100 mL of methanol. The analysis was performed at a constant temperature (22 ± 1 °C) and humidity (47 ± 1%) [23]. The compounds were identified by comparing the peak (retention time) of each sample with the peak of the standard compound (i.e., pure compound) [24].

2.9. Physiochemical Analysis of Soil

Soil samples were taken from the donor's extract-treated weed pots at the termination of the experiment. All soil samples were taken randomly from three points in each pot, mixed uniformly, and kept inside zipper plastic bags. After removing fine roots, stones, and plant organic residue by hand, each soil sample was air-dried at ambient temperature. Following that, samples were pulverized, strained through a 2 mm mesh sieve, and homogeneously blended for additional soil physicochemical analysis. The potentials of hydrogen ions (pH), water holding capacity (WHC), soil texture, nitrogen (N), phosphorus (P), potassium (K), organic carbon (OC), organic matter (OM), calcium (Ca), and magnesium (Mg) of all soil samples were determined. The pH was determined using a digital meter (Labtronics instrument LT-49) [25]. A hydrometer was used to measure soil texture [26], while the WHC was estimated by the gravimetric method [27]. Total nitrogen was calculated by the Kjeldhal method, as described in [28]. A UV-vis spectrophotometer (Labtronics instrument LT-2800, Labtronics, Panchkula, India) was used to calculate the phosphorus concentration [29]. A Systronics flame photometer-128 was used to measure potassium [30]. The calcium, magnesium, organic carbon, and organic matter contents were measured using an atomic absorption spectrophotometer (Analyst-800) [31,32]. The standard ranges of the physiochemical parameters of soil, as described by Krishi Vigyan Kendra (KVK) (Delhi, India), for good agricultural soil are mentioned in Table 1.

Table 1. The standard ranges of the physiochemical parameters of soil, as described by Krishi Vigyan Kendra.

Parameters	Standard Range
pH	6.5–8.2
Organic Carbon	0.5–0.75%
Total Nitrogen	240–480 kg/ha
Available Phosphorus	28–56 kg/ha
Available Potassium	140–280 kg/ha
Calcium	3.36–6.72 kg/ha
Magnesium	2.24–4.48 kg/ha

2.10. Statistical Analysis

A statistical analysis of the experimental data was performed using the Analysis of Variance (ANOVA). The level of significance was taken as 0.05. The null hypothesis was rejected if the p-value of the test statistic (F) was less than or equal to 0.05. The Tukey's post hoc test was used for pair-wise comparisons whenever the null hypothesis was rejected.

3. Results and Discussion

3.1. Seed Germination

Different concentrations of *M. koenigii* aqueous extract (MAKE) significantly ($p \leq 0.05$) affected the germination percentage (GP) of *C. arietinum* and *T. aestivum*. In the case of *C. arietinum*, the germination percentage was moderate at 2% and 4% concentrations of extract, i.e., around 60–70%, but it started decreasing with increasing the concentration of extract, from 50% at 6% to 0–10% at 8% and 10% concentrations. In *T. aestivum*, the germination percentage was found to be 90–100% at a 2% concentration of extract, whereas it was observed to be 10–20% at 4% and 6% concentrations, and 0% at 8% and 10% concentrations. The values of germination percentage in *C. arietinum* and *T. aestivum* pure culture were close to 100% at the control level (Table 2).

The application of *M. koenigii* extracts showed that at 2% and 4% concentrations, the plumule and radicle lengths of both crops were similar to those of the control. In comparison, they were very low at 6%, 8%, and 10% concentrations. This implies that higher concentrations of extract slowed plant growth, which is unacceptable for crop yield. We also determined that the germination inhibition percentages (GIPs) of *C. arietinum* and *T. aestivum* were significantly influenced ($p \leq 0.01$) by the apparent allelopathic action of the MAKE concentrations (Table 2). GIP was not obtained in the controls. The minimum inhibition was recorded at a 2% concentration, while the maximum was at 8% and 10%.

Table 2. The number of germinated seeds, germination percentage, germination inhibition percentage, and radicle and plumule length in two important crop plants (*C. arietinum* and *T. aestivum*) under control and treated situations. Under treated conditions, different concentrations of *M. koenigii* aqueous extract (MKAE) were used at different temperatures in the Petri dish experiment.

Parameters	Plant Name	Treatments					
		Control	2%	4%	6%	8%	10%
Germinated seeds	*C. arietinum*	10.00 [a]	9.33 [ab]	7.67 [c]	5.00 [d]	0.33 [e]	0.67 [f]
	T. aestivum	10.00 [a]	10.00 [a]	2.00 [b]	0.67 [d]	0.00 [e]	0.00 [e]
Germination (%)	*C. arietinum*	100.00 [a]	93.30 [ab]	76.70 [c]	50.00 [d]	3.30 [e]	6.70 [f]
	T. aestivum	100.00 [a]	100.00 [a]	20.00 [b]	6.70 [c]	0.00 [e]	0.00 [e]
Germination Inhibition (%)	*C. arietinum*	0.00 [a]	0.00 [a]	23.30 [b]	50.00 [c]	96.70 [d]	93.30 [e]
	T. aestivum	0.00 [a]	0.00 [a]	80.00 [b]	93.30 [c]	100.00 [cd]	100.00 [cd]
Radicle length (cm)	*C. arietinum*	17.53 [a]	5.50 [b]	5.27 [bc]	4.20 [d]	0.67 [e]	1.23 [f]
	T. aestivum	11.13 [a]	6.73 [b]	4.30 [c]	0.10 [d]	NG	NG
Plumule length (cm)	*C. arietinum*	2.00 [a]	0.90 [b]	0.57 [c]	0.37 [cd]	0.10 [e]	0.07 [ef]
	T. aestivum	11.70 [a]	9.40 [b]	7.83 [c]	NG	NG	NG

Differences between the mean of each row superscripted by the same letter were not statistically significant ($p \leq 0.05$). NG—No growth.

Although the allelochemicals were shown either to be stimulatory or inhibitory in the first tranche of experiments, i.e., Petri plate experiments (PPEs), the results revealed that higher concentrations (6%, 8%, and 10%) of aqueous extracts significantly inhibited the germination of both crops, i.e., *C. arietinum* and *T. aestivum*. Lawan et al. [33], Vashishth et al. [34] and Khattak et al. [35] have also reported similar findings. The allelochemicals found in aqueous extracts of *Jatropha curcas* were found to be stimulatory to wheat at lower concentrations, while at higher concentrations, the extracts were either inhibitory or did not affect wheat plant growth compared with the control. This could have been because the biological activity of allelochemicals is concentration dependent [36]; however, these findings are contrary to those of Bora et al. [37], whose study reported the allelopathic effect of *Acacia auriculiformis* leaf extracts on the seed germination of certain crops. Eyini et al. [38] reported the allelopathic impact of *Bambusa arundinacea* on *Arachis hypogaea* and concluded that the aqueous extracts of weeds hindered the germination of crop seeds. Higher concentrations of the *Acacia auriculiformis* extract cause significant reductions in the root length of *Brassica juncea*, *C. arietinum*, *Phaseolus mungo*, *Raphanus sativus*, and *Vigna unguiculata* [39]. Kruse et al. [40] reported unfavorable results, as the germination rate and growth of *Triticum durum* were retarded due to the toxic effects of the phenolic compounds present in leachates (from the leaf, root, and litter) of *Populus nigra* at high concentrations (75 g/L). The findings of Hachani et al. [41] suggested that *Populus nigra* extracts, on the other hand, were more phyto-inhibitory than *Casuarina glauca* extracts. This could have been due to the high concentration of polyphenols found in all *Populus nigra* extracts. This helps to explain the allelopathic effects of chemical substances released by *C. glauca* and *P. nigra* on *T. durum* germination and early seedling growth.

3.2. Cash Crop Growth and Physiological Parameters

The shoot lengths of *C. arietinum* ranged from 19.68 cm to 20.20 cm, including the control specimens. Table 3 shows that at 1% concentration, the shoot length was better than under the control conditions, while in *T. aestivum*, it ranged from 14.67 to 15.30 cm, including in the control specimens. The total plant dry mass ranged from 0.6 to 1.2 gm for both crops, including in the control specimens, and both these parameters, as estimated for plant morphological growth, were not significantly ($p \leq 0.05$) affected by the application of extract. A comparison of means revealed that there was no significant difference in chlorophyll a (Chl a) and chlorophyll b (Chl b) levels between the control and MAKE-treated plant leaves.

Table 3. Shoot height, dry mass, photosynthetic pigments (Chlorophyll a, b, total chlorophyll, and carotenoids), net photosynthetic rate (Pn), stomatal conductance (gs), transpiration rate, and chlorophyll fluorescence (Fv/Fm) in two important crop plants (*C. arietinum* and *T. aestivum*) under control and treated conditions.

Parameters	Plant Name	Treatments					
		Control	1%	2%	3%	4%	5%
Shoot length (cm)	*C. arietinum*	20.00 ^{a}	20.20 ^{ab}	19.63 ^{c}	20.17 ^{bd}	19.68 ^{ce}	19.93 ^{cef}
	T. aestivum	15.30 ^{a}	14.70 ^{b}	14.73 ^{bc}	14.67 ^{d}	14.90 ^{e}	15.00 ^{f}
Total plant dry mass (gm)	*C. arietinum*	1.1 ^{a}	1.0 ^{ab}	0.9 ^{bc}	0.7 ^{d}	0.6 ^{de}	0.7 ^{d}
	T. aestivum	1.2 ^{a}	1.1 ^{ab}	1.1 ^{ab}	0.9 ^{c}	0.6 ^{d}	0.7 ^{e}
Chl a (mg g^{-1} FW)	*C. arietinum*	6.766 ^{a}	6.748 ^{ab}	6.749 ^{abc}	6.668 ^{d}	6.631 ^{e}	6.614 ^{ef}
	T. aestivum	1.810 ^{a}	1.616 ^{b}	1.571 ^{c}	1.522 ^{d}	1.497 ^{e}	1.508 ^{ef}
Chl b (mg g^{-1} FW)	*C. arietinum*	3.526 ^{a}	3.883 ^{b}	3.773 ^{c}	3.737 ^{d}	3.641 ^{e}	3.582 ^{f}
	T. aestivum	5.071 ^{a}	6.304 ^{b}	5.970 ^{c}	5.966 ^{cd}	5.975 ^{cde}	5.884 ^{f}
Total Chl (mg g^{-1} FW)	*C. arietinum*	18.493 ^{a}	19.739 ^{b}	19.681 ^{c}	19.565 ^{d}	19.333 ^{e}	19.188 ^{e}
	T. aestivum	7.652 ^{a}	8.725 ^{b}	7.913 ^{c}	7.130 ^{d}	6.754 ^{e}	6.638 ^{ae}
Car (mg g^{-1} FW)	*C. arietinum*	2.116 ^{a}	1.968 ^{b}	2.010 ^{c}	1.996 ^{d}	2.033 ^{ce}	2.048 ^{ef}
	T. aestivum	1.458 ^{a}	2.127 ^{b}	2.172 ^{bc}	2.224 ^{d}	2.277 ^{e}	2.311 ^{e}
Pn (μmol m^{-2} s^{-1})	*C. arietinum*	30.00 ^{a}	29.98 ^{ab}	29.98 ^{ab}	29.97 ^{abc}	29.95 ^{bcd}	29.94 ^{de}
	T. aestivum	18.02 ^{a}	18.03 ^{ab}	18.02 ^{a}	18.00 ^{c}	17.99 ^{cd}	17.97 ^{e}
gs (mol m^{-2} s^{-1})	*C. arietinum*	0.25 ^{a}	0.24 ^{ab}	0.22 ^{bc}	0.22 ^{bc}	0.21 ^{cd}	0.20 ^{de}
	T. aestivum	0.28 ^{a}	0.27 ^{ab}	0.27 ^{ab}	0.25 ^{c}	0.24 ^{cd}	0.21 ^{e}
E (mmol m^{-2} s^{-1})	*C. arietinum*	5.72 ^{a}	5.71 ^{ab}	5.69 ^{bc}	5.68 ^{cd}	5.68 ^{cd}	5.65 ^{e}
	T. aestivum	4.93 ^{a}	4.92 ^{ab}	4.89 ^{c}	4.87 ^{cd}	4.85 ^{de}	4.81 ^{f}
Fv/Fm	*C. arietinum*	0.789 ^{a}	0.789 ^{a}	0.788 ^{ab}	0.787 ^{abc}	0.788 ^{ab}	0.786 ^{bcd}
	T. aestivum	0.820 ^{a}	0.821 ^{ab}	0.820 ^{a}	0.818 ^{ac}	0.817 ^{cd}	0.817 ^{cd}

NG—No growth. Differences between the mean of each row superscripted by the same letter were not statistically significant ($p \leq 0.05$).

After examining the total chlorophyll data, it was found that stress conditions significantly impacted the content. However, there was no significant difference between the control and MAKE-treated specimens. Similar results were observed for carotenoids, photosynthetic rate (Pn), stomatal conductance (gs), and transpiration rate (Table 3). The Fv/Fm ratio is a suitable indicator for evaluating PSII in plants exposed to environmental stress [17]. The Fv/Fm test is intended to allow as much light energy as possible to enter the fluorescence pathway. A saturating light source reduces the maximum number of reaction centers [42]. The Fv/Fm ratio typically decreases as plant stress increases, because fewer open reaction centers become available. Similar to other physiological parameters, no difference was noticed statistically in the Fv/Fm ratio of control and treated crop plants (Table 3).

The increased proline levels in response to environmental stress indicate a plant's stress response. The accumulation of proline in plant cells under stress conditions serves as a protective mechanism, shielding from toxicity, as reported by Rahbarian et al. [43]. Proline also plays a significant role in osmoregulation, membrane stability, detoxification, cytosol pH adjustment, and the maintenance of cell enzyme structure [44]. As no significant difference was reported in the means comparison of control and MKAE-treated plants, it can be concluded that plants were not facing any kind of stress condition. A comparison of mean data on leaf protein content showed that there was no significant effect on either cash crop at lower concentrations of MKAE treatments (Table 4).

Table 4. Contents of proline and soluble protein in two important crop plants (*C. arietinum* and *T. aestivum*) under control and treated conditions.

Biochemical Features	Plant Name	Treatments					
		Control	1%	2%	3%	4%	5%
Proline (mg g^{-1} FW)	*C. arietinum*	1.08 ^{a}	1.09 ^{ab}	1.07 ^{abc}	1.05 ^{cd}	1.02 ^{de}	1.01 ^{ef}
	T. aestivum	1.17 ^{a}	1.17 ^{a}	1.16 ^{ab}	1.15 ^{abc}	1.13 ^{cd}	1.12 ^{cde}
Soluble Protein (mg g^{-1} DW)	*C. arietinum*	56.24 ^{a}	56.24 ^{a}	56.23 ^{a}	56.21 ^{ab}	56.20 ^{abc}	56.19 ^{abcd}
	T. aestivum	48.45 ^{ab}	48.48 ^{a}	47.47 ^{bc}	46.79 ^{de}	47.17 ^{cd}	46.72 ^{def}

Differences between the mean of each row superscripted by the same letter were not statistically significant ($p \leq 0.05$).

There was no significant difference observed in superoxidase dismutase activity (SOD) for control and MKAE-treated plants. SOD constitutes the first line of defense against reactive oxygen species (ROS). A mean data comparison showed that there was no significant difference observed in the activity of PER (Peroxidase) and CAT (Catalase) between all MKAE concentrations tested and the control groups. PER catalyzes peroxide reduction and generates ROS. Under water stress conditions, plants produce reactive oxygen species (ROS) that can act as secondary messengers in signal transduction pathways at low concentrations [45]. However, high concentrations of ROS can cause damage to the membrane, photosystems (PSI and PSII), photosynthetic pigments, and enzymes [43]. Chloroplasts and mitochondria are the main sites for electron transfer cycles in plant cells and are particularly vulnerable to ROS production [46]. CAT is responsible for preventing the accumulation of cellular organelles and protecting these organelles and tissue from destruction by peroxide, as several metabolic reactions continuously synthesize peroxide [45] (Table 5).

Table 5. Antioxidant enzyme activities in two important crop plants (*C. arietinum* and *T. aestivum*) under control and treated conditions.

Biochemical Features	Plant Name	Treatments					
		Control	1%	2%	3%	4%	5%
SOD (Uμg^{-1} protein)	*C. arietinum*	1.03 ^{a}	1.02 ^{ab}	1.02 ^{ab}	1.00 ^{abc}	1.00 ^{abc}	1.00 ^{abc}
	T. aestivum	2.56 ^{a}	2.51 ^{ab}	2.50 ^{ab}	2.48 ^{bc}	2.47 ^{bc}	2.45 ^{cd}
PER (Uμg^{-1} protein)	*C. arietinum*	0.374 ^{a}	0.373 ^{ab}	0.373 ^{ab}	0.371 ^{abc}	0.369 ^{cd}	0.368 ^{de}
	T. aestivum	0.22 ^{a}	0.20 ^{ab}	0.19 ^{bc}	0.18 ^{cd}	0.15 ^{e}	0.15 ^{e}
CAT (Uμg^{-1} protein)	*C. arietinum*	0.308 ^{a}	0.308 ^{a}	0.307 ^{ab}	0.304 ^{bc}	0.302 ^{cd}	0.300 ^{de}
	T. aestivum	0.28 ^{a}	0.28 ^{a}	0.27 ^{ab}	0.26 ^{abc}	0.23 ^{cd}	0.22 ^{cde}

Differences between the mean of each row superscripted by the same letter were not statistically significant ($p \leq 0.05$).

3.3. Number of Weed Plants Died

No effect of the first dose on *A. arvensis* plants was observed after fifteen days, as they were green, fresh, and healthy at all aqueous extract concentrations. All the weed plants followed the same trend after the first treatment dose. The average plant height of *A. arvensis* ranged between 14.67 and 15.33 cm (compared to the plant height of the control, i.e., 14.93 cm), 0.80 cm to 1.33 cm (control 1.60 cm) for *L. didymum*, 8.70 cm to 9.73 cm including the control for *P. annua*, and between 9.73 cm and 10.00 cm, compared to the control value of 10.27 cm, for *V. sativa*. Seven days after the second dose, all *A. arvensis* plants died at 1%, while flowers and leaves started drooping at 2, 3, 4, and 5%. Approximately 50% of plants died at 2%, 25% at 3% and 4%, and a few plants died at 5% treatment concentration, whereas all of the control plants were green and healthy with flowers. One week after the second treatment dose, the plants of *L. didymum* began to droop at 1% and 2%, but not at 3%, 4%, or 5%. After a fortnight, about 50% of plants died at 1%, and a few plants died at 2% and 3% concentrations, whereas all of the control plants were green and healthy. The second dose did not affect the treatments at 4% and 5% concentrations. In the case of *P. annua*, the mean heights ranged from 9.40 to 10.30 cm (including the control), and budding started in all of the plants after the second dose of treatment. None of the plants drooped or died, which indicated that the second dose had no visual effect. One week after the second dose of treatment, the plants of *V. sativa* started drooping at 1, 2, 3, and 4% concentrations. No significant impact was seen at 5% concentration, while after a fortnight, all the plants died at 1, 2, and 3% and a few at 4% and 5%, while under control conditions, all plants were green and healthy (Figure 2). Hence, it can be concluded that the inhibition percentage (IP) is zero under control conditions and at all concentrations of *P. annua*. The IP of *A. arvensis* was 100% at 1% treatment, 50% at 2%, and too low to be measured at 3%, 4%, and 5%. In the case of *L. didymum*, the IP was 50% at 1% and 2%, 25% at 3%, and zero at 4% and 5% treatments, whereas in *V. sativa*, the IP was 100% at 1, 2, and 3%, 50% at 4% and very low at 5% treatment (Figure 2). When the concentration of an herbicide increases but its herbicidal action decreases, it is known as the herbicide "dilution effect". This phenomenon appears to contradict the basic idea that larger quantities of a chemical should have a greater influence. However, various scientific concepts, like antagonistic effect [47,48], selectivity and non-target effects [49,50], and physiological responses [51,52], can assist in understanding why this happens.

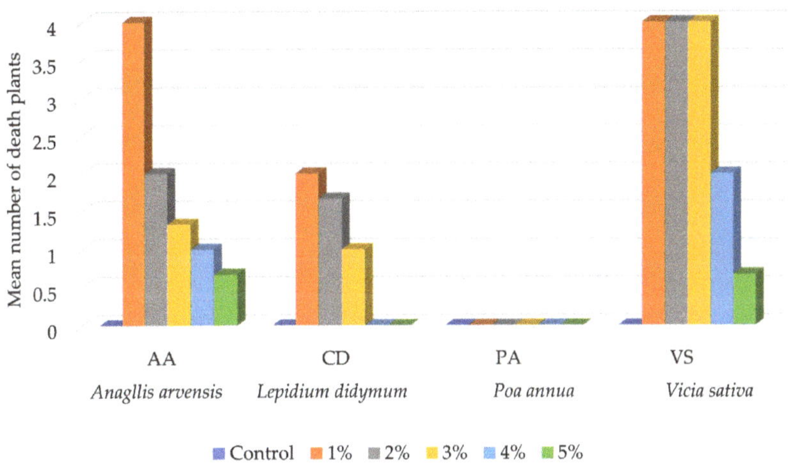

Figure 2. Mean number of weed death plants following treatment with *M. koenigii* aqueous extract.

Several studies have reported that the germination and growth of cash crops are significantly inhibited by various plant extracts or leachates, with some exceptions. It is therefore important to determine the donor plant extract concentration at which it supports the growth of recipient crop plants and inhibits the growth of recipient weed plants, as such findings are more likely to fulfill the above-mentioned requirements to make a bioherbicide. Bansal [53] found similar results to the current study, reporting that extracts from *Lantana camara* inhibited the seed germination and seedling growth of various weed species, including *Echinochloa colonum*, *Digitaria sanguinalis*, *Panicum psilopodium*, and *Commelina benghalensis*, which are common weeds associated with rice. However, there was no significant effect on the seed germination of rice. Water extracts of sorghum with sunflower and sorghum, sunflower, and mulberry were combined and used as an herbicide to control various weed species such as wild oats, little seed canary grass, labsquarters, and swine cress in wheat. The extracts were sprayed at different concentrations (12, 15, 18, and 21 L/ha) at two different growth stages of wheat (40 and 55 days after sowing). The study also included a control treatment, where synthetic herbicide (Affinity 50 WP) and a weekly check were used for comparison. Applying an increasing dose of water extract (12–21 L/ha) in all treatment combinations resulted in a significant reduction in total weed density and dry biomass production of narrow and broad-leaved weeds. Applying a mixture of sorghum, sunflower, and mulberry water extracts at 18 L/ha 40 + 55 days after sowing resulted in an 87.14% decrease in total weed dry matter and a 19.5% increase in wheat grain yield. However, it was found that applying sorghum + sunflower water extracts at 12 L/ha 40 + 55 days after sowing was more cost-effective than the other treatments [54].

3.4. HPTLC Analysis of Allelopathic Phytochemicals

An HPTLC analysis of the extracts revealed the presence of quercetin, benzoic acid, and kaempferol compounds (Figure 3). The outcomes from our HPTLC analysis contributed to our comprehension of the allelopathic capabilities of *M. koenigii* and offer scientific proof of its viability as a natural herbicide for eco-friendly weed management. The detection of allelopathic components such as benzoic acid (1.95%), kaempferol (3.28%), and quercetin (2.92%) additionally reinforced the idea that utilizing allelochemicals holds promise for ecologically sound and efficient weed management. This approach lessens the need to use synthetic herbicides, encouraging more sustainable agricultural methods. Various researchers have reported allelopathic activity against various plant species in these compounds. Compared to the control, quercetin significantly inhibited the radicle growth of *Pasteuria ramosa* seedlings [55]. Additionally, at 0.06 mM, p-hydroxybenzoic acid inhibited *Anabaena flosaquae* and *Chlorella pyrenoidosa* growth [56] and the activity of kaempferol was also reported by Cid-Ortega and Monroy-Rivera [57] and Li et al. [58].

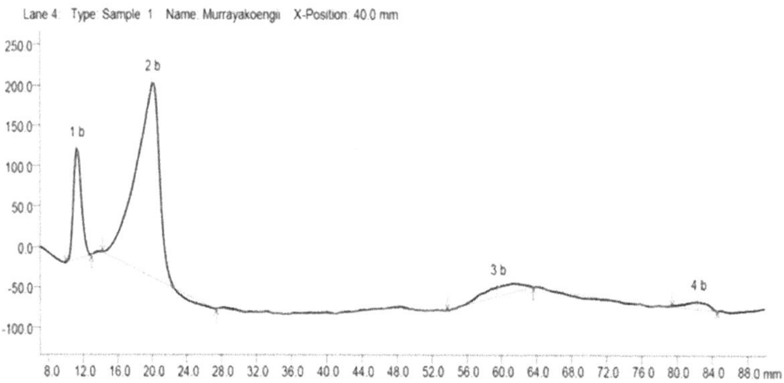

Figure 3. HPTLC analysis graph of *M. koenigii* aqueous extract. (Arrow represents major peak 1b: benzoic acid, 2b: kaempferol, 3b: quercetin).

3.5. Physiochemical Parameters of Soil

The pH of the soil ranged from 7.23 to 7.37 in *A. arvensis* pots, 7.20 to 7.33 in *L. didymum* pots, 7.20 to 7.37 in *P. annua* pots, and 7.23 to 7.37 in *V. sativa* pots (Figure S5a). The moisture content in the soil of *A. arvensis* pots was 16.87–17.20%, 16.97–17.23% in the soil of *L. didymum* pots, 16.93–17.17% in the soil of *P. annua* pots, and 17.03–17.20% in the soil of *V. sativa* pots (Figure S5b). The water holding capacity ranged from 46.29 to 46.42 in *A. arvensis* pots, 47.23 to 47.33 in *L. didymum* pots, 47.27 to 47.41 in *P. annua* pots, and 46.36 to 46.52 in *V. sativa* pots (Table 6) (Figure S5c). The pH of the control soil was recorded as

7.20 for *A. arvensis*, 7.22 for *L. didymum*, 7.21 for *P. annua*, and 7.24 for *V. sativa*. Meanwhile, the moisture content and WHC of the control were 17.12% and 46.40% for *A. arvensis*, 17.08% and 46.39% for *L. didymum*, 17.13% and 46.35% for *P. annua*, and 17.18% and 46.31% for *V. sativa*. All of the parameters were within KVK's standard ranges. This means that the statistical analysis did not find a significant difference between the treatments and the control group.

Table 6. Impact of *M. koenigii* extract on the physical parameters of loam soil.

Plant Name	Type of Plant	Treatment	pH	Moisture Content (%)	Water Holding Capacity (%)
C. arietinum	Crop	Control	7.38 ± 0.07	17.10 ± 0.26	46.39 ± 0.04
	Crop	1%	7.23 ± 0.03	17.23 ± 0.22	47.34 ± 0.08
	Crop	2%	7.40 ± 0.06	17.20 ± 0.17	47.46 ± 0.04
	Crop	3%	7.40 ± 0.10	17.27 ± 0.15	47.32 ± 0.11
	Crop	4%	7.20 ± 0.06	16.87 ± 0.27	47.35 ± 0.08
	Crop	5%	7.27 ± 0.09	17.10 ± 0.15	47.36 ± 0.01
T. aestivum	Crop	Control	7.32 ± 0.07	17.10 ± 0.26	46.40 ± 0.04
	Crop	1%	7.30 ± 0.12	17.23 ± 0.22	46.38 ± 0.07
	Crop	2%	7.23 ± 0.07	16.93 ± 0.28	46.47 ± 0.05
	Crop	3%	7.33 ± 0.07	16.93 ± 0.24	46.28 ± 0.07
	Crop	4%	7.27 ± 0.03	16.97 ± 0.23	46.28 ± 0.02
	Crop	5%	7.33 ± 0.03	17.17 ± 0.20	46.39 ± 0.08
A. arvensis	Weed	Control	7.20 ± 0.07	17.12 ± 0.26	46.40 ± 0.04
	Weed	1%	7.27 ± 0.12	16.87 ± 0.22	46.29 ± 0.02
	Weed	2%	7.33 ± 0.03	16.87 ± 0.18	46.38 ± 0.11
	Weed	3%	7.37 ± 0.07	17.20 ± 0.21	46.33 ± 0.09
	Weed	4%	7.23 ± 0.07	17.03 ± 0.20	46.41 ± 0.08
	Weed	5%	7.30 ± 0.06	16.90 ± 0.25	46.42 ± 0.12
L. didymum	Weed	Control	7.22 ± 0.07	17.08 ± 0.26	46.39 ± 0.04
	Weed	1%	7.20 ± 0.06	17.23 ± 0.15	47.32 ± 0.02
	Weed	2%	7.20 ± 0.06	17.10 ± 0.20	47.33 ± 0.06
	Weed	3%	7.33 ± 0.03	17.07 ± 0.20	47.29 ± 0.07
	Weed	4%	7.30 ± 0.12	17.17 ± 0.09	47.33 ± 0.07
	Weed	5%	7.27 ± 0.09	16.97 ± 0.27	47.23 ± 0.04
P. annua	Weed	Control	7.21 ± 0.07	17.13 ± 0.26	46.35 ± 0.04
	Weed	1%	7.37 ± 0.03	17.07 ± 0.22	47.41 ± 0.07
	Weed	2%	7.27 ± 0.12	17.00 ± 0.15	47.27 ± 0.06
	Weed	3%	7.20 ± 0.06	17.10 ± 0.21	47.36 ± 0.06
	Weed	4%	7.23 ± 0.07	17.17 ± 0.28	47.31 ± 0.05
	Weed	5%	7.30 ± 0.06	16.93 ± 0.19	47.36 ± 0.10
V. sativa	Weed	Control	7.24 ± 0.07	17.18 ± 0.26	46.31 ± 0.04
	Weed	1%	7.27 ± 0.09	17.07 ± 0.15	46.41 ± 0.01
	Weed	2%	7.33 ± 0.09	17.03 ± 0.20	46.36 ± 0.03
	Weed	3%	7.23 ± 0.07	17.13 ± 0.27	46.52 ± 0.01
	Weed	4%	7.30 ± 0.06	17.07 ± 0.17	46.41 ± 0.08
	Weed	5%	7.37 ± 0.07	17.20 ± 0.21	46.43 ± 0.09

Total organic carbon and organic matter were found to constitute 0.51–0.52% (Figure S5d) and 0.88–0.89% (Figure S5e) in all pots, respectively. In the *A. arvensis* pots, total nitrogen ranged from 324.05 to 342.87 kg/ha, 319.87–330.33 kg/ha in pots of *L. didymum*, 326.14–340.78 kg/ha in *P. annua* pots, and 328.23–340.78 kg/ha in *V. sativa* pots (Figure S5f). The available phosphorus ranged from 43.49 to 45.55 kg/ha in *A. arvensis pots*, 42.75 to 44.80 kg/ha in *L. didymum* pots, 44.24 to 46.85 kg/ha in *P. annua* pots, and 42.56 to 44.80 kg/ha in *V. sativa* pots (Figure S5g). The potassium content was found to be 171.81–172.63 kg/ha in *A. arvensis* pots, 171.81–172.63 kg/ha in *L. didymum* pots, 171.51–172.55 kg/ha in *P. annua* pots, and 172.03–172.44 kg/ha in *V. sativa* pots (Figure S5h).

The calcium and magnesium contents ranged from 5.79 to 6.53 kg/ha and 2.24–3.03 kg/ha in *A. arvensis* pots, 6.46–6.72 kg/ha and 2.32–2.81 kg/ha in *L. didymum* pots, 6.03–6.73 and 2.27–3.33 kg/ha in *P. annua* pots, and 6.24–6.62 kg/ha (Figure S5i) and 2.42–3.03 kg/ha (Figure S5j) in *V. sativa* pots, respectively (Table 7). The results of our analysis showed that there was no statistically significant difference between all treatments and the control. Additionally, all the parameters were found to be within the ranges prescribed by Krishi Vigyan Kendra.

Table 7. Impact of *M. koenigii* extract on the chemical parameters of soil.

Plant Name	Type of Plant	Treatment	TOC (%)	OM (%)	TN (kg/ha)	AP (kg/ha)	K (kg/ha)	Ca (kg/ha)	Mg (kg/ha)
C. arietinum	Crop	Control	0.52 ± 0.00	0.88 ± 0.00	332.24 ± 9.11	42.32 ± 0.37	172.26 ± 0.36	6.52 ± 0.00	2.96 ± 0.00
	Crop	1%	0.51 ± 0.00	0.89 ± 0.00	340.78 ± 4.18	45.17 ± 0.19	172.55 ± 0.38	6.05 ± 0.00	2.69 ± 0.00
	Crop	2%	0.51 ± 0.00	0.88 ± 0.00	328.23 ± 2.09	44.24 ± 0.86	172.14 ± 0.45	6.81 ± 0.00	2.02 ± 0.00
	Crop	3%	0.51 ± 0.00	0.89 ± 0.00	326.14 ± 9.58	46.11 ± 0.19	172.18 ± 0.21	6.27 ± 0.00	2.99 ± 0.00
	Crop	4%	0.51 ± 0.00	0.88 ± 0.00	326.14 ± 9.58	45.55 ± 1.78	172.33 ± 0.13	6.83 ± 0.00	2.09 ± 0.00
	Crop	5%	0.52 ± 0.00	0.89 ± 0.00	332.42 ± 7.24	47.04 ± 0.32	172.33 ± 0.21	6.03 ± 0.00	2.84 ± 0.00
T. aestivum	Crop	Control	0.52 ± 0.00	0.89 ± 0.00	334.60 ± 9.11	43.38 ± 0.37	171.52 ± 0.36	6.53 ± 0.00	2.28 ± 0.00
	Crop	1%	0.52 ± 0.00	0.89 ± 0.00	334.51 ± 4.18	45.36 ± 1.71	172.82 ± 0.13	6.33 ± 0.00	2.09 ± 0.00
	Crop	2%	0.52 ± 0.00	0.89 ± 0.00	324.05 ± 4.18	44.43 ± 1.04	171.88 ± 0.25	6.23 ± 0.00	2.84 ± 0.00
	Crop	3%	0.51 ± 0.00	0.89 ± 0.00	334.51 ± 5.53	46.85 ± 0.37	171.92 ± 0.17	6.71 ± 0.00	2.94 ± 0.00
	Crop	4%	0.51 ± 0.00	0.89 ± 0.00	332.42 ± 9.58	42.19 ± 0.19	172.29 ± 0.10	6.47 ± 0.00	2.06 ± 0.00
	Crop	5%	0.51 ± 0.00	0.89 ± 0.00	328.23 ± 9.11	45.73 ± 0.67	171.77 ± 0.26	6.07 ± 0.00	3.48 ± 0.00
A. arvensis	Weed	Control	0.52 ± 0.00	0.89 ± 0.00	336.60 ± 9.11	43.31 ± 0.37	172.22 ± 0.36	6.53 ± 0.00	2.96 ± 0.00
	Weed	1%	0.51 ± 0.00	0.89 ± 0.00	324.05 ± 4.18	43.49 ± 0.19	171.81 ± 0.17	6.41 ± 0.00	2.66 ± 0.00
	Weed	2%	0.52 ± 0.00	0.89 ± 0.00	326.14 ± 10.86	45.55 ± 0.67	172.63 ± 0.31	6.54 ± 0.00	2.24 ± 0.00
	Weed	3%	0.51 ± 0.00	0.89 ± 0.00	342.87 ± 2.09	44.80 ± 1.41	172.63 ± 0.16	5.79 ± 0.00	2.62 ± 0.00
	Weed	4%	0.52 ± 0.00	0.89 ± 0.00	328.23 ± 11.64	44.61 ± 0.67	172.03 ± 0.30	6.53 ± 0.00	2.99 ± 0.00
	Weed	5%	0.52 ± 0.00	0.89 ± 0.00	324.05 ± 8.36	45.36 ± 1.71	172.07 ± 0.10	6.28 ± 0.00	3.03 ± 0.00
L. didymum	Weed	Control	0.51 ± 0.00	0.88 ± 0.00	335.20 ± 9.11	42.11 ± 0.37	171.12 ± 0.36	6.64 ± 0.00	2.66 ± 0.00
	Weed	1%	0.51 ± 0.00	0.89 ± 0.00	319.87 ± 7.24	42.75 ± 0.19	172.33 ± 0.44	6.72 ± 0.00	2.32 ± 0.00
	Weed	2%	0.51 ± 0.00	0.89 ± 0.00	326.14 ± 3.62	43.68 ± 0.32	171.96 ± 0.26	6.59 ± 0.00	2.62 ± 0.00
	Weed	3%	0.51 ± 0.00	0.89 ± 0.00	330.33 ± 8.36	44.80 ± 0.86	171.81 ± 0.17	6.46 ± 0.00	2.59 ± 0.00
	Weed	4%	0.51 ± 0.00	0.88 ± 0.00	319.87 ± 7.24	44.43 ± 1.04	172.22 ± 0.42	6.69 ± 0.00	2.81 ± 0.00
	Weed	5%	0.51 ± 0.00	0.88 ± 0.00	326.14 ± 13.06	42.93 ± 0.49	172.63 ± 0.23	6.57 ± 0.00	2.51 ± 0.00
P. annua	Weed	Control	0.52 ± 0.00	0.89 ± 0.00	332.71 ± 9.11	44.11 ± 0.37	172.63 ± 0.36	6.43 ± 0.00	2.74 ± 0.00
	Weed	1%	0.52 ± 0.00	0.89 ± 0.00	330.33 ± 9.11	46.85 ± 0.19	171.92 ± 0.20	6.03 ± 0.00	2.27 ± 0.00
	Weed	2%	0.52 ± 0.00	0.89 ± 0.00	332.42 ± 3.62	44.24 ± 0.32	172.55 ± 0.43	6.44 ± 0.00	2.47 ± 0.00
	Weed	3%	0.51 ± 0.00	0.88 ± 0.00	326.14 ± 3.62	46.29 ± 1.04	171.92 ± 0.20	6.77 ± 0.00	2.54 ± 0.00
	Weed	4%	0.51 ± 0.00	0.89 ± 0.00	328.23 ± 5.53	44.61 ± 1.31	172.26 ± 0.19	6.52 ± 0.00	2.89 ± 0.00
	Weed	5%	0.51 ± 0.00	0.89 ± 0.00	340.78 ± 5.53	44.61 ± 0.67	171.51 ±0.10	6.68 ± 0.00	3.33 ± 0.00
V. sativa	Weed	Control	0.51 ± 0.00	0.88 ± 0.00	338.38 ± 9.11	43.62 ± 0.37	171.54 ± 0.36	6.62 ± 0.00	2.78 ± 0.00
	Weed	1%	0.51 ± 0.00	0.89 ± 0.00	330.33 ± 7.54	44.80 ± 0.97	172.03 ± 0.17	6.53 ± 0.00	2.96 ± 0.00
	Weed	2%	0.52 ± 0.00	0.89 ± 0.00	340.78 ± 5.53	45.55 ± 0.67	172.44 ± 0.37	6.24 ± 0.00	2.54 ± 0.00
	Weed	3%	0.51 ± 0.00	0.89 ± 0.00	328.23 ± 8.36	44.24 ± 0.97	172.33 ± 0.38	6.60 ± 0.00	2.42 ± 0.00
	Weed	4%	0.52 ± 0.00	0.89 ± 0.00	334.51 ± 11.06	42.56 ± 0.32	172.07 ± 0.27	6.35 ± 0.00	2.89 ± 0.00
	Weed	5%	0.51 ± 0.00	0.89 ± 0.00	330.33 ± 4.18	43.68 ± 1.17	172.07 ± 0.37	6.32 ± 0.00	3.03 ± 0.00

TOC: total organic carbon, OM: organic matter, TN: total nitrogen, AP: available phosphorous, K: potassium, Ca: calcium, Mg: magnesium.

The ideal pH range for plant growth is 5.5–6.5, because this allows for optimal availability of nutrients [59]. Additionally, most soil microbes also thrive in this pH range, because plants grow well in this range and produce more root exudates that can serve as a carbon source for the survival and proliferation of microbes [59]. Total organic carbon influences many soil characteristics, including color, nutrient holding capacity (cation and anion exchange capacity), nutrient turnover, and stability, influencing water relations, aeration, and workability [60]. Singh et al. [61] detected pHs in a range of 8.5–9.0, organic carbon contents in a range of 0.9–2.5 (%), nitrogen contents in a range of 0.11–0.22 (g/kg), phosphorus contents in a range of 0.42–13 (g/kg), potassium contents in a range of 0.19–3.9 (g/kg), and calcium contents in a range of 0.64–1.65 (g/kg) in the soil of a botanical garden and an agricultural field. Their findings were close to those of the present study.

4. Conclusions

Wheat is an important crop that is grown worldwide. Still, it suffers from heavy weed infestations, which can result in the loss of 20–40% of a crop's production [62]. Weed problems in wheat vary by region due to differences in agroclimatic conditions and management practices. Similarly, chickpea is one of the most important pulse crops worldwide, but it has low productivity, of which weeds are a major cause. Weeds reduce crop yield by 24–63% on average [63]. Chickpea is a poor weed competitor due to its slow initial growth and vegetative cover. The present study demonstrated that a 1% concentration of *M. koenigii* aqueous extract could kill *A. arvensis*, *L. didymum*, and *V. sativa* without significantly affecting the germination, growth or photosynthetic and biochemical properties of cash crops (*C. arietinum* and *T. aestivum*) or the physiochemical properties of soil. These findings show that curry leaf extract treatment has a remarkable species-specific sensitivity in terms of seed germination and plant development and that crops are more resistant than weeds. A good understanding of the intricacies of the curry leaf weed–crop interactions can facilitate the development of bioherbicides to suppress weeds.

Supplementary Materials: The following supporting information can be downloaded at: https://www.mdpi.com/article/10.3390/agriculture13091678/s1, Figure S1. *C. arietinum* and *T. aestivum* seeds treated with *M. koenigii* aqueous extract (MKAE); Figure S2: *T. aestivum* and *C. arietinum* plants treated *M. koenigii* aqueous extract (MKAE); Figure S3: *A. arvensis* and *L. didymum* plants treated with *M. koenigii* aqueous extract (MAKE); Figure S4: *P. annua* and *V. sativa* plants treated with *M. koenigii* aqueous extract (MKAE); Figure S5: Graph showing impact of *M. koenigii* aqueous extract on physiochemical parameters of soil (AA—*A. arvensis*, LD—*L. didymium*, PA—*P. annua* and VS—*V. sativa*).

Author Contributions: Conceptualization, D.S.V. and A.B.; Data curation, R.K.B., S.K.A., L.A.A.-S., P.K. and A.H.; Formal analysis, D.S.V.; Funding acquisition, S.K.A. and L.A.A.-S.; Investigation, D.S.V.; Methodology, D.S.V., A.B. and R.K.B.; Project administration, S.K.A. and L.A.A.-S.; Resources, A.B.; Software, L.A.A.-S., P.K. and A.H.; Supervision, A.B.; Validation, R.K.B., S.K.A., P.K. and A.H.; Visualization, R.K.B., P.K. and A.H.; Writing—original draft, D.S.V.; Writing—review & editing, A.B., R.K.B., S.K.A., L.A.A.-S., P.K. and A.H. All authors have read and agreed to the published version of the manuscript.

Funding: This research publication was funded by Princess Nourah bint Abdulrahman University Researchers Supporting Project number (PNURSP2023R365), Princess Nourah bint Abdulrahman University, Riyadh, Saudi Arabia. This research publication was funded by the Deanship of Scientific Research at King Khalid University through Large Groups Project under grant number (R.G.P.2/77/44).

Institutional Review Board Statement: Not applicable.

Informed Consent Statement: Not applicable.

Data Availability Statement: Not applicable.

Acknowledgments: The infrastructural support provided by the Head of the Department of Environmental Science, Graphic Era (Deemed to be University), Dehradun, India is gratefully acknowledged. Deepti Singh Vashishth wishes to thank Ashish Kumar Arya and Rinky Yadav for their support. The authors extend their appreciation to Princess Nourah bint Abdulrahman University Researchers Supporting Project number (PNURSP2023R365), Princess Nourah bint Abdulrahman University, Riyadh, Saudi Arabia. The authors extend their appreciation to the Deanship of Scientific Research at King Khalid University for funding this work through Large Groups Project under grant number (R.G.P.2/77/44). All individuals included in this section have consented to the acknowledgement.

Conflicts of Interest: The authors declare no conflict of interest.

References

1. Chauvel, B.; Guillemin, J.P.; Gasquez, J.; Gauvrit, C. History of chemical weeding from 1944 to 2011 in France: Changes and evolution of herbicide molecules. *Crop Prot.* **2012**, *42*, 320–326. [CrossRef]
2. Carballido, J.; Pérez, R.M.; Rodríguez, L.A.; Agüera, V.J. Field sprayer for inter-and intra-row weed control: Performance and labour savings. *Span. J. Agric. Res.* **2013**, *11*, 642–651. [CrossRef]
3. Bachheti, A.; Sharma, A.; Bachheti, R.K.; Husen, A.; Pandey, D.P. Plant Allelochemicals and Their Various Applications. In *Co-Evolution of Secondary Metabolites. Reference Series in Phytochemistry*; Mérillon, J.M., Ramawat, K., Eds.; Springer: Cham, Switzerland, 2020.
4. Bhat, M.A. Evaluation of Allelopathic Effect of Aqueous Leaf Extract of Dried Leaves of *Tecoma stans* L. on Seed Germination and Biochemical Changes in *Vigna radiata* L. (Green Gram). Doctoral Dissertation, Pondicherry University, Pondicherry, India, 2017.
5. Xie, X.; He, Z.; Chen, N.; Tang, Z.; Wang, Q.; Cai, Y. The roles of environmental factors in regulation of oxidative stress in plant. *BioMed Res. Int.* **2019**, *2019*, 9732325. [CrossRef]
6. Gulzar, A.; Siddiqui, M.B. Allelopathic effect of *Calotropis procera* (Ait.) R. Br. on growth and antioxidant activity of *Brassica oleracea* var. botrytis. *J. Saudi Soc. Agric. Sci.* **2017**, *16*, 375–382. [CrossRef]
7. Chaves, M.M.; Flexas, J.; Pinheiro, C. Photosynthesis under drought and salt stress: Regulation mechanisms from whole plant to cell. *Ann. Bot.* **2009**, *103*, 551–560. [CrossRef]
8. Tarafdar, M.; Bahadur, V.; Rana, S.; Singh, R.K. A Review: Abiotic Stress on Transpiration, Stomatal Diffusive Resistance and Photosynthetic Rate. *Pharma Innov. J.* **2022**, *11*, 1632–1635.
9. Barman, K.K.; Varshney, J.G. Impact of herbicides on soil environment. *Indian J. Weed Sci.* **2008**, *40*, 10–17.
10. Rose, M.T.; Cavagnaro, T.R.; Scanlan, C.A.; Rose, T.J.; Vancov, T.; Kimber, S.; Van Zwieten, L. Impact of herbicides on soil biology and function. *Adv. Agron.* **2016**, *136*, 133–220.
11. Saini, S.C.; Reddy, G.B.S. A review on curry leaves (*Murraya koenigii*): Versatile multi-potential medicinal plant. *Am. J. Phytomed. Clin. Ther.* **2015**, *3*, 363–368.
12. Zain, N.M.; Yew, O.H.; Sahid, I.; Seng, C. Potential of napier grass (*Pennisetum purpureum*) extracts as a natural herbicide. *Pak. J. Bot.* **2013**, *45*, 2095–2100.
13. Li, J.; Chen, L.; Chen, Q.; Miao, Y.; Peng, Z.; Huang, B.; Guo, L.; Liu, D.; Du, H. Allelopathic effect of *Artemisia argyi* on the germination and growth of various weeds. *Sci. Rep.* **2021**, *11*, 4303. [CrossRef] [PubMed]
14. Arnon, D.I. Copper enzymes in isolated chloroplasts. Polyphenoloxidase in *Beta vulgaris*. *Plant Physiol.* **1949**, *24*, 1–15. [CrossRef] [PubMed]
15. Husen, A. Growth characteristics, physiological and metabolic responses of teak (*Tectona grandis* Linn. f.) clones differing in rejuvenation capacity subjected to drought stress. *Silvae Genet.* **2010**, *59*, 124–136. [CrossRef]
16. Bates, L.S.; Waldren, R.A.; Teare, I.D. Rapid determination of free proline for water-stress studies. *Plant Soil* **1973**, *39*, 205–207. [CrossRef]
17. Hosseinzadeh, S.R.; Amiri, H.; Ismaili, A. Nutrition and biochemical responses of chickpea (*Cicer arietinum* L.) to vermicompost fertilizer and water deficit stress. *J. Plant Nutr.* **2018**, *40*, 2259–2268. [CrossRef]
18. Lowry, O.H.; Rosebrough, N.J.; Farr, A.L.; Randall, R.J. Protein measurement with the Folin phenol reagent. *J. Biol. Chem.* **1951**, *193*, 265–275. [CrossRef]
19. Prochazkova, D.; Sairam, R.K.; Srivastava, G.C.; Singh, D.V. Oxidative stress and antioxidant activity as the basis of senescence in maize leaves. *Plant Sci.* **2001**, *161*, 765–771. [CrossRef]
20. Dhindsa, R.S.; Plumb-Dhindsa, P.; Thorpe, T.A. Leaf senescence: Correlated with increased levels of membrane permeability and lipid peroxidation, and decreased levels of superoxide dismutase and catalase. *J. Exp. Bot.* **1981**, *32*, 93–101. [CrossRef]
21. Husen, A. Clonal propagation of *Dalbergia sissoo* Roxb. and associated metabolic changes during adventitious root primordium development. *New For.* **2008**, *36*, 13–27. [CrossRef]
22. Chandlee, J.M.; Scandalios, J.G. Analysis of variants affecting the catalase developmental program in maize scutellum. *Theor. Appl. Genet.* **1984**, *69*, 71–77. [CrossRef]
23. Shivarkar, R.; Bhise, S.; Gupta, R.M. HPTLC fingerprinting and antimicrobial activity of *Azadirachta indica* leaf extracts. *Int. J. Biomed. Adv. Res.* **2018**, *9*, 235–240.
24. Rodrıguez-Delgado, M.A.; Malovana, S.; Perez, J.P.; Borges, T.; Montelongo, F.G. Separation of phenolic compounds by high-performance liquid chromatography with absorbance and fluorimsetric detection. *J. Chromatogr. A* **2001**, *912*, 249–257. [CrossRef] [PubMed]
25. Kashyap, T.; Athokpam, H.S.; Devi, N.S.; Venu, N. Forms and status of phosphorus in acid soils of Kakching district, Manipur (India). *Pharma Innov. J.* **2021**, *10*, 407–410.
26. Agbeshie, A.A.; Adjei, R.; Anokye, J.; Banunle, A. Municipal waste dumpsite: Impact on soil properties and heavy metal concentrations, Sunyani, Ghana. *Sci. Afr.* **2020**, *8*, e00390.
27. Estefan, G. *Methods of Soil, Plant, and Water Analysis: A Manual for the West Asia and North Africa Region*; International Center for Agricultural Research in the Dry Areas (ICARDA): Beirut, Lebanon, 2013; Volume 3, pp. 65–119.
28. Goyal, K.; Singh, N.; Jindal, S.; Kaur, R.; Goyal, A.; Awasthi, R. *Kjeldahl Method in book of Advanced Techniques of Analytical Chemistry*; Bentham Science: Sharjah, United Arab Emirates, 2022; Volume 1, p. 105.

29. McDowell, R.W.; Condron, L.M. Estimating phosphorus loss from New Zealand grassland soils. New Zealand. *J. Agric. Res.* **2004**, *47*, 137–145.
30. Doruk, K.; Venkatakrishnan, D. Evaluation of nutritional quality attributes and their inter-relationship in maize. *J. Pharmacogn. Phytochem.* **2022**, *11*, 296–297.
31. Lalrammawia, K.; Buragohain, A.; Kakki, B.; Zote, L.; Marak, N.K.; Lalmuanpuii, R.; Kumar, N.S.; Jahau, L.; Sudarshan, M.; Chaligava, O.; et al. Determination of multi elements in tobacco plant of northeast India by neutron activation analysis and atomic absorption spectrometry. *Biol. Trace Elem. Res.* **2021**, *24*, 1–16. [CrossRef]
32. Ou-Zine, M.; Symanczik, S.; Rachidi, F.; Fagroud, M.; Aziz, L.; Abidar, A.; Mäder, P.; Achbani, E.H.; Haggoud, A.; Abdellaoui, M.; et al. Effect of organic amendment on soil fertility, mineral nutrition, and yield of majhoul date palm cultivar in Drâa-Tafilalet region, Morocco. *J. Soil Sci. Plant Nutr.* **2021**, *21*, 1745–1758. [CrossRef]
33. Lawan, S.A.; Suleiman, K.; Iortsuun, D. Effects of allelochemicals of some eucalyptus species on germination and radicle growth of *Arachis hypogea*. *Bayero J. Pure Appl. Sci.* **2011**, *4*, 59–62. [CrossRef]
34. Vashishth, D.S.; Bachheti, A.; Bachheti, R.K.; Husen, A. Allelopathic effect of *Callistemon viminalis's* leaves extract on weeds, soil features, and growth performance of wheat and chickpea plants. *J. Plant Interact.* **2023**, *18*, 2248172. [CrossRef]
35. Khattak, A.; Ullah, F.; Wazir, S.M.; Shinwari, Z.K. Allelopathic potential of *Jatropha curcas* L. leaf aqueous extracts on seedling growth of wheat. *Pak. J. Bot.* **2015**, *47*, 2449–2454.
36. Einhellig, F.A. Mechanisms and modes of action of allelochemicals. In *The Science of Allelopathy*; Putnam, A.R., Tang, C., Eds.; John Wiley and Sons, Inc.: Hoboken, NJ, USA, 1986.
37. Bora, I.P.; Singh, J.; Borthakur, R. Allelopathic effect of leaf extracts of *Acacia auriculiformis* on seed germination of some Agricultural crops. *Ann. For.* **1999**, *7*, 143–146.
38. Eyini, M.; Joyakumar, M.; Pannirselvam, S. Allelopathic effect of bamboo leaf extracts on the seedling of groundnut. *Trop. Ecol.* **1989**, *30*, 138–141.
39. Rafiqul Hoque, A.T.M.; Ahmed, R.; Uddin, M.B.; Hossain, M.K. Allelopathic effect of different concentrations of water extracts of *Acacia auriculiformis* leaf on some initial growth parameters of five common agricultural crops. *Pak. J. Agric.* **2003**, *2*, 92–100. [CrossRef]
40. Kruse, M.; Strandberg, M.; Strandberg, B. *Ecological Effects of Allelopathic Plants: A Review*; NERI Technical Report 315; National Environmental Research Institute: Silkeborg, Denmark, 2000; p. 66.
41. Hachani, C.; Abassi, M.; Lazhar, C.; Lamhamedi, M.S.; Béjaoui, Z. Allelopathic effects of leachates of *Casuarina glauca* Sieb. Ex Spreng. and *Populus nigra* L. on germination and seedling growth of *Triticum durum* Desf. Under laboratory conditions. *Agrofor. Syst.* **2019**, *93*, 1973–1983. [CrossRef]
42. Baker, N.R. Chlorophyll fluorescence: A probe of photosynthesis in vivo. *Annu. Rev. Plant Biol.* **2008**, *59*, 89–113. [CrossRef]
43. Rahbarian, R.; Khavari-Nejad, R.; Ganjeali, A.; Bagheri, A.; Najafi, F. Drought stress effects on photosynthesis, chlorophyll fluorescence and water relations in tolerant and susceptible chickpea (*Cicer arietinum* L.) genotypes. *Acta Biol. Cracoviensia Ser. Bot.* **2011**, *53*, 47–56. [CrossRef]
44. Bian, S.; Jiang, Y. Reactive oxygen species, antioxidant enzyme activities and gene expression patterns in leaves and roots of Kentucky bluegrass in response to drought stress and recovery. *Sci. Hortic.* **2009**, *120*, 264–270. [CrossRef]
45. Eraslan, F.; Inal, A.; Savasturk, O.; Gunes, A. Changes in antioxidative system and membrane damage of lettuce in response to salinity and boron toxicity. *Sci. Hortic.* **2007**, *114*, 5–10. [CrossRef]
46. Armand, N.; Amiri, H.; Ismaili, A. Interaction of methanol spray and water-deficit stress on photosynthesis and biochemical characteristics of *Phaseolus vulgaris* L. cv. Sadry. *Photochem. Photobiol.* **2016**, *92*, 102–110. [CrossRef]
47. Matzenbacher, F.D.O.; Kalsing, A.; Dalazen, G.; Markus, C.; Merotto, A., Jr. Antagonism is the predominant effect of herbicide mixtures used for imidazolinone-resistant barnyardgrass (Echinochloa crus-galli) control. *Planta Daninha* **2015**, *33*, 587–597. [CrossRef]
48. Zhang, J.; Hamill, A.S.; Weaver, S.E. Antagonism and synergism between herbicides: Trends from previous studies. *Weed Technol.* **1995**, *9*, 86–90. [CrossRef]
49. Jugulam, M.; Shyam, C. Non-target-site resistance to herbicides: Recent developments. *Plants* **2019**, *8*, 417. [CrossRef] [PubMed]
50. Neve, P.; Powles, S. High survival frequencies at low herbicide use rates in populations of Lolium rigidum result in rapid evolution of herbicide. *Herdity* **2005**, *95*, 485–492. [CrossRef]
51. Sachdev, S.; Ansari, S.A.; Ansari, M.I.; Fujita, M.; Hasanuzzaman, M. Abiotic stress and reactive oxygen species: Generation, signaling, and defense mechanisms. *Antioxidants* **2021**, *10*, 277. [CrossRef] [PubMed]
52. Maurya, A.K. Oxidative Stress in Crop Plants. In *Agronomic Crops: Stress Responses and Tolerance*; Hasanuzzaman, M., Ed.; Springer: Singapore, 2020; pp. 349–380.
53. Bansal, G.L. Allelopathic effect of Lantana camara on rice and associated weeds under the mid-hill conditions of Himachal Pradesh, India. In Proceedings of the Workshop on Allelopathy in Rice; Olofsdotter, M., Ed.; International Rice Research Institute: Manila, Philippines, 1998; pp. 133–138.
54. Naeem, M.; Cheema, Z.A.; Ihsan, M.Z.; Hussain, Y.; Mazari, A.; Abbas, H.T. Allelopathic effects of different plant water extract on yield and weeds of wheat. *Planta Daninha* **2018**, *36*, e018177840. [CrossRef]
55. Golisz, A.; Lata, B.; Gawronski, S.W.; Fujii, Y. Specific and total activities of allelochemicals identified in buckwheat. *Weed Biol. Manag.* **2007**, *7*, 164–171. [CrossRef]

56. Zhang, T.T.; Zheng, C.Y.; Hu, W.; Xu, W.W.; Wang, H.F. The allelopathy and allelopathic mechanism of phenolic acids on toxic *Microcystis aeruginosa*. *J. Appl. Phycol.* **2010**, *22*, 71–77. [CrossRef]
57. Cid-Ortega, S.; Monroy-Rivera, J.A. Extraction of kaempferol and its glycosides using supercritical fluids from plant sources: A review. *Food Technol. Biotechnol.* **2018**, *56*, 480–493. [CrossRef]
58. Li, J.; Ye, Y.; Huang, H.; Dong, L. Kaempferol-3-O-β-D-glucoside, a potential allelochemical isolated from *Solidago Canadensis*. *Allelopath. J.* **2011**, *28*, 259–266.
59. Msimbira, L.A.; Smith, D.L. The roles of plant growth promoting microbes in enhancing plant tolerance to acidity and alkalinity stresses. *Front. Sustain. Food Syst.* **2020**, *4*, 106. [CrossRef]
60. Ezeji, E.U.; Anyanwu, C.O.; Ukwandu, N.C.D. Evaluation of Heavy Metals and Physicochemical Parameters of Oil-contaminated Soil and Water Samples from Bonny, South-South, Nigeria. *Arch. Curr. Res. Int.* **2019**, *19*, 1–11. [CrossRef]
61. Singh, S.; Singh, J.; Vig, A.P. Earthworm as ecological engineers to change the physico-chemical properties of soil: Soil vs. vermicast. *Ecol. Eng.* **2016**, *90*, 1–5. [CrossRef]
62. Indian Council of Agricultural Research. *Weed management in Wheat*; Directorate of Weed Research: Jabalpur, India, 2009.
63. Nath, C.P.; Kumar, N.; Hazra, K.K.; Praharaj, C.S.; Singh, S.S.; Dubey, R.P.; Sharma, A.R. Topramezone: A selective post-emergence herbicide in chickpea for higher weed control efficiency and crop productivity. *Crop Prot.* **2021**, *150*, 105814. [CrossRef]

agriculture

MDPI

Article

Estimation of Air Pollutant Emissions by Tractor Utilization in Korea

Myoung Ho Kim [1,2,3] and Seong Min Kim [1,2,3,4,*]

1 Department of Bioindustrial Machinery Engineering, College of Agriculture and Life Sciences, Jeonbuk National University, Jeonju 54896, Republic of Korea; myoung59@jbnu.ac.kr
2 Department of Agricultural Machinery Engineering, Graduate School, Jeonbuk National University, Jeonju 54896, Republic of Korea
3 Institute for Agricultural Machinery & ICT Convergence, Jeonbuk National University, Jeonju 54896, Republic of Korea
4 Department of Agricultural Convergence Technology, Graduate School, Jeonbuk National University, Jeonju 54896, Republic of Korea
* Correspondence: smkim@jbnu.ac.kr; Tel.: +82-63-270-2617

Abstract: The utilization of tractors is essential because of both the shortage of agricultural labor forces and the rapid aging of farmers in Korean agriculture. Tractors can be classified into two types: a walking tractor called as a power tiller, and a riding tractor. In this study, diesel emission inventories of tractors were established and variations of air pollutants were analyzed with each other using 2011 and 2019 survey data in Korea. Agricultural tractors including walking and riding types, which were categorized into three levels via the rated output power, were the main sources of diesel emissions. The emission inventory including CO, NOx, SOx, TSP(PM_{10}), $PM_{2.5}$, VOCs, and NH_3 were established using a Tier 3 methodology. The total amount of emissions using agricultural tractors was decreased by about 13% from 2011 to 2019. The number of walking tractors were decreased by about 19% in 8 years; on the other hand, that of riding tractors were increased by about 12%. However, the emission reduction is about 48% for walking tractors, and the emission increment is about 5% for riding tractors. Thus, the total emission from agricultural tractors was decreased by about 16% in those periods. It is due to the decrease of 21% and 15% in the hours of use of walking and riding tractors, respectively, in 2019. Walking tractors mainly emit air pollutants from spraying and transporting. Riding tractors emit about 61% of the total air pollutants mainly from soil preparation and transporting operations. The geographic information system (GIS) was used to visualize the distribution of air pollutants in Korea. High-emission generating regions and the changes of emissions over 8 years were clearly seen in the GIS analysis. High air-pollutant emitting regions are mainly located in the western and southern regions of Korea, which have plenty of arable areas compared to other regions in Korea.

Keywords: agricultural tractors; diesel emission inventory; air pollutants; geographic information system

1. Introduction

In 2019, the total cultivated acreage in Korea was 1,643,465 ha, and the amounts of agricultural production were 4375×10^3, 8186×10^3, 2206×10^3, and 65×10^3 tons for grain crops, vegetables, fruits, and specialty crops, respectively [1,2]. Due to the decline of agricultural labor forces and the rapid aging of farmers, the utilization of a variety of agricultural machinery is expanding rapidly in Korea. Compared to a decade ago, Korean farmers are now using bigger and more powerful tractors [3]. The mechanization rate of rice farming is more than 99%, and that of the other major agricultural works in open field farming is about 62% on average in 2019 [4].

The tractor is a multi-purpose vehicle that performs major agricultural operations while driving with various implements such as a plow, a rotary, or a baler, and it is frequently used in Korean agriculture [5,6]. Tractors are used to perform many agricultural practices such as tillage, harrowing, fertilizer, compost spreading, transportation in almost every agricultural sector of grain, vegetable, and fruit production, and livestock husbandry. Tractors usually use diesel as a fuel and emit a lot of pollutant substances which are primary and secondary sources of particulate matter (PM) and other air pollutants. PM is considered as one of the most concerning air pollutants due to its effect on human health and the environment both in urban and rural areas [7–13]. The non-road mobile machinery sector including tractors largely contributes to the emissions of PM_{10} and $PM_{2.5}$, being responsible for 7.4% and 16.4% of the total emissions, respectively [14].

An emission inventory can show the temporal and spatial distributions and changes in pollutants in a certain area over a period of time. Currently, the global level non-road machinery activities and emission data are difficult to obtain, and the development of emission models and inventories is still being undertaken. Only the NONROAD model, developed by the U.S. Environmental Protection Agency (EPA), has been widely used [15].

Mobile sources can exhaust many kinds of pollutants including sulfur dioxide (SO_2), nitrogen oxides (NOx), total hydrocarbons (THC), carbon monoxide (CO), and particulate matter (PM) during the process of fuel combustion. These pollutants can cause direct or indirect adverse influences on air pollution [16–18], human health [19,20], and climate change [21–23]. For example, emitted PM could directly increase the atmospheric particulate matter with a diameter below 2.5 μm ($PM_{2.5}$) concentrations; as important fine particle precursors, SO_2 and NOx can transform into sulfate and nitrate, making a secondary contribution to $PM_{2.5}$ [24,25].

In this study, refined diesel emission inventories for agricultural tractors including two-wheeled driving (2WD) and four-wheeled driving (4WD) tractors were established and variations of air pollutants were analyzed with each other using 2011 and 2019 data in Korea by categorizing tractors in terms of their rated powers. Also, the emission characteristics of seven air pollutants from various field operations were analyzed. In addition, the spatial distribution of the amount of eight pollutants was visualized via the geographic information system (GIS) and was investigated on a country scale.

2. Materials and Methods

2.1. Calculation of Air Pollutant Amounts Emitted from Agricultural Tractor Operation

The emitted amounts of seven air pollutant substances due to tractor operation were calculated via the method of NIER [26]. The formula is shown in Equation (1):

$$E_{i,j,k} = \sum\{N_{i,k} \times HP_i \times LF \times HRS_i \times EF_{i,j}\} \quad (1)$$

where $E_{i,j,k}$ is the total amount of air pollutants emitted from a specific region (kg/yr); $N_{i,k}$ is the number of machinery in the specific region (unit); HP_i is the average rated power of the tractor (kw); LF is the load factor (=0.48); HRS_i is the average annual activity of the tractor (hr/yr); $EF_{i,j}$ is the emission factor (kg/(kWh-unit)); i is the farm tractor type (walking and riding) (i = 1, ⋯, 4); j is the type of air pollutant (j = 1, ⋯, 8); and k is the region (k = 1, ⋯,10).

The NIER handbook gives the emission factors for CO, NOx, TSP, $PM_{2.5}$, VOCs, and NH_3, but the following equation is used to calculate the emission factor for SOx [21].

$$EF_i = FF_i\ (g/kWh\text{-}unit)/1000 \times l \times (\text{Fuel sulfur weight }\%/100) \quad (2)$$

where EF_i is the emission factor (kg/(kWh-unit)); FF_i is the fuel factor (g/(kWh-unit)); l is the constant (=2.0) (grams of SOx formed from one gram of sulfur); and i is the farm tractor type (i = 1, ⋯, 4).

2.2. Average Rated Power and Number of Agricultural Tractors

From the agricultural machinery statistics by the Ministry of Agriculture, Food and Rural Affairs (MAFRA) of Korea, the number of agricultural tractors was categorized into two groups: a power tiller called as a walking (two-wheeled) tractor and a riding (four-wheeled) tractor. The riding tractors are further divided into three groups via their rated engine power—small, medium, and large. Table 1 shows the average rated power and relevant statistics of agricultural tractors used in Korea in 2011 and 2019 [3].

Table 1. Average rated power and number of agricultural tractors used in Korea.

Tractor Type	Engine Size (Rated Power Range)	ARP [4] (kW)	Unit (ea)	
			2011	2019
Walking Tractor	-	6.7	666,897	544,005
Riding Tractor	S [1] ($S \leq 29.4$ kW)	23.0 [5]	73,901	73,604
	M [2] ($29.4 \text{ kW} \leq M < 44.1$ kW)	39.0 [5]	142,856	150,283
	L [3] ($44.1 \text{ kW} \leq L$)	52.1 [5]	51,114	74,793
	Sub Total	-	267,871	298,680

[1] S: Small, [2] M: Medium, [3] L: Large, [4] ARP: Average rated power. [5] ARP is defined as a weighted average value and is calculated from the number of tractors and their rated engine power.

2.3. Average Annual Operating Hours of Agricultural Tractors

The average annual operating hours of agricultural tractors are given in Table 2 according to tractor types as well as some typical agricultural practices. The survey data by the Rural Development Administration (RDA) in Korea on the utilization of agricultural machinery were used to identify the average annual operating hours of agricultural tractors [27,28].

Table 2. Average annual operating hours of agricultural tractors.

Tractor Type	Operation Type	Average Annual Activity (hr/yr)	
		2011	2019
Walking Tractor	TL [1]	3.7	1.4
	HW [2]	6.1	3.8
	PP [3]	9.1	10.5
	SY [4]	25.2	19.9
	TP [5]	56.5	34.3
	Others	0.2	2.3
	Total	100.8	72.2
Riding Tractor	TL [1]	21.1	18.1
	HW [2]	31.5	44.6
	LL [6]	11.7	14.1
	FS [7]	6.9	8.2
	CS [8]	16.6	5.7
	LD [9]	13.2	23.4
	BL [10]	6.9	2.7
	TP [5]	29.1	14.0
	Others	13.9	9.1
	Total	153.7	139.9

[1] TL: Tilling, [2] HW: Harrowing, [3] PP: Pumping, [4] SY: Spraying, [5] TP: Transporting, [6] LL: Leveling, [7] FS: Fertilizer spreading, [8] CS: Compost spreading, [9] LD: Loading, [10] BL: Baling.

2.4. Emission and Fuel Factors of Agricultural Tractors

Table 3 shows the emission and fuel factors required to calculate the amount of seven air pollutants emitted from agricultural tractors [26].

Table 3. Emission factors of six air pollutants and fuel factors for SOx of agricultural tractors.

Tractor Type	Size	Emission Factor (kg/kWh-Unit)						Fuel Factor (g/kWh-Unit)
		CO	NOx	TSP [4]	$PM_{2.5}$	VOCs	NH_3	SOx
Walking Tractors		6.80	13.60	1.36	1.251	2.04	0.00004	0.00542
Riding Tractors	S [1]	2.48	7.84	0.39	0.359	0.48	0.00003	0.00538
	M [2]							0.00530
	L [3]							0.00530

[1] S: Small, [2] M: Medium, [3] L: Large. [4] TSP includes PM_{10}.

2.5. Geological Distribution of Air Pollutant Emissions from Agricultural Tractors

To calculate the domestic spatial distribution of the total air pollutant emissions from tractors, an open source geographic information system (GIS) software (QGIS, Version 3.10, qgis.org) was used. The total of nine provinces and one total metropolitan city (TMC), which combines and represents eight metropolitan cities, were analyzed in the study. Based on the geocoded residential addresses, individual exposure to the various variables was assessed when applying the GIS program. QGIS is a widely-used, open-source GIS visualization tool that allows users to produce, edit, visualize, and analyze spatial data. It supports the vector and raster format, as well as the database format and its functionalities.

3. Results and Discussion

3.1. Total Air Pollutant Emissions from Agricultural Tractors

The air pollutant emission inventory for agricultural tractors in 2011 and 2019 in Korea was calculated. Tables 4 and 5 show that the amount of air pollutants emitted from various agricultural practices from agricultural tractors. The two main sources of air pollutants emitted from walking tractors are transporting and spraying. They emit about 81% of emissions from walking tractors. Riding tractors are heavily used in soil preparation, which emits about 42% of air pollutants, including tilling, harrowing, and leveling, for seeding and planting in Korea. Figure 1 shows the changes of cultivated land and the calculated total emissions in agricultural tractors categorized into four groups. The cultivated land for agriculture was decreased about 7% from 1,698,000 ha to 1,581,000 ha, from 2011 to 2019. However, the total amount of emissions using agricultural tractors was decreased about 13% during those periods. The main reason is that the number of power tillers were decreased by about 19% in 8 years; on the other hand, that of farm tractors were increased by about 12%. However, the emission reduction is about 48% for power tillers, and the emission increment is about 5% for farm tractors. Thus, the total emissions from agricultural tractors was decreased by about 16% in those periods. It is due to the decrease of 21% and 15% in the hours of use of walking and riding tractors, respectively, in 2019. However, the total amount of air pollutants emitted from large-size riding tractors was increased by about 33% in 2019. Recently, Korean farmers are showing a tendency of using large-size riding tractors to reduce working hours in the fields.

3.2. Amounts of Air Pollutant Emissions from Various Agricultural Tractor Operations

Figures 2 and 3 show the calculated amounts of average air pollutant emissions from various agricultural operations of walking and riding tractors. Walking tractors emit air pollutants mainly from spraying and transporting operations. The transporting operation is responsible for 2639 Mg and 1429 Mg of air pollutant emissions in 2011 and 2019, respectively. Riding tractors emit air pollutants mainly from soil preparation and transporting operations. About 61% of the total air pollutants are emitted from the two operations. Particularly, the air pollutant emission amount from the loading operation was nearly doubled in 2019 compared to that in 2011. Korean farmers have made frequent use of riding tractors to load and unload heavy materials recently.

Table 4. Calculated amounts of air pollutants emitted from agricultural tractor operations in 2011. unit: Mg/yr.

Machinery		Operation Type	CO	NOx	SOx ($\times 10^2$)	TSP	$PM_{2.5}$	VOCs	NH_3 ($\times 10$)
Walking Tractor (Power Tiller)		TL [1]	53.9	107.8	2.88	10.8	9.9	16.2	3.17
		HW [2]	88.9	177.7	4.75	17.8	16.4	26.7	5.23
		PP [3]	132.6	265	7.08	26.5	24.4	39.8	7.80
		SY [4]	367	734	19.61	73.4	67.5	110.1	21.6
		TP [5]	823	1646	44.0	164.6	151.4	246.9	48.4
		Others	2.91	5.83	0.16	0.58	0.54	0.87	0.017
		Sub Total	1469	2940	78.4	296	270	441	86.2
Riding Tractor	Small	TL [1]	42.7	135.1	6.21	6.7	6.2	8.3	5.17
		HW [2]	63.9	202	9.28	10.0	9.2	12.4	7.73
		LL [6]	23.7	74.9	3.44	3.7	3.4	4.6	2.87
		FS [7]	14.0	44.2	2.03	2.2	2.0	2.7	1.69
		CS [8]	33.6	106.3	4.89	5.3	4.9	6.5	4.07
		LD [9]	26.7	84.5	3.89	4.2	3.9	5.2	3.23
		BL [10]	19.6	62.1	2.86	3.1	2.8	3.8	2.38
		TP [5]	58.9	186.3	8.57	9.3	8.5	11.4	7.13
		Others	28.2	89.0	4.09	4.43	4.07	5.45	3.41
		Sub Total	311	984	45.3	49.0	45.1	60.3	37.7
	Medium	TL [1]	140.0	442	20.0	22.0	20.2	27.1	16.9
		HW [2]	209	661	29.9	32.9	30.3	40.5	25.3
		LL [6]	77.6	245	11.1	12.2	11.2	15.0	9.38
		FS [7]	45.7	145.0	6.55	7.19	6.62	8.85	5.53
		CS [8]	110.0	348	15.8	17.3	15.9	21.3	13.30
		LD [9]	87.5	277	12.5	13.8	12.7	16.9	10.60
		BL [10]	64.3	203	9.21	10.1	9.31	12.4	7.78
		TP [5]	193.0	610	27.6	30.3	27.9	37.3	23.3
		Others	92.1	291	13.2	14.5	13.3	17.8	11.10
		Sub Total	1019	3220	145.9	160.3	147.5	197.3	123.3
	Large	TL [1]	66.9	211	9.57	10.5	9.68	12.9	8.09
		HW [2]	99.9	316	14.3	15.7	14.5	19.3	12.10
		LL [6]	37.1	117.0	5.31	5.83	5.37	7.18	4.48
		FS [7]	21.9	69.1	3.13	3.44	3.16	4.23	2.64
		CS [8]	52.6	166.0	7.53	8.27	7.61	10.2	6.36
		LD [9]	41.8	132.0	5.99	6.58	6.05	8.90	5.06
		BL [10]	30.7	97.2	4.40	4.83	4.45	5.95	3.72
		TP [5]	92.2	291	13.20	14.50	13.30	17.80	11.20
		Others	44.0	139.0	6.31	6.93	6.38	8.52	5.33
		Sub Total	487	1540	69.7	76.6	70.5	94.3	58.9
	Sub Total		1818	5746	261	286	267	352	220
Total			3290	8683	339	580	537	792	306

[1] TL: Tilling, [2] HW: Harrowing, [3] PP: Pumping, [4] SY: Spraying, [5] TP: Transporting, [6] LL: Leveling, [7] FS: Fertilizer spreading, [8] CS: Compost spreading, [9] LD: Loading, [10] BL: Baling.

3.3. Geological and Temporal Distribution of Air Pollutant Emissions from Agricultural Tractors

The amount of air pollutants emitted by Korean agricultural tractors was calculated on a regional level. Tables 6–9 show the calculated air pollutant emissions of 10 regions in Korea. The total emissions of CO, NOx, SOx, TSP, $PM_{2.5}$, VOCs, and NH_3 from walking tractors in 2011 were 1468 Mg, 2937 Mg, 0.77 Mg, 293 Mg, 270 Mg, 442 Mg, and 8.7 Mg, respectively. Also, the total emissions of CO, NOx, SOx, TSP, $PM_{2.5}$, VOCs, and NH_3 from walking tractors in 2019 were 857 Mg, 1617 Mg, 0.44 Mg, 189 Mg, 158 Mg, 257 Mg, and 5.0 Mg, respectively. In 2019, the walking tractor emissions of Gyeongsangbuk-do, Jeollanam-do, and Gyeongsangnam-do account for 23%, 16%, and 13% of the total emissions. The total emissions of CO, NOx, SOx, TSP, $PM_{2.5}$, VOCs, and NH_3 from riding tractors were 1816 Mg, 5744 Mg, 2.6 Mg, 287 Mg, 266 Mg, 352 Mg, and 22 Mg, respectively, in 2011 and were 1906 Mg, 6028 Mg, 2.76 Mg, 301 Mg, 276 Mg, 369 Mg, and 23.3 Mg, respectively, in 2019. Gyeongsangbuk-do and Jeollanam-do are major regions emitting about 30% and 32% of the total air pollutants utilizing walking and riding tractors in 2019.

Table 5. Calculated amounts of air pollutants emitted from agricultural tractor operations in 2019. unit: Mg/yr.

Machinery		Operation Type	CO	NOx	SOx ($\times 10^2$)	TSP	$PM_{2.5}$	VOCs	NH_3 ($\times 10$)
Walking Tractor (Power Tiller)		TL [1]	16.64	33.3	0.90	3.33	3.06	4.99	0.98
		HW [2]	40.4	80.8	2.20	8.08	7.43	12.12	2.38
		PP [3]	124.8	250	6.70	25.0	23.0	37.4	7.34
		SY [4]	237	473	12.6	47.3	43.5	71.0	13.9
		TP [5]	408	815	21.8	81.5	75.0	122.3	24.0
		Others	27.3	54.7	1.50	5.47	5.03	8.20	1.61
		Sub Total	853	1707	45.6	170.7	157.0	256	50.2
Riding Tractor	Small	TL [1]	36.5	115.4	5.30	5.74	5.29	7.07	4.42
		HW [2]	90.0	284	13.10	14.15	13.02	17.4	10.88
		LL [6]	28.4	89.9	4.10	4.47	4.12	5.51	3.44
		FS [7]	16.54	52.3	2.40	2.60	2.39	3.20	2.00
		CS [8]	11.50	36.3	1.700	1.808	1.664	2.23	1.390
		LD [9]	47.2	149.2	6.90	7.42	6.83	9.14	5.71
		BL [10]	5.45	17.22	0.800	0.856	0.788	1.054	0.660
		TP [5]	28.2	89.3	4.10	4.44	4.09	5.47	3.42
		Others	18.36	58.0	2.70	2.89	2.66	3.55	2.22
		Sub Total	282	892	41.0	44.4	40.8	54.6	34.1
	Medium	TL [1]	126.2	399	18.10	19.85	18.27	24.3	15.27
		HW [2]	311	983	44.50	48.91	45.0	60.20	37.62
		LL [6]	98.3	311	14.10	15.46	14.23	19.03	11.89
		FS [7]	57.2	180.8	8.20	8.9	8.28	11.07	6.92
		CS [8]	39.8	125.7	5.70	6.25	5.75	7.69	4.81
		LD [9]	163.2	516	23.4	25.7	23.6	31.6	19.74
		BL [10]	18.83	59.5	2.70	2.96	2.73	3.64	2.28
		TP [5]	97.6	309	14.00	15.35	14.13	18.90	11.81
		Others	63.5	201	9.10	9.98	9.19	12.28	7.68
		Sub Total	976	3080	139.7	153.4	141.2	188.8	118.0
	Large	TL [1]	83.9	265	12.00	13.18	12.15	16.24	10.15
		HW [2]	207	654	29.6	32.52	29.9	40.0	25.01
		LL [6]	65.4	207	9.40	10.28	9.46	12.65	7.91
		FS [7]	38.0	120.2	5.40	5.98	5.50	7.36	4.60
		CS [8]	26.4	83.5	3.80	4.16	3.83	5.12	3.20
		LD [9]	108.5	343	15.50	17.06	15.71	21.0	13.12
		BL [10]	12.52	39.6	1.800	1.969	1.812	2.42	1.510
		TP [5]	64.9	205	9.30	10.21	9.40	12.56	7.85
		Others	42.2	133.4	6.00	6.64	6.11	8.17	5.10
		Sub Total	649	2050	92.9	102.0	93.9	125.5	78.5
	Sub Total		1906	6030	274	300	276	369	231
Total			2760	7730	319	470	433	625	281

[1] TL: Tilling, [2] HW: Harrowing, [3] PP: Pumping, [4] SY: Spraying, [5] TP: Transporting, [6] LL: Leveling, [7] FS: Fertilizer spreading, [8] CS: Compost spreading, [9] LD: Loading, [10] BL: Baling.

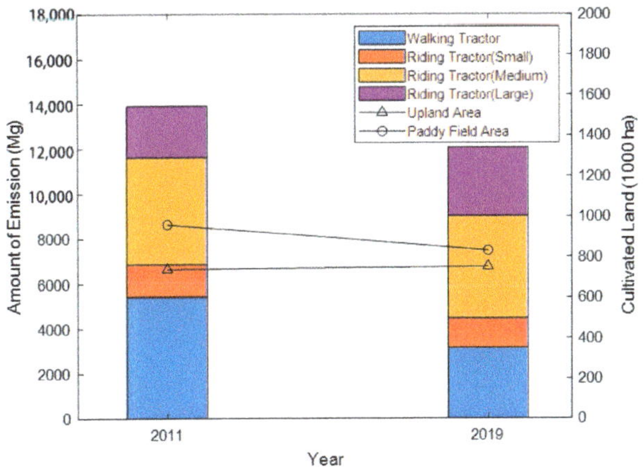

Figure 1. Changes in cultivated acreage and calculated amounts of total air pollutant emissions from walking and riding tractors.

Table 6. Calculated air pollutant emissions of walking tractor by region in Korea (2011).

Region	Emission (Mg/yr)							
	CO	NOx	SOx (×10^2)	TSP	$PM_{2.5}$	VOCs	NH_3 (×10)	Total
CHB [1]	100	199	5	20	18	30	6	368
CHN [2]	192	384	10	38	35	58	11	708
GAW [3]	85	171	5	17	16	26	5	316
GYB [4]	306	613	16	61	56	92	18	1128
GYG [5]	129	258	7	26	24	39	8	477
GYN [6]	195	390	10	39	36	58	11	719
JEB [7]	133	265	7	27	24	40	8	490
JEJ [8]	32	65	2	6	6	10	2	119
JEN [9]	232	464	12	46	43	70	14	857
TMC [10]	64	128	3	13	12	19	4	236
Total	1468	2937	77	293	270	442	87	5419

[1] CHB: Chungcheongbuk-do, [2] CHN: Chungcheongnam-do, [3] GAW: Gangwon-do, [4] GYB: Gyeongsangbuk-do, [5] GYG: Gyeonggi-do, [6] GYN: Gyeongsangnam-do, [7] JEB: Jeollabuk-do, [8] JEJ: Jeju-do, [9] JEN: Jeollanam-do, [10] TMC: Total of eight metropolitan cities.

Table 7. Calculated air pollutant emissions of walking tractor by region in Korea (2019).

Region	Emission (Mg/yr)							
	CO	NOx	SOx (×10^2)	TSP	$PM_{2.5}$	VOCs	NH_3 (×10)	Total
CHB [1]	58	116	3	12	11	18	3	215
CHN [2]	100	109	5	20	18	30	6	278
GAW [3]	50	100	3	10	9	15	3	184
GYB [4]	195	389	10	39	36	57	11	717
GYG [5]	64	127	3	13	12	19	4	235
GYN [6]	113	225	6	23	21	34	7	417
JEB [7]	76	151	4	15	14	23	4	279
JEJ [8]	23	46	1	5	4	7	1	85.1
JEN [9]	135	270	7	27	25	41	8	499
TMC [10]	43	84	2	25	8	13	3	173
Total	857	1617	44	189	158	257	50	3083

[1] CHB: Chungcheongbuk-do, [2] CHN: Chungcheongnam-do, [3] GAW: Gangwon-do, [4] GYB: Gyeongsangbuk-do, [5] GYG: Gyeonggi-do, [6] GYN: Gyeongsangnam-do, [7] JEB: Jeollabuk-do, [8] JEJ: Jeju-do, [9] JEN: Jeollanam-do, [10] TMC: Total of eight metropolitan cities.

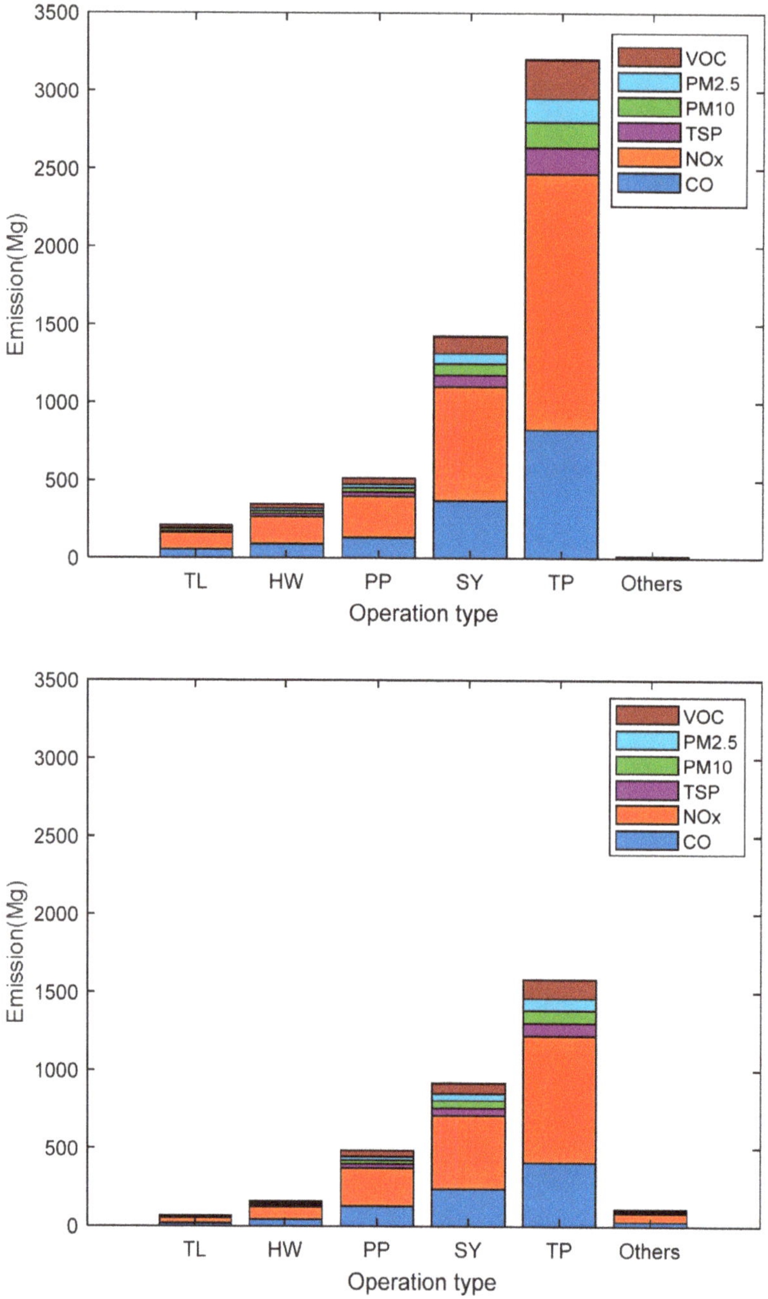

Figure 2. Calculated amounts of average air pollutant substances emitted from various agricultural operations by walking tractors in 2011 (**top**) and 2019 (**bottom**). TL: Tilling, HW: Harrowing, PP: Pumping, SY: Spraying, TP: Transporting.

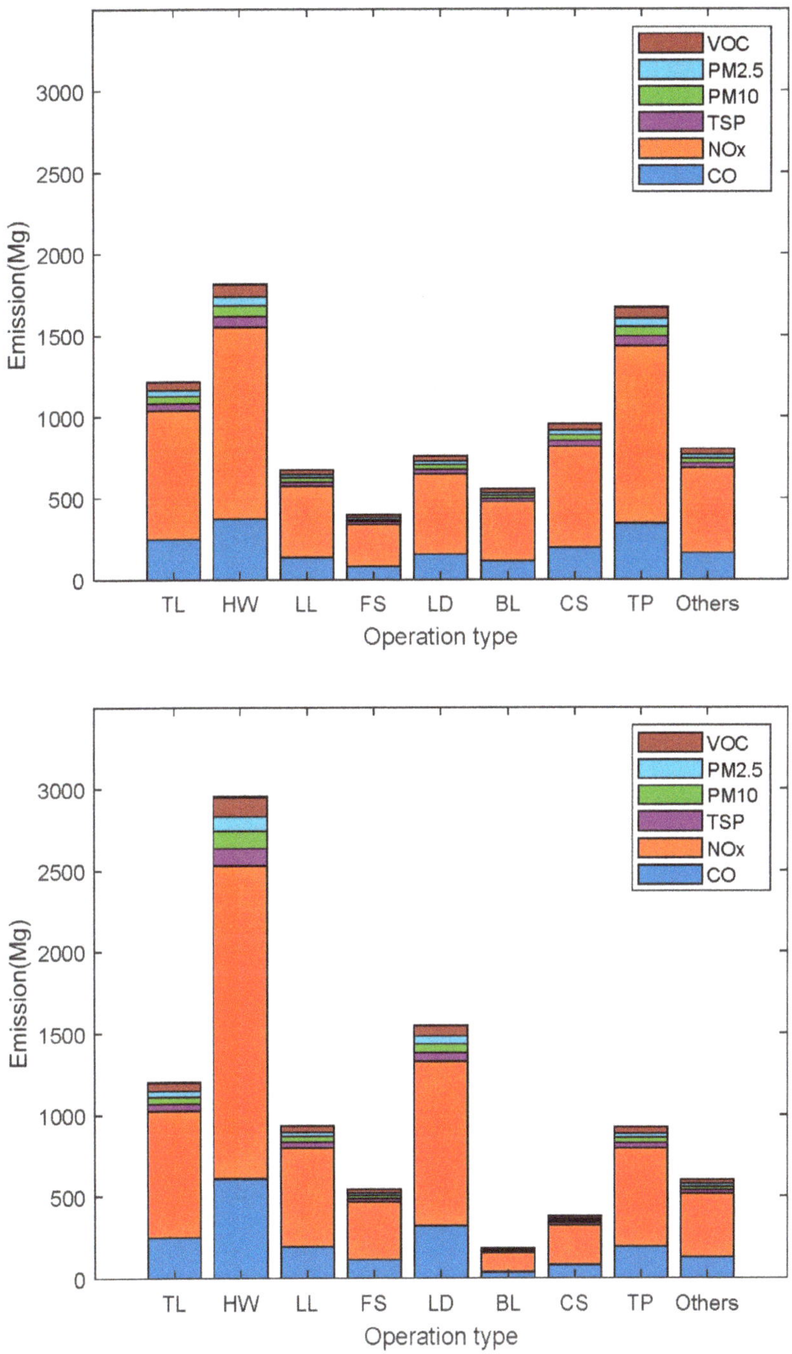

Figure 3. Calculated amounts of average air pollutant substances emitted from various agricultural operations by riding tractors in 2011 (**top**) and 2019 (**bottom**). TL: Tilling, HW: Harrowing, LL: Leveling, FS: Fertilizer spreading, LD: Loading, BL: Baling, CS: Compost spreading, TP: Transporting.

Table 8. Calculated air pollutant emissions of riding tractor by region in Korea (2011).

Region	Emission (Mg/yr)							
	CO	NOx	SOx ($\times10^2$)	TSP	$PM_{2.5}$	VOCs	NH_3 ($\times$10)	Total
CHB [1]	108	342	16	17	16	21	13	505
CHN [2]	253	801	36	40	38	49	31	1184
GAW [3]	127	402	18	20	18	25	15	594
GYB [4]	270	854	39	43	39	52	33	1262
GYG [5]	259	819	37	41	38	50	31	1210
GYN [6]	207	655	30	33	31	40	25	969
JEB [7]	210	664	30	33	30	41	25	981
JEJ [8]	30	94	4	5	4	6	4	139
JEN [9]	273	864	39	43	41	53	33	1278
TMC [10]	79	249	11	12	11	15	10	357
Total	1816	5744	260	287	266	352	220	8489

[1] CHB: Chungcheongbuk-do, [2] CHN: Chungcheongnam-do, [3] GAW: Gangwon-do, [4] GYB: Gyeongsangbuk-do, [5] GYG: Gyeonggi-do, [6] GYN: Gyeongsangnam-do, [7] JEB: Jeollabuk-do, [8] JEJ: Jeju-do, [9] JEN: Jeollanam-do, [10] TMC: Total of eight metropolitan cities.

Table 9. Calculated air pollutant emissions of riding tractor by region in Korea (2019).

Region	Emission (Mg/yr)							
	CO	NOx	SOx ($\times10^2$)	TSP	$PM_{2.5}$	VOCs	NH_3 ($\times$10)	Total
CHB [1]	113	358	16	18	16	22	14	529
CHN [2]	256	810	37	40	37	50	31	1196
GAW [3]	137	433	20	22	20	27	17	641
GYB [4]	312	986	45	49	45	60	38	1456
GYG [5]	230	727	33	36	33	45	28	1074
GYN [6]	219	694	32	35	32	42	27	1025
JEB [7]	228	720	33	36	33	44	28	1064
JEJ [8]	32	103	5	5	5	6	4	151.5
JEN [9]	291	919	42	46	42	56	35	1358
TMC [10]	88	278	13	14	13	17	11	411
Total	1906	6028	276	301	276	369	233	8906

[1] CHB: Chungcheongbuk-do, [2] CHN: Chungcheongnam-do, [3] GAW: Gangwon-do, [4] GYB: Gyeongsangbuk-do, [5] GYG: Gyeonggi-do, [6] GYN: Gyeongsangnam-do, [7] JEB: Jeollabuk-do, [8] JEJ: Jeju-do, [9] JEN: Jeollanam-do, [10] TMC: Total of eight metropolitan cities.

The spatial distribution of the total amount of tractor air pollutant emissions in Korea were visualized at the regional level using a GIS technique, as shown in Figures 4 and 5. Figure 4a shows the emissions in 2011 and Figure 4b shows the emissions in 2019 from walking tractors. Emissions from walking tractors are decreased nationwide except in Chungcheongbuk-do during those periods. Regions with high emissions are mainly southern regions such as Gyeongsangbuk-do, Jeollanam-do, and Gyeongsangnam-do. In the case of Gyeongsangbuk-do compared to other regions, it emits about 22.7% of the total air pollutants in 2019. Figure 5a,b shows the emissions from riding tractors in 2011 and 2019. Gyeongsangbuk-do and Jeollanam-do emit about 31.6% of the total air pollutants in 2019. Figures 4 and 5 show that the emission of the walking tractor is decreased, but the emission of the riding tractor is increased in the Gyeongsangbuk-do region. In 2011, riding tractor emissions were in about the same range in Chungcheongnam-do, Gyeongsangbuk-do, Gyeonggi-do, and Jeollanam-do, but in 2019, the emission range in Gyeongsangbuk-do, Jeollanam-do, Chungcheongbuk-do, and Jeollabuk-do increased by one level. These high-emission areas are mainly located in the western and southern regions of Korea, which have plenty of arable areas compared to other regions in Korea. However, riding tractor emissions are much lower in Gangwon-do and Chungcheongbuk-do. This area is mainly mountainous, so there is little agricultural area to produce agricultural products.

Metropolitan cities with a high population generate less emissions from agricultural tractors compared to other regions.

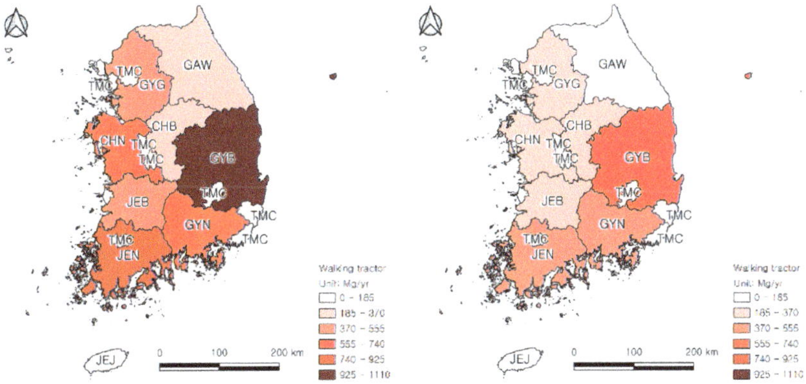

(**a**) (**b**)

Figure 4. Spatial distribution of total emissions from walking tractors in 2011 (**a**) and 2019 (**b**). CHB: Chungcheongbuk–do, CHN: Chungcheongnam–do, GAW: Gangwon–do, GYB: Gyeongsangbuk–do, GYG: Gyeonggi–do, GYN: Gyeongsangnam–do, JEB: Jeollabuk–do, JEJ: Jeju–do, JEN: Jeollanam–do, TMC: Total of eight metropolitan cities.

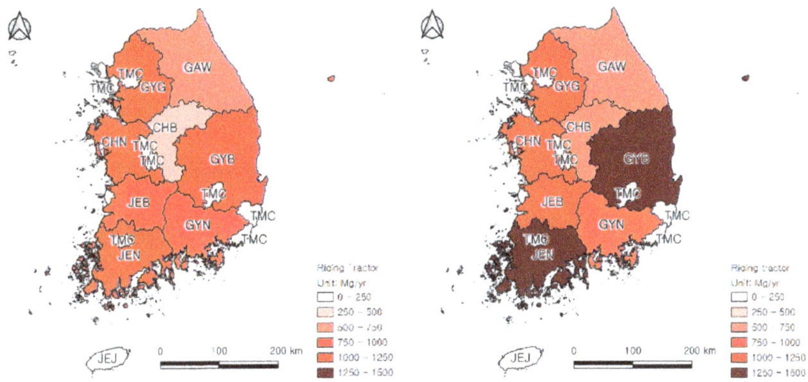

(**a**) (**b**)

Figure 5. Spatial distribution of total emissions from riding tractors in 2011 (**a**) and 2019 (**b**). CHB: Chungcheongbuk–do, CHN: Chungcheongnam–do, GAW: Gangwon–do, GYB: Gyeongsangbuk–do, GYG: Gyeonggi–do, GYN: Gyeongsangnam–do, JEB: Jeollabuk–do, JEJ: Jeju–do, JEN: Jeollanam–do, TMC: Total of eight metropolitan cities.

4. Conclusions

Agricultural tractors in Korea were categorized into four levels via the rated engine power and their air pollutants' emission characteristics were studied using 2011 and 2019 survey data. Calculations showed that, from 2011 to 2019, the emission amounts of all seven air pollutant substances decreased. NOx was the dominant pollutant substance in emission quantity, and then CO, VOCs, TSP, and $PM_{2.5}$ followed. In 2019, the calculated emission amounts of NOx, CO, VOCs, TSP, and $PM_{2.5}$ from all agricultural tractors in Korea were 7730, 2760, 625, 470, and 433 Mg, respectively.

Changes in tractor holding and the annual working hours from 2011 to 2019 resulted in the reduction in the air pollutants' emission amount by about 48% for power tillers and an increment of about 5% for riding tractors. Overall, the total air pollutant emissions from agricultural tractor use decreased by about 13% in Korea from 2011 to 2019.

Walking tractors emit air pollutants mainly from spraying and transporting operations. They are responsible for 81% of the total air pollutant emissions. Riding tractors, on the other hand, are heavily used in soil preparation work (tilling, harrowing, and leveling) for seeding and planting, and it accounts for about 42% of the total air pollutant emissions from riding tractors.

Emissions from walking tractors in 2019 decreased in all regions except one when comparing with those in 2011, but emissions from riding tractors increased or at least was steady nationwide. From 2011 to 2019, the total amount of emissions from tractors in Gyeongsangbuk-do, Jeollanam-do, and Jeollabuk-do went up by one level. These provinces are located in the western and southern part of Korea, where there are plenty of arable areas compared to other regions in Korea.

Author Contributions: Conceptualization, S.M.K.; software, S.M.K.; validation, M.H.K. and S.M.K.; investigation, M.H.K. and S.M.K.; resources, M.H.K. and S.M.K.; writing—original draft preparation, S.M.K.; writing—review and editing, M.H.K. and S.M.K.; visualization, S.M.K.; supervision, S.M.K.; project administration, S.M.K.; funding acquisition, S.M.K. All authors have read and agreed to the published version of the manuscript.

Funding: This work was carried out with the support of "Study on Particulate Matter Outbreak Source Characteristics during Agricultural Practice and Inventory Integration (Project No. PJ01428301)" Rural Development Administration, Korea, and was also supported by the research funds of Jeonbuk National University in 2022.

Institutional Review Board Statement: Not applicable.

Data Availability Statement: The datasets generated and/or analyzed during this study are available from the corresponding author upon reasonable request.

Conflicts of Interest: The authors declare no conflict of interest. The funders had no role in the design of the study; in the collection, analyses, or interpretation of the data; in the writing of the manuscript; or in the decision to publish the results.

References

1. MAFRA. *2019 Agricultural Area Survey*; Ministry of Agriculture, Food and Rural Affairs: Sejong, Republic of Korea, 2020.
2. MAFRA. *2019 Crop Production Survey*; Ministry of Agriculture, Food and Rural Affairs: Sejong, Republic of Korea, 2020.
3. MAFRA. *2019 Agricultural Machinery Holdings Survey*; Ministry of Agriculture, Food and Rural Affairs: Sejong, Republic of Korea, 2020.
4. Han, G.-G.; Jeon, J.-H.; Cho, Y.-J.; Kim, M.-H.; Kim, S.-M. Analysis of Air Pollutant Emissions for Mechanized Rice Cultivation in Korea. *Agriculture* **2021**, *11*, 1208. [CrossRef]
5. Kim, J.G.; Park, J.S.; Choi, K.J.; Lee, D.K.; Shin, M.S.; Oh, J.Y.; Nam, J.S. Analysis of Agricultural Tractor Transmission using Actual Farm Workload. *J. Korean Soc. Manuf. Process Eng.* **2020**, *19*, 42–48. [CrossRef]
6. Lee, N.G.; Kim, Y.J.; Beak, S.M.; Moon, S.P.; Park, S.U.; Choi, Y.S.; Cho, C.H. Analysis of Traction of Agricultural Tractor According to Soil Condition. *J. Drive Control* **2020**, *17*, 133–140.
7. Lewis, P.; Frey, H.C.; Rasdorf, W. Development and use of emissions inventories for construction vehicles. *Transp. Res. Rec.* **2009**, *2123*, 46–53. [CrossRef]
8. Millstein, D.E.; Harley, R.A. Revised estimates of construction activity and emissions: Effects on ozone and elemental carbon concentrations in southern California. *Atmos. Environ.* **2009**, *43*, 6328–6635. [CrossRef]
9. Lelicveld, J.; Evans, J.; Fnais, M.; Giannadaki, D.; Pozzer, A. The contribution of outdoor air pollution sources to premature mortality on a global scale. *Nature* **2015**, *525*, 367–371. [CrossRef]
10. Douglas, P.; Robertson, S.; Gay, R.; Hansell, A.L.; Gant, T.W. A systematic review of the public health risks of bioaerosols from intensive farming. *Int. J. Hyg. Environ. Health* **2018**, *221*, 134–173. [CrossRef]
11. Giannadaki, D.; Giannakis, E.; Pozzer, A.; Lelieveld, J. Estimating health and economic benefits of reductions in air pollution from agriculture. *Sci. Total Environ.* **2018**, *622–623*, 1304–1316. [CrossRef]
12. Giannakis, E.; Kushta, J.; Giannadaki, D.; Georgiou, G.K.; Bruggeman, A.; Lelieveld, J. Exploring the economy-wide effects of agriculture on air quality and health: Evidence from Europe. *Sci. Total Environ.* **2019**, *663*, 889–900. [CrossRef]
13. Yang, W.H. Changes in Air Pollutant Concentrations Due to Climate Change and the Health Effect of Exposure to Particulate Matter. *Korean Stud. Inf. Serv. Syst.* **2019**, *269*, 20–31.
14. NIER. *2017 National Air Pollutants Emission*; National Institute of Environmental Research: Sejong, Republic of Korea, 2020; p. 19.

15. Hou, X.; Tian, J.; Song, C.; Wang, J.; Zhao, J.; Zhang, X. Emission inventory research of typical agricultural machinery in Beijing, China. *Atmos. Environ.* **2019**, *216*, 116903. [CrossRef]
16. Kim, H.H.; Byun, G.R.; Choi, Y.S.; Kim, S.R.; Kim, S.Y.; Lee, J.T. Effects of long-term exposure to air pollution on all-cause mortality and cause-specific mortality in seven major cities of South Korea: Korean national health and nutritional examination surveys with mortality follow-up. *Environ. Res.* **2021**, *192*, 110290. [CrossRef]
17. Yu, G.E.; Park, S.S. Chemical characterization and source apportionment of $PM_{2.5}$ at an urban site in Gwangju, Korea. *Atmos. Pollut. Res.* **2021**, *12*, 101092. [CrossRef]
18. Choi, S.W.; Kim, T.K.; Lee, H.K.; Kim, H.C.; Hna, J.H.; Lee, K.B.; Lim, E.H.; Shin, S.H.; Jin, H.A.; Cho, E.G.; et al. Analysis of the National Air Pollutant Emission Inventory (CAPSS 2016) and the Major Cause of Change in Republic of Korea. *Asian J. Atmos. Environ.* **2020**, *14*, 422–445. [CrossRef]
19. Bae, S.H.; Kwon, H.J. Current State of Research on the Risk of Morbidity and Mortality Associated with Air Pollution in Korea. *Yonsei Med. J.* **2019**, *60*, 243–256. [CrossRef]
20. Tran, H.; Kim, J.y.; Kim, D.U.; Choi, M.Y.; Choi, M.H. Impact of air pollution on cause-specific mortality in Korea: Results from Bayesian Model Averaging and Principle Component Regression approaches. *Sci. Total. Environ.* **2018**, *636*, 1020–1031. [CrossRef]
21. Anenberg, S.C.; Balakrishnan, K.; Jetter, J.; Masera, O.; Mehta, S.; Moss, J.; Ramanathan, V. Cleaner Cooking Solutions to Achieve Health, Climate, and Economic Cobenefits. *Environ. Sci. Technol.* **2013**, *47*, 3944–3952. [CrossRef]
22. Kapadia, Z.Z.; Spracklen, D.V.; Arnold, S.R.; Borman, D.J.; Mann, G.W.; Pringle, K.J.; Monks, S.A.; Reddington, C.L.; Benduhn, F.; Rap, A.; et al. Impacts of aviation fuel sulfur content on climate and human health. *Atmos. Chem. Phys.* **2016**, *16*, 10521–10541. [CrossRef]
23. Kodros, J.K.; Scott, C.E.; Farina, S.C.; Lee, Y.H.; L'Orange, C.; Volckens, J.; Pierce, J.R. Uncertainties in global aerosols and climate effects due to biofuel emissions. *Atmos. Chem. Phys.* **2015**, *15*, 8577–8596. [CrossRef]
24. Lang, J.L.; Zhou, Y.; Chen, D.S.; Xing, X.F.; Wei, L.; Wang, X.T.; Zhao, N.; Zhang, Y.Y.; Guo, X.R.; Han, L.H.; et al. Investigating the contribution of shipping emissions to atmospheric $PM_{2.5}$ using a combined source apportionment approach. *Environ. Pollut.* **2017**, *229*, 557–566. [CrossRef]
25. Wu, Y.Y.; Gu, B.J.; Erisman, J.W.; Reis, S.; Fang, Y.Y.; Lu, X.H.; Zhang, X.M. $PM_{2.5}$ pollution is substantially affected by ammonia emissions in China. *Environ. Pollut.* **2016**, *218*, 86–94. [CrossRef] [PubMed]
26. NIER. *A Handbook of Method for National Air Pollutant Emission Estimation*; National Institute of Environmental Research: Sejong, Republic of Korea, 2013.
27. RDA. *2011 Survey on the Utilization of Agricultural Machinery and Farmwork Mechanization Rate*; Rural Development Administration: Jeonju, Republic of Korea, 2012.
28. RDA. *2019 Survey on the Utilization of Agricultural Machinery and Farmwork Mechanization Rate*; Rural Development Administration: Jeonju, Republic of Korea, 2020.

agriculture

MDPI

Article

The Content, Uptake and Bioaccumulation Factor of Copper and Nickel in Grass Depending on Zinc Application and Organic Fertilization

Beata Kuziemska *, Andrzej Wysokinski and Paulina Klej

Faculty of Agrobioengineering and Animal Husbandry, Institute of Agriculture and Horticulture, Siedlce University of Natural Sciences and Humanities, 08110 Siedlce, Poland; andrzej.wysokinski@uph.edu.pl (A.W.); t.paula@wp.pl (P.K.)

* Correspondence: beata.kuziemska@uph.edu.pl

Abstract: Different content of Zn in the soil and organic fertilization can affect micronutrient uptake by plants. A pot experiment was carried out to determine the impact of increasing Zn application rates, i.e., 200, 400 and 600 Zn mg·kg^{-1}, in combination with bovine and chicken manures and mushroom substrate on Cu and Ni content, uptake and bioaccumulation factor in cocksfoot (*Dactylis glomerata* L.). Control objects without Zn and organic fertilizers and after application of only different Zn doses and only organic fertilizers were also tested. Application of Zn at 400 and 600 mg·kg^{-1} significantly decreased the content, uptake and value of bioaccumulation factor of Cu in the grass. Different Zn doses were not found to influence the content and bioaccumulation factor of Ni in cocksfoot, but application at 200 Zn mg·kg^{-1} increased Ni uptake. Chicken manure increased the content and bioaccumulation factor of Cu and Ni in the test plant, and all the organic fertilizers increased their uptake. Cocksfoot showed no tendency to excessive bioaccumulation of Cu and Ni.

Keywords: bovine manure; chicken manure; cocksfoot; heavy metals; mushroom substrate

1. Introduction

To obtain high yields of good quality in agricultural production, it is necessary not only to provide an adequate supply of macroelements to crop plants, but to meet their demand for microelements as well. Essential microelements in plant nutrition include Zn, Cu, Ni and others (e.g., B, Fe, Mn, Mo) [1–3]. Their source for plants can be mineral and organic fertilizers or waste substances used in fertilization and industrial human activity [4]. However, taken up by plants in excessive amounts, these metals can have toxic effects, disturbing metabolic processes and decreasing yields [5–8].

Zn is a catalytic and structural cofactor of many enzymes, influencing the activity of enzymes such as hydrogenase and carbonic anhydrase [9,10]. It also performs crucial structural functions in the domains of proteins which interact with other molecules [11,12]. In addition, Zn increases crop resistance to drought [13]. Zn deficiency inhibits the activity of enzymes involved in photosynthesis [14] and also reduces accumulation of amino acids and protein synthesis. An excessive amount of this metal inhibits photosynthesis and chlorophyll biosynthesis, resulting in reduced whole mass of plants [15,16].

Cu is a catalyst in respiration and photosynthesis and plays an important role in hormonal signalling and cell wall metabolism [17,18]. Cu ions are cofactors in many enzymes, e.g., cytochrome c oxidase, Cu/Zn superoxide dismutase (SOD), amine oxidase, polyphenol oxidase, plastocyanin and laccase [19]. Cu deficiency disturbs metabolism, reduces stress resistance, damages enzyme systems and, in effect, reduces yields [20]. At high concentrations in tissues, Cu inhibits plant growth, disturbs photosynthetic electron transport and causes chlorosis, necrosis and dwarfism [21,22].

Ni is an element that activates urease, which catalyses hydrolysis of urea in plants [23,24]. It is also a component of other metalloenzymes, such as NiFe hydrogenase, superoxide dismutase, carbon monoxide dehydrogenase, methyl-coenzyme M reductase, hydrogenase and RNase A, acetyl-coenzyme-A synthase. Ni deficiency reduces urease activity, disturbs N assimilation and limits scavenging of superoxide free radicals. Excessive amounts of this metal adversely affect nutrient absorption by the roots, disturb plant metabolism, and inhibit photosynthesis and transpiration [25,26].

The uptake of heavy metals into plants' roots occurs mainly through two pathways: the apoplastic pathway (passive diffusion) and the symplastic pathway (active transport against electrochemical potential gradients and concentrations across the plasma membrane [27]. Uptake of Zn, Ni and Cu by plants depends on numerous soil factors, including pH, content of organic matter, redox conditions and the antagonistic or synergistic interactions between these elements [28,29].

Studies carried out by many researchers have shown that nearly half of the arable soils in the world have a low content of available Zn, resulting in a deficiency of this element in crop plants [2,28,30–33]. Application of this microelement in the form of mineral and organic fertilizer is the most practical means of preventing the effects of its deficiencies [34]. However, a high content of Zn in the soil may reduce its biological activity and cause the accumulation of this metal in plants [8]. At the same time, it can reduce the uptake of other elements by plants, in relation to which it is an antagonist. In such conditions, the organic fertilizers used may be a source of, e.g., Cu and Ni for plants.

The presented studies show that the content of zinc in Polish soils may range from 12 to 632 mg $Zn \cdot kg^{-1}$ of soil, which was taken into account when determining its doses in the experiment [35]. Therefore, a study was undertaken to determine the impact of varied levels of Zn applied together with organic materials, e.g., bovine and chicken manures or mushroom substrate, on the content, uptake and bioaccumulation factor of Cu and Ni in cocksfoot.

Grasses occupy a special place among fodder crops. When adequately fertilized, they have high yield potential for dry matter, protein and carbohydrates. Cocksfoot is one of the most important fodder crops in Europe and in the world [36,37]. It attains high yields and is disease-resistant [36], as well as highly resistant to drought [38]. Due to its high quality as fodder, it is used for production of hay and silage and for grazing [36]. The National List of varieties in Poland includes eight cocksfoot cultivars: Amera, Bepra, Berta, Crown Royale, Dika, Minora, Trerano and Tukan [39]. The Amera cultivar, the test plant in the present study, is the earliest variety of cocksfoot, with early onset of growth, a high rate of spring growth and rapid regrowth. It produces high yields of palatable and digestible biomass.

It was hypothesized that increasing doses of Zn reduce the content and bioaccumulation factor of Cu and Ni in cocksfoot, while organic fertilization causes their increase.

2. Materials and Methods

2.1. Experimental Design

A pot experiment was carried out in the years 2014–2016 in a greenhouse in Siedlce (52°10′ N, 22°17′ E, 155 m a.s.l.).

Pots with a capacity of 10 L were filled with 12 kg of soil. The soil was Luvisols consisting of 71% sand, 24% silt and 5% clay.

The experiment was planned in a randomized design in triplicate, with two factors:

I—Zn application amount

II—organic fertilizer.

The experimental design is shown in Table 1.

Zn and organic fertilizer were applied once, only in the first year of the study, two weeks before sowing of the grass. Zn was applied as an aqueous solution of $ZnSO_4 \cdot 5H_2O$. Organic fertilizers were applied in amounts that introduced 2 g $C_{org} \cdot kg^{-1}$ to the soil. In three years of the study, the test plant was the grass cocksfoot (*Dactylis glomerata* L.), the Amera cultivar, which was sown in the first decade of May of each year. The green parts of the grass were collected every 30 days, four times a year. During the growing period, the soil moisture in the pots was kept at 60–70% of the total water capacity.

Table 1. Design of the pot experiment.

Organic Fertilizer	Dose of Zn [mg·kg^{-1} of Soil]			
	0	200	400	600
Without organic fertilization (CO)	1	2	3	4
Bovine manure (BM)	5	6	7	8
Chicken manure (ChM)	9	10	11	12
Mushroom substrate (MS)	13	14	15	16

1—no organic fertilizer and without Zn applied; 2—200 mg Zn; 3—400 mg Zn; 4—600 mg Zn; 5—only BM; 6—BM + 200 mg Zn; 7—BM + 400 mg Zn; 8—BM + 600 mg Zn; 9—only ChM; 10—ChM + 200 mg Zn; 11—ChM + 400 mg Zn; 12—ChM + 600 mg Zn; 13—only MS; 14—MS + 200 mg Zn; 15—MS + 400 mg Zn; 16—MS + 600 mg Zn.

2.2. Laboratory Analyses

2.2.1. Soil

Before starting the experiment, the soil was analysed for pH by potentiometric titration in 1 mol·dm^{-3} of KCl solution [40]; organic C by a modification of the Tyurin method [41]; total N content by CHNS (automatic CHN analyser with an IDC detector, Series II 2400, Perkin-Elmer, Valencia, CA, USA); and content of total K, P, Zn, Cu and Ni by ACP-AES (Optima 3200 RL spectrometer, Perkin-Elmer, Waltham, MA, USA), following wet mineralization in a mixture of concentrated HCl and HNO_3 (3:1 ratio) [42]. Available K and P were determined by the Egner–Riehm method [43]. Selected properties of soil are shown in Table 2.

Table 2. Some soil properties before the start of the experiment.

pH_{KCl}	$C_{organic}$	N_{total}	P_{total}	K_{total}	$P_{available}$	$K_{available}$	Cu_{total}	Zn_{total}	Ni_{total}
		g·kg^{-1}					mg·kg^{-1}		
6.65	16.40	1.52	0.52	1.22	176	108	13.50	56.60	9.85

2.2.2. Organic Fertilizers

The organic materials used in the experiment (bovine and chicken manures and mushroom substrate) were analysed for dry matter (DW at 105 °C) organic C by the Tyurin method [41]; total N content by CHNS; and content of P, K, Zn, Cu and Ni by ICP-AES following dry mineralization of the samples at 500 °C [42]. The results of the chemical analyses are given in Table 3.

Table 3. Chemical composition of organic fertilizers.

Organic Fertilizers	Dry Matter (%)	$C_{organic}$	N	P	K	Cu	Zn	Ni
			g·kg^{-1} DM				mg·kg^{-1} DM	
BM	19.6	405.1	23.90	5.38	15.28	5.97	60.28	11.16
ChM	27.8	167.3	13.50	8.44	9.32	42.98	190.80	32.15
MS	30.4	319.3	24.20	6.22	17.48	15.61	117.50	11.20

BM—bovine manure, ChM—chicken manure, MS—mushroom substrate.

The contents of dry matter, total N and K were highest in the mushroom substrate, organic C content was highest in the bovine manure, and the contents of P, Zn, Cu and Ni were highest in chicken manure.

2.2.3. Plant Material

The aerial parts of cocksfoot were analysed for content of Cu and Ni by inductively coupled plasma atomic emission spectroscopy (ICP-AES). Prior to the analysis, the samples were subjected to dry mineralization at 450 °C and dissolved in 10% HCl solution [44].

2.3. Calculations

Uptake of Cu and Ni by cocksfoot was calculated according to the formulas given by Bokhari et al. [45]:

$$Cu_{UP} = Y \times Cu_{plant}/1000, \tag{1}$$

where:

Cu_{UP}—Cu uptake by grass (accumulation), Cu mg·pot^{-1}
Y—yield of grass, g·pot^{-1}
Cu_{plant}—Cu content in grass, Cu mg·kg^{-1}.

$$Ni_{UP} = Y \times Ni_{plant}/1000, \tag{2}$$

where:

Ni_{UP}—Ni uptake by grass (accumulation), Ni mg·pot^{-1}
Y—yield of grass, g·pot^{-1}
Ni_{plan}—Ni content in grass, Ni mg·kg^{-1}.

In addition, the bioaccumulation factor of Cu and Ni [46,47] was calculated according to the following formulas:

$$BF_{Cu} = Cu_{plant}/Cu_{soil}, \tag{3}$$

where:

BF_{Cu}—Bioaccumulation factor of Cu
Cu_{plant}—Cu content in grass, Cu mg·kg^{-1}
Cu_{soil}—total Cu content in the soil, Cu mg·kg^{-1}

$$BF_{Ni} = Ni_{plant}/Ni_{soil}, \tag{4}$$

where:

BF_{Ni}—Bioaccumulation factor of Ni
Ni_{plant}—Ni content in grass, Ni mg·kg^{-1}
Ni_{soil}—total Ni content in the soil, Ni mg·kg^{-1}.

2.4. Statistical Analysis

The content, uptake and bioaccumulation factor obtained in subsequent years of the experiment were analysed by analysis of variance using Statistica 13 PL software (ver. 13.1, StatSoft, Tulsa, OK, USA).

$$y_{ijk} = \mu + a_i + b_j + c_k + ab_{ij} + ac_{ik} + bc_{jk} + abc_{ijk} + e_{ijkp} \tag{5}$$

where μ—mean from all treatments; a_i—effect of Zn level; b_j—effect of organic fertilizer type; c_k—effect of year (third source of variation); ab_{ij}—interaction of Zn level and fertilizer type; ac_{ik}—interaction of Zn level and year; bc_{jk}—interaction of fertilizer type and year; abc_{ijd}—interaction of Zn level and fertilizer type and year; e_{ijkp}—sampling error.

The total Cu and Ni uptake over a period of 3 years was also calculated in this application according to the equation:

$$y_{ijk} = \mu + a_i + b_j + ab_{ij} + e_{ijp} \tag{6}$$

where μ—mean from all treatments; a_i—effect of Zn level; b_j—effect of organic fertilizer type; ab_{ij}—interaction of Zn level and fertilizer type; e_{ijp}—sampling error.

The significance of the experimental factors was determined on the basis of the Fisher–Snedecor distribution. Least significant difference (LSD) values were calculated by Tukey's test. The calculations were performed for a significance level of $p = 0.05$. In addition, Pearson's linear correlation coefficient was calculated.

3. Results

The content of Cu in the biomass of cocksfoot ranged from 2.35 to 5.24 mg·kg^{-1} DM (on average, 3.34 Cu mg·kg^{-1} DM) (Table 4). It was significantly reduced by application of Zn at 400 and 600 mg·kg^{-1}, by 10.6% and 12.0%, respectively, relative to the control. The content of Cu in the grass was not significantly affected by application of 200 Zn mg·kg^{-1}, bovine manure or mushroom substrate. The highest content of this metal was noted in the cocksfoot fertilized with chicken manure; it was 32.5% higher than in the plants from the control treatment and 24.2% and 28.2% higher, respectively, than following application of mushroom substrate and bovine manure. The Cu content in the biomass of the grass significantly decreased in successive years of the experiment. In the second and third years, it was 7.8% and 31.7% lower, respectively, than in the first year of the experiment.

Table 4. Cu content in grass biomass (Cu mg·kg^{-1} DM).

Organic Fertilizers	Years	Doses of Zn [mg·kg^{-1}]				Means
		0	200	400	600	
CO	1st	3.68	3.44	3.24	3.43	3.44
	2nd	3.42	3.29	3.15	3.01	3.22
	3rd	2.46	2.43	2.35	2.35	2.40
	means	3.18	3.05	2.91	2.93	3.02 a
BM	1st	4.12	3.98	3.75	3.36	3.80
	2nd	3.66	3.38	3.25	3.23	3.38
	3rd	2.54	2.50	2.40	2.46	2.48
	means	3.44	3.29	3.13	3.02	3.22 a
ChM	1st	5.24	4.96	4.14	4.08	4.61
	2nd	4.72	4.26	4.08	3.86	4.23
	3rd	3.38	3.12	3.00	3.15	3.16
	means	4.45	4.11	3.74	3.70	4.00 b
MS	1st	3.85	3.61	3.41	3.28	3.54
	2nd	3.57	3.40	3.26	3.18	3.35
	3rd	2.50	2.49	2.45	2.47	2.48
	means	3.31	3.17	3.04	2.98	3.12 a
Means for Zn dose		3.59 b	3.41 ab	3.21 a	3.16 a	3.34
Means for years	1st	4.22	4.00	3.64	3.54	3.85 c
	2nd	3.84	3.58	3.44	3.32	3.55 b
	3rd	2.72	2.64	2.55	2.61	2.63 a

CO—without organic fertilization, BM—bovine manure, ChM—chicken manure, MS—mushroom substrate. a,b,c—means investigated factors with different letters (in the columns for organic fertilization and for years but in the row for Zn doses) are significantly different.

All experimental factors also significantly influenced Cu uptake by cocksfoot, calculated as the mean from the three years of the experiment (Table 5) and the total for the 3-year cycle (Figures 1 and 2). In the first year of the study, Cu accumulation was highest following application of 200 Zn mg·kg^{-1} and lowest following application of 600 Zn mg·kg^{-1}. In the subsequent years of the experiment, increasing application of Zn did not significantly affect Cu uptake by the test plant. The organic fertilizers, irrespective of their origin, increased Cu uptake by cocksfoot. In all years of the study, its accumulation by the plants was significantly the highest following application of chicken manure. Cu uptake by the grass decreased in successive years. In the second and third years of the experiment, it was 57.0% and 31.4%, respectively, of the amount taken up in the first year.

Table 5. Cu uptake by grass (Cu mg·pot^{-1}).

Organic Fertilizers	Years	Doses of Zn [mg·kg^{-1}]				Means
		0	200	400	600	
CO	1st	0.056	0.063	0.042	0.040	0.050 C
	2nd	0.042	0.037	0.028	0.038	0.036 B
	3rd	0.022	0.020	0.018	0.019	0.020 A
	means	0.040	0.040	0.029	0.032	0.035 a
BM	1st	0.104	0.116	0.099	0.071	0.098 C
	2nd	0.053	0.051	0.042	0.041	0.047 B
	3rd	0.032	0.030	0.025	0.028	0.029 A
	means	0.063	0.066	0.056	0.047	0.058 c
ChM	1st	0.141	0.142	0.110	0.093	0.122 C
	2nd	0.077	0.072	0.061	0.057	0.067 B
	3rd	0.034	0.036	0.026	0.030	0.032 A
	means	0.084	0.084	0.066	0.060	0.073 d
MS	1st	0.083	0.091	0.075	0.058	0.076 C
	2nd	0.047	0.038	0.052	0.046	0.046 B
	3rd	0.027	0.028	0.024	0.025	0.026 A
	means	0.052	0.052	0.050	0.043	0.049 b
Means for Zn dose		0.060 b	0.060 b	0.050 a	0.0454 a	0.054
Means for years	1st	0.096 C	0.103 D	0.081 B	0.065 A	0.086 c
	2nd	0.055 A	0.050 A	0.046 A	0.045 A	0.049 b
	3rd	0.029 A	0.029 A	0.023 A	0.026 A	0.027 a

CO—without organic fertilization, BM—bovine manure, ChM—chicken manure, MS—mushroom substrate. a,b,c—means investigated factors with different letters (in the columns for organic fertilization and for years but in the row for Zn doses) are significantly different. A,B,C,D—means the interaction with different letters in the rows of the table are significantly different.

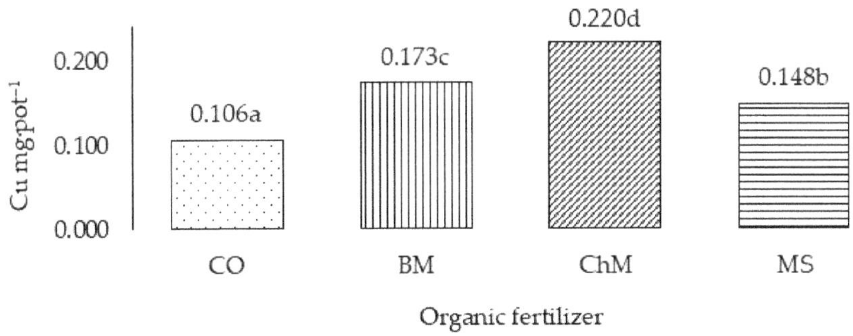

Figure 1. Total Cu uptake by grass in three years (Cu mg·pot^{-1})—variability caused by different organic fertilizers. CO—without organic fertilization, BM—bovine manure, ChM—chicken manure, MS—mushroom substrate. a,b,c,d—means investigated factors with different letters are significantly different.

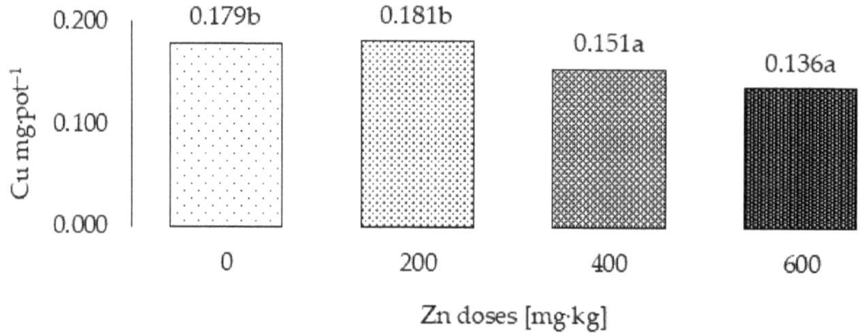

Figure 2. Total Cu uptake by grass in three years (Cu mg·pot^{-1})—variability caused by different zinc doses. a,b—means investigated factors with different letters are significantly different.

Ni content in tested grass ranged from 1.34 to 6.40 Ni mg·kg^{-1} DM (on average, 3.20 Ni mg·kg^{-1} DM) and did not depend on the amount of Zn applied to the soil (Table 6). It was significantly increased by application of chicken manure, but only in the first year of the study. Mushroom substrate and bovine manure were not shown to significantly influence this parameter. The content of Ni in grass significantly decreased in successive years of the experiment. In the second and third years of the experiment, it was 43.1% and 62.7% lower, respectively, than in the first year.

Table 6. Ni content in grass biomass (Ni mg·kg^{-1} DM).

Organic Fertilizers	Years	Doses of Zn [mg·kg^{-1}]				Means
		0	200	400	600	
CO	1st	4.67	4.38	4.31	4.51	4.47 A
	2nd	2.46	2.95	3.19	3.12	2.93 A
	3rd	1.54	1.42	1.44	1.34	1.44 A
	means	2.89	2.92	2.98	2.99	2.95 a
BM	1st	4.80	4.89	4.87	5.12	4.92 A
	2nd	2.89	3.08	2.64	2.79	2.85 A
	3rd	1,70	1.77	1.99	1.89	1.84 A
	means	3.13	3.25	3.17	3.27	3.20 a
ChM	1st	6.13	6.40	5.79	5.52	5.96 B
	2nd	2.91	2.98	3.21	3.26	3.09 A
	3rd	1.65	2.04	2.01	1.72	1.86 A
	means	3.56	3.81	3.67	3.50	3.63 b
MS	1st	4.78	4.89	5.01	4.97	4.91 A
	2nd	2.05	2.80	2.86	2.80	2.63 A
	3rd	1.59	1.50	1.52	1.47	1.52 A
	means	2.80	3.06	3.13	3.08	3.02 a
Means for Zn dose		3.10	3.26	3.24	3.21	3.20
Means for years	1st	5.09	5.14	4.99	5.03	5.06 c
	2nd	2.58	2.95	2.98	2.99	2.88 b
	3rd	1.62	1.68	1.74	1.61	1.66 a

CO—without organic fertilization, BM—bovine manure, ChM—chicken manure, MS—mushroom substrate. a,b,c—means investigated factors with different letters (in the columns for organic fertilization and for years but in the row for Zn doses) are significantly different. A,B—means the interaction with different letters in the rows of the table are significantly different.

Uptake of Ni by cocksfoot, calculated as the mean from the 3 years of the study (Table 7) and the total for the 3-year cycle (Figures 3 and 4), was significantly dependent on the level of Zn applied to the soil, organic fertilizer, and the year of the experiment. In the 3-year cycle, following application of Zn at 200 mg·kg^{-1} of soil, the plants accumulated 13.7% more Ni than the plants from the control treatment and 14.4% and 27.5% more, respectively, than following application of 400 and 600 Zn mg·kg^{-1}. No significant differences were noted in accumulation of Ni in the plants from the treatments without Zn application or with application at 400 and 600 Zn mg·kg^{-1}. Bovine and chicken manures and mushroom substrate increased uptake of nickel by the grass by 63.2%, 107.2% and 39.6%, respectively, relative to the control treatment. Significantly, the highest uptake of this metal was noted following application of chicken manure. It was 105.5% higher than in the control plants and 20.4% and 42.7% higher, respectively, than following application of bovine manure and mushroom substrate. During the three years of the experiment, Ni uptake by grass was highest in the first year and lowest in the third year. Uptake in the second and third year was 34.5% and 15.0%, respectively, of the level of uptake in the first year.

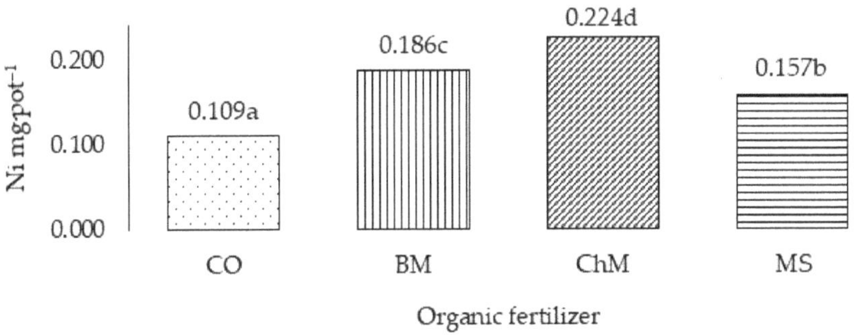

Figure 3. Total Ni uptake by grass in three years (Ni mg·pot^{-1})—variability caused by different organic fertilizers. CO—without organic fertilization, BM—bovine manure, ChM—chicken manure, MS—mushroom substrate. a,b,c,d—means investigated factors with different letters are significantly different.

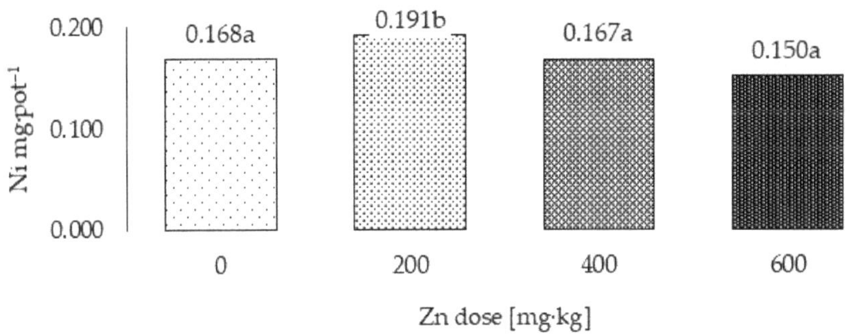

Figure 4. Total Ni uptake by grass in three years (Ni mg·pot^{-1})—variability caused by different zinc doses. a,b—means investigated factors with different letters are significantly different.

The application of 400 and 600 Zn mg·kg^{-1} reduced the factor of Cu bioaccumulation in cocksfoot (Figure 5). There was no significant effect of all doses of Zn on the bioaccumulation factor of Ni in this grass.

Table 7. Ni uptake by grass (Ni mg·pot^{-1}).

Organic Fertilizers	Years	Doses of Zn [mg·kg^{-1}] 0	200	400	600	Means
CO	1st	0.071	0.079	0.056	0.052	0.065 A
	2nd	0.031	0.033	0.028	0.039	0.033 A
	3rd	0.014	0.012	0.011	0.011	0.012 A
	means	0.038	0.041	0.032	0.034	0.036 a
BM	1st	0.122	0.144	0.129	0.108	0.126 C
	2nd	0.041	0.047	0.034	0.035	0.039 AB
	3rd	0.020	0.021	0.021	0.022	0.021 A
	means	0.061	0.071	0.061	0.055	0.062 c
ChM	1st	0.164	0.183	0.154	0.127	0.157 D
	2nd	0.048	0.050	0.048	0.048	0.048 B
	3rd	0.017	0.024	0.017	0.016	0.019 A
	means	0.076	0.085	0.073	0.064	0.075 d
MS	1st	0.102	0.123	0.109	0.087	0.105 B
	2nd	0.027	0.031	0.046	0.040	0.036 A
	3rd	0.017	0.017	0.015	0.015	0.016 A
	means	0.049	0.057	0.057	0.047	0.052 b
Means for Zn dose		0.056 a	0.064 b	0.055 a	0.050 a	0.056
Means for years	1st	0.115 B	0.132 C	0.112 B	0.094 A	0.113 c
	2nd	0.037 A	0.040 A	0.039 A	0.041 A	0.039 b
	3rd	0.017 A	0.018 A	0.016 A	0.016 A	0.017 a

CO—without organic fertilization, BM—bovine manure, ChM—chicken manure, MS—mushroom substrate. a,b,c,d—means investigated factors with different letters (in the columns for organic fertilization and for years but in the row for Zn doses) are significantly different. A,B,C,D—means the interaction with different letters in the rows of the table are significantly different.

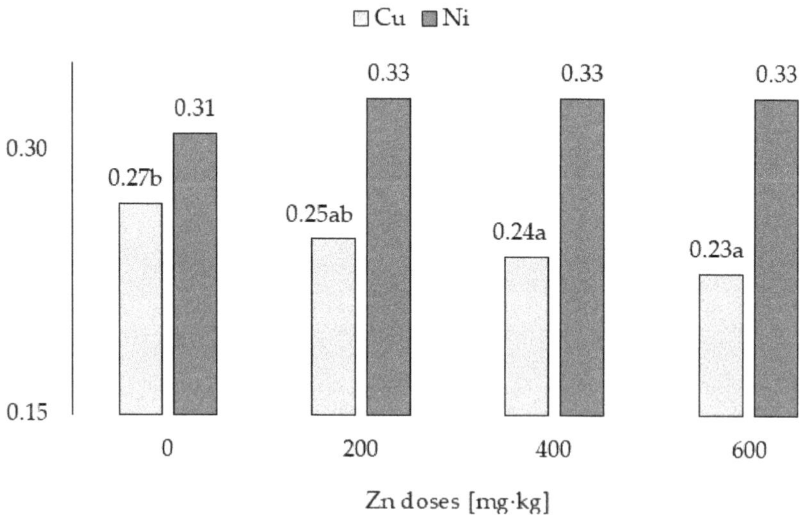

Figure 5. Bioaccumulation factor of Cu and Ni in grass—variability caused by different zinc doses. a,b for Cu—means investigated factors with different letters are significantly different.

Mushroom substrate and bovine manure did not significantly change the bioaccumulation factor of Cu and Ni, whereas chicken manure increased its value (Figure 6).

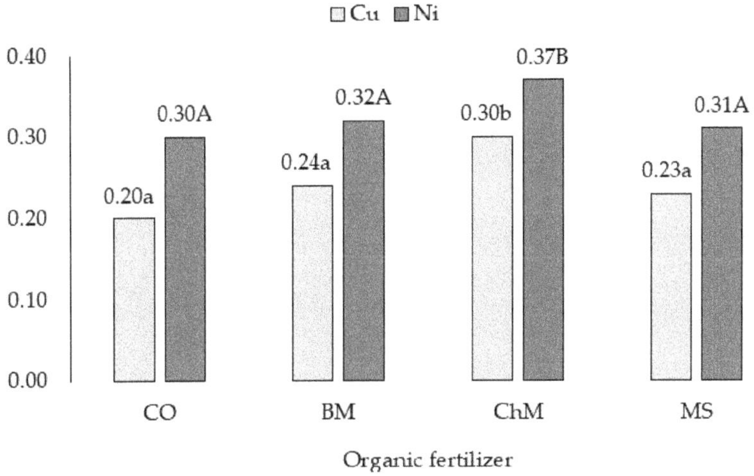

Figure 6. Bioaccumulation factor of Cu and Ni in grass—variability caused by different organic fertilizers. CO—without organic fertilization, BM—bovine manure, ChM—chicken manure, MS—mushroom substrate. a,b for Cu and A,B for Ni—means investigated factors with different letters are significantly different.

In the subsequent years of the study, the bioaccumulation factor of Cu and Ni decreased (Figure 7).

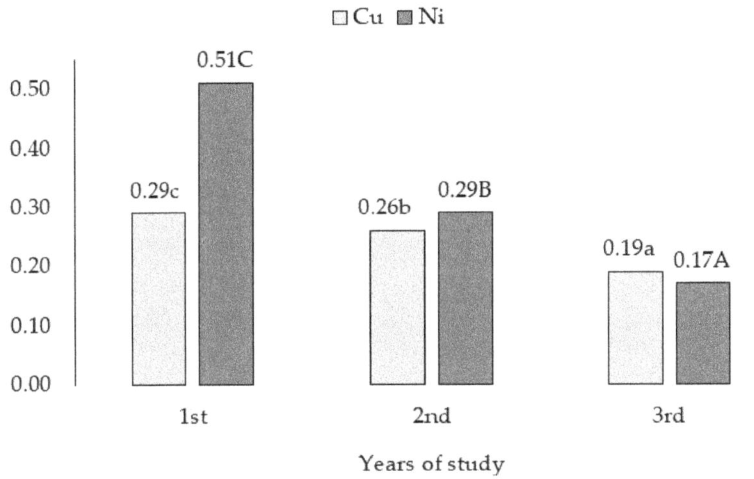

Figure 7. Bioaccumulation factor of Cu and Ni in grass—variability caused by years of study. a,b,c for Cu and A,B,C for Ni—means investigated factors with different letters are significantly different.

The correlation coefficients showed no clear effect of increasing Zn application (200, 400 and 600 Zn mg·kg^{-1}) on the content and uptake of Cu and Ni by grass (Table 8). They also showed no significant relationship between the content of these two metals and the content of Zn in the grass given by Kuziemska et al. [48]. At the same time, they indicated a significant relationship between the uptake and content of Cu and Ni and the yield of this plant reported in the study cited above.

Table 8. Linear correlation coefficients between some properties of cultivated grass.

Properties	Cockfoot Yield	Zn_Content	Zn_Uptake	Cu_Content	Cu_Uptake	Ni_Content	Ni_Uptake
Zn dose	ns	0.946 *	0.851 *	ns	ns	ns	ns
Cockfoot yield	-	ns	ns	0.645 *	0.908 *	0.662	0.944 *
Zn_content	ns	-	0.837 *	ns	ns	ns	ns
Zn_uptake	ns	0.837 *	-	ns	ns	ns	ns
Cu_content	0.645 *	ns	ns	-	0.901 *	0.792 *	0.798 *
Cu_uptake	0.908 *	ns	ns	0.901 *	-	0.801 *	0.963 *
Ni_content	0.662 *	ns	ns	0.792 *	0.801 *	-	0.862 *
Ni_uptake	0.944	ns	ns	0.798 *	0.963 *	0.862	-

*—is important, $p < 0.05$. ns—not significant.

4. Discussion

The biological value of crops is influenced not only by their content of macronutrients and microelements, but also by their proportions. Interactions between elements in crops occur when the supply of one element affects the absorption and use of other elements. Interaction of this type is most common when one element is present in the soil in excessive amounts. Antagonistic and synergistic relationships between plant nutrients cause disturbances in their chemical composition. According to Rietra et al. [49], knowledge of the interactions between nutrients can help to optimize fertilization strategies to achieve high yields of good biological value.

In the present study, the average Cu content in the cocksfoot biomass was 3.34 Cu mg·kg^{-1} DM. Łukowski et al. [50] reported content of this metal ranging from 2.2 to 21 Cu mg·kg^{-1} DM in various grass species cultivated in Poland. In a study by Gawryluk et al. [51], the average Cu content in grasses ranged from 8.6 to 16.2 Cu mg·kg^{-1} DM. Zn application at 400 and 600 mg·kg^{-1} reduced Cu content, uptake and bioaccumulation factor in cocksfoot. Decreases in Cu content is indicative of antagonistic relationships between these metals. Chaudhry et al. [52] investigated the effect of various levels of application of Zn and Cu on their uptake by rice (*Oryza sativa* L.) and rice yield. They showed that Zn application reduced the content of Cu in the biomass of the plant. Similar tendencies were obtained by Imtiaz et al. [53] for the relationship between Zn and Cu. In a pot experiment, they tested the effect of increasing Zn application (0, 5, 10, 15 and 20 Zn mg·ml^{-1}) on the content of Cu in wheat. The authors found that Zn, irrespective of the rate of application, decreased the content of Cu in the test plant. Zhao et al. [54] tested the effect of Zn, Mn and Cu on plant growth and accumulation of these elements in *Phytolacca americana*. They showed that excessive Zn application leads to chlorosis, disturbs elemental homeostasis and reduces the Cu concentration in the plant. Antagonistic relationships between Zn and Cu have also been reported by Abou Seeda et al. [3], Mousavi et al. [55] and Liščáková et al. [56].

Contrasting results for the interrelationships between Zn and Cu were obtained by Malinowska et al. [57], who tested the content of these metals in selected plants growing along a motorway. They found no significant correlations between their content in the plants. The results obtained by the cited authors and the results of our own research indicate that the relationship between the content of Zn and Cu in plants probably depends on their species.

The present study showed no significant effect of bovine manure or mushroom substrate on the content and bioaccumulation factor of Cu by cocksfoot. Application of chicken manure, with which the largest quantity of this metal was applied to the soil, significantly increased its content in the biomass of the grass. In addition, all of the organic fertilizers increased Cu uptake by the test plant. Sumner [58] and Wuana and Okieimen [59] described bovine manure and chicken manure as potential sources of Cu for plants. At the same time, Wuana and Okieimen [59] reported that Cu is added to livestock fodder as a growth promoter, which may result in an increase in its content in animal waste and in the biomass of fertilized crops. Velusami et al. [60] tested the fertilizer value of mushroom substrates and showed that they are a valuable source of macro- and microelements, including Cu, for plants. The correlation coefficients indicate that the increase in Cu accumulation by cocksfoot under the influence of chicken manure, bovine manure and mushroom substrate is associated with stimulation of yield by these fertilizers, discussed in a previous paper [48].

The average content of Ni in cocksfoot was 3.20 Ni mg·kg^{-1} DM. Wowkowicz et al. [61] tested the content of heavy metals in crops grown on permanent grassland in the vicinity of Warsaw. The Ni content in the grasses ranged from 0.8 to 2.28 Ni mg·kg^{-1} DM. Contrasting results were presented by Łukowski et al. [50], who found highly varied content of this metal in fodder grasses, ranging from 4.55 to 39.60 Ni mg·kg^{-1} DM.

The present study showed no significant effect of increasing soil application of Zn (200, 400 and 600 Zn mg·kg^{-1}) on the content and bioaccumulation factor of Ni in the biomass of cocksfoot. The results do not indicate an interaction between Zn and Ni. Previous studies also showed no antagonistic or synergistic relationships between these metals in cocksfoot cultivation [62,63]. Contrasting results were obtained by Dalir et al. [64], who studied the relationship between the uptake of Zn and Ni by spring wheat grown in a water medium containing varied concentrations of both metals. The authors found that the interactions were dependent on their content in the medium. Zn at low concentrations did not affect Ni uptake, but higher Zn content in the medium reduced Ni uptake by wheat. Antagonistic relationships between Zn and Ni in cultivation of *Arabidopsis thaliana* (L.) were also reported by Nishida et al. [65]. At the same time, application of 200 Zn mg·kg^{-1} soil caused a significant increase in the uptake of Ni in the test plant. Analysis of the correlation coefficients indicates that this is linked to the increase in the yield of cocksfoot.

In our study, the highest amount of Ni was applied to the soil with chicken manure, which caused a significant increase in its content and bioaccumulation factor in the cocksfoot biomass. Earlier research on the chemical composition of selected organic fertilizers (pig, bovine and chicken manure) showed that chicken manure was the best source of Ni for plants [66]. The organic fertilizers, irrespective of their origin, increased uptake of this metal by the test plant. Analysis of the correlation coefficients indicates that this is directly linked to their positive effect on the yield of cocksfoot. The values of the bioaccumulation factor provide information about the ability of plants to absorb elements from the soil. In research conducted by Łukowski et al. [50], the average bioaccumulation factor of Cu and Ni in Polish grasses was 0.8 and 1.8, respectively. Small values (much lower than 1) of this factor for Cu and Ni obtained in our own research indicate that cocksfoot does not tend to excessive accumulation of these heavy metals [46].

5. Conclusions

Zn applied to the soil at 200 mg·kg^{-1} did not influence the content, uptake and bioaccumulation factor of Cu in cocksfoot, whereas higher application rates—400 and 600 Zn mg·kg^{-1}—significantly reduced them. Decrease in the copper content indicates antagonistic relationships between Zn and Cu. Application of Zn, irrespective of the application rate, did not affect the content and bioaccumulation factor of Ni in the grass, but it increased its uptake at 200 Zn mg·kg^{-1}, as a result of increasing the yield. Bovine manure and mushroom substrate did not affect the contents and bioaccumulation factor of Cu or Ni in the grass. The chicken manure application, with which the largest amount of Cu and Ni was introduced, resulted in a significant increase in their content and bioaccumulation coefficient in tested grass. All of the organic fertilizers increased the uptake of Ni and Cu by the biomass of the test grass. Uptake of both of these metals was highest in the grass fertilized with chicken manure. Low values of bioaccumulation factors of Cu and Ni indicate a low potential of cocksfoot for their excessive accumulation.

Author Contributions: Conceptualization, B.K., A.W. and P.K.; methodology, B.K., P.K. and A.W.; resources, B.K., P.K. and A.W.; writing—original draft preparation, B.K. and A.W.; writing—review and editing, B.K. and A.W.; visualization, A.W. and B.K. All authors have read and agreed to the published version of the manuscript.

Funding: This research was funded by the Polish Ministry of Science and Higher Education, grant numbers 314/12/S, 105/14/MN and 158/23/B.

Institutional Review Board Statement: Not applicable.

Data Availability Statement: Not applicable.

Conflicts of Interest: The authors declare no conflict of interest. The funders had no role in the design of the study; in the collection, analyses or interpretation of data; in the writing of the manuscript; or in the decision to publish the results.

References

1. Fageria, N.K.; Baligar, V.C.; Clark, R.B. Micronutrients in Crop Production Sparks. In *Advances in Agronomy*; Donald, L., Ed.; Academic Press: Cambridge, MA, USA, 2002; Volume 77, pp. 185–268. [CrossRef]
2. Alloway, B.J. Micronutrients and crop production: An introduction. In *Micronutrient Deficiencies in Global Crop Production*; Alloway, B.J., Ed.; Springer: Dordrecht, The Netherlands, 2008; pp. 1–39. [CrossRef]
3. Abou Seeda, M.A.; Abou El-Nour, E.A.A.; Yassen, A.A.; Gad Mervat, M.; Sahar, M.Z. Interaction of copper, zinc, and their importance in plant physiology: Review, Acquisition and Transport. *Middle East J. Appl. Sci.* **2020**, *10*, 407–434. [CrossRef]
4. Kopittke, P.M.; Menzies, N.W.; de Jonge, M.D.; McKenna, B.A.; Donner, E.; Webb, R.I.; Paterson, D.J.; Howard, D.L.; Ryan, C.G.; Glover, C.J.; et al. In situ distribution and speciation of toxic copper, nickel, and zinc in hydrated roots of cowpea. *Plant Physiol.* **2011**, *156*, 663–673. [CrossRef] [PubMed]
5. Arif, N.; Yadav, V.; Singh, S.; Singh, S.; Ahmad, P.; Mishra, R.K.; Sharma, S.; Tripathi, D.K.; Dubey, N.K.; Chauhan, D.K. Influence of high and low levels of plant-beneficial heavy metal ions on plant growth and development. *Front. Environ. Sci.* **2016**, *4*, 69. [CrossRef]
6. Alengebawy, A.; Abdelkhalek, S.T.; Qureshi, S.R.; Wang, M.-Q. Heavy metals and pesticides toxicity in agricultural soil and plants: Ecological risks and human health implications. *Toxics* **2021**, *9*, 42. [CrossRef]
7. Angulo-Bejarano, P.I.; Puente-Rivera, J.; Cruz-Ortega, R. Metal and metalloid toxicity in plants: An overview on molecular Aspects. *Plants* **2021**, *10*, 635. [CrossRef]
8. Kaur, H.; Garg, N. Zinc toxicity in plants: Review. *Planta* **2021**, *253*, 129. [CrossRef]
9. Hambidge, M.; Cousins, R.J.; Costello, R.B. Zinc and health: Current status and future directions. *J. Nutr.* **2020**, *130*, 1437S–1446S. [CrossRef]
10. Hafeez, B.; Khanif, Y.; Saleem, M. Role of zinc in plant nutrition—A review. *Am. J. Exp. Agric.* **2013**, *3*, 374–391. [CrossRef]
11. Ma, D.; Sun, D.; Wang, C.; Ding, H.; Qin, H.; Hou, J.; Huang, X.; Xie, Y.; Guo, T. Physiological responses and yield of wheat plants in zinc-Mediated alleviation of drought stress. *Front. Plant Sci.* **2017**, *8*, 860. [CrossRef]
12. Cabot, C.; Martos, S.; Llugany, M.; Gallego, B.; Tolrà, R.; Poschenrieder, C.A. Role for zinc in plant defense against pathogens and herbivores. *Front. Plant Sci.* **2019**, *10*, 1171. [CrossRef]
13. Umair Hassan, M.; Aamer, M.; Umer Chattha, M.; Haiying, T.; Shahzad, B.; Barbanti, L.; Nawaz, M.; Rasheed, A.; Afzal, A.; Liu, Y.; et al. The critical role of zinc in plants facing the drought stress. *Agriculture* **2020**, *10*, 396. [CrossRef]
14. Brown, P.H.; Cakmak, I.; Zhang, Q. Form and function of zinc plants. In *Zinc in Soils and Plants: Developments in Plant and Soil Sciences*; Robson, A.D., Ed.; Springer: Dordrecht, The Netherlands, 1993; Volume 55, pp. 93–106. [CrossRef]
15. Stoyanova, Z.; Doncheva, S. The effect of zinc supply and succinate treatment on plant growth and mineral uptake in pea plant. *Braz. J. Plant Physiol.* **2002**, *14*, 111–116. [CrossRef]
16. Balafrej, H.; Bogusz, D.; Triqui, Z.E.A.; Guedira, A.; Bendaou, N.; Smouni, A.; Fahr, M. Zinc hyperaccumulation in plants: A review. *Plants* **2020**, *9*, 562. [CrossRef] [PubMed]
17. Yamasaki, H.; Pilon, M.; Shikanai, T. How do plants respond to copper deficiency? *Plant Signal Behav.* **2008**, *3*, 231–232. [CrossRef] [PubMed]
18. Wairich, A.; De Conti, L.; Lamb, T.I.; Keil, R.; Neves, L.O.; Brunetto, G.; Sperotto, R.A.; Ricachenevsky, F.K. Throwing copper around: How plants control uptake, distribution, and accumulation of copper. *Agronomy* **2022**, *12*, 994. [CrossRef]
19. Aguirre, A.J.; Meyers, R.M.; Weir, B.A.; Vazquez, F.; Zhang, C.Z.; Ben-David, U.; Cook, A.; Ha, G.; Harrington, W.F.; Doshi, M.B.; et al. Genomic copy number dictates a gene-independent cell response to CRISPR/Cas9 targeting. *Cancer Discov.* **2016**, *6*, 914–929. [CrossRef]
20. Hippler, F.W.R.; Boaretto, R.M.; Quaggio, J.A.; Mattos Júnior, D. Copper in citrus production: Required but avoided. *Citrus Res. Technol.* **2017**, *38*, 99–106. [CrossRef]
21. Yruela, I. Copper in plants. *Braz. J. Plant. Physiol.* **2005**, *17*, 185–268. [CrossRef]
22. Pietrini, F.; Carnevale, M.; Beni, C.; Zacchini, M.; Gallucci, F.; Santangelo, E. Effect of different copper levels on growth and morpho-physiological parameters in giant reed (*Arundo donax* L.) in semi-hydroponic mesocosm experiment. *Water* **2019**, *11*, 1837. [CrossRef]
23. Fabiano, C.C.; Tezotto, T.; Favarin, J.L.; Polacco, J.C.; Mazzafera, P. Essentiality of nickel in plants: A role in plant stresses. *Front. Plant Sci.* **2015**, *6*, 754. [CrossRef]
24. Harasim, P.; Filipek, T. Nickel in the environment. *J. Elem.* **2015**, *20*, 525–534. [CrossRef]

25. Ahmad, M.S.; Ashraf, M. Essential roles and hazardous effects of nickel in plants. *Rev. Environ. Contam. Toxicol.* **2011**, *214*, 125–167. [CrossRef]
26. Shahzad, B.; Tanveer, M.; Rehman, A.; Cheema, S.A.; Fahad, S.; Rehman, S.; Sharma, A. Nickel; whether toxic or essential for plants and environment—A review. *Plant Physiol. Biochem.* **2018**, *132*, 641–651. [CrossRef]
27. Yan, A.; Wang, Y.; Tan, S.N.; Mohd Yusof, M.L.; Ghosh, S.; Chen, Z. Phytoremediation: A promising approach for revegetation of heavy metal-polluted land. *Front. Plant Sci.* **2020**, *11*, 359. [CrossRef] [PubMed]
28. Prasad, R.; Shivay, Y.; Kumar, D. Interactions of zinc with other nutrients in soils and plants—A Review. *Indian J. Fert.* **2016**, *12*, 16–26.
29. Fan, T.T.; Wang, Y.J.; Li, C.B.; He, J.Z.; Gao, J.; Zhou, D.M.; Friedman, S.P.; Sparks, D.L. Effect of organic matter on sorption of Zn on soil: Elucidation by wien effect measurements and EXAFS spectroscopy. *Environ. Sci. Technol.* **2016**, *50*, 2931–2937. [CrossRef]
30. Naik, S.K.; Das, D.K. Relative performance of chelated zinc and zinc sulphate for lowland rice (*Oryza sativa* L.). *Nutr. Cycl. Agroecosyst.* **2008**, *81*, 219–227. [CrossRef]
31. Sharma, A.; Patni, B.; Shankhdhar, D.; Shankhdhar, S.C. Zinc—An indispensable micronutrient. *Physiol. Mol. Biol. Plants.* **2013**, *19*, 11–20. [CrossRef] [PubMed]
32. Singh, S.; Parihar, P.; Singh, R.; Singh, V.P.; Prasad, S.M. Heavy metal tolerance in plants: Role of transcriptomics, proteomics, metabolomics, and ionomics. *Front. Plant Sci.* **2016**, *6*, 1143. [CrossRef]
33. Doolette, C.L.; Read, T.L.; Li, C.; Scheckel, K.G.; Donner, E.; Kopittke, P.M.; Schjoerring, J.K.; Lombi, E. Foliar application of zinc sulphate and zinc EDTA to wheat leaves: Differences in mobility, distribution, and speciation. *J. Exp. Bot.* **2018**, *69*, 4469–4481. [CrossRef]
34. Kumar, S.; Verma, G.; Dhaliwal, S.S. Zinc phasing and fertilization effects on soil properties and some agromorphological parameters in maize–wheat cropping system. *Commun. Soil Sci. Plant Anal.* **2022**, *53*, 453–462. [CrossRef]
35. Arasimowicz, M.; Niemiec, M.; Wiśniowska-Kielian, B. Zinc, copper and chromium content in soils and needles of the scots pine (*Pinus silvestris* L.) from the Krakow agglomeration terrain. *Ecol. Chem. Eng. A* **2010**, *17*, 1543–1552.
36. Last, L.; Widmer, F.; Fjellstad, W.; Stoyanova, S.; Kölliker, R. Genetic diversity of natural orchardgrass (*Dactylis glomerata* L.) populations in three regions in Europe. *BMC Genet.* **2013**, *14*, 102. [CrossRef] [PubMed]
37. Šiaudinis, G.; Jasinskas, A.; Karčauskienė, D.; Šarauskis, E.; Lekavičienė, K.; Repšienė, R. The dependence of cocksfoot productivity of liming and nitrogen application and the assessment of qualitative parameters and environmental impact using biomass for biofuels. *Sustainability* **2020**, *12*, 8208. [CrossRef]
38. Zhouri, L.; Kallida, R.; Shaimi, N.; Barre, P.; Volaire, F.; Gaboun, F.; Douaik, A.; Fakiri, M. Characterization of cocksfoot (*Dactylis glomerata* L.) population for growth traits and summer dormancy. *J. Mater. Environ. Sci.* **2017**, *8*, 4378–4384. [CrossRef]
39. National Register of Crop Varieties. Available online: https://coboru.gov.pl/pl/szczegoly_odmiany?nrodm=1990 (accessed on 25 May 2023).
40. Stępień, M.; Bodecka, E.; Gozdowski, D.; Wijata, M.; Groszyk, J.; Studnicki, M.; Sobczyński, G.; Rozbicki, J.; Samborski, S. Compatibility of granulometric groups determined based on standard BN-78/9180-11 and granulometric groups according to PTG 2008 and USDA texture classes. *Soil Sci. Ann.* **2018**, *69*, 223–233. [CrossRef]
41. Barančíková, G.; Makovníková, J. Comparison of two methods of soil organic carbon determination. *Pol. J. Soil Sci.* **2015**, *48*, 48–56. [CrossRef]
42. Adamczyk, D.; Jankiewicz, B. Effects of thiuram on uptake of copper, zinc and manganese by *Valeriana officinalis*. *Pol. J. Environ. Stud.* **2008**, *17*, 823–826.
43. Egnér, H.; Riehm, H.; Domingo, W.R. Studies on chemical soil analysis as a basis for assessing the nutrient status of soils. II. Chemical extraction methods for phosphorus and potassium determination. *K. Lantbrukshögskolans Ann.* **1960**, *26*, 199–215.
44. Malinowska, E.; Jankowski, K.; Wiśniewska-Kadżajan, B.; Skrzyczyńska, J.; Sosnowski, J. Cobalt and arsenic concentration in herbs growing in field pond areas in Poland. *Appl. Ecol. Environ. Res.* **2018**, *16*, 3805–3814. [CrossRef]
45. Bokhari, S.H.; Ahmad, I.; Mahmood-Ul-Hassan, M.; Mohammad, A. Phytoremediation potential of *Lemna minor* L. for heavy metals. *Int. J. Phytoremediation* **2016**, *18*, 25–32. [CrossRef]
46. Netty, S.; Wardiyati, T.; Maghfoer, M.D.; Handayanto, E. Bioaccumulation of nickel by five wild plant species on nickel contaminated soil. *IOSR J. Eng.* **2013**, *3*, 1–6. [CrossRef]
47. Aladesanmi, O.T.; Oroboade, J.G.; Osisiogu, C.P.; Osewole, A.O. Bioaccumulation factor of selected heavy metals in Zea mays. *J. Health Pollut.* **2019**, *9*, 191207. [CrossRef] [PubMed]
48. Kuziemska, B.; Klej, P.; Wysokinski, A.; Jaremko, D.; Pakuła, K. Yielding and bioaccumulation of zinc by cocksfoot under conditions of different doses of this metal and organic fertilization. *Agronomy* **2022**, *12*, 686. [CrossRef]
49. Rietra, R.P.J.J.; Heinen, M.; Dimkpa, C.O.; Prem, S.; Bindraban, P.S. Effects of nutrient antagonism and synergism on yield and fertilizer use efficiency. Commun. *Soil Sci. Plant Anal.* **2017**, *48*, 1895–1920. [CrossRef]
50. Łukowski, A.; Wiater, J.; Dymko, A. Bioaccumulation of heavy metals in forage grasses. *Ecol. Eng. Environ. Technol.* **2017**, *18*, 149–158. [CrossRef]
51. Gawryluk, A.; Wyłupek, T.; Wolański, P. Assessment of Cu, Pb and Zn content in selected species of grasses and in the soil of the roadside embankment. *Ecol. Evol.* **2020**, *10*, 9841–9852. [CrossRef]
52. Chaudhry, F.M.; Sharif, M.; Latif, A.; Qureshi, R.H. Zinc-copper antagonism in the nutrition of rice (*Oryza sativa* L.). *Plant Soil* **1973**, *38*, 573–580. [CrossRef]

53. Imtiaz, M.; Alloway, B.J.; Shah, K.H.; Siddiqui, S.H.; Memon, M.Y.; Aslam, M.; Khan, P. Zinc nutrition of wheat: II: Interaction of zinc with other trace elements. *Asian J. Plant Sci.* **2003**, *2*, 156–160. [CrossRef]
54. Zhao, H.; Wu, L.; Chai, T.; Zhang, Y.; Tan, J.; Ma, S. The effects of copper, manganese and zinc on plant growth and elemental accumulation in the manganese-hyperaccumulator *Phytolacca americana*. *J. Plant Physiol.* **2012**, *169*, 1243–1252. [CrossRef]
55. Mousavi, S.R.; Galavi, M.; Rezaei, M. The interaction of zinc with other elements in plants: A review. *Indian J. Adv. Chem. Sci.* **2012**, *4*, 1881–1884.
56. Liščáková, P.; Nawaz, A.; Molnárová, M. Reciprocal effects of copper and zinc in plants. *Int. J. Environ. Sci. Technol.* **2022**, *19*, 9297–9312. [CrossRef]
57. Malinowska, E.; Jankowski, K.; Wiśniewska-Kadżajan, B.; Sosnowski, J.; Kolczarek, R.; Jankowska, J.; Ciepiela, G.A. Content of zinc and copper in selected plants growing along a motorway. *Bull. Environ. Contam. Toxicol.* **2015**, *95*, 638–643. [CrossRef] [PubMed]
58. Sumner, M.E. Beneficial use of effluents, wastes, and biosolids. *Commun. Soil Sci. Plant Anal.* **2008**, *31*, 1701–1715. [CrossRef]
59. Wuana, R.A.; Okieimen, F.E. Heavy metals in contaminated soils: A review of sources, chemistry, risks and best available strategies for remediation. *ISRN Ecol.* **2011**, *2011*, 402647. [CrossRef]
60. Velusami, B.; Jordan, S.N.; Curran, T.; Grogan, H. Fertiliser characteristics of stored spent mushroom substrate as a sustainable source of nutrients and organic matter for tillage, grassland and agricultural soils. *Ir. J. Agric. Food Res.* **2021**, *60*, 1–11. [CrossRef]
61. Wowkonowicz, P.; Malowaniec, B.; Niesiobędzka, K. Heavy metals in soil and plants on grassland around Warsaw. *Environ. Prot. Natur. Resour.* **2011**, *49*, 309–319.
62. Kuziemska, B.; Kalembasa, D.; Kalembasa, S. Effect of liming and organic materials on content of selected metals in of cocksfoot grown in soil contaminated with nickel. *Acta Agroph.* **2014**, *21*, 293–304.
63. Kuziemska, B.; Kalembasa, S.; Kalembasa, D. Effect of liming and use of organic materials on the contents of cooper adn zinic in of cock's-foot grown in soil contaminated with nickel. *Acta Agroph.* **2017**, *24*, 263–270.
64. Dalir, N.; Tandy, S.; Gramlich, A.; Khoshgoftarmanesh, A.; Schulin, R. Effects of nickel on zinc uptake and translocation in two wheat cultivars differing in zinc efficiency. *Environ. Exp. Bot.* **2017**, *134*, 96–101. [CrossRef]
65. Nishida, S.; Kato, A.; Tsuzuki, C.; Yoshida, J.; Mizuno, T. Induction of nickel accumulation in response to zinc deficiency in *Arabidopsis thaliana*. *Int. J. Mol. Sci.* **2015**, *16*, 9420–9430. [CrossRef] [PubMed]
66. Kuziemska, B.; Jaremko, D.; Wysokiński, A. Content of macroelements and fractions of nickel in selected organic fertilizers. *Fresenius Env. Bull.* **2019**, *28*, 1417–1422.

agriculture

Article

The Environmental Tax Scheme in China's Large-Scale Pig Farming: Balancing Economic Burden and Responsibility

Tiemei Yan, Tong Zhang and Zhanguo Zhu *

College of Economics and Management, Nanjing Agricultural University, No. 1 Weigang, Nanjing 210095, China; 2021106054@stu.njau.edu.cn (T.Y.); 2019206025@njau.edu.cn (T.Z.)
* Correspondence: zgzhu@njau.edu.cn

Abstract: China has implemented an environmental protection tax for large-scale agricultural farming to address environmental pollution caused by livestock and poultry breeding. Studying the environmental management of large-scale pig farming is crucial for controlling agricultural pollution. However, the economic impact and effectiveness of the current tax design, including responsibility sharing, tax rates, and taxpayers, are unclear. This study aims to address these gaps. Firstly, two types of taxation principles, production-based and consumption-based, are established, and their effects on pollution emissions distribution between pig production and marketing areas are comparatively analyzed. Secondly, the economic impact of consumption-based environmental taxes is estimated from inter-provincial and rural–urban perspectives under the current tax mechanism. Thirdly, a new alternative tax rate aligned with inter-provincial pig consumption levels is proposed based on the consumption principle. By comparing alternative tax systems, the potential impact of the current environmental tax on the regional economic burden is analyzed. The results highlighted that a production-based tax system resulted in 83% of regions experiencing varying degrees of implicit emission transfers, and a consumption-based tax system helped coordinate the environmental economic burden between pig production and marketing regions. Additionally, a linear relationship between tax rates and pork consumption expenditure not only alleviated the overall economic burden, leading to tax savings in 30% of regions, but also increased the national environmental revenue from the pig farming industry, resulting in a remarkable 147% rise in overall environmental tax revenue. These findings provide theoretical support for adjusting responsibility and economic burden through environmental tax modifications, facilitating the establishment of a compensation mechanism for the benefits between pig production and marketing areas.

Keywords: environmental tax design; inequality; pig industry; consumption-based pollution policy

1. Introduction

With economic growth and social structural changes, meat has become an increasingly significant source of protein. In order to meet the rising demand, the global meat supply is projected to expand, reaching 377 million tons by 2031. This growth will be accompanied by an increase in livestock inventories [1]. As the proportion of large-scale livestock farming continues to rise, livestock farming emissions have increased the carrying pressure on ecosystems, posing potential risks to the ecological environment [2]. Freshwater systems are interconnected and the impact of agricultural pollution, such as livestock production, on freshwater systems is complex [3]. Agricultural water pollution has become a focal point of concern for policymakers in most OECD countries [4]. While these increased outputs have helped meet the demand for food, they have also given rise to significant environmental threats. In many countries, agriculture is now the primary source of water pollution, surpassing cities and industries; it is worth noting that water pollution caused by organic matter from livestock farming is currently more widespread than organic pollution originating from urban areas [5].

Pig farming is a crucial sector within the livestock industry. Due to the large volume and high density of pig manure accumulation, large-scale pig farming easily leads to environmental pollution [6]. China stands as the world's largest producer and consumer of swine, consuming nearly 50% of global pork production [7]. However, China is currently facing serious environmental pollution problems stemming from pig farming, with pig manure alone exceeding 600 million tons, accounting for approximately one-third of the total amount of animal manure. In such circumstances, the introduction of environmental taxes has become an inevitable choice, leading to long-term transformations in the agricultural sector [8].

In order to address environmental challenges, the implementation of environmental taxes has been recognized as a significant policy tool [9,10]. Over the past few decades, many countries have introduced environmental tax reforms to tackle these issues but the majority of environmental tax policies in most countries have primarily focused on industrial sectors such as fossil fuel consumption [11–14]. However, China's environmental protection tax targets not only industrial activities but also large-scale livestock farming, which provides a new research perspective for the reform of environmental tax policies. Compared to industrial pollution with concentrated point-source emissions, agricultural pollution involves dispersed non-point-source emissions, making pollution control more challenging. Thus, research on environmental taxes in the agricultural sector holds significant importance for controlling ecological pollution. The effectiveness of environmental tax reform depends on the rational design of the tax system. This study took China's large-scale pig farming industry as an example and conducted a theoretical exploration of the operational mechanism of environmental taxes under different tax principles, and then analyzed the economic impact of environmental taxes on pollution control in the production and sales areas of pig farming. Moreover, this paper examined the fairness and effectiveness of the existing environmental tax system, considering aspects such as tax rates and tax subjects. The findings provided theoretical support for a progressive reform of environmental taxes and offered new insight for policymakers to adopt agriculture-specific environmental tax policies that are incentive-driven and yield positive environmental benefits.

Our contribution can be summarized in three aspects. Firstly, we examined the mechanism design and economic impacts of environmental taxes in the large-scale pig farming industry, addressing the existing institutional research gap in the agricultural sector concerning environmental taxes. Secondly, from the perspective of compensating interests between production and consumption regions, we employed a framework that considers the process of tax collection and allocation in these regions. We demonstrated the theoretical operation mechanism of environmental taxes under different taxation principles and, by incorporating "pre-measures" of differential tax rates and "post-measures" of transfer payments, we proposed a novel alternative taxation scheme. Thirdly, our work facilitates the establishment of a compensatory mechanism for production and consumption regions by providing a fresh perspective. We discovered that the proposed alternative taxation scheme not only helps in harmonizing the environmental and economic burdens between pig production and consumption regions, but also achieves a better balance in environmental governance responsibilities between these regions. Policymakers can promote the development of the compensatory mechanism from an environmental protection tax perspective.

2. Background

2.1. The Evolution of Environmental Tax in China

With the increasing adoption of carbon taxes globally, China has also begun utilizing environmental policies to combat pollution issues. In 2003, China implemented pollution discharge fees, requiring all industrial enterprises that directly discharge pollutants into the environment to pay fees. In order to further enhance pollution control measures, China enacted the "Environmental Protection Tax Law of the People's Republic of China" in 2018, marking a significant legislative reform in the field of ecological environment and representing China's first independent green tax specifically targeting environmental protection. According to China's Environmental Protection Tax Law, the producers directly emitting taxable pollutants into the environment are considered taxpayers for environmental protection tax. The tax is primarily calculated and levied based on the number of pollutant emissions, pollutant equivalent values, pollutant equivalent numbers, and tax rate standards. In the agricultural sector, China's environmental protection tax is only applicable to large-scale livestock and poultry farms with a stock size exceeding 50 head of cattle, 500 head of pigs, 5000 birds chickens or ducks. The taxation mainly targets water pollutants in the livestock and poultry farming industry. The specific applicable tax rates are determined and adjusted by the provincial, autonomous regional, and municipal governments in consideration of the local environmental carrying capacity. The range of tax rates varies from CNY 1.4 to CNY 14 per pollutant equivalent.

2.2. The Theoretical Analysis of Environmental Taxation Based on Large-Scale Pig Farming

China experiences significant inter-provincial economic inequality, highlighting the need to consider not only the revenue generated from environmental taxes but also the potential impact on regional economies when implementing measures to address environmental pollution. In line with China's environmental tax law, pig farming operations with a size exceeding 500 heads are required to pay the environmental protection tax [15]. In China, due to the imbalanced production and sales of pigs across different regions, it is common to divide major production areas and major consumption areas. In order to meet the demand for swine supply in various provinces, the major production areas not only cater to local consumption but also employ measures such as "pig redistribution" to transport pigs from major production areas to major consumption areas, thus addressing the issue of uneven distribution of pig resources. Based on the pig production layout, the current specific taxation methods are depicted in Figure 1. Figure 1 illustrates the current environmental tax mechanism. In Figure 1, Region A represents the area of pig production. Pigs produced in Region A can be divided into two components: one component is sold locally, consisting of M heads, which generates R_M' environmental tax; the other component is sold in other regions, consisting of N heads. As per the existing tax system, all emissions throughout the production chain are attributed to the province of origin. Consequently, the polluter in Region A is responsible for bearing the environmental tax of $R_M + R_N$, which is retained by Region A and utilized for environmental governance purposes in due course.

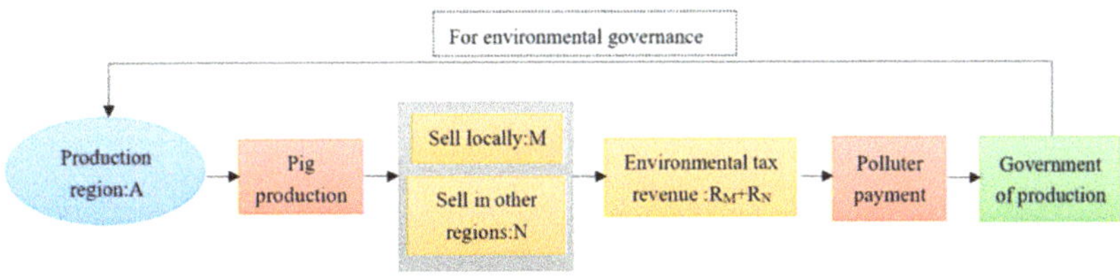

Figure 1. The environmental tax mechanism based on local sales and sales in other regions of pigs.

However, the tax mechanism imposed on large-scale pig farming may potentially contribute to economic inequality among the main pig producers and purchasers, while the effectiveness of pollution mitigation remains uncertain. Firstly, in 2016, China issued a guideline on pig production development; the primary pig-producing regions are concentrated in the central and western areas, with major purchasers predominantly located in eastern areas. This production distribution in pig farming leads to an unequal consumption of environmental resources among regions. Moreover, within the context of regional uneven development, imposing an environmental tax on the main producing areas tends to burden less developed provinces with excessive pollutant emissions and pollution tax payments. Meanwhile, the main pig-consuming provinces tend to be net importers of pig products at a significantly lower cost, as depicted in Figure 2. In Figure 2, Region A represents the pig production region, while Region B represents the pig consumption region. Under the current pig production layout, Region B tends to externalize pollution to Region A, thereby increasing the intensity of pollution control measures and tax burdens on Region A, ultimately constraining its development prospects.

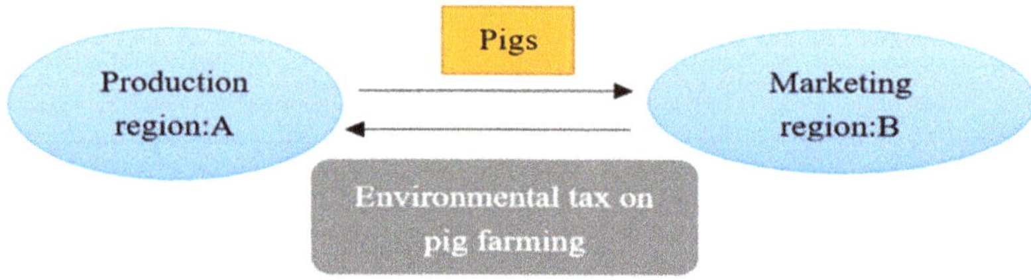

Figure 2. Environmental tax sharing mechanism between pig production and marketing regions.

Secondly, under the environmental tax system, despite the local ownership of tax revenue, the central and western provinces often face a disparity in the valuation of environmental resources compared to the eastern region, resulting in lower tax revenue that is inadequate to cover the higher costs associated with environmental governance [16]. Lastly, the current tax mechanism frequently encounters challenges related to high implementation costs, including the need for extensive information monitoring and the prevalence of local government protectionism. These factors often contribute to inefficiencies in the law enforcement process [17].

Although the primary objective of the environmental tax is to address environmental issues, the design of the tax mechanism, at the very least, should not exacerbate income inequality [18]. In recent years, there has been an increasing scholarly discourse on China's environmental tax. Many researchers argue that the current tax system largely inherits the structure of pollution discharge fees, and the underlying asymmetry between the benefits and costs for polluting entities remains a fundamental issue in environmental problems [19]. It is important to acknowledge that, under different realistic constraints, the polluter-pays principle may not always be the optimal choice. Lu et al. [20] advocate for a perspective of shared environmental responsibility, suggesting that consuming provinces should also bear the relevant emission responsibility. In comparison with a single-polluter-pays principle, a series of studies reveal that tax deviations among regions and the distribution of tax benefits should be thoroughly considered, with the introduction of a consumer-pays principle for taxation [17].

Inter-provincial tax differences serve as a significant means to achieve inter-regional compensation [20]. The current environmental tax system on pig farming employs differential tax rates as ex-ante measures to mitigate potential adverse distributional effects. However, challenges remain regarding the unreasonable distribution of regional tax burdens. Building upon ex-ante measures, a novel alternative tax mechanism has been proposed, which incorporated insights from the transfer payment approach discussed by Liang and Wei [21], Alton et al. [22], and Wang et al. [23] to propose a novel alternative tax mechanism, as depicted in Figure 3. In Figure 3, Region A represents the province of pig production, while Region B represents the province of pig consumption. The key distinction lies in the allocation of emissions to the final consumption area rather than the place of origin. In this scenario, the environmental tax R_M generated from locally sold pigs is paid by the consumer in Region A and utilized for environmental governance within the same region. Simultaneously, the consumer in Region B bears the environmental tax R_N resulting from the consumption of M-head pigs as environmental management fees, which are subsequently transferred to Region A through financial transfer payments.

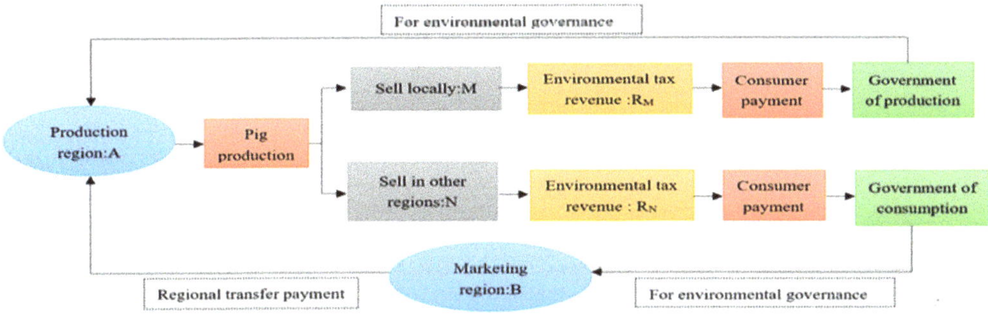

Figure 3. The adjusted environmental tax mechanism based on local sales and sales in other regions of pigs.

Under this proposed alternative tax system, the challenges associated with the current tax mechanism can be addressed. Firstly, the alternative tax takes into account the issue of conflicting interests between pig production and marketing regions. By altering the taxation principles, the unfair tax burden between production and marketing in pig farming has been rectified, as illustrated in Figure 4. In this graph, when compared to Figure 1, Region B not only consumes pigs but also assumes responsibility for the emissions resulting from the production process. This adjustment significantly reduces the costs and pressures faced by pig production areas.

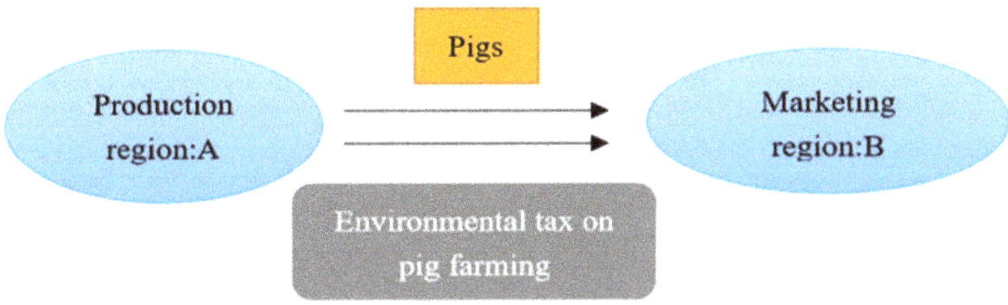

Figure 4. The new mechanism for sharing environmental tax between pig production and marketing regions.

Secondly, in comparison to the mid-west production areas, the eastern consumption regions demonstrate a preference for paying higher prices to access favorable environmental resources, resulting in a higher expected environmental tax revenue. Through the transfer of payments, the eastern region can allocate tax revenue to the governments of the central-western regions. This approach aids in augmenting the funding sources for environmental governance and alleviating the environmental governance pressures faced by the production areas. Lastly, transitioning the burden of environmental tax on pigs from producers to consumers will contribute to reducing the costs associated with environmental tax regulation while enhancing the efficiency of tax collection. Additionally, the increased funding can facilitate the diversification of environmental investments, thereby improving pollution control efficiency in the production areas. To enhance the efficiency of combating environmental pollution, it is essential to establish a compensation mechanism between production areas and marketing areas. This mechanism should guide the existing environmental tax system towards a framework that embodies both fairness and efficiency simultaneously.

Currently, the tax principles solely target behaviors that directly discharge pollutants into the environment, without considering the emissions resulting from pig consumption, as well as the implicit emissions and tax transfers that occur during the consumption process [17]. To address this gap, this paper aimed to assess the potential impact of environmental taxes on pollutant emissions from large-scale pig farming in mainland China. Our analysis focused on the year 2020, utilizing available data, and specifically examined the distortion in environmental tax burdens between pig production and marketing regions. Moreover, this paper compared the distributional effects of tax principles based on the polluter-pays and consumer-pays approaches. These comparisons offer valuable insights and novel perspectives for the establishment of an environmental compensation mechanism between pig production and marketing areas within the existing tax system. The structure of the remaining sections in our paper is outlined as follows. In Section 2, we provide a comprehensive literature review on environmental taxes implemented under different principles. In Section 3, we elucidate our methodological approach, including the strategy and data utilized for our analysis. The results obtained from various scenarios are presented in Section 4. Finally, in Section 5, we summarize the research findings, draw conclusions, and discuss the policy implications derived from our study.

3. Literature Review

The calculation and allocation of emissions among stakeholders play a crucial role in the implementation of environmental taxes. Two widely adopted methods are production-based and consumption-based environmental taxes [24]. Production-based environmental taxes have gained popularity in many countries due to their simplicity in tax procedures [25]. However, such taxes can result in emission leakage effects, with pollutants flowing into regions with less stringent emission regulations [26]. On the other hand, consumption-based environmental taxes may undermine emitters' incentives to reduce emissions, potentially limiting the effectiveness of emission reductions [25]. Scholars have compared the effects of these two tax schemes in various scenarios. Ren et al. [24] discovered that, under a production-based emission tax, higher demand uncertainty may incentivize manufacturers to invest more in emissions reduction. However, increasing the emission tax under both tax principles could potentially reduce manufacturers' investments in emission reductions. Schmutzler and Goulder [27] argued that a production-based tax is reasonable if monitoring is comprehensive and costs are sufficiently low. However, in cases of high monitoring costs, limited emission reduction options, and close output substitutes, the introduction of a consumption-based environmental tax is preferable. While the production-based principle is likely to remain dominant, the consumption-based principle offers valuable insights for climate mitigation policies [28]. Edjabou and Smed [29] observed that the agricultural sector meets the requirements for implementing consumption-based environmental taxes; they conducted scenario simulations and discovered that using consumption-based taxes to promote climate-friendly diets could be a cost-effective policy tool for reducing agricultural emissions. Existing studies have demonstrated the incentivizing role of environmental tax policies in achieving sustainable development goals for growth and development [30]. Research on agricultural output taxes also indicates that environmental tax policies are equally applicable in the context of agricultural emission reduction.

On the one hand, establishing public support for ensuring effective and sustainable environmental taxes is of paramount importance [31]. Imposing taxes on agricultural output implies taxing consumption. Smed [32] and Säll [33] have studied the impact of environmental taxation policies on social welfare and consumption burden from various aspects, including social costs, consumer surplus, and changes in dietary patterns. Smed [32] discussed the anticipated effects of Denmark's taxation on saturated fats in food based on pre-tax data and the estimated price elasticity of consumer response to the market. It was found that this tax would lead to a reduction of approximately 8% in saturated fat consumption. Meanwhile, Säll [33] employed a consumer welfare approach to investigate the distributional impact of Sweden's environmental tax on meat. The results indicated that, following the introduction of the tax, in order to maintain the utility level of consumption, the middle-income group would require compensation equal to 0.81% of total expenditure, the lowest-income group would need compensation equivalent to 0.99% of income, and the highest-income group would require 0.51% of compensation.

On the other hand, the rationality of the environmental tax system design is equally important. The tax system design often involves a trade-off between efficiency and fairness [23]. Currently, economic research on environmental tax system design is primarily focused on industrial sectors [17,25,34] and not the agricultural sector, which is of significant importance for evaluating the potential implementation effects of environmental tax policies in agriculture. Wang et al. [17] analyzed the distributional impacts of consumption-based environmental taxes based on multi-regional input–output (MRIO) tables and a unified tax calculation method. They found that, while industrial and corporate entities are the taxed subjects, the economic burden of taxation is passed on to consumers through price shifts, resulting in regional inequalities. Thus, attention must be given to potential economic implications when designing environmental taxes. Xue et al. [25] developed a tax model based on production and consumption and argued that effective environmental policies should ensure equal marginal abatement costs for each economic entity's final unit of emissions, and they suggested incorporating consumption-based environmental responsibilities into the tax model framework. Nielsen et al. [34], through constructing a fixed-effects model, conducted a comparative analysis of different environmental tax systems and found that differentiating tax rates based on the products' environmental harm and substantially raising product prices help to address deficiencies in tax design. These studies contribute valuable insights to the progressive reform of environmental tax systems in the agricultural sector.

The agricultural sector holds significant potential for providing substantial short-term ecological relief at relatively low costs, ranking as the second-largest contributor to this potential for mitigation [35]. The environmental policy effects of this sector are crucial for ensuring stable agricultural production and supply, as well as promoting regional sustainability. Compared to the previous research, our study focused on the intensive-scale pig farming within the agricultural sector and constructed an appropriate environmental taxation model. By taking into account the unique production layout of pig farming in China, we examined the regional distributional effects of environmental taxes on pig farming from the perspective of compensating the interests of production and consumption regions. Additionally, we simulated the implementation effects under different taxation scenarios, offering valuable theoretical insights for the progressive reform of environmental taxation.

4. Materials and Methods

The analysis in this study consisted of the following steps. First, we collected production and consumption data for large-scale pig farming in mainland China. By applying the appropriate accounting method and emission coefficients, we obtained both production-based and consumption-based emission data for large-scale pig farming. Next, we calculated the net transfer of pollutant emissions across regions based on the distribution of pig farming emissions. Using the consumption-based emission data and the environmental tax rates specific to each region, we calculated the revenue generated from the consumption-based environmental tax for large-scale pig farming. Finally, we proposed two alternative tax mechanisms and compared them to the original tax mechanism, then analyzed the differences between the alternative and original tax mechanisms, and assessed their respective impacts on regional economic inequality. Additionally, due to data limitations, we did not include the Tibet Autonomous Region in the analysis.

4.1. Calculation of Production-Based Environmental Tax on Large-Scale Pig Farming

We primarily referred to the "Manual of Production and Emission Accounting Methods and Coefficients for Statistical Surveys of Emission Sources" and the "Agricultural Pollution Source Production and Discharge Factor Manual" (available at https://www.mee.gov.cn/xxgk2018/xxgk/xxgk01/202106/t20210618_839512.html (accessed on 11 June

2021)) published by the Ministry of Ecology and Environment of the People's Republic of China in 2021. The environmental tax on pig farming was applicable only to water pollutant emissions from farms with a stock size exceeding 500. We utilized the emission coefficients provided for large-scale pig farming as reference data. According to the "Agricultural Pollution Source Production and Discharge Factor Manual", water pollutants from livestock farming mainly include chemical oxygen demand (COD), ammonia nitrogen (NH3-N), total nitrogen (TN), and total phosphorus (TP). However, total nitrogen (TN) is not considered a taxable water pollutant under China's environmental tax law. Therefore, we focused on the three primary water pollutants: chemical oxygen demand (COD), ammonia nitrogen (NH3-N), and total phosphorus (TP). The specific emission coefficients can be found in Table A1. To estimate the production of large-scale pig farming, we multiplied the proportion of farms with over 500 pigs slaughtered per year, as announced by the Ministry of Agriculture and Rural Affairs, with the pig production of each region. The calculation formula for water pollutant emissions from large-scale pig farms is as follows:

$$Q_j = \left(q \times e_j\right) \times 10^{-3} \tag{1}$$

$$Q = \sum_{j}^{n} Q_j \tag{2}$$

where Q represents the total water pollutant emissions from large-scale pig farming in the region, Q_j denotes the emissions of the j-th pollutant from large-scale pig farming in the region (unit: tons), q represents the number of pigs slaughtered in large-scale pig farming exceeding 500 (unit: heads), and e_j represents the emission coefficient of the j-th pollutant in large-scale pig farming in the region (unit: kg/head).

To calculate the environmental tax revenue on large-scale pig farming based on production emissions, we incorporated the pollutant equivalent values and the tax rate table for taxable water pollutants (Tables A2 and A3). The conversion formula was as follows:

$$TAX = \sum_{j}^{n} \mathrm{TAX}_j = \sum_{j}^{n} AP_j \cdot R = \sum_{j}^{n} \frac{Q_j}{W_j} \cdot R \tag{3}$$

where TAX_j represents the tax revenue generated from the j-th water pollutant, AP_j denotes the equivalent value of the j-th water pollutant, Q_j represents the emissions of the j-th water pollutant, W_j signifies the pollution equivalent value of the j-th water pollutant, and R represents the tax rate ranging from CNY 1.4 to CNY 14 per pollution equivalent in the respective region.

4.2. Calculation of Consumption-Based Environmental Tax on Large-Scale Pig Farming

Based on the methodologies employed by Wang et al. [17] and Xue et al. [25], we assumed that environmental taxes would eventually be transferred to consumers through tax transfer mechanisms, resulting in consumers bearing the tax burden based on the producer's tax rate. To estimate pig consumption in each region, we utilized data on per capita pig consumption (measured in kilograms per person) and population size (measured in ten thousand people) obtained from the China Statistical Yearbook. By multiplying the proportion of pigs slaughtered in farms with over 500 heads and the total pig consumption, we estimated the consumption volume of large-scale pig farming in each region. It should be noted that not all pig consumption is sourced from farms with a scale of over 500 heads within each region. The "Agricultural Cost and Benefit Yearbook" distinguishes medium-scale pig farms as those with 100–1000 pigs and large-scale pig farms as those with over 1000 pigs. Since the average weight of pigs is influenced by the number of farrowings, in order to accurately estimate the weight of pigs on farms with over 500 heads (in kilograms per pig), we calculated the ratio of the number of farms with 500–1000 pigs slaughtered to the number of farms with more than 500 pigs slaughtered, and the ratio of farms with over 1000 pigs to the number of farms with more than 500 pigs slaughtered. These ratios were used as the weights for medium-scale and large-scale pig farms, respectively. The pig weight on farms with over 500 heads can then be estimated by taking the weighted average. Finally, to estimate the number of pigs consumed from all large-scale pig farming in each region (in ten thousand heads), we divided each region's pig consumption by the pig weight on farms with over 500 heads. The calculation formula was as follows:

$$Q^s = \frac{consume \times per_{More\ than\ 500}}{scale_1 \times per_{500\sim1000} + scale_2 \times per_{More\ than\ 1000}} \tag{4}$$

where Q^s represents the consumption of pigs from large-scale farms with over 500 heads in each region (unit: 10,000 heads), consume denotes the total consumption of pigs in each region (unit: 10,000 tons), $per_{More\ than\ 500}$ indicates the proportion of large-scale farms with over 500 heads among all pig farms in terms of annual output, $scale_1$ denotes the average pig production of medium-scale farms (unit: kg/head), $per_{500\sim1000}$ represents the ratio of pig farms with 500 to 1000 heads to all pig farms with over 500 heads, $scale_2$ represents the average pig production of large-scale farms (unit: kg/head), and $per_{More\ than\ 1000}$ is the ratio of pig farms with over 1000 heads to all pig farms with over 500 heads. Based on the above, the formula for calculating the water pollutant emissions from pig consumption is as follows:

$$Q_j{}^s = (Q^s \times e_j) \times 10^{-3} \tag{5}$$

$$Q^* = \sum_j^n Q_j{}^s \tag{6}$$

where Q^* represents the total water pollutant emissions from pig consumption in each region (in ten thousand tons), $Q_j{}^s$ represents the j-th water pollutant emission from pig consumption based on over 500-head scale pig farming in each region (in ten thousand tons), Q^s is the consumption of pigs from 500 or more-scale pig farming in each region (in ten thousand heads), and e_j represents the emission coefficient of the j-th pollutant from large-scale pig farming in the region (in kilograms per head). We have adopted the approach proposed by Wang et al. [17] and assumed that, regardless of where these pigs are produced, each region applies its current tax rate to all the products it consumes. The formula for calculating the environmental tax based on pig consumption was as follows:

$$TAX^* = \sum_j^n \mathrm{TAX}_j{}^s = \sum_j^n AP_j{}^s \cdot R = \sum_j^n \frac{Q_j{}^s}{W_j} \cdot R \tag{7}$$

where TAX^* represents the environmental tax revenue based on the consumption of pigs in each region, $\mathrm{TAX}_j{}^s$ denotes the tax revenue from the j-th water pollutant, $AP_j{}^s$ represents the equivalent value of the j-th water pollutant, $Q_j{}^s$ signifies the emission of the j-th water pollutant, W_j represents the pollution equivalent value of the j-th water pollutant, and R represents the original official tax rate ranging from CNY 1.4 to CNY 14 per pollution equivalent.

4.3. Calculation of Alternative Environmental Tax on Large-Scale Pig Farming

In this study, we proposed a new tax approach for pig consumption, which takes into account the pig consumption levels in different regions. The new tax rates in each region were determined based on a linear correlation with pig consumption expenditure. The region with the highest pig consumption expenditure was assigned the highest tax rate of CNY 14 per unit of pollutant, while the region with the lowest pig consumption expenditure was assigned the lowest tax rate of CNY 1.4 per unit of pollutant. The tax rates for intermediate regions were determined through differential interpolation.

$$TAX' = \sum_{j}^{n} \mathrm{TAX}_j^s = \sum_{j}^{n} AP_j^s \cdot R' = \sum_{j}^{n} \frac{Q_j^s}{W_j} \cdot R' \tag{8}$$

where TAX' represents the environmental tax revenue under the new tax rate and R' represents the updated tax rate determined based on the pig consumption expenditure of each region. The specific values of the new tax rates can be found in Table A4.

5. Results

5.1. Comparison of Pollutant Emission Distribution Based on Production and Consumption

As depicted in Figure 1, using Formulas (1) to (2) and (5) to (6), this paper calculated the regional pollutant emissions from pig production and consumption separately for the year 2020. Following the approach of Xue et al. [25], we determined the net emissions transfer by subtracting the production-based emissions from the consumption-based emissions. If the result is greater than zero, it indicates the presence of hidden net pollution transfer in that region. In Figure 5, among the analyzed regions, Guangdong, Beijing, Tianjin, Shanghai, and Zhejiang remained unaffected by pollution transfer, while other regions received water pollution in amounts of 2.842×10^4 t, 4.301×10^3 t, 7.015×10^3 t, 3.813×10^3 t, and 3.748×10^3 t, respectively. The remaining 25 regions experienced varying degrees of hidden pollution net transfer. Among them, Hunan had the highest level of hidden pollution net transfer. As a major pig production province, Hunan contributed the highest production-based pollutant emissions of 3.23×10^5 t, accounting for 15.81% of the national total. Henan and Hubei followed closely, bearing 13.21% and 12.54% of the hidden net emission transfer, respectively. Therefore, it is evident that the developed coastal regions reap the greatest benefits from cross-provincial pig production and marketing, while the developing inland regions suffer from the loss of ecological benefits. This indicates that cross-provincial pig production and marketing exhibit a certain level of inequality. Consequently, the design of an environmental tax must take into account the equitable sharing of environmental governance responsibilities between the production and marketing regions.

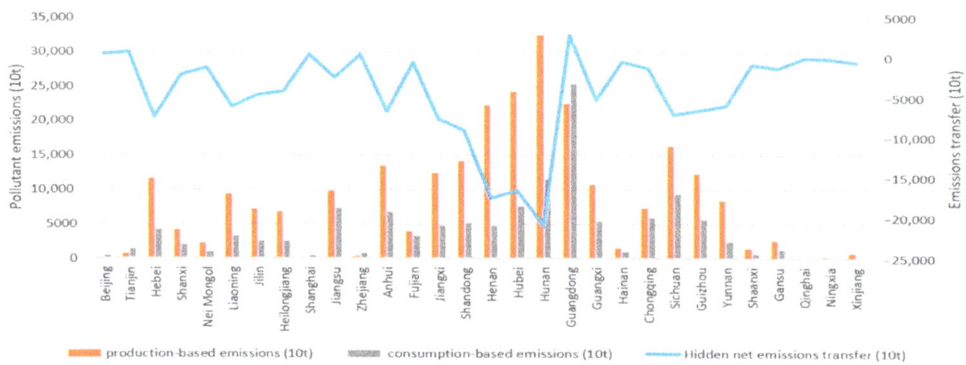

Figure 5. Distribution of pollutant emissions based on pig production and marketing in 2020.

5.2. Estimation of Environmental Tax Revenue Based on Pig Consumption under the Current Tax Rates

Figure 6 illustrates the regional environmental tax revenue generated from pig consumption in 2020, considering the current tax rates. Among all regions, Guangdong exhibited the highest tax revenue from live pig consumption, reaching CNY 747 million, which accounted for 18.2% of the total tax revenue. Jiangsu followed with 10.4% of the tax revenue. In contrast, regions with a smaller population of the Hui ethnic group, such as

Ningxia, had the lowest tax revenue from live pig consumption. Qinghai also demonstrated relatively low tax revenue, amounting to only CNY 1.1 million. The highest estimated tax revenue based on consumption was approximately 600 times higher than the lowest tax revenue.

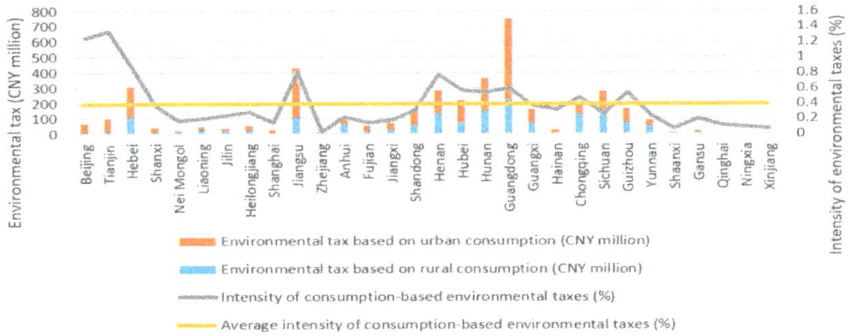

Figure 6. Consumption-based environmental tax revenue by region in 2020 under the current tax system.

Regarding urban and rural areas, with the exception of Liaoning, Jilin, Anhui, Chongqing, Sichuan, Gansu, Yunnan, Qinghai, and Zhejiang, two-thirds of the provinces exhibited higher pig consumption in urban areas than in rural areas. Notably, pollutant emissions caused by urban consumption were significantly higher than those in rural areas. Tax revenue generated from urban consumption constituted approximately 60.9% of the total tax revenue. This trend was particularly evident in Beijing, Tianjin, Shanghai, and Xinjiang, where tax revenue from urban consumption reached as high as 80%. These findings indicated that, under the current tax mechanism, the consumption-based tax system aligned better with the pig consumption patterns in urban and rural areas.

In addition, we employed the method proposed by Wang et al. [17] to measure the tax intensity using the ratio of consumption-based tax revenue to pig consumption expenditures. As depicted in Figure 7, the national average tax intensity was merely 0.39% under the current tax rate, which was higher than the tax intensity observed in the majority of regions. Tianjin exhibited the highest tax intensity at 1.33%, followed by Beijing at 1.24%. With the exception of regions characterized by low live pig consumption expenditures and environmental tax rates, such as Shaanxi, Qinghai, Ningxia, and Xinjiang, developed and coastal regions such as Zhejiang, Fujian, and Anhui exhibited significantly higher pig consumption expenditures than the national average. However, their tax rates and tax intensities were considerably lower than the national average, particularly Zhejiang, with a tax intensity of only 0.02%. Furthermore, Figure 7 illustrates a linear simulation relationship between the consumption-based environmental tax intensity and pig consumption expenditure under the current tax rate in 2020. The graph displays a slight positive correlation, suggesting that the current rate did not align well with the regional level of pig consumption. This indicated the need for further adjustments to the environmental tax mechanism under the replacement scheme, considering the regional pig consumption capacity. The tax rate mechanism based on consumption also required further improvement.

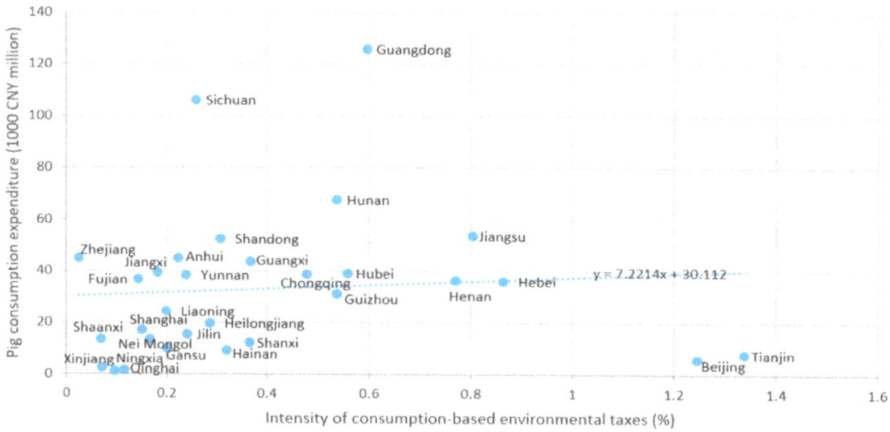

Figure 7. Linear simulation of consumption-based environmental tax intensity and expenditure on pig consumption under the current tax rate.

5.3. *Estimation of Environmental Tax Revenue Based on Pig Consumption under the New Tax Rate*

Currently, there exists a reversed interest between pig-producing and marketing areas in China. This is primarily influenced by distribution costs and regional advantages. Pig farming is predominantly concentrated in East China, Central China, Southwest China, and South China, while the Yangtze River Delta, Pearl River Delta, and Bohai Rim economic circle serve as the main markets for pigs. Pigs often flow from the main producing areas to the main consuming areas at lower prices. This situation poses challenges for the safety and stability of the pig supply chain. The producing areas face high pressures from swine diseases, environmental pollution, and input costs, leading to financial losses. On the other hand, the main consuming areas directly benefit from the deep processing and distribution of pigs. In order to address this imbalance, we investigated the impact of each region adjusting its environmental tax rate based on its pork consumption level on the economic burden among regions. Our focus was to examine the overall changes in regional environmental tax and assess the economic burden on both the pig-producing and marketing areas. To do so, we designed a new consumption-based tax rate based on the pig consumption levels (pig consumption expenditure) in each region (refer to Table A4). Figure 8 illustrates the tax revenue generated from pig consumption under the new tax rate, providing a comparison of three different tax mechanisms.

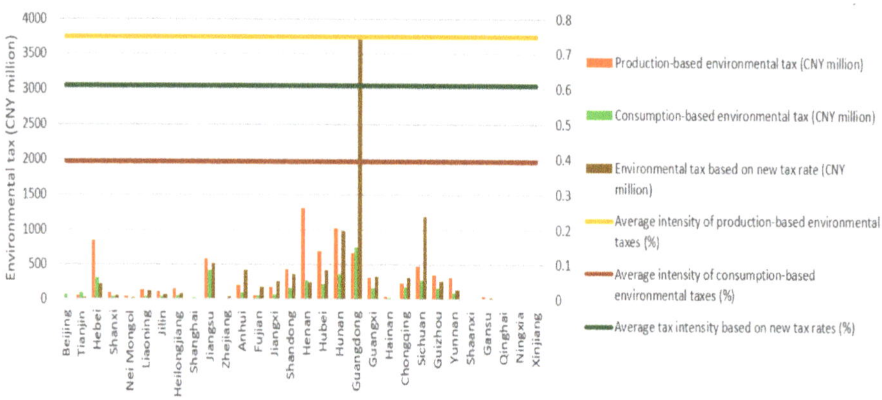

Figure 8. Consumption-based tax revenue under the new tax rate.

Under the new environmental tax rate, regions with higher pig consumption expenditures, such as Anhui, Guangdong, Jiangsu, and Zhejiang, witnessed significant increases in their consumption-based environmental taxes. Specifically, compared to the previous tax rate, the consumption-based environmental taxes rose by 131%, 125%, 613%, and 131% in these regions, respectively. Nationally, the average environmental tax revenue for pigs increased by 285%. In Figure 8, the green bar represents the environmental tax intensity based on consumption under the new tax rate, the red bar depicts the environmental tax intensity based on consumption under the original tax rate, and the yellow bars illustrate the environmental tax intensity based on production under the original tax rate. The national average intensity under the three different tax mechanisms is 0.61, 0.39, and 0.74, respectively, with corresponding national environmental tax revenues for pigs of CNY 1.012×10^4, 0.842×10^4, and 0.409×10^4 million, respectively. Compared to the current tax mechanism, the consumption-based environmental tax mechanism gradually reduces the tax burden on pig-producing areas such as Hebei, Henan, and Hainan, while increasing the tax burden on regions with higher pig consumption expenditures. Under the new tax mechanism, the national average environmental burden for pig farming was dispersed and reduced, leading to a 147% increase in overall environmental tax revenue for pig farming.

In addition, Figure 9 illustrates the linear simulation relationship between the environmental tax intensity based on consumption and pig consumption expenditure under the new tax rate. The figure clearly demonstrates a significant positive correlation between the environmental tax intensity and pig consumption expenditure, indicating that the new environmental tax mechanism effectively aligns with the pig consumption level in each region and helps balance the economic burden between pig-producing and pig-marketing areas.

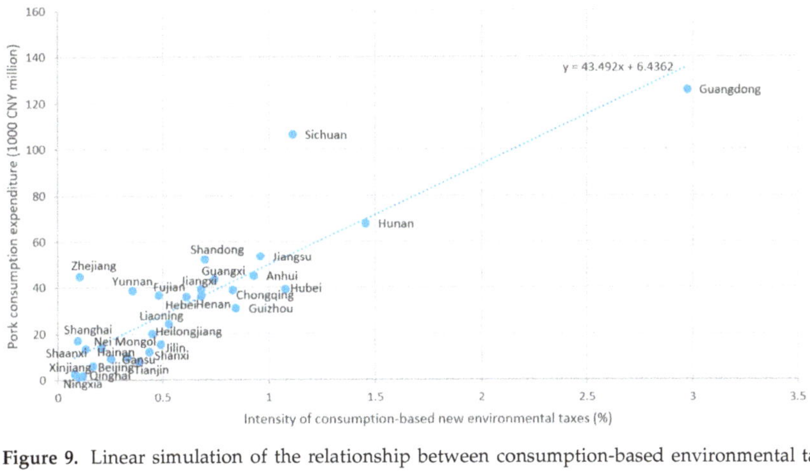

Figure 9. Linear simulation of the relationship between consumption-based environmental tax intensity and pig consumption expenditure under the new tax rate.

5.4. Comparison of Three Environmental Tax Mechanisms on Pig Farming

To compare the disparities between the existing tax mechanism and the alternative tax system, we adopted the methodology outlined by Wang et al. [17]. In line with their approach, we referred to the underpaid environmental taxes resulting from the new alternative tax system as "environmental tax savings", while any additional environmental taxes paid were labeled as "tax losses". Initially, we calculated the discrepancy between the environmental taxes based on pig consumption and those based on pig production by subtracting the former from the latter. This enabled comparison of the tax mechanisms based on production and consumption using the original tax rate. The findings of this analysis are presented in Figure 10.

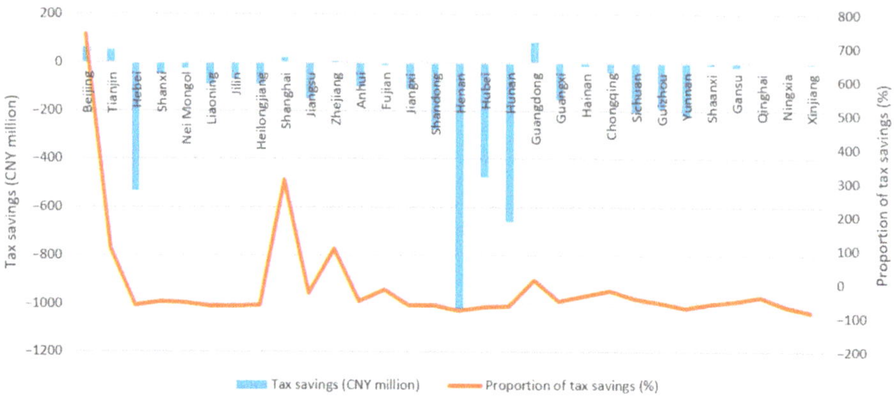

Figure 10. Comparison between production-based and consumption-based tax mechanisms under the original tax rate.

Contrasting the environmental tax mechanism based on pig consumption, only five pig-production and economically developed regions, namely Beijing, Tianjin, Shanghai, Zhejiang, and Guangdong, achieved tax savings under the original tax rate, while other regions experienced varying degrees of tax losses. Among them, Beijing had the highest proportion of pig tax savings, at 739%, followed by Zhejiang with a savings rate of 104%. Guangdong, as the province with the highest pig consumption expenditure, saved a tax revenue of up to CNY 83.8 million, with a savings rate of 12.63%. Conversely, Henan, Yunnan, and Hunan, as major pig supply regions in China, experienced tax losses of 78.6%, 70.3%, and 64.4% respectively. The current tax mechanism exacerbated the unequal economic burden between pig-producing and marketing regions.

Next, we compared the new tax mechanism with the original tax mechanism and the results are presented in Figure 11. When the alternative tax mechanism under the new tax rate was used as the standard, nine regions, including Zhejiang, Shanghai, Anhui, and Guangdong, achieved tax savings. The national average tax savings rate was 37.3%. However, 70% of the regions experienced tax losses, indicating that the national environmental tax level under the original system was relatively low compared to the alternative tax system. Moreover, the proportion of regions experiencing tax savings and losses differed significantly, potentially leading to an unequal regional economic burden under a lower environmental tax intensity.

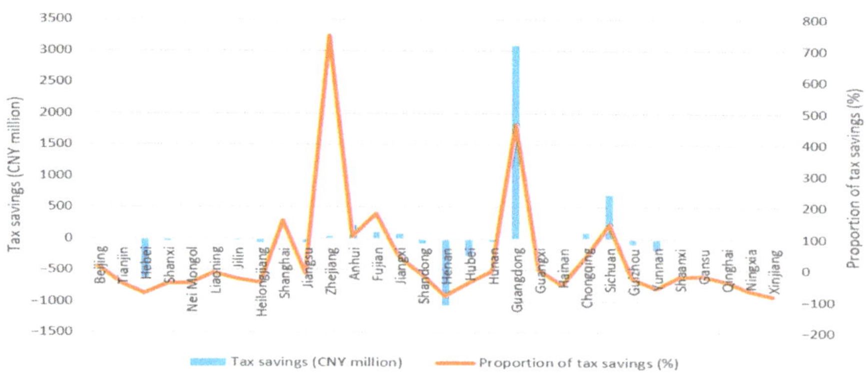

Figure 11. Comparison between the original production-based tax and the new consumption-based tax mechanism.

6. Discussion

The implementation of the environmental protection tax law in China has elevated the level of green taxation and expedited the process of green tax reform. While the environmental tax primarily targeted the industrial sector, agricultural pollution originating from non-point sources posed a more discreet and diffuse challenge. Pollution generated by large-scale livestock farming has emerged as a significant environmental issue and the introduction of an environmental tax specifically targeting this sector could serve as a crucial measure for controlling agricultural pollution in China.

According to the current distribution of pig production in China, the producers and polluters are mainly concentrated in the central and western regions, while the consumers are predominantly located in the eastern regions. Considering the existing development disparities in China, where the east was economically prosperous compared to the west, levying taxes on pig producers in the central and western regions would lead to increased production costs for the producers and additional environmental governance burdens for the local governments. Consequently, this would further exacerbate the developmental inequality between swine production and consumption regions. Our research findings indicated that, under the prevailing environmental tax system, the major beneficiaries of cross-provincial pig production and consumption were the regions of consumption, while the regions of production experienced varying degrees of profit loss. Wang et al. [17] quantified the potential impact of environmental taxes on air pollutant emissions in different provinces of mainland China. They discovered that developed provinces had relatively low tax intensities, while less developed provinces had relatively high tax intensities. However, our research results slightly differ. We found that some developed provinces, such as Jiangsu, had significantly higher tax intensities than the average, while some less developed provinces, such as Shaanxi, Qinghai, and Ningxia, had considerably lower tax intensities than the average. This discrepancy could be attributed to the different research scopes selected. Wang et al. [17] based their study on data from a Multi-Regional Input–Output tabel, reflecting the general patterns of environmental taxes in various industries. In contrast, our research primarily focused on the large-scale pig farming industry, which exhibited distinct production distribution characteristics. The tax intensity was not only correlated with the level of economic development but also with the regional pig production capacity. Consequently, our study better illustrates the features of environmental taxes in the large-scale pig farming industry and is more suitable for analyzing the perspective of compensating interests between pig production and consumption regions.

Meanwhile, despite the suggestions put forth by scholars such as Wang et al. [23], Wang et al. [17], and Xue et al. [25] regarding the design of environmental taxes based on different criteria, including tax rates, taxable objects, and taxation principles, there is a lack of further discussion on the specific taxation procedures and operational mechanisms. In this study, taking the large-scale pig farming industry as an example, we comprehensively examined the entire process of environmental tax collection and distribution, thereby enhancing the feasibility of environmental tax reform.

7. Conclusions

Pig farming is a traditional industry in China and plays a crucial role in ensuring the well-being of the population. In this study, we examined large-scale pig farming as a case study and proposed two alternative environmental taxation mechanisms within the existing tax framework. By comparing the current taxation system with the alternative approaches, we observed that, under the current tax regime, 83% of regions bore varying degrees of potential implicit emission transfers, leading to an increased burden on environmental governance in pig farming areas. Furthermore, the environmental tax burden borne by pig production and consumption areas did not align appropriately with their respective pig consumption expenditures.

Based on our research findings, we proposed two main adjustments during the transition from the existing tax system to an alternative tax mechanism. Firstly, there should be a shift in the tax collection focus from production-based pollutant emissions to consumption-based pollutant emissions. Secondly, it is essential to adjust the environmental tax rates to align with the consumption expenditure of pork in different regions. The current environmental tax system, though straightforward, may exacerbate inequalities between production and sales regions due to its inefficiency. In contrast, the alternative tax mechanism took into account the conflicting interests between pig production and sales regions, promoting fairness and enhancing the effectiveness of environmental tax collection through information transparency and public oversight. This ensured the revenue generated from environmental governance and held the relevant entities responsible for environmental pollution control. Therefore, the environmental tax system for pig farming needs to gradually guide the tax structure towards efficiency and fairness.

Currently, our research has only conducted technical calculations on the economic impact of environmental taxes on the large-scale farming industry. We did not delve deeper into the subject using econometric models, which represents a limitation of this study and an area that can be further explored in the future. Nevertheless, despite this limitation, our study provided some novel insights into the reform and development of environmental protection taxes in the agricultural sector. Additionally, we offered theoretical support for adjusting environmental governance responsibilities and economic burdens between the production and sales regions of the pig industry through the modification of tax policies.

The objective of our study was to effectively alleviate agricultural production pollution through the reform of environmental taxation. In future policy implementation, policymakers can increase environmental tax rates to raise the cost of pollution and thereby discourage polluting behaviors. In addition, based on the research findings by Işık et al. [36], renewable energy can contribute to reducing environmental cleanup costs and minimizing pollution emissions. Therefore, in future practices, it is also worth considering the recycling of fecal waste generated by the agricultural farming sector, for example, by implementing agricultural-farming integration. By employing external constraints through environmental taxes and internal conversion of agricultural pollutants, more favorable outcomes in addressing pollution in the agricultural sector may be achieved. In future research, further investigations can be conducted based on different types of agricultural pollution (such as soil pollution and air pollution) and variations in livestock farm scales (livestock quantities) among different regions. This approach will help to better delineate the applicability of environmental tax reforms in the agricultural sector. Simultaneously, expansions of this work could consider the specific allocation of environmental tax burdens between producers and consumers in the context of large-scale pig farming. Given the unique position of agricultural products such as pork in the national economy, a deeper investigation into the distributional effects of environmental tax policies on producers and consumers in the agricultural sector using welfare analysis and other methods is warranted.

Author Contributions: Conceptualization, T.Y. and Z.Z.; data curation, T.Y.; formal analysis, T.Y. and T.Z.; funding acquisition, Z.Z.; writing—original draft preparation, T.Y.; writing—review and editing, T.Y. and T.Z. All authors have read and agreed to the published version of the manuscript.

Funding: The study is funded by the General Project of National Social Science Foundation of China (grant number 20BGL177) and the Center for agricultural industrial chain management and development of China.

Institutional Review Board Statement: Not applicable.

Data Availability Statement: The data presented in this study are available upon request from the corresponding author.

Conflicts of Interest: The authors declare no conflict of interest.

Appendix A

This appendix contains four tables that display information on emission coefficients for large-scale pig farming, pollution equivalent values of taxable water pollutants, equivalent tax amount of water pollution discharge by region, and official tax rate and new tax rate values.

Table A1. Emission coefficients for large-scale pig farming.

	COD	NH3-N	TP
Beijing	4.3178	0.1094	0.0677
Tianjin	10.5351	0.1707	0.1572
Hebei	6.2497	0.1276	0.0951
Shanxi	9.7331	0.1615	0.1452
Nei Mongol	5.2113	0.1049	0.0801
Liaoning	7.6073	0.1081	0.1177
Jilin	9.4502	0.13	0.1458
Heilongjiang	6.6873	0.0913	0.1039
Shanghai	2.0527	0.0556	0.0356
Jiangsu	8.8285	0.2761	0.1764
Zhejiang	1.0047	0.0498	0.0227
Anhui	10.2912	0.2673	0.2016
Fujian	5.7742	0.1736	0.1163
Jiangxi	8.43	0.2388	0.1671
Shandong	6.7607	0.2539	0.1376
Henan	8.0811	0.1076	0.1432
Hubei	12.7263	0.145	0.2234
Hunan	11.6476	0.1428	0.2051
Guangdong	12.9476	0.1512	0.2271
Guangxi	7.1333	0.1027	0.1276
Hainan	6.7538	0.0892	0.1207
Chongqing	9.6497	0.1467	0.143
Sichuan	5.9272	0.1034	0.0887
Guizhou	12.7358	0.1691	0.1871
Yunnan	4.1215	0.0795	0.0627
Shaanxi	2.6206	0.0296	0.0533
Gansu	7.5284	0.0536	0.1505
Qinghai	3.8713	0.0385	0.0932
Ningxia	3.718	0.0324	0.076
Xinjiang	2.5414	0.0283	0.0562

Table A2. Pollution equivalent values of taxable water pollutants.

	Contamination Equivalent Value
COD	1
NH3-N	0.8
TP	0.25

Table A3. Equivalent tax amount of water pollution discharge by region.

	COD	NH3-N	TP
Beijing	14.0	14.0	14.0
Tianjin	7.5	7.5	1.4
Hebei	7	7	5.6
Shanxi	2.1	2.1	2.1
Nei Mongol	2.1	2.1	2.1
Liaoning	1.4	1.4	1.4
Jilin	1.4	1.4	1.4

Table A3. *Cont.*

	COD	NH3-N	TP
Heilongjiang	2.1	2.1	2.1
Shanghai	5	4.8	1.4
Jiangsu	5.6	5.6	5.6
Zhejiang	1.4	1.4	1.4
Anhui	1.4	1.4	1.4
Fujian	1.5	1.5	1.4
Jiangxi	1.4	1.4	1.4
Shandong	3	3	1.4
Henan	5.6	5.6	5.6
Hubei	2.8	2.8	1.4
Hunan	3.5	3.5	3.5
Guangdong	2.8	2.8	2.8
Guangxi	2.8	2.8	2.8
Hainan	2.8	2.8	2.8
Chongqing	3	3	3
Sichuan	2.8	2.8	2.8
Guizhou	2.8	2.8	2.8
Yunnan	3.5	3.5	3.5
Shaanxi	1.4	1.4	1.4
Gansu	1.4	1.4	1.4
Qinghai	1.4	1.4	1.4
Ningxia	1.4	1.4	1.4
Xinjiang	1.4	1.4	1.4

Table A4. Official tax rate and new tax rate values.

	Original Tax Rate	New Tax Rate
Beijing	14.0	1.9
Tianjin	5.5	2.1
Hebei	6.5	4.9
Shanxi	2.1	2.5
Nei Mongol	2.1	2.6
Liaoning	1.4	3.7
Jilin	1.4	2.8
Heilongjiang	2.1	3.3
Shanghai	3.7	3.0
Jiangsu	5.6	6.7
Zhejiang	1.4	5.8
Anhui	1.4	5.8
Fujian	1.5	5.0
Jiangxi	1.4	5.2
Shandong	2.5	6.6
Henan	5.6	4.9
Hubei	2.3	5.2
Hunan	3.0	8.1
Guangdong	2.8	14.0
Guangxi	2.8	5.7
Hainan	2.8	2.2
Chongqing	3.0	5.2
Sichuan	2.8	12.0
Guizhou	2.8	4.4
Yunnan	3.5	5.2
Shaanxi	1.4	2.6
Gansu	1.4	2.3
Qinghai	1.4	1.4
Ningxia	1.4	1.4
Xinjiang	1.4	1.5

References

1. OECD/FAO. *OECD-FAO Agricultural Outlook 2022–2031*; OECD Publishing: Paris, France, 2022; pp. 190–191. [CrossRef]
2. Jia, W.; Qin, W.; Zhang, Q.; Wang, X.; Ma, Y.; Chen, Q. Evaluation of crop residues and manure production and their geographical distribution in China. *J. Clean. Prod.* **2018**, *188*, 954–965. [CrossRef]
3. Moss, B. Water pollution by agriculture. *Philos. Trans. R. Soc. B Biol. Sci.* **2008**, *363*, 659–666. [CrossRef]
4. Parris, K. Impact of agriculture on water pollution in OECD countries: Recent trends and future prospects. *Int. J. Water Resour. Dev.* **2011**, *27*, 33–52. [CrossRef]
5. Mateo-Sagasta, J.; Zadeh, S.M.; Turral, H. (Eds.) *More People, More Food, Worse Water?: A Global Review of Water Pollution from Agriculture*; FAO: Rome, Italy; International Water Management Institute (IWMI): Colombo, Sri Lanka, 2018; CGIAR Research Program on Water, Land and Ecosystems (WLE); 224p. Available online: https://hdl.handle.net/10568/96991 (accessed on 20 June 2018).
6. Schaffner, M.; Bader, H.P.; Scheidegger, R. Modeling the contribution of pig farming to pollution of the Thachin River. *Clean Techn. Environ. Policy* **2010**, *12*, 407–425. [CrossRef]
7. Zhang, S.; Wu, X.; Han, D.; Hou, Y.; Tan, J.; Kim, S.W.; Li, D.; Yin, Y.; Wang, J. Pork production systems in China: A review of their development, challenges and prospects in green production. *Front. Agric. Sci. Eng.* **2021**, *8*, 15–24. [CrossRef]
8. Yao, W. Emission trading mechanism in pig farming pollution control: An empirical study of Zhejiang Province, China. *Environ. Sci. Pollut. Res.* **2021**, *28*, 30007–30018. [CrossRef] [PubMed]
9. Morley, B. Empirical evidence on the effectiveness of environmental taxes. *Appl. Econ. Lett.* **2012**, *19*, 1817–1820. [CrossRef]
10. Shahzad, U. Environmental taxes, energy consumption, and environmental quality: Theoretical survey with policy implications. *Environ. Sci. Pollut. Res.* **2020**, *27*, 24848–24862. [CrossRef]
11. Mark, P.; Stephen, S. *The European Carbon Tax: An Assessment of the European Commission's Proposals*; IFS Report, No. R39; Institute for Fiscal Studies (IFS): London, UK, 1991; ISBN 1873357125. [CrossRef]
12. Wier, M.; Birr-Pedersen, K.; Jacobsen, H.K.; Klok, J. Are CO_2 taxes regressive? Evidence from the Danish experience. *Ecol. Econ.* **2005**, *52*, 239–251. [CrossRef]
13. Beck, M.; Rivers, N.; Wigle, R.; Yonezawa, H. Carbon tax and revenue recycling: Impacts on households in British Columbia. *Resour. Energy Econ.* **2015**, *41*, 40–69. [CrossRef]
14. Khastar, M.; Aslani, A.; Nejati, M. How does carbon tax affect social welfare and emission reduction in Finland? *Energy Rep.* **2020**, *6*, 736–744. [CrossRef]
15. Ministry of Justice Website. The Implementation Regulations of the People's Republic of China on the Environmental Protection Tax Law. 2018. Available online: https://www.gov.cn/zhengce/2020-12/27/content_5574446.htm (accessed on 25 December 2017).
16. Wang, J.; Lin, J.; Feng, K.; Liu, Y.; Jiao, X.; Ni, R.; Du, M.; Hubacek, K. Towards reducing inter-city economic inequality embedded in China's environmental protection tax law. *Environ. Res. Lett.* **2021**, *16*, 124007. [CrossRef]
17. Wang, J.; Lin, J.; Feng, K.; Liu, P.; Du, M.; Ni, R.; Chen, L.; Kong, H.; Weng, H.; Liu, M.; et al. Environmental taxation and regional inequality in China. *Sci. Bull.* **2019**, *64*, 1691–1699. [CrossRef] [PubMed]
18. Carattini, S.; Carvalho, M.; Fankhauser, S. Overcoming public resistance to carbon taxes. *Wiley Interdiscip. Rev. Clim. Change* **2018**, *9*, e531. [CrossRef] [PubMed]
19. Wang, K.; Wang, J.; Hubacek, K.; Mi, Z.; Wei, Y.-M. A cost–benefit analysis of the environmental taxation policy in China: A frontier analysis-based environmentally extended input–output optimization method. *J. Ind. Ecol.* **2020**, *24*, 564–576. [CrossRef]
20. Lu, Y.; Wang, Y.; Zhang, W.; Hubacek, K.; Bi, F.; Zuo, J.; Jiang, H.; Zhang, Z.; Feng, K.; Liu, Y.; et al. Provincial air pollution responsibility and environmental tax of China based on interregional linkage indicators. *J. Clean. Prod.* **2019**, *235*, 337–347. [CrossRef]
21. Liang, Q.M.; Wei, Y.M. Distributional impacts of taxing carbon in China: Results from the CEEPA model. *Appl. Energy* **2012**, *92*, 545–551. [CrossRef]
22. Alton, T.; Arndt, C.; Davies, R.; Hartley, F.; Makrelov, K.; Thurlow, J.; Ubogu, D. Introducing carbon taxes in South Africa. *Appl. Energy* **2014**, *116*, 344–354. [CrossRef]
23. Wang, Q.; Hubacek, K.; Feng, K.; Wei, Y.; Liang, Q. Distributional effects of carbon taxation. *Appl. Energy* **2016**, *184*, 1123–1131. [CrossRef]
24. Ren, J.; Chen, X.; Hu, J. The effect of production-versus consumption-based emission tax under demand uncertainty. *Int. J. Prod. Econ.* **2020**, *219*, 82–98. [CrossRef]
25. Xue, J.; Zhu, D.; Zhao, L.; Li, L. Designing tax levy scenarios for environmental taxes in China. *J. Clean. Prod.* **2022**, *332*, 130036. [CrossRef]
26. Van Asselt, H.; Brewer, T. Addressing competitiveness and leakage concerns in climate policy: An analysis of border adjustment measures in the US and the EU. *Energy Policy* **2010**, *38*, 42–51. [CrossRef]
27. Schmutzler, A.; Goulder, L.H. The choice between emission taxes and output taxes under imperfect monitoring. *J. Environ. Econ. Manag.* **1997**, *32*, 51–64. [CrossRef]
28. Peters, G.P. From production-based to consumption-based national emission inventories. *Ecol. Econ.* **2008**, *65*, 13–23. [CrossRef]
29. Edjabou, L.D.; Smed, S. The effect of using consumption taxes on foods to promote climate friendly diets–The case of Denmark. *Food Policy* **2013**, *39*, 84–96. [CrossRef]

30. Shafi, M.; Ramos-Meza, C.S.; Jain, V.; Salman, A.; Kamal, M.; Shabbir, M.S.; Rehman, M.U. The dynamic relationship between green tax incentives and environmental protection. *Environ. Sci. Pollut. Res.* **2023**, *30*, 32184–32192. [CrossRef]
31. Muhammad, I.; Mohd Hasnu, N.N.; Ekins, P. Empirical Research of Public Acceptance on Environmental Tax: A Systematic Literature Review. *Environments* **2021**, *8*, 109. [CrossRef]
32. Smed, S. Financial Penalties on Foods: The Fat Tax in Denmark. *Nutr. Bull.* **2012**, *37*, 142–147. [CrossRef]
33. Säll, S. Environmental food taxes and inequalities: Simulation of a meat tax in Sweden. *Food Policy* **2018**, *74*, 147–153. [CrossRef]
34. Nielsen, H.Ø.; Konrad, M.T.H.; Pedersen, A.B.; Gyldenkærne, S. Ex-post evaluation of the Danish pesticide tax: A novel and effective tax design. *Land Use Policy* **2023**, *126*, 106549. [CrossRef]
35. IPCC. *Climate Change 2022: Impacts, Adaptation, and Vulnerability. Contribution of Working Group II to the Sixth Assessment Report of the Intergovernmental Panel on Climate Change*; Pörtner, H.-O., Roberts, D.C., Tignor, M., Poloczanska, E.S., Mintenbeck, K., Alegría, A., Craig, M., Langsdorf, S., Löschke, S., Möller, V., et al., Eds.; Cambridge University Press: Cambridge, UK; New York, NY, USA, 2022; 3056p. [CrossRef]
36. Işık, C.; Simionescu, M.; Ongan, S.; Radulescu, M.; Yousaf, Z.; Rehman, A.; Alvarado, R.; Ahmad, M. Renewable energy, economic freedom and economic policy uncertainty: New evidence from a dynamic panel threshold analysis for the G-7 and BRIC countries. *Stoch Environ. Res. Risk Assess.* **2023**. [CrossRef] [PubMed]

agriculture

MDPI

Article

Integrated Ecological Risk Assessment of the Agricultural Area under a High Anthropopressure Based on Chemical, Ecotoxicological and Ecological Indicators

Agnieszka Klimkowicz-Pawlas *, Bożena Smreczak and Barbara Maliszewska-Kordybach

Department of Soil Science Erosion and Land Protection, Institute of Soil Science and Plant Cultivation—State Research Institute, Czartoryskich 8, 24-100 Puławy, Poland; bozena@iung.pulawy.pl (B.S.); bkord@poczta.onet.pl (B.M.-K.)
* Correspondence: agnes@iung.pulawy.pl

Abstract: Agricultural land is often located close to highly urbanised/industrialised areas and is subject to continuous anthropogenic pressure associated with the emission of many pollutants, ultimately deposited in the soil. Most studies on ecological risk assessment have only analysed the total contaminants' concentration, which does not reflect their bioavailability or toxicity and often leads to an overestimation of risk. Therefore, in our study, we used an interdisciplinary approach, whereby the final conclusions about the risk in a given area are based on the integration of detailed data from chemical, ecotoxicological and ecological analysis. The research was carried out on agricultural land exposed to high levels of anthropopression for more than 100 years. Chemical measurements comprised both the total and bioavailable PAH content. A battery of bio-assays describing effects on soil retention and habitat function was used for ecotoxicity testing, and ecological indicators included enzymatic activity, respiration, microbial biomass, carbon mineralisation and nitrification. The integrated IntRisk index ranged from 0.19 to 0.94, and this was mainly due to high values of the chemical risk index, while the ecotoxicological and ecological results indicated no or low risk. The majority of the area (almost 90%) had acceptable risk levels, no/low risk (IntRisk < 0.5) at 57% of the sites and medium risk at 28% of the area. Very high unacceptable risk (IntRisk 0.77–0.94) was only at three sampling sites. The integration of data from a set of 15 indicators allowed us to derive quantitative risk indexes and delineate the limited area which needs additional action.

Keywords: agricultural soils; bioassays; ecological indicators; soil contamination; ecological risk assessment; Triad approach; lines of evidence; integrated risk

1. Introduction

Soils are fundamental and non-renewable natural resources providing goods and services essential for ecosystems and human life. The condition of soils is significantly affected by ongoing anthropogenic pressures associated with agricultural (use of organic and mineral fertilisers, pesticides) and non-agricultural activities (industrial emissions), which result in the release of a range of contaminants (e.g., polycyclic aromatic hydrocarbons, PAHs) that ultimately reach the soil. This may affect soil biodiversity, thereby reducing resistance to stress, influencing soil retention and habitat functions and, finally, deteriorating soil quality, fertility and productivity [1]. The Intergovernmental Technical Panel on Soil has identified soil contamination as one of the most worrying threats to soil health and the provision of key soil ecosystem functions and services [2]. Therefore, the protection of agricultural soils is one of the most important issues in current European policy, as reflected by many soil-related initiatives and strategies, such as the European Green Deal, the Biodiversity Strategy and the new European Union Soil Strategy. Given the objectives of the Green Deal, it is crucial to identify contaminated sites, and the final decision to remediate them should be based on site-specific risk assessment procedures to determine when risk is unacceptable.

A lot of studies on ecological risk assessment are focused on the prediction of potential risk based on the chemical analysis of the total pollutants concentrations and chemical risk indexes derived from such data [3–7]. Various indexes were used, e.g., the hazard quotient and hazard index [6,7], contamination factor and pollution load index [7], toxic units for individual PAHs and PAHs mixtures [6,7] and toxic equivalent concentration [5–7]. In each of these cases, the total contaminant content was compared with the generic reference values from the applicable soil quality standards. The final result of the assessment often depended on the type of chemical index [6,7], the model used for the calculation [6] and the assessment criterion adopted [5,6]. This traditional approach is not sufficient for fully assessing the response of soil organisms, does not provide information on the bioavailability of pollutants and their actual toxicity to soil biota and often leads to an overestimation or underestimation of risk [8,9]. Therefore, chemical analyses should be complemented by the results of ecotoxicological and ecological studies, preferably using organisms representing different groups of biological organisation.

Triad is one of the weight-of-evidence methods in which conclusions are based on the quantitative analysis of three independent assessment tools (lines of evidence, LoEs), i.e., chemical measurements of pollutant concentrations, results of ecotoxicological testing and ecological indicators [10]. Combining the three LoEs assessments is necessary because no single line of evidence provides comprehensive information, whereas the combined efforts and assessment of results from three independent disciplines provide a pragmatic reduction in the conceptual uncertainty associated with risk assessment. So far, the Triad method has mainly been used to assess the risk of sites contaminated by metals [11–20]. There are almost no reports on other groups of contaminants; only a few studies on the use of this method for the risk assessment of soils contaminated with organic pollutants can be found in the literature [21,22]. This method usually consists of several tiers and is recommended for site-specific risk assessment [10,23]. In practice, Tier 1 comprises screening studies to determine whether further stages of ecological risk assessment are required at a site. The screening phase is often based on a generic assessment, i.e., a simple comparison of environmental concentrations of pollutants with available threshold values [11,21,22], as well as on simple biotests, which, at this stage, should be quick, sensitive, easy to perform and inexpensive [12,22,23]. Higher tiers of this assessment (refined and detailed assessment), levels 2 and 3, are carried out to verify and reduce uncertainty and should take into account a wider set of more relevant and sophisticated indicators.

This research is the next step forward in the ecological risk assessment study focusing on the ecotoxicological and ecological effects in the agricultural area long-time-contaminated with organic pollutants (PAHs). Our previous studies characterised the contamination status of the area [22,24], a screening risk assessment based on chemical indexes [6] and simple bioassays implementation [22]. The aim of the current study was the deployment of a detailed integrated risk assessment procedure including a multidisciplinary approach based on chemical, ecotoxicological and ecological lines of evidence. We hypothesised that organic pollutants present in soils at high concentrations may deteriorate their retention and habitat functions and that the inclusion of parameters characterising the availability of pollutants in the assessment increases the reliability of the procedure. To the best of our knowledge, this study presents the first application of a higher-tier Triad approach to the ecological risk assessment of agricultural land with a long history of PAH contamination.

2. Materials and Methods

2.1. Study Area and Sampling Campaigns

The study was carried out in the area located in the south-western part of the Silesian region in Czerwionka, Rybnik district, Poland. The territory is very diverse; it includes agricultural and urban parts as well as industrial and post-industrial sites (Figure 1).

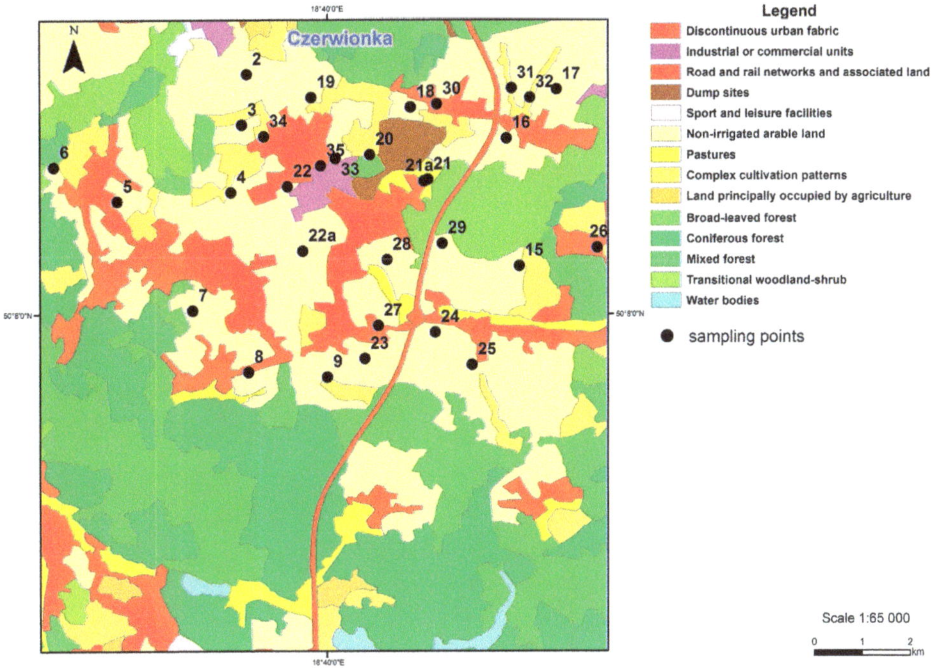

Figure 1. Map of the study area (land cover according to CORINE Land Cover 2018 database).

The high anthropopressure in this region results from diverse diffuse sources: the long-time coke production, former coal mining activity, fossil fuel combustion, waste recovery and road transport. Moreover, high values were found for anthropogenic indices such as total dust emissions (6827 kg year^{-1} km^{-2}) and dust emissions from industrial sources (1084 kg year^{-1} km^{-2}). The detailed characteristics of the region can be found elsewhere [22,24]. The climate in the region is influenced by three colliding subtropical, Arctic and continental air masses. The mean annual precipitation and temperature during the sampling period were 719 mm and 8.9 °C, respectively. The wind distribution was dominated by south-westerly winds with an average speed of 3.8 m s^{-1}. Soil samples were collected from agricultural land after the growing season during two sampling campaigns; initially, soil samples were collected to characterise the contamination status of the area (115 km^2; one sampling site per ca. 4.8 km^2) and to conduct a screening risk assessment [6,22]. Tier 1 assessment [22] showed that the acceptable risk was exceeded, or the uncertainty was too high in 46% of investigated soils, and it was concluded that, for 11 sampling sites, the risk analysis needed to be refined with a broader battery of tests. Therefore, the next phase—the second campaign (current research)—focused on a region of possible high risk limited to the northern and central part of the Czerwionka region (38 km^2). Finally, 31 soil samples (1 sample per 1.2 km^2) from the 0–0.3 m layer were collected after the growing season; each sample (approximately 5 kg) consisted of 6 subsamples taken from a 1 m^2 area. After initial manual mixing and homogenisation in the field, the samples were divided in the laboratory. Part of the sample was then air-dried and prepared for chemical analyses. Directly after collection, fresh soil samples were sieved through a 2 mm sieve, and ecotoxicological and ecological analyses were successively performed. If analyses could not be carried out immediately, the samples were stored under ISO 10381-6 standard conditions (temperature 4 ± 2 °C, in the dark with access to air) for a maximum of 7 days.

2.2. Soil Characteristic and Chemical Analysis

Soils were characterised in terms of texture (laser diffraction method), total organic carbon content (TOC, sulfochromic oxidation method), total carbon and nitrogen content (TC and TN, dry combustion) and pH in 1 mol L^{-1} KCl (potentiometric method) (Table 1). The total content of 16 PAHs (16PAH) was determined by gas chromatography–mass spectrometry (GC-MS) using the Agilent GC-MS system (Agilent Technologies, Santa Clara, CA, USA) after the extraction of soil samples with dichloromethane in an ASE200 Accelerated Solvent Extractor (Dionex Co., Sunnyvale, CA, USA) [22,25]. Moreover, the bioavailable fraction of the 16 PAH compounds (16PAH bioav) was measured by GC-MS after extraction with Tenax-TA as a solid-phase sorbent according to Smreczak [26].

Table 1. Summary statistics for the selected physical-chemical properties and PAH concentration (n = 31).

Indicator	Mean	Median	SD	Min	Max	CV%
Sand (%)	72.2	70.7	7.4	58.1	90.6	10
Silt (%)	25.6	26.8	6.8	9.2	39.0	27
Clay (%)	2.1	2.06	0.7	0.2	3.5	33
TOC (g kg^{-1})	17.4	11.5	20.0	5.2	106.9	115
TC (g kg^{-1})	24.9	14.6	32.7	6.3	155.0	131
pH_{KCl}	5.4	5.4	0.9	3.9	7.2	16
TN (g kg^{-1})	1.0	0.7	0.9	0.2	4.1	88
16PAH bioav. (μg kg^{-1})	97.1	33.6	239.6	22.6	1183.8	247
16PAH (μg kg^{-1}) [a]	12,620	1412	56,461	311	316,085	447

[a] Adapted from Smreczak [26] and Klimkowicz-Pawlas and Debaene [6].

Soil physical-chemical properties (including quality assurance (QA)/quality control (QC) parameters) were described in detail by Klimkowicz-Pawlas et al. [22,24] and Ukalska-Jaruga et al. [25]. The study area was dominated by loamy sand soils (sand 72%, silt 26%, clay 2%, on average; CV = 10, 27 and 33%, respectively) with acidic reaction (mean pH_{KCl} value equal to 5.4). TOC and TN contents were in the ranges of 5.2–106.9 g kg^{-1} (CV = 115%) and 0.2–4.1 g kg^{-1} (CV = 88%), respectively. Considerable differences were observed for the total (CV = 447%) and bioavailable PAH concentration (CV = 247%). 16PAHs averaged 12,620 μg kg^{-1} and reached levels as high as 316,085 μg kg^{-1}, but the content of bioavailable PAHs was significantly lower (Table 1).

The reference site was selected taking into account the following criteria: no contamination, pedological properties (organic carbon and nitrogen content, pH and texture) similar to those of the tested soils, the same land use and similar biological properties [15,27]. The reference site had the following characteristics: loamy sand texture (70% sand, 28% silt, 2% clay), TOC content of 12.9 g kg^{-1}, pH of 5.4 and TN contents of 0.6 g kg^{-1} and 408 μg Σ16 PAHs kg^{-1}. A more detailed analysis of the difficulties involved in selecting a reference site is given elsewhere [22,28,29].

2.3. Soil Ecotoxicity Testing with Bioassays Battery (Ecotoxicological Line of Evidence)

Ecological risk was assessed towards organisms from different taxonomic groups: microorganisms, invertebrates and plants (Tables S1 and S3). Soil elutriates were prepared according to the ISO 21268-1 [30] leaching procedure: soils were mixed with 0.001 mol L^{-1} $CaCl_2$ in a ratio of 1:2 (w/v), shacked through 24 h (20 ± 2 °C, 125 rpm) and filtered with the 0.45 μm syringe filters. We applied two kinds of tests: a water-phase to assess the soil retention function and a solid-phase to assess the soil habitat function.

2.3.1. Water-Phase Ecotoxicity Testing

An acute toxicity assay with the luminescent bacteria *Aliivibrio fischeri* was performed according to the Microtox 81.9% Screening test procedure [31]. Duplicated samples of soil leachates were incubated with bacteria at 15 °C for 15 min, and then the light output of the samples was recorded with a Microtox Model 500 analyser and analysed using the MicrotoxOmni software. The results were expressed as a percentage of the toxic effect (PE).

Rapidtoxkit is an indicator test for detecting the presence of toxic substances using crustacean larvae *Thamnocephalus platyurus* [22,32,33], which take up their food by filtration from the surrounding solution. The rate and level of filtration decrease or, in extreme cases, stop completely after the exposure of *T. platyurus* to toxic substances. Prior to the test, the cysts were rehydrated, transferred to a hatching container and incubated for 34 h at 25 ± 2 °C under continuous illumination (3000–4000 lux). *T. platyurus* larvae were then exposed to soil elutriates (or a culture medium as a control) for 1 h (25 °C, in the dark), followed by feeding with artificial food (suspension of red latex microspheres) for 30 min. The presence or absence of coloured microspheres in the digestive tract of the crustacean larvae was observed under a stereomicroscope, and the results were expressed as the mean percentage inhibition of particle uptake.

The Phytotestkit was conducted in flat test plates using two plant species: white mustard (*Sinapis alba* L.) and garden cress (*Lepidium sativum* L.), for which germination inhibition and root length inhibition were determined. The test was performed according to the procedure recommended by the manufacturer [34]. The lower part of the plate (21 × 15.5 × 0.8 cm) was filled with a foam pad and covered with a white thick cellulose filter, on which 25 mL of soil elutriate (in 0.001 mol L^{-1} $CaCl_2$) was slowly spread in order to hydrate the filter paper completely. Then, the lower part of the plate was covered with a black paper filter, and 10 seeds of the test plants were placed on it. The test plates were incubated vertically at 25 ± 2°C in darkness for 3 days. A photo of the plate was taken using a digital camera, and the root length was measured in the 'Image Tools' image analysis program ver. 3.0. The test control was a 0.001 mol L^{-1} $CaCl_2$ solution used to prepare the soil extracts. The test was carried out in three replicates for each of the soil/plant combinations tested.

2.3.2. Solid-Phase Ecotoxicity Testing

A direct contact toxicity assay with the ostracods *Heterocypris incongruens* was performed according to ISO 14371 [35]. The assay was carried out in the six-well polystyrene multi-well plates. Ten neonates of the *H. incongruens* hatched from cysts were added to each well of the microplate containing the tested soil sample (2 g), standard nutrient medium (2 mL) and algal suspension (1 mL) as food. After 6 days of exposure to contaminated soil samples (25 ± 2 °C, in the darkness), the percentage mortality and the growth of the *H. incongruens* were determined and compared to those of the control (non-toxic sediment).

Two plant growth tests were carried out: seed germination and an early growth assay with white mustard [36] and a screening test for the emergence of lettuce seedlings [37]. Approximately 90 mL of the soil sample was placed on the bottom part of the test plate, hydrated with distilled water to 100% of the water holding capacity and covered with a paper filter. Ten seeds of the same plant species (*Sinapis alba* L. or *Lactuca sativa* L.) were sown in one line at the same distance from one another; the test was carried out in triplicate. After closing with a transparent cover, the test plates were placed vertically in a holder and incubated at 25 °C for 3 days (*S. alba*) or 5 days (*L. sativa*). After incubation, a photo was taken with a digital camera, and the length of the roots was measured with the use of the image analysis program (ImageTool, ver. 3.0). The test control was OECD soil, the standard recommended for use in the plant growth test [36].

2.4. Soil Microbial and Biochemical Parameters (Ecological Line of Evidence)

Soil enzymatic activity and respiration, microbial biomass, nitrification and carbon mineralisation were considered as indicators within the ecological line of evidence (Tables S1 and S3).

Dehydrogenases activity (DH) was assessed after 24 h of incubation of the soil samples (37 ± 2 °C) with 3% triphenyltetrazolium chloride (TTC) as an electron acceptor [38]. The resulting triphenylformazan (TPF) was extracted from soils with ethanol and determined colorimetrically (wavelength of 485 nm) using a Lambda 45 UV-VIS spectrophotometer (Perkin Elmer, Waltham, MA, USA).

The basal (BR) and substrate-induced (SIR) soil respiration were analysed according to the ISO 16072 and ISO 14240-1 methods, respectively. Field-moist soil subsamples were incubated for 24 h (BR) or 6 hours (SIR) at 20 ± 2 °C in darkness. Glucose at 10 g kg^{-1} was applied as a substrate in SIR measurements. CO_2 released during incubation (in both BR and SIR analyses) was entrapped in 0.05 mol L^{-1} NaOH and precipitated as barium carbonate by adding 0.5 mol L^{-1} $BaCl_2$ solution. The non-consumed NaOH was titrated with HCl in the presence of phenolphthalein as an indicator. BR and SIR were expressed as a µg of released CO_2 per gram of dry weight of soil per hour. The microbial biomass (MB) was based on the substrate-induced respiration ISO 14240-1 method and was expressed in µg of microbial C per g of dry weight of soil.

The carbon mineralisation test (CMIN) was carried out according to the OECD 217 [39] standard. Soil samples (100 g) of 60% water holding capacity were incubated for 28 days (20 ± 2 °C, darkness). The soil moisture was checked daily by the weighting of each experimental pot, and the water content was replenished as needed. C mineralisation was assessed at the beginning of the experiment (time 0 days) and after 28 days of soil incubation. At each time interval, 20 g of the soil subsamples was amended with 200 mg of glucose powder, and CMIN was calculated by CO_2 evolution after the next 24 h, as described in the ISO 16072 method.

The nitrification potential (NIT) was determined according to the method described by Maliszewska-Kordybach et al. [40]. Briefly, soil samples were incubated (24 h, 125 rpm, 20 ± 2 °C) with a mineral medium containing ammonium sulphate as the substrate, potassium phosphate buffer and sodium chlorate (to prevent the further oxidation of NO_2^- to NO_3^-). The sample slurry was then mixed with 4 mol L^{-1} KCl and filtered, and the colour reagent (containing sulphanilamide and N-(1-naphthyl)ethylenediamine dihydrochloride) was added. The amount of NO_2^- formed by the nitrification process was determined after 1 h using a Lambda 45 UV-VIS spectrophotometer (Perkin Elmer, Waltham, MA, USA) at 543 nm and expressed as µg NO_2^- per gram of soil dry weight per 24 h. All microbiological and biochemical measurements were made in triplicate, and each series of analyses included control samples without soil material. The accuracy of the biological methods, expressed in terms of the relative standard deviation, ranged from 5 to 25% depending on the parameter analysed.

2.5. Statistical Analysis and Integrated Risk Calculation

Statistical parameters—mean, median, standard deviation (SD), range (Min and Max) and coefficient of variation (CV)—were analysed using the Statgraphics Centurion software (ver. XVIII, Statpoint Technologies). Differences between the response of test organisms and ecological parameters in contaminated and reference soil were assessed via a one-way ANOVA variance analysis, followed by the Duncan's test ($p \leq 0.05$).

The method described by Jensen et al. [23] and Niemeyer et al. [14] was used to calculate risk, including scaling the results for each singular parameter, integrating the scaled results for each assessment line (chemical—IR_Chem, ecological—IR_Ecol and ecotoxicological—IR_Ecotox) and calculating an overall integrated risk index (IntRisk) (Table S2). All parameters were normalised with respect to the reference soil and expressed on a scale from zero (no effect) to one (maximal ecosystem effect). In the chemical line of evidence, the toxic pressure (TP′) for each PAH (expressed as a potentially affected fraction of species) was calculated according to Equation (1) [23].

$$TP' = \frac{1}{1 + exp\left\{\frac{logMPC - logPAH}{0.4}\right\}} \quad (1)$$

The PAH content was related to the maximum permissible concentration (MPC) given by Verbruggen [41].

Next, the scaled TP′ were integrated into one TP index according to the formula [6,23,28]:

$$TP_{16PAH} = 1 - (1 - TP'_1)(1 - TP'_2) \ldots (1 - TP'_n) \quad (2)$$

The chemical, ecotoxicological and ecological lines of evidence included two, seven and six indicators (mentioned in Sections 2.2–2.4 and Tables S1 and S3), respectively, and IR_Chem, IR_Ecotox and IR_Ecol were calculated as follows [14,19,23,29]:

$$IR_{Chem} \text{ or } IR_{Ecotox} \text{ or } IR_{Ecol} = 1 - \left(10^{\,average\ of\ \log(1-A)}\right) \quad (3)$$

where A is the scaled value of each chemical, ecotoxicological or ecological parameter.

In the final step, the integrated risk index was determined according to Equation (4) [23,42], and uncertainty was assessed based on SD calculation for three separate IR indexes.

$$IntRisk = 1 - \left(10^{[\log(1-IR_{\_Chem})+\log(1-IR_{\_Ecotox})+\log(1-IR_{\_Ecol})]/n}\right) \quad (4)$$

3. Results and Discussion

3.1. Chemical Line of Evidence

The site-specific toxic pressure of the 16PAH mixture, reflecting the potentially affected fraction of species, was calculated based on a response addition model [3,23] using the total measured concentration of PAHs (Table 1) and MPC values reported by Verbruggen [41]. Maximum permissible concentrations were derived from the no-observed-effect concentration (NOEC) or 10% effect concentration (EC_{10}) value of the most sensitive terrestrial species and should be protective for all species [3,41]. For 30% of the sampling sites [6], TP_{16PAH} values ranged from 0.01 to 0.20 and were below the lowest threshold (0.25) according to the classification given by Jensen et al. [23] and Dagnino et al. [28], indicating no risk. Furthermore, the Σ16PAH content in these soils ranged from 311 to 591 μg kg^{-1} and was at the level recorded in the reference soil; such contaminant content <600 μg kg^{-1} corresponds to the class of uncontaminated soils according to the classification of Maliszewska-Kordybach [43]. The permissible limits (100 or 200 μg kg^{-1}) for 10 compounds from Polish soil guidelines [44] were also not exceeded, and these nine sampling points were therefore excluded from further stages of the assessment.

The TP_{16PAH} values in the remaining area were rather high, ranging from 0.36 to 1.00 (Table 6); 63% of the soils can be classified as medium and high risk, with TP values of 0.56–0.74 and 0.86–1.00, respectively. The high toxic pressure is due to the high total concentration of PAHs (956–316,085 μg kg^{-1} (Table S3). As highlighted in previous studies by Semenzin et al. [45], Hong et al. [18] and Grassi et al. [46], such a conservative approach considering only the total content of contaminants may lead to a biased risk assessment and should only be used at screening stages. According to Ortega-Calvo et al. [8] and Peijnenburg [9], determining the bioavailability of contaminants is crucial for a proper and realistic assessment of contaminated sites. In our study, the bioavailable PAHs ranged from 22.6 to 1184 μg kg^{-1} (Table 1), representing only 0.4 to 7.5% of the total 16PAHs. It was in line with the previous data of Smreczak [26] and Ukalska-Jaruga and Smreczak [47], who reported that the potentially bioavailable PAHs in historically contaminated soils did not exceed 8%. Therefore, both total and bioavailable 16PAH concentrations were used in our study for risk calculation.

3.2. Ecotoxicity Analysis

The battery of bioassays with various organisms was utilised to assess the toxicity of soils and soil elutriates (Tables 2 and S1). Liquid-phase tests with soil elutriates were applied to obtain information about the fraction of contaminants potentially reaching the groundwater by the water path (retention function of soils), whereas solid-phase tests were used to assess the soil habitat function [4,14]. The response of the test organisms varied significantly according to the type of samples tested and the type of test species (Table 2, Figure 2).

Table 2. Response of organisms (PE%) for water-phase and solid soil samples—ecotoxicological line of evidence (n = 31).

Organism	Endpoint	Mean	Median	SD	Min	Max	CV%
			Water-phase (retention function)				
A.fischeri	LUI_Scr	13.9	13.7	15.5	−10.9	45.7	111
T. platyurus	IF	25.2	33.6	21.9	−10.0	56.7	87
L. sativum	IRG_LEP	7.9	6.8	12.7	−9.2	39.8	161
S. alba	IRG_SIN	−4.6	−8.6	22.1	−48.1	37.8	482
			Solid-phase (habitat function)				
H. incongruens	GI	30.0	30.2	17.0	2.5	63.8	57
S. alba	IRG_SINS	15.6	11.1	18.5	−12.6	60.6	118
L. sativa	IRG_LAC	21.5	19.1	30.9	−24.9	72.9	144

CV%—variation coefficient; LUI_Scr—luminescence inhibition of *A. fischeri*; IF—inhibition of food ingestion by *T. platyurus*; IRG_SIN and IRG_LEP—inhibition of root growth of *S. alba* and *L. sativum*, respectively (in soil leachates); GI—growth inhibition of *H. incongruens*; IRG_SINS and IRG_LAC—inhibition of root growth of *S. alba* and *L. sativa*, respectively (soil samples).

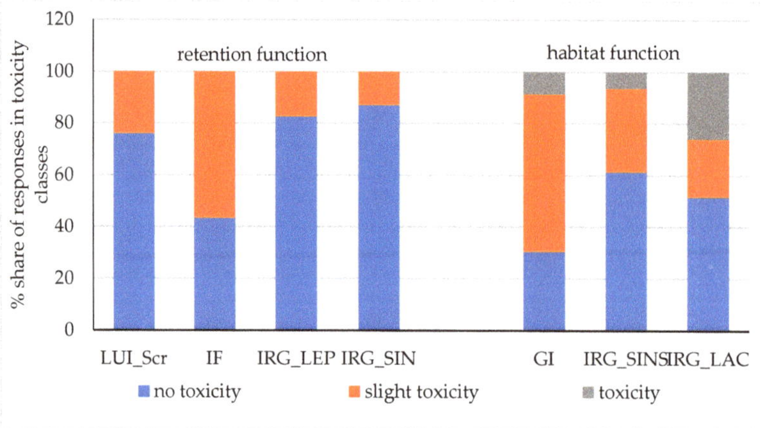

Figure 2. Classification of organisms' responses to soil leachates and solid samples.

The soil and soil leachates toxicity was evaluated using the toxicity assessment method developed by Persoone et al. [48], where PE < 20% means no toxic effect, PE from 20 to 50% means a low toxic sample, PE in the range of 50–100% means a toxic sample and PE = 100% means a very toxic sample. In general, a weaker response of the organisms was observed for the water-phase tests. The sensitivity of the organisms can be ordered as follows: crustacean *T. platyurus* > bacteria *A. fischeri* > plants *L. sativum* = *S. alba* (Table 2, Figure 2).

Food uptake by T. platyurus was inhibited by an average of 25.2%; for most samples, it was within the range of up to 50%, indicating slight toxicity [48]. The *A. fischeri* luminescence inhibition was at a level of 14%; for a majority of samples, no toxic effect was observed, and only 24% of samples exhibited low toxicity (Figure 1), which was correlated with soil pH_{KCl} ($r = -0.40$, $p < 0.05$). Both plants from the Phytotestkit (*L. sativum* and *S. alba*) test showed a similar response, averaging 7.9% (IRG_LEP) and −4.6% (IRG_SIN), respectively, with the stimulation of root growth more frequently observed in mustard. For extracts from several samples (17, 18 and 35), the inhibition of root growth was approximately 30–40%.

Significantly higher effects were observed in the solid-phase tests (Table 2, Figure 2). The most sensitive was *H. incongruens*; no crustacean mortality was observed, while the growth inhibition ranged from 2.5 to 64% and averaged 30%. Only seven soil samples were non-toxic; the remaining samples were low-toxic (61% of sampling points) or toxic (9% of soils), corresponding to toxicity classes II and III, according to Persoone et al. [48]. The inhibition of plant root elongation (IRG_SINS and IRG_LAC) was 15.6% and 21.5% for white mustard and lettuce, respectively (Table 2). For both plants, 50–60% of the samples were non-toxic (Figure 2), with root growth stimulation effects often observed. Lettuce was the more sensitive plant; based on the IRG_LAC endpoint, 23% of the samples can be classified as slightly toxic, and 26% of the samples can be classified as toxic.

The no toxicity or low toxicity for test organisms observed in our study (Table 2, Figure 2), particularly in the aqueous phase tests, indicates the limited availability and leaching of PAHs from the soil system and thus a low risk of their transfer to the water environment. Soil contamination in the Czerwionka region is mainly the result of long-term deposition from several sources, i.e., coking plant, hard coal mining activity, storage of mine wastes and bituminous substances production [22,24,25]. Several authors [25,26,47,49–52] indicated that, in historically contaminated soils, PAHs compounds are subjected to ageing processes leading to significant PAH sequestration or binding, resulting in the reduction in their mobility, extractability and availability. Soil organic matter plays a major role in these processes [25,47,50,51]. Although most of the soils from the region were characterised by a low TOC content (Table 1), according to a previous study by Ukalska-Jaruga et al. [25], these soils showed a significant contribution of humin (up to 123 g kg^{-1}) and black carbon (up to 45 g kg^{-1}) fractions, further limiting the desorption and bioavailability/bioaccessibility of PAHs.

3.3. Ecological Indicators

Soil contamination can affect both the abundance and activity of soil microorganisms, which play a key role in many soil biochemical processes (e.g., decomposition of organic compounds, organic matter synthesis, nutrient cycling, soil detoxification, etc.) essential for the soil ecological functions [40,53,54]. Furthermore, due to their direct contact with the soil, soil microorganisms can respond quickly to changes caused by natural and anthropogenic factors and can therefore be a direct measure of soil function [55,56]. Our study included parameters reflecting the soil's functional diversity: enzymatic activity (DH), soil respiration (BR and SIR), microbial biomass (MB), nitrification (NIT) and carbon mineralisation (CMIN) (Tables 3 and S3), indicators commonly used in both soil quality and contaminant exposure assessments [14,22,53,57,58].

Table 3. Statistical evaluation of biological indicators used to calculate risk within the ecological line of evidence (n = 31).

Indicator	Unit	Mean	Median	SD	Min	Max	CV%
DH	µg TPF g^{-1} d.w. 24 h^{-1}	33.0	32.6	13.7	13.1	86.9	41
NIT	µg NO_2^- g^{-1} d.w. 24 h^{-1}	3.0	2.4	2.4	0.8	13.7	80
BR	µg CO_2 g^{-1} d.w. h^{-1}	2.9	2.5	1.7	0.3	7.8	58
SIR	µg CO_2 g^{-1} d.w. h^{-1}	10.6	8.4	6.1	2.6	26.3	57
MB	µg Cmic g^{-1} d.w.	21.9	17.5	12.4	5.7	54.0	56
CMIN	µg CO_2 g^{-1} d.w. h^{-1}	15.9	14.0	9.3	1.9	38.2	58

Overall, it can be concluded that all soils were characterised by relatively low biological activity, which, for most indicators, did not vary significantly between sampling sites (CV 41–58%). Slightly greater variability (CV = 80%) was found for the nitrification potential, which reflects the activity of the ammonia-oxidising bacteria responsible for the first stage of the nitrification process and is considered one of the most sensitive microbiological parameters responding strongly to the natural factors [59] and the presence of pollutants such as PAHs [40] or metals [14]. NIT values ranged from 0.8 to 13.7 µg NO_2^- g^{-1}

d.w. $24h^{-1}$, with a median value of 2.4 μg NO_2^- g^{-1} d.w. 24 h^{-1}, and were strongly related to both soil properties (r for TOC = 0.72, $p < 0.000$; r for $pH_{KCl} = 0.67$, $p < 0.000$) and the pollutants content (r for 16PAHs = 0.82, $p < 0.000$). For the other parameters (BR, SIR, MB), these relationships were much weaker, and the correlation coefficients with TOC, soil pH and PAH content were 0.36–0.54, 0.39–0.61 and 0.35–0.45 ($p < 0.05$), respectively. The dehydrogenase activity, intracellular enzymes related to the oxidative phosphorylation process, was 33 μg TPF g^{-1} d.w. 24 h^{-1}, on average (Table 3). According to Chakravarty et al. [7], low enzymatic activity (e.g., dehydrogenases, urease, phosphatase) in contaminated soils could be attributed to the poor microbial growth and deficiency of available substrates. Soil microbial respiration and soil microbial biomass are indicators closely related to organic matter degradation and synthesis; soil respiration reflected soil carbon availability to microorganisms, and MB is an important reservoir of nutrients in ecosystems [53,60]. Sites with a high microbial biomass can store and recycle more nutrients [53]; both soil respiration and biomass, in our study, were rather low at the levels of 2.5 μg CO_2 g^{-1} d.w. h^{-1} (for BR) and 17.5 μg Cmic g^{-1} d.w. (MB), indicating that these functions are impaired. It was also confirmed by the low carbon mineralisation (CMIN) rate (Table 3). Under stress conditions (e.g., presence of pollutants), microorganisms usually exhibit a lower metabolic efficiency and need more energy for survival, resulting in less TOC incorporation into the microbial biomass [53].

3.4. Integrated Environmental Risk Assessment

The results from all analyses (chemical, ecotoxicological and ecological) were used to determine the environmental risks associated with PAHs in soils. Initially, the results were normalised to the reference soil, and risk indices were calculated for the individual indicators measured (Tables 4, 5 and S2); the results were then integrated across the different lines of evidence (IR_Chem, IR_Ecotox, IR_Ecol in Tables 6 and S2), with the combined risk for retention (IR_reten) and habitat function (IR_habit) additionally calculated for the ecotoxicological assessment line (Table 4). Finally, after considering all available data, an integrated risk index (IntRisk) was calculated for each of the study sites (Equation (4), Tables 6 and S2).

Table 4. Scaled and corrected risk values (RI) calculated for single ecotoxicological test endpoints and combined risk for retention (IR_reten) and habitat function (IR_habit), respectively.

	Retention Function					Habitat Function			
Sample	**LUI_Scr**	**IF**	**IRG_SIN**	**IRG_LEP**	**IR_Reten**	**GI**	**IRG_SINS**	**IRG_LAC**	**IR_Habit**
3	0.00	0.00	0.11	0.09	0.05	0.00	0.00	0.48	0.20
4	0.00	0.00	0.06	0.16	0.06	0.26	0.00	0.00	0.09
5	0.00	0.00	0.00	0.07	0.02	0.34	0.21	0.60	0.32
6	0.20	0.00	0.00	0.01	0.06	0.59	0.00	0.30	0.67
7	0.10	0.00	0.00	0.00	0.03	0.03	0.04	0.59	0.28
8	0.24	0.00	0.00	0.07	0.08	0.18	0.22	0.50	0.25
16	0.07	0.00	0.21	0.16	0.11	0.26	0.02	0.73	0.34
17	0.01	0.00	0.29	0.20	0.14	0.21	0.11	0.57	0.26
18	0.07	0.00	0.16	0.29	0.14	0.28	0.04	0.67	0.55
19	0.27	0.00	0.00	0.00	0.08	0.17	0.00	0.00	0.05
20	0.07	0.00	0.00	0.00	0.02	0.37	0.05	0.35	0.34
21	0.18	0.00	0.00	0.00	0.05	0.36	0.18	0.00	0.15
21a	0.04	0.00	0.00	0.07	0.03	0.17	0.17	0.14	0.38
22a	0.00	0.00	0.00	0.02	0.01	0.00	0.00	0.02	0.01
26	0.39	0.00	0.00	0.00	0.12	0.37	0.14	0.00	0.38
27	0.00	0.00	0.15	0.26	0.11	0.00	0.34	0.00	0.10
29	0.00	0.00	0.00	0.00	0.00	0.58	0.38	0.39	0.37
30	0.00	0.00	0.00	0.04	0.01	0.24	0.28	0.14	0.70
32	0.00	0.00	0.06	0.19	0.07	0.35	0.43	0.00	0.22
33	0.00	0.00	0.00	0.02	0.00	0.00	0.37	0.57	0.28
35	0.00	0.00	0.38	0.40	0.22	0.06	0.23	0.00	0.08

Explanation of abbreviations under Table 2.

Table 5. Scaled and corrected risk values (RI) calculated for single biological activity indicators according to the ecological line of evidence.

Sample	DH	NIT	BR	SIR	MB	CMIN
3	0.20	0.62	0.13	0.50	0.49	0.01
4	0.38	0.45	0.06	0.10	0.10	0.61
5	0.23	0.58	0.00	0.54	0.53	0.00
6	0.10	0.78	0.00	0.31	0.30	0.67
7	0.46	0.68	0.41	0.26	0.25	0.44
8	0.12	0.44	0.09	0.33	0.32	0.00
16	0.00	0.00	0.00	0.00	0.00	0.00
17	0.01	0.02	0.36	0.00	0.00	0.04
18	0.10	0.52	0.84	0.71	0.70	0.59
19	0.00	0.44	0.00	0.00	0.00	0.00
20	0.09	0.52	0.00	0.07	0.07	0.13
21	0.00	0.23	0.42	0.00	0.00	0.00
21a	0.00	0.55	0.40	0.36	0.36	0.18
22a	0.44	0.00	0.00	0.00	0.00	0.00
26	0.00	0.61	0.00	0.00	0.00	0.47
27	0.06	0.00	0.00	0.00	0.00	0.43
29	0.00	0.00	0.00	0.00	0.00	0.05
30	0.21	0.72	0.00	0.13	0.13	0.74
32	0.33	0.00	0.00	0.19	0.19	0.28
33	0.37	0.00	0.00	0.00	0.00	0.30
35	0.26	0.15	0.00	0.00	0.00	0.00

Table 6. Toxic pressure for total (TP_{16PAH}) and bioavailable PAHs ($TP_{16PAH\ bioav.}$), risk indexes (IR) for individual LoEs and final integrated risk index (IntRisk).

Sample	TP_{16PAH}	$TP_{16PAH\ bioav.}$	IR_Chem	IR_Ecotox	IR_Ecol	IntRisk	SD	Evaluation *
3	0.39	0.01	0.23	0.12	0.36	0.24	0.12	no risk
4	0.59	0.01	0.36	0.07	0.32	0.26	0.16	low risk
5	0.67	0.00	0.43	0.18	0.36	0.33	0.13	low risk
6	0.63	0.01	0.40	0.44	0.44	0.43	0.03	low risk
7	0.36	0.00	0.20	0.14	0.44	0.27	0.16	low risk
8	0.98	0.00	0.85	0.17	0.23	0.55	0.38	moderate risk
16	0.99	0.00	0.92	0.23	0.00	0.61	0.48	moderate risk **
17	0.71	0.00	0.46	0.20	0.08	0.27	0.19	low risk
18	0.89	0.01	0.67	0.38	0.63	0.58	0.16	moderate risk
19	0.74	0.01	0.49	0.06	0.09	0.24	0.24	no risk
20	0.97	0.00	0.83	0.19	0.17	0.52	0.38	moderate risk
21	0.92	0.00	0.72	0.10	0.12	0.40	0.35	low risk
21a	0.99	0.00	0.92	0.22	0.33	0.65	0.37	moderate risk
22a	0.65	0.00	0.41	0.01	0.09	0.19	0.21	no risk
26	1.00	0.08	0.98	0.26	0.23	0.77	0.42	high risk **
27	0.56	n.d.	0.56	0.10	0.10	0.29	0.27	low risk
29	0.69	0.07	0.46	0.21	0.01	0.25	0.23	no risk
30	0.86	0.10	0.64	0.45	0.40	0.51	0.13	moderate risk
32	0.59	n.d.	0.59	0.14	0.18	0.34	0.25	low risk
33	1.00	0.53	1.00	0.15	0.13	0.94	0.50	high risk **
35	1.00	0.08	1.00	0.15	0.07	0.94	0.51	high risk **

n.d.—not determined; SD—standard deviation; * evaluation of risk level based on Jensen et al. [23] and Dagnino et al. [28]; ** SD value > 0.40 indicates high uncertainty and no acceptable risk.

The combined IR_Chem index included the total and bioavailable fraction of the 16PAH content and was slightly lower compared to TP based on total PAHs only (Table 6). High chemical risk values (0.67–1.00) were found in nine sampling sites (8, 16, 18, 20, 21, 21a, 26, 33 and 35) (Table 6). On the contrary, the risk index ($TP_{16PAH\ bioav.}$) calculated for the bioavailable fraction was at a level below 0.1 (no-risk), and for one sampling site (sample 33), it was equal to 0.53. This sample had an extremely high 16PAH content (316,085 μg kg^{-1}, Table S3) and, thus, high 16PAH bioav. (1184 μg kg^{-1}); however, 16PAH bioav. accounted for only 0.4% of the total PAH content. Although the consideration of bioavailability in environmental risk analyses is recommended by other authors [8–10,46,50], so far, only a few papers describing the Triad method have considered availability analysis, and this has been for metals only [12,14,20,61].

As described in Section 3.2, organisms reacted differently to the presence of contaminants, resulting in different risk index values in relation to ecotoxicological parameters (Table 4). The risk values for the soil retention function (IR_reten) did not exceed 0.22, thus indicating no risk. When considering single parameters, low risk was found in only two (RI for LUI_Scr and IRG_SIN) or three cases (RI for IRG_LEP). Interestingly, for *T. palyurus*, characterised by the highest sensitivity (Figure 2), the risk index was 0.00, undoubtedly related to the higher toxicity found in the reference soil; similar observations were also reported by Gutierrez et al. [15] and Niemeyer et al. [14], indicating that the selection of a reference site is a critical point in the risk assessment process.

The RI values for tests describing the effect of pollutants on the soil's habitat function were significantly different; IR_habit was three times higher and resulted mainly from high indices for *H. incongruens* and *L. sativa* (Table 4), for which the highest toxicity was observed. The risk index value finally calculated for all seven ecotoxicological endpoints (IR_Ecotox) was in the range of 0.01–0.45 (Table 6) and indicated no risk in 80% of the sampling sites; only for four sampling points (6, 18, 26 and 30) could the risk be described as low.

Upon analysing the risk indices for the ecological line of evidence (Table 6), no risk was found for 50% of the samples (IR_Ecol from 0.00 to 0.23). The remaining area showed low risk (0.32–0.44), and only soil 18 showed medium risk (IR_Ecol = 0.63), closely linked to the lower activity of NIT, BR, SIR, MB and CMIN in this case (Table 5), due to the acidic soil reaction ($pH_{KCl} = 4.1$).

The integrated IntRisk (calculated from the three independent LoEs using Equation (4)) ranged from 0.19 to 0.94 (Table 6). The risk level was determined according to the classification given by Jensen et al. [23] and Dagnino et al. [28], according to the following: IntRisk $\leq$ 0.25 no risk; 0.25 < IntRisk $\leq$ 0.50 low risk; 0.50 < IntRisk $\leq$ 0.75 moderate risk and 0.75 < IntRisk $\leq$ 1.00 high risk. Furthermore, in order to determine the uncertainty of the results obtained, the SD between LoEs was calculated, assuming that IntRisk values with a standard deviation greater than 0.4 [23] indicate high uncertainty in the risk assessment and unacceptable risk.

The lowest IntRisk below 0.25 (no risk) was found in sampling points 3, 19, 22a and 29, located at a distance of about 5–8 km from the coking plant (the main pollution source in the area), with 16PAHs from 989 to 1620 μg kg^{-1}. A low risk level (IntRisk 0.26–0.43) was associated with eight locations (4, 5, 6, 7, 17, 21, 27 and 32) unevenly distributed over the study area with a similar 16PAH concentration (not exceeding 1500 μg kg^{-1}), and due to the low discrepancy in the results between the three lines (SD values < 0.4), the risk should be considered acceptable. For the other six soils, the IntRisk ranged from 0.51 to 0.65 (moderate risk) and was mainly linked with high IR_Chem values; based on adopted criteria [23,28], risk in five sampling sites (8, 18, 20, 21a and 30) can be assessed as acceptable. Only for one site (sample 16 located near the coal recovery plant) were the differences between three LoEs high, indicating that additional evaluation in this area is needed. The IntRisk had very high and unacceptable values (0.77 and 0.94, respectively) in only three sampling points (26, 33 and 35), representing less than 10% of the study area. These sites were in close vicinity to the coke plant (site 33 approx. 400 m and site 35 about 100 m from the source) and to the asphalt processing plant (site 26). In addition, extremely high concentrations of 16 PAHs were recorded at these points (Table S3): 7604 μg kg^{-1} (site 26), 316,085 μg kg^{-1} (site 33) and 21,768 μg kg^{-1} (site 35). However, it should be emphasised that the ecological and ecotoxicological risk indices in this sampling point were low and did not exceed 0.25. In the previous study of Klimkowicz-Pawlas et al. [22], describing the Tier 1 assessment, almost 50% of the area was classified under medium and high risk. Moreover, high differences between chemical, ecological and ecotoxicological RI indexes were observed.

In our detailed risk assessment, a total of 15 different indicators giving information on the PAH content, bioavailability, toxicity and soil biota activity were included. This comprehensive approach is very valuable, as previous studies mostly reported a screening risk assessment [12,17,20,22] and analysed only chemical and ecotoxicological data [4,17,20,61], and only a few papers [14,19] described higher tiers of the Triad procedure. Most often, these studies focused on post-industrial sites contaminated by metals [11–13,17–19], and only a few papers focused on sites contaminated by PAHs [21,22]. A very important issue in the Triad methodology is the weighting of results from different disciplines. Equal weighting is recommended by default [10,46], and this approach was used in our study. However, when we assume that chemical analyses (despite taking into account the bioavailability of PAHs) are less important than the results of the ecotoxicological and ecological tests and we adopt the different weights respectively, e.g., 1, 1.5 and 2 suggested by Dagnino et al. [28] or Terekhova et al. [13], the final result of the risk assessment may be quite different, especially in cases where the contribution of IR_Chem to the total IntRisk is very high. This also applies to our study, where the unacceptable risks found at sampling points 26, 33 and 35 are mainly due to extremely high chemical indices (Table 6). Recalculating the data with different weights gave a lower IntRisk index (0.41, 0.32 and 0.31 for sites 26, 33 and 35 respectively), within the range of low risk values. Nevertheless, due to the high uncertainty of the results (SD > 0.4), the risk is still unacceptable.

4. Conclusions

The detailed assessment using the Triad approach was conducted for the agricultural area contaminated with organic pollutants, with PAHs as an example. A comprehensive set of 15 indicators was analysed and subsequently scaled and integrated to obtain risk indicators for the individual assessment routes and the final integrated IntRisk. The values for IR_Chem, IR_Ecotox and IR_Ecol were 0.20–1.00, 0.00–0.45 and 0.00–0.63, respectively, while IntRisk ranged from 0.19 to 0.94 and was mainly dependent on high chemical risk index values. Almost 90% of the study area had no or low risk (IntRisk was < 0.5). Very high unacceptable risk (IntRisk 0.77–0.94) was found at only three sampling sites located in close proximity (100–400 m) to a coke plant or asphalt processing plant. Our study revealed that the integration of data on the organism's response and ecological indicators into the risk assessment process provides a realistic assessment and allows for deriving quantitative risk indexes and for delineating the area of the high risk which needs additional action, e.g., further monitoring, changes in land use or remediation.

Several important points should be highlighted, which are crucial for the final outcome of an environmental risk assessment. Our research confirms that the risk assessment cannot be based solely on the total content of contaminants; the determination of the available fraction of compounds should also be included in the chemical line of evidence. In addition, a more realistic assessment is provided by the consideration of toxicity tests and ecological indicators. However, it has to be kept in mind that no species is sensitive to all contaminants, so bioassays representing different taxonomic groups of organisms should be included in the studies. A certain challenge is the selection of a suitable reference site, which should be uncontaminated and have pedological characteristics similar to those of contaminated sites. Sometimes, however, despite meeting these criteria, higher toxicity or lower biological activity may be observed for the reference site, which may be related to the presence of other contaminants not considered in the study. Another issue is the use of the appropriate weighting of results obtained using equal weights (assuming an equal influence of different LoEs on the final result) or different weights (assigning higher weights to indicators that better reflect the state of the soil ecosystem). In each case, the final value of the risk index may be different.

However, it is important to emphasise that the calculation of integrated risk indexes provides a more comprehensive and realistic assessment, enables the prioritisation of contaminated areas and allows for a decision on the actions needed at a site. Such a methodology can be used in the future to assess other pollutants that are a threat to agricultural lands, including emerging contaminants such as pharmaceuticals or microplastics.

Supplementary Materials: The following supporting information can be downloaded at: https://www.mdpi.com/article/10.3390/agriculture13071353/s1, Table S1: Indicators used for the characterisation of chemical, ecotoxicological and ecological lines of evidence; Table S2: General scheme of the Triad procedure for detailed assessment in our study; Table S3: Raw data for chemical, ecotoxicological and ecological indicators used for integrated risk calculation.

Author Contributions: Conceptualisation, A.K.-P. and B.M.-K.; methodology, A.K.-P.; formal analysis, A.K.-P.; investigation, A.K.-P.; data curation, A.K.-P. and B.S.; writing—original draft preparation, A.K.-P.; writing—review and editing, A.K.-P. and B.S.; visualisation, A.K.-P.; supervision, A.K.-P. and B.M.-K.; project administration, A.K.-P.; funding acquisition, A.K.-P. All authors have read and agreed to the published version of the manuscript.

Funding: The study was carried out under the statutory projects of IUNG-PIB no. 4.1.2 (Ecological risk assessment of agricultural soils contaminated with organic pollutants) and no. 4.07 (Application of ecological risk assessment procedure for the evaluation of soil retention function for chemical pollutants).

Institutional Review Board Statement: Not applicable.

Data Availability Statement: All data are available from the authors of the publication.

Acknowledgments: The authors are grateful to the staff of the Soil Science Department at IUNG, especially Urszula Pasternak for the technical assistance and Magdalena Łysiak for the map preparation.

Conflicts of Interest: The authors declare no conflict of interest.

Abbreviations

16PAH	total content of 16 polycyclic aromatic hydrocarbons
16PAH bioav	content of the bioavailable fraction of 16 polycyclic aromatic hydrocarbons
BR	basal soil respiration
CMIN	carbon mineralisation
CV	coefficient of variation
DH	dehydrogenases activity
EC_{10}	10% effect concentration
GC-MS	gas chromatography–mass spectrometry
GI	growth inhibition of *H. incongruens*
IF	inhibition of food ingestion by *T. platyurus*
IntRisk	overall integrated risk index
IR_Chem	integrated risk for the chemical line of evidence
IR_Ecol	integrated risk for the ecological line of evidence
IR_Ecotox	integrated risk for the ecotoxicological line of evidence
IR_habit	combined risk for soil habitat function
IR_reten	combined risk for soil retention function
IRG_LAC	inhibition of root growth of *L. sativa* in solid soil samples
IRG_LEP	inhibition of root growth of *L. sativum* in soil leachates
IRG_SIN	inhibition of root growth of *S. alba* in soil leachates
IRG_SINS	inhibition of root growth of *S. alba* in solid soil samples
LoEs	lines of evidence
LUI_Scr	luminescence inhibition of *A. fischeri*
Max	maximum value
MB	microbial biomass
Min	minimum value
MPC	maximum permissible concentration
NIT	nitrification potential

NOEC	no-observed-effect concentration
QA/QC	quality assurance/quality control
r	correlation coefficient
RI	risk values for single parameters
SD	standard deviation
SIR	substrate-induced soil respiration
TC	total carbon content
TN	total nitrogen content
TOC	total organic carbon content
TP	toxic pressure of 16 PAH mixture
TP′	toxic pressure for individual PAH compounds

References

1. Rodríguez-Eugenio, N.; McLaughlin, M.; Pennock, D. *Soil Pollution: A Hidden Reality*; Food and Agriculture Organization of the United Nations: Rome, Italy, 2018; p. 142.
2. FAO; ITPS. *Status of the World's Soil Resources (SWSR)—Main Report*; Food and Agriculture Organization of the United Nations; Intergovernmental Technical Panel on Soils: Rome, Italy, 2015; p. 648.
3. Cachada, A.; da Silva, E.F.; Duarte, A.C.; Pereira, R. Risk assessment of urban soils contamination: The particular case of polycyclic aromatic hydrocarbons. *Sci. Total Environ.* **2016**, *551–552*, 271–284. [CrossRef]
4. Lors, C.; Ponge, J.-F.; Damidot, D. Environmental hazard assessment by the Ecoscore system to discriminate PAH-polluted soils. *Environ. Sci. Pollut. Res.* **2018**, *25*, 26747–26756. [CrossRef] [PubMed]
5. Wu, J.; Li, K.; Ma, D.; Yu, N.; Chai, C. Contamination, source identification, and risk assessment of polycyclic aromatic hydrocarbons in agricultural soils around a typical coking plant in Shandong, China. *Hum. Ecol. Risk Assess.* **2018**, *24*, 225–241. [CrossRef]
6. Klimkowicz-Pawlas, A.; Debaene, G. Screening risk assessment of agricultural areas under a high level of anthropopressure based on chemical indexes and VIS-NIR spectroscopy. *Molecules* **2020**, *25*, 3151. [CrossRef]
7. Chakravarty, P.; Chowdhury, D.; Deka, H. Ecological risk assessment of priority PAHs in crude oil contaminated soil and its impacts on soil biological properties. *J. Hazard. Mater.* **2022**, *437*, 129325. [CrossRef]
8. Ortega-Calvo, J.-J.; Harmsen, J.; Parsons, J.R.; Semple, K.T.; Aitken, M.D.; Ajao, C.; Eadsforth, C.; Galay-Burgos, M.; Naidu, R.; Oliver, R.; et al. From bioavailability science to regulation of organic chemicals. *Environ. Sci. Technol.* **2015**, *49*, 10255–10264. [CrossRef] [PubMed]
9. Peijnenburg, W.J.G.M. Implementation of Bioavailability in Prospective and Retrospective Risk Assessment of Chemicals in Soils and Sediments. In *Bioavailability of Organic Chemicals in Soil and Sediment. The Handbook of Environmental Chemistry*; Ortega-Calvo, J.J., Parsons, J.R., Eds.; Springer: Cham, Switzerland, 2020; Volume 100, pp. 391–422.
10. *ISO 19204:2017*; Soil Quality—Procedure for Site-Specific Ecological Risk Assessment of Soil Contamination (TRIAD Approach). ISO: Geneva, Switzerland, 2017.
11. Niemeyer, J.C.; Moreira-Santos, M.; Nogueira, M.A.; Carvalho, G.M.; Ribeiro, R.; da Silva, E.M.; Sousa, J.P. Environmental risk assessment of a metal-contaminated area in the Tropics. Tier I: Screening phase. *J. Soil. Sediment* **2010**, *10*, 1557–1571. [CrossRef]
12. Ribé, V.; Aulenius, E.; Nehrenheim, E.; Martell, U.; Odlare, M. Applying the Triad method in a risk assessment of a former surface treatment and metal industry site. *J. Hazard. Mater.* **2012**, *207–208*, 15–20. [CrossRef] [PubMed]
13. Terekhova, V.A.; Pukalchik, M.A.; Yakovlev, A.S. The Triad approach to ecological assessment of urban soils. *Eurasian Soil Sci.* **2015**, *47*, 952–958. [CrossRef]
14. Niemeyer, J.C.; Moreira-Santos, M.; Ribeiro, R.; Rutgers, M.; Nogueira, M.A.; da Siva, E.M.; Sousa, P.J. Ecological risk assessment of metal-contaminated area in the tropics. Tier II: Detailed assessment. *PLoS ONE* **2015**, *10*, 1–25. [CrossRef]
15. Gutiérrez, L.; Garbisu, C.; Ciprián, E.; Becerril, J.M.; Soto, M.; Etzebarria, J.; Madariaga, J.M.; Antigüedad, I.; Epelde, L. Application of ecological risk assessment based on a novel TRIAD-tiered approach to contaminated soil surrounding a closed non-sealed landfill. *Sci. Total Environ.* **2015**, *514*, 49–59. [CrossRef] [PubMed]
16. Da Silva, M.B.; Abrantes, N.; Patinha, C.; da Silva, E.F.; Marques, J.C.; Gonçalves, F.; Pereira, R. Soil ecotoxicological screening (tier 1) for a diffuse-contaminated drainage area surrounding a lacustrine ecosystem in the Centre of Portugal. *J. Soil Sediment* **2018**, *18*, 189–204. [CrossRef]
17. Son, J.; Kim, J.-G.; Hyun, S.; Cho, K. Screening level ecological risk assessment of abandoned metal mines using chemical and ecotoxicological lines of eviudence. *Environ. Pollut.* **2019**, *249*, 1081–1090. [CrossRef] [PubMed]
18. Hong, Y.K.; Yoon, D.H.; Chae, M.J.; Ko, B.K.; Kim, S.C. Ecological risk assessment of heavy metal-contaminated soil using the triad approach. *J. Soil Sediment* **2021**, *21*, 2732–2743. [CrossRef]
19. Kim, D.; Kwak, J.I.; Hwang, W.; Lee, Y.; Lee, Y.S.; Kim, J.-I.; Hong, S.; Hyun, S.; An, Y.-J. Site-specific ecological risk assessment of metal-contaminated soils based on the TRIAD approach. *J. Hazard. Mater.* **2022**, *434*, 128883. [CrossRef] [PubMed]
20. Pereira, J.L.; Pereira, P.; Padeiro, A.; Gonçalves, F.; Amaro, E.; Leppe, M.; Verkulich, S.; Hughes, K.A.; Peter, H.-U.; Canário, J. Environmental hazard assessment of contaminated soils in Antarctica: Using a structured tier 1 approah to inform decision-making. *Sci. Total Environ.* **2017**, *574*, 443–454. [CrossRef] [PubMed]

21. Gworek, B.; Baczewska-Dąbrowska, A.H.; Kalinowski, R.; Górska, E.B.; Rekosz-Burlaga, H.; Gozdowski, D.; Olejniczak, I.; Graniewska, M.; Dmuchowski, W. Ecological risk assessment for land contaminated by petrochemical industry. *PLoS ONE* **2018**, *13*, e0204852. [CrossRef]
22. Klimkowicz-Pawlas, A.; Maliszewska-Kordybach, B.; Smreczak, B. Triad-based screening risk assessment of the agricultural area exposed to the long-term PAHs contamination. *Environ. Geochem. Health* **2019**, *41*, 1369–1385. [CrossRef]
23. Jensen, J.; Mesman, M.; Bierkens, J.; Loibner, A.; Rutgers, M.; Bogolte, T.; Celis, R.; Dirven-van Breemen, E.M.; Erlacher, E.; Ehlers, C.; et al. *Ecological Risk Assessment of Contaminated Land-Decision Support System for Site Specific Investigation*; National Institute for Public Health and Environment (RIVM): Bilthoven, The Netherlands, 2006; No. 711701047.
24. Klimkowicz-Pawlas, A.; Smreczak, B.; Ukalska-Jaruga, A. The impact of selected soil organic matter fractions on the PAH accumulation in the agricultural soils from areas of different anthropopressure. *Environ.Sci. Pollut. Res.* **2017**, *24*, 10955–10965. [CrossRef]
25. Ukalska-Jaruga, A.; Smreczak, B.; Klimkowicz-Pawlas, A. Soil organic matter composition as a factor affecting the accumulation of polycyclic aromatic hydrocarbons. *J. Soil Sediment* **2019**, *19*, 1890–1900. [CrossRef]
26. Smreczak, B. *Bioaccessibility of Polycyclic Aromatic Hydrocarbons (PAHs) in Soils*; Monografie i Rozprawy Naukowe; IUNG-PIB: Puławy, Poland, 2018; Volume 56, p. 110. (In Polish)
27. *ISO 15799:2019*; Soil Quality—Guidance on the Ecotoxicological Characterization of Soils and Soil Materials. ISO: Geneva, Switzerland, 2019.
28. Dagnino, A.; Sforzini, S.; Dondero, F.; Fenoglio, S.; Bona, E.; Jensen, J.; Viarengo, A. A „Weight-of-Evidence" approach for the integration of environmental "Triad" data to assess ecological risk and biological vulnerability. *Integr. Environ. Assess.* **2008**, *4*, 314–326. [CrossRef]
29. Sorvari, J.; Schultz, E.; Haimi, J. Assessment of ecological risks at former landfill site using TRIAD procedure and multicriteria analysis. *Risk Anal.* **2013**, *33*, 203–219. [CrossRef]
30. *ISO 21268-1:2019*; Leaching Procedure for Subsequent Chemical and Ecotoxicological Testing of Soil and Soil Materials. Part 1: Batch Test Using a Liquid to Soil Ratio of 2 L/kg. ISO: Geneva, Switzerland, 2019.
31. Microbics Corporation. *Microtox Manual Toxicity Testing Handbook*; Microbics Corporation: Carlsband, CA, USA, 1992.
32. Nałęcz-Jawecki, G.; Szczęsny, Ł.; Solecka, D.; Sawicki, J. Short ingestion test as alternative proposal for conventional range finding assays with *Thamnocephalus platyurus* and *Brachionus calyciflorus*. *Int. J. Environ. Sci. Technol.* **2011**, *8*, 687–694. [CrossRef]
33. Szara, M.; Baran, A.; Klimkowicz-Pawlas, A.; Tarnawski, M. Ecotoxicological characteristics and ecological risk assessment of trace elements in the bottom sediments of the Rożnów reservoir (Poland). *Ecotoxicology* **2020**, *29*, 45–57. [CrossRef] [PubMed]
34. Phytotestkit. *Phytotoxkit Liquid Samples for Determination of the Direct Effects of Chemicals on Seed Germination and Early Growth of Plants*; MicroBioTest Inc.: Nazareth, Belgium, 2012; p. 22.
35. *ISO 14371:2012*; Water Quality. Determination of Freshwater Sediment Toxicity to Heterocypris incongruens (Crustacea Ostracoda). ISO: Geneva, Switzerland, 2012.
36. *ISO 18763:2016*; Determination of the Toxic Effects of Pollutants on Germination and Early Growth of Higher Plants. ISO: Geneva, Switzerland, 2016.
37. *ISO 17126:2005*; Determination of the Effects of Pollutants on Soil Flora—Screening Test for Emergence of Lettuce Seedlings (*Lactuca sativa* L.). ISO: Geneva, Switzerland, 2016.
38. Casida, L.E.; Klein, D.A.; Santoro, T. Soil dehydrogenase activity. *Soil Sci.* **1964**, *98*, 371–376. [CrossRef]
39. OECD. *Test No. 217: Soil Microorganisms: Carbon Transformation Test*; OECD Guidelines for the Testing of Chemicals, Section 2; OECD Publishing: Paris, France, 2000. [CrossRef]
40. Maliszewska-Kordybach, B.; Klimkowicz-Pawlas, A.; Smreczak, B.; Janusauskaite, D. Ecotoxicological effect of phenanthrene on nitrifying bacteria in soils of different properties. *J. Environ. Qual.* **2007**, *36*, 1635–1640. [CrossRef] [PubMed]
41. Verbruggen, E.M.J. *Environmental Risk Limits for Polycyclic Aromatic Hydrocarbons (PAHs)—For Direct Aquatic, Benthic, and Terrestrial Toxicity*; National Institute for Public Health and the Environment (RIVM): Bilthoven, The Netherlands, 2012; No. 607711007.
42. Crevecoeur, S.; Debacker, V.; Joaquim-Justo, C.; Gobert, S.; Scippo, M.-L.; Dejonghe, W.; Martin, P.; Thome, J.-P. Groundwater quality assessment of one former industrial site in Belgium using a TRIAD-like approach. *Environ. Pollut.* **2011**, *159*, 2461–2466. [CrossRef]
43. Maliszewska-Kordybach, B. Polycyclic aromatic hydrocarbons in agricultural soils in Poland: Preliminary proposals for criteria to evaluate the level of soil contamination. *Appl. Geochem.* **1996**, *11*, 121–127. [CrossRef]
44. Dz, U. Regulation of the Minister of the Environment of 1 September 2016 on the Conduct of the Assessment of Contamination of the Surface of the Earth. 2016. Available online: http://isap.sejm.gov.pl/isap.nsf/DocDetails.xsp?id=WDU20160001395 (accessed on 5 May 2023). (In Polish)
45. Semenzin, E.; Critto, A.; Rutgers, M.; Marcomini, A. Integration of bioavailability, ecology and ecotoxicology by three lines of evidence into ecological risk indexes for contaminated soil assessment. *Sci. Total Environ.* **2008**, *389*, 71–86. [CrossRef] [PubMed]
46. Grassi, G.; Lamy, I.; Pucheux, N.; Ferrari, B.J.D.; Faburé, J. State of the art of Triad-based ecological risk assessment: Current limitations and needed implementations in the case of soil diffuse contamination. *Front. Environ. Sci.* **2022**, *10*, 878238. [CrossRef]
47. Ukalska-Jaruga, A.; Smreczak, B. The impact of organic matter on Polycyclic Aromatic Hydrocarbon (PAH) availability and persistence in soils. *Molecules* **2020**, *25*, 2470. [CrossRef] [PubMed]

48. Persoone, G.; Marsalek, B.; Blinova, I.; Törökne, A.; Zarina, D.; Manusadzianas, L.; Nalecz-Jawecki, G.; Tofan, L.; Stepanova, N.; Tothova, L.; et al. A practical and user-friendly toxicity classification system with microbiotests for natural waters and wastewaters. *Environ. Toxicol.* **2003**, *18*, 395–402. [CrossRef] [PubMed]
49. Čvančarová, M.; Křesinová, Z.; Cajthaml, T. Influence of the bioaccessible fraction of polycyclic aromatic hydrocarbons on the ecotoxicity of historically contaminated soil. *J. Hazard. Mater.* **2013**, *254–255*, 116–124. [CrossRef]
50. Umeh, A.C.; Duan, L.; Naidu, R.; Semple, K.T. Residual hydrophobic organic contaminants in soil: Are they a barrier to risk-based approaches for managing contaminated land? *Environ. Int.* **2017**, *98*, 18–34. [CrossRef]
51. Yu, L.; Duan, L.; Naidu, R.; Semple, K.T. Abiotic factors controlling bioavailability and bioaccessibility of polycyclic aromatic hydrocarbons in soil: Putting together a bigger picture. *Sci. Total Environ.* **2018**, *613–614*, 1140–1153. [CrossRef]
52. Cipullo, S.; Prpich, G.; Campo, P.; Coulon, F. Assessing bioavailability of complex mixtures in contaminated soils: Progress made and research need. *Sci. Total Environ.* **2018**, *615*, 708–723. [CrossRef]
53. Niemeyer, J.C.; Lolata, G.B.; de Carvalho, G.M.; Da Silva, E.M.; Sousa, J.P.; Nogueira, M.A. Microbial indicators of soil health as tools for ecological risk assessment of a metal contaminated site in Brazil. *Appl. Soil Ecol.* **2012**, *59*, 96–105. [CrossRef]
54. Bhaduri, D.; Sihi, D.; Bhowmik, A.; Verma, B.C.; Munda, S.; Dari, B. A review on effective soil health bio-indicators for ecosystem restoration and sustainability. *Front. Microbiol.* **2022**, *13*, 938481. [CrossRef]
55. Asensio, V.; Guala, S.D.; Vega, F.L.; Covelo, E.F. A soil quality index for reclaimed mine soils. *Environ. Toxicol. Chem.* **2013**, *32*, 2240–2248. [CrossRef]
56. Garrigues, E.; Corson, M.S.; Angers, D.A.; van der Werf, H.M.G.; Walter, C. Soil quality in Life Cycle Assessment: Towards development of an indicator. *Ecol. Indic.* **2012**, *18*, 434–442. [CrossRef]
57. Muhlbachova, G.; Sagova-Mareckova, M.; Omelka, M.; Szakova, J.; Tlust, P. The influence of soil organic carbon on interactions between microbial parameters and metal concentrations at log-term contaminated site. *Sci Total Environ.* **2015**, *502*, 218–223. [CrossRef]
58. Gospodarek, J.; Rusin, M.; Barczyk, G.; Nadgórksa-Socha, A. The effect of petroleum-derived substances and their bioremediation on soil enzymatic activity and soil invertebrates. *Agronomy* **2021**, *11*, 80. [CrossRef]
59. Suszek-Łopatka, B.; Maliszewska-Kordybach, B.; Klimkowicz-Pawlas, A.; Smreczak, B. The drought and high wet soil condition impact on PAH (phenathrene) toxicity towards nitrifying bacteria. *J. Hazard. Mater.* **2019**, *368*, 274–280. [CrossRef] [PubMed]
60. Creamer, R.E.; Hannula, S.E.; Van Leeuwen, J.P.; Stone, D.; Rutgers, M.; Schmelz, R.M.; de Ruiter, P.C.; Hendriksen, N.B.; Bolger, T.; Bouffaud, M.L.; et al. Ecological network analysis reveals the inter-connection between soil biodiversity and ecosystem function as affected by land use across Europe. *Appl. Soil Ecol.* **2016**, *97*, 112–124. [CrossRef]
61. Karjalainen, A.-M.; Kilpi-Koski, J.; Väisänen, A.O.; Penttinen, S.; van Gestel, C.A.; Penttinen, O.-P. Ecological risks of an old wood impregnation mill: Application of the Triad approach. *Integr. Environ. Assess. Manag.* **2009**, *5*, 379–389. [CrossRef]

agriculture

MDPI

Article

Bioremediation of Battery Scrap Waste Contaminated Soils Using Coco Grass (*Cyperus rotundus* L.): A Prediction Modeling Study for Cadmium and Lead Phytoextraction

Arwa A. AL-Huqail [1], Mostafa A. Taher [2], Ivan Širić [3], Madhumita Goala [4], Bashir Adelodun [5,6], Kyung Sook Choi [6], Piyush Kumar [7], Vinod Kumar [8], Pankaj Kumar [8,*] and Ebrahem M. Eid [9,*]

1 Department of Biology, College of Science, Princess Nourah bint Abdulrahman University, P.O. Box 84428, Riyadh 11671, Saudi Arabia
2 Biology Department, Faculty of Science and Arts, King Khalid University, Mohail Assir 61321, Saudi Arabia
3 Faculty of Agriculture, University of Zagreb, Svetosimunska 25, 10000 Zagreb, Croatia
4 Department of Environment Science, Graphic Era (Deemed to be University), Dehradun 248002, India
5 Department of Agricultural and Biosystems Engineering, University of Ilorin, PMB 1515, Ilorin 240103, Nigeria
6 Department of Agricultural Civil Engineering, Kyungpook National University, Daegu 41566, Republic of Korea
7 Department of Science, Vivek College of Education, Moradabad Road, Bijnor 246701, India
8 Agro-Ecology and Pollution Research Laboratory, Department of Zoology and Environmental Science, Gurukula Kangri (Deemed to be University), Haridwar 249404, India
9 Botany Department, Faculty of Science, Kafrelsheikh University, Kafr El-Sheikh 33516, Egypt
* Correspondence: rs.pankajkumar@gkv.ac.in (P.K.); ebrahem.eid@sci.kfs.edu.eg (E.M.E.)

Abstract: With the increasing demand for electronic devices that use batteries, e-waste is also becoming a major threat to the environment. Battery e-waste contains hazardous heavy metals that affect the health of the soil ecosystem. Thus, the present study evaluates the cadmium (Cd) and lead (Pb) phytoextraction potential of coco grass (*Cyperus rotundus* L.) grown in soils contaminated with battery scrap waste (BSW). Pot experiments were conducted to grow *C. rotundus* under different treatments (0%: control, T1: 1%, T2: 2%, T3: 3%, and T4: 4%) of BSW mixed with soil (w/w). The results showed that BSW mixing significantly ($p < 0.05$) increased the physicochemical properties and heavy metal (Cd and Pb) content in the soil. BSW mixing resulted in a reduction in growth and biochemical traits of *C. rotundus* and an increase in oxidative stress enzymes with an increase in BSW dose. The Pearson correlation studies also showed that soil HM concentration had a negative influence on the growth and biochemical parameters of *C. rotundus*. The bioaccumulation and translocation factor analysis showed that *C. rotundus* was a hyperaccumulator plant with a maximum accumulation of Cd and Pb (38.81 and 109.06 mg·kg^{-1}) in root parts followed by the whole plant (277.43 and 76.10 mg·kg^{-1}) and shoot (21.30 and 22.65 mg·kg^{-1}) parts. Moreover, predictive models based on multiple linear regression (MLR) and artificial neural network (ANN) approaches were developed for Cd and Pb uptake by *C. rotundus*. Mathematical modeling results showed that soil properties were useful to construct quality MLR and ANN models with good determination coefficient ($R^2 > 0.98$), model efficiency (ME > 0.99), and low root mean square error (RMSE < 5.72). However, the fitness results of the ANN models performed better compared with those of the MLR models. Overall, this study presents an efficient and sustainable strategy to eradicate hazardous HMs by growing *C. rotundus* on BSW-contaminated soils and reducing its environmental and health consequences.

Keywords: hazardous waste; heavy metals; mathematical models; phytoremediation; soil pollution; waste management

1. Introduction

Batteries are devices that convert stored chemical energy into electrical energy, which can be used to power electronic devices [1]. Different types of batteries, such as nickel-cadmium

(Ni-Cd), lead-acid (Pb-acid), lithium-ion (Li-ion), zinc-carbon (Zn-C), etc., are used in various sectors, including commercial, communication, transportation, and household [2]. It is estimated that the global battery market shared approximately USD 104.31 billion in 2022 with a >15% of compound annual growth rate (CAGR). The revenue of the battery market is expected to hit USD 329.84 billion by 2030 [3]. India imported more than 0.62 million tons of Ni-Cd-based ores during 2017–2018 to meet its battery production demands [4]. However, the management of spent batteries has continued to become a serious environmental issue since the efficiency of these batteries is reduced after a certain period and the batteries need to be discarded. In the last decades, rapid industrialization and urbanization have created devastating impacts by releasing hazardous wastes into our environment [5]. Spent BSW is one of the most dangerous hazardous wastes that pollute soil, water, and air [6]. Being hazardous waste, battery scrap waste (BSW) has become a global issue with the rapid growth of electronic devices and electric vehicles worldwide.

With an increasing demand for batteries, the problem of BSW management is also becoming threatening. Developing countries often lack the necessary infrastructure to sustainably manage BSW, which leads to increased environmental and health risks (Guerrero et al., 2013). Thus, significant portions of BSW, including metal plates, plastic, straps, intercell connectors, spent electrolytes (acids and bases), etc., end up in landfill rather than being properly recycled [7]. Unsafe disposal and poor recycling of BSW can also lead to several toxic heavy metals (HMs), such as Cd, Pb, Ni, Co, and Li, entering the soil and groundwater continuum [8]. Elevated levels of hazardous HMs in soil and groundwater can inflict significant health damage to humans, animals, plants, and soil microbes [9,10]. In particular, Cd and Pb are reported to cause several respiratory issues, kidney damage, bone osteoporosis, cardiovascular effects, and reproductive and developmental problems, among many others. [11,12]. Plants also experience severe effects of high Cd and Pb stress, such as reduced growth and development, chlorosis, leaf damage, altered mineral balance, oxidative stress, impaired fertilization, and accumulation in edible parts [13,14]. Recent studies have shown that soils impacted by BSW have an extremely high content of Cd and Pb, which requires urgent attention to prevent further movement to upper trophic levels, including human beings [15,16].

Environmentally friendly decontamination of soils impacted by hazardous HMs has become crucial to avoid any health consequences. In recent times, phytoremediation has emerged as one of the best approaches for the efficient removal of several HMs using hyperaccumulator plant species [17]. Several authors reported successful reclamation of soils contaminated with HMs, including those affected by BSW [18], landfill leachate [19], coal mining [20], and copper smelting [21], using hyperaccumulator plants. Such plants respond to physiological stress by sequestering HMs into their vegetative parts, chelating HMs using specialized chelators and transporter molecules, and producing antioxidant enzymes [22,23]. Therefore, these features of plant phytoremediation make them ideal candidates for the bio-extraction of HMs from contaminated soils. Moreover, the efficiency of phytoremediation systems can be enhanced by the use of several mathematical models [24]. The pollutant removal process of plants can be efficiently optimized by regulating several intrinsic and extrinsic factors. For this, different types of mathematical models, such as logistic, multiple linear regression (MLR), artificial neural networks (ANN), etc., are widely used [25].

Cyperus rotundus L., also known as coco grass, nut grass, nut sedge, or java grass, belongs to the family Cyperaceae and is a perennial plant widely found in Africa, South Asia, and central Europe. It is a small plant, considered a weed, that generally reproduces rapidly in moist soils to take over soil fertilizers [26]. It propagates by tubers, basal bulbs, division, and seeds. The tubers sprout into numerous rhizomes and expand vertically, forming several roots which radiate horizontally [27]. Tubers are found mostly in the upper 45 cm of the soil surface and can remain unaffected under severe droughts for several years. It is estimated that *C. rotundus* can produce over 10–30 million tubers/ha in one season, which may negatively affect crop production [28]. However, evidence shows that tubers of *C. rotundus* are consumed as a medicinal and staple food by several tribal people in India and Africa [29,30]. Apart from its culinary uses, *C. rotundus* is also utilized for the phytoextraction of HMs from contaminated soils [31]. *C. rotundus* can tolerate high HM stress, thereby providing a solution to the contaminated soil management. Significant bio-extraction of HMs, including Cd, Pb, Zn, and Ni, was reported by Nwaichi et al. [32].

Considering the aforementioned, this study aimed to investigate the Cd and Pb phytoextraction potential of *C. rotundus* grown on BSW-contaminated soils. Laboratory experiments were conducted to understand the effect of different BSW loading rates on growth, biochemical and enzyme response, and HMs accumulation by *C. rotundus*.

2. Materials and Methods

2.1. Experimental Materials

Loam soil was collected from agricultural lands of rural Haridwar, Uttarakhand, India (29°53′13.6″ N and 78°08′15.2″ E), which was not previously exposed to any industrial or urban activities. The experimental soil was identified as having a loamy texture (clay: 19%, silt: 42%, and sand: 38%) based on the composition of clay, silt, and sand present in it. The mixed type of battery scrap waste (BSW) containing Pb-acid and Ni-Cd batteries was collected from a local scraper near Nakur town, Saharanpur district, Uttar Pradesh, India (29°55′38.6″N and 77°18′58.4″E). The acid and outer case of the batteries were carefully removed and internal parts were dismantled. Then, selected internal components were separated (cells, plates, connectors, and terminal posts) shredded, and converted into fine powder form using a portable industrial shredder with the help of a local scrapper. On the other hand, juvenile and healthy rhizomes (2–3 cm in length) of coco grass (*C. rotundus*) were also collected from the soil sampling site by digging 15 cm below the surface. The tubers were washed with tap water to remove adhering soil and collected in 100 gm capacity zip-locking aerated polythene bags and transported to the laboratory for further experiments.

2.2. Experimental Design and Operation

For the present study, phytoremediation experiments were conducted using rectangular plastic pots (35 × 20 × 15 cm; length × width × depth). Each pot was filled with 10 kg arable soil and mixed with an appropriate dose (w/w) of BSW—i.e., 0% (no BSW mixing), 1, 2, 3, and 4%, respectively. Figure 1 shows the design of phytoremediation experiments using *C. rotundus*. The soil was loosened to a depth of 8–10 cm in order to promote quick propagation. For each pot, a total of five identical rhizomes were sown in small holes at a depth of 5 cm. The holes were covered with soil and pots were equally watered using a borewell water supply to maintain consistent moisture. Tubers were sown on 1 March 2022 and experiments were terminated after 45 days, i.e., on 14 April 2022. For this, pots were placed under the open sky with a direct sunlight period of 10 h and natural environmental conditions. The tubers were allowed to sprout, and seedlings emerged within 15 days, followed by 30 days of vegetative plant growth. The plants were irrigated twice a week with borewell water, and other emerging weeds were immediately removed to avoid any experimental inference.

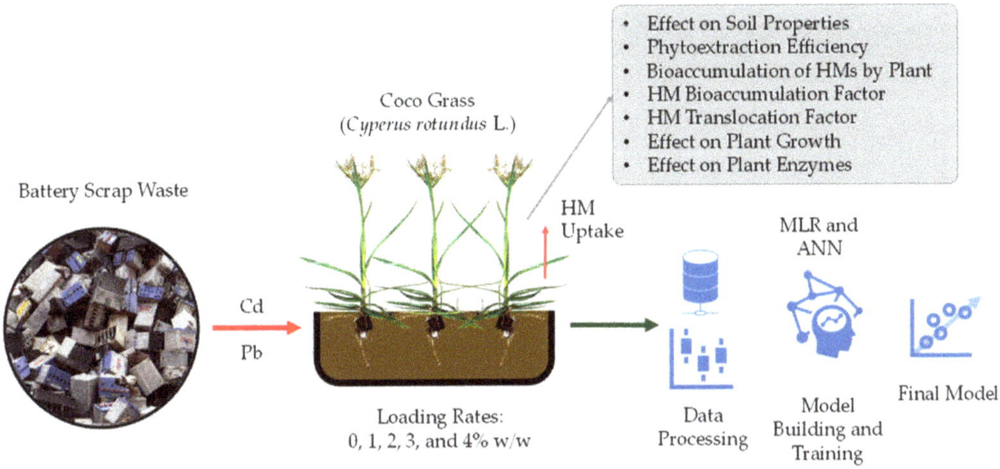

Figure 1. Experimental design for phytoextraction of Cd and Pb from soils contaminated with battery scrap waste using coco grass (*C. rotundus*).

2.3. Analytical and Instrumental Methods

The arable soil used in this study was analyzed before and after BSW mixing, as well as after phytoextraction experiments. Purposely, the pH of the soil was analyzed using a microprocessor-based digital meter (1611, ESICO International, Parwanoo, India). Cation exchange capacity (CEC) was measured as previously outlined by Kriti et al. [33]. Organic matter (OM) of soil was determined using Walkley and Black method [34]. Redox potential (mV) was measured using a meter. The texture of the soil was determined based on the percent composition of clay, silt, and sand content as previously outlined by Bharti et al. [35]. On the other hand, the contents of cadmium (Cd) and lead (Pb) in experimental soil and *C. rotundus* parts (shoot and root) were determined based on inductively coupled plasma optical emission spectroscopy (ICP-OES: 7300 DV, Perkin Elmer, Waltham, MA, USA). For this, total and bioavailable Cd and Pb contents were determined by two different extraction methods, i.e., di-acid and ethylenediaminetetraacetic acid (EDTA)-based digestion, respectively. For total metal content, a di-acid extraction solution with a 1:3 ratio containing $HClO_4$ and HNO_3 was used to digest soil samples at 180 °C for 2 h. For bioavailable heavy metal extraction, a 0.05 M EDTA solution was used to digest samples for 12 h on a mechanical shaker at room temperature [36]. The instrument was calibrated as per the manufacturer's instructions using standard reference materials. The detection limits of ICP-OES for Cd and Pb determination were 0.01 and 0.10 ppb, respectively. Appropriate quality control measures, such as replicate analysis, cross-validation, and certified reference materials, were ensured for the accuracy and precision of the results.

2.4. Data Analysis

In this study, the Cd and Pb removal efficiency and removal rate of *C. rotundus* grown on BSW-amended soils were computed using the following Equations (1) and (2):

$$\text{Removal efficiency (\%)} = [(\text{HMi} - \text{HMf})/\text{HMi}] \times 100 \quad (1)$$

$$\text{Removal Rate } (\text{mg}\cdot\text{kg}^{-1}\text{day}^{-1}) = \text{HMp}/\text{t} \quad (2)$$

where HMi and HMf refer to initial and final HM concentrations in arable soil (before and after experiments), while HMp indicates heavy metal concentration in *C. rotundus* after t time (45 days). The Cd and Pb uptake potential of *C. rotundus* was computed using the bioaccumulation factor (BAF) index [37] as given in the following Equation (3):

$$\text{Bioaccumulation factor (BAF)} = \text{HMp}/\text{HMs} \tag{3}$$

where, HMp and HMs denote the ratio of HM concentration in plant and experimental soil, respectively. The ability of *C. rotundus* to mobilize HM from its root region to shoot was calculated using the translocation factor [38] as per the following Equation (4):

$$\text{Translocation factor (Tf)} = \text{HMsh}/\text{HMro} \tag{4}$$

where, HMsh and HMro refer to the HM concertation in the shoot and root parts of *C. rotundus*, respectively.

Two different modeling approaches, i.e., multiple linear regression (MLR) and artificial neural network (ANN), were adopted to predict the amount of Cd and Pb accumulated by *C. rotundus*. Both MLR and ANN have widely accepted tools for monitoring HM in phytoextraction experiments [39,40]. MLR establishes a linear relationship between the input and output factors. In this study, the selected soil properties, such as pH, CEC, OM, and HM concentration, were taken as input (independent) variables while the total amount of heavy metal accumulated by *C. rotundus* was considered an output (dependent) variable. The following MLR function (Equation (5)) was used to compute the Cd and Pb uptake amount (y: $\text{mg}\cdot\text{kg}^{-1}$):

$$\text{y} = \text{a} + (\text{b} \times \text{pH}) + (\text{c} \times \text{CEC}) + (\text{d} \times \text{OM}) + (\text{e} \times \text{HMs}) \tag{5}$$

Here, a is the regression intercept, while b, c, d, and e are the regression coefficients for pH, CEC, OM, and HMs as independent variables, respectively.

On the other hand, the ANN approach was also used to predict HM uptake by *C. rotundus*. For this purpose, a neural network was created comprising three different processing layers, i.e., input, hidden, and output layers. For this, the input layer is comprised of four nodes or neurons (n = 4) referring to the same independent variables as in the case of the MLR model. The hidden layer had ten neurons (n = 10) while the output layer had only one neuron (n = 1), indicating the predicted response (y) in terms of effective HM uptake by *C. rotundus*. A widely used logistic model activation function was used for model training. This is a sigmoid-type function that can take in any number of inputs (x) and constrain the output to be between 0 and 1 [41]. The form of the model is given in Equation (6):

$$f(x) = 1/(1 + \exp(-x)) \tag{6}$$

The whole dataset was categorized into three groups (training: 70%; testing: 15%; and validation: 15%). The learning rate of the model was adjusted to 0.001 with maximum iterations of 10,000 and epochs of 100. Figure 2 shows the overall architecture of the proposed ANN with different numbers of input/output layers and their respective neurons. Using specific validation tools, such as measured vs. predicted response values, coefficient of determination (R^2), Nash–Sutcliffe model efficiency (ME) coefficient, and root mean square error (RMSE), the created MLR and ANN models were evaluated for the goodness of fit and accuracy [42]. The following Equations (7)–(9) were used for the computation of R^2, ME, and RMSE:

$$R^2 = 1 - \frac{\sum_{i=1}^{n} \left(y_{\hat{obs}} - \overline{y_{pre}}\right)^2}{\sum_{i=1}^{n} \left(y_{obs} - \overline{y_{mean}}\right)^2} \tag{7}$$

$$\text{ME} = 1 - \frac{\sum_{i=1}^{n} \left(y_{obs} - y_{pre}\right)^2}{\sum_{i=1}^{n} \left(y_{obs} - \overline{y_{mean}}\right)^2} \tag{8}$$

$$\text{RMSE} = \sqrt{\frac{\sum_{i=1}^{n} \left(y_{obs} - y_{pre}\right)^2}{n}} \tag{9}$$

where, y_{obs}, y_{pre}, and y_{mean} are experimental, predicted, and mean values of HM content in *C. rotundus*.

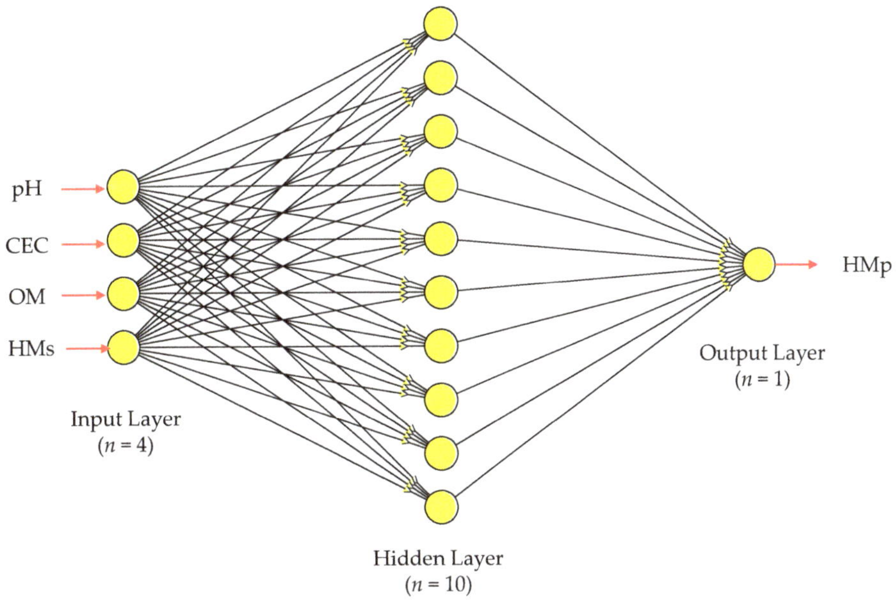

Figure 2. Design of the artificial neural network depicting different neurons and layers (Shi et al. 2023).

2.5. Software and Statistics

All experiments were performed in triplicate ($n = 3$). The data obtained in this study were analyzed using Pearson correlation and Tukey's post-hoc test to understand the significant impacts of BSW mixing on the soil as well as on *C. rotundus* properties. The level of statistical significance was $p < 0.05$. The data were analyzed using OriginPro (Version 2023, OriginLab Corp., Northampton, MA, USA) and the "*nntool*" function of MATLAB (Version 2021, MathWorks, Natick, MA, USA) software packages.

3. Results and Discussion

3.1. Impact of Battery Waste on Soil Properties and Cd-Pb Bioavailability

Table 1 depicts the impact of BSW mixing on the physicochemical and heavy metal properties of soil used for the cultivation of *C. rotundus*. The results show that BSW mixing had a significant ($p < 0.05$) impact on selected soil properties. In particular, it was observed that the initial pH (7.23) of arable soil was significantly ($p < 0.05$) increased after BSW mixing in different ratios, i.e., 7.55–8.35. This increase was for different BSW treatments, including T1, T2, T3, and T4 as follows: 4.42, 5.53, 9.40, and 15.49%, respectively. The increase in soil pH might be associated with the mixing of electrolytes (usually 30% NaOH) present in several Ni-Cd batteries. However, Pb-acid batteries also contain H_2SO_4, which might decrease the pH of soil [33]. Instead, the increased pH might be due to a larger proportion of Ni-Cd batteries in the BSW used in this study. Likewise, the CEC of soil was significantly ($p < 0.05$) increased by 5.13, 7.45, 8.92, and 10.51% in different BSW treatments, such as T1, T2, T3, and T4, respectively. The ability to retain and exchange ions is influenced by basic conditions, which can alter soil CEC [43]. However, the content of OM was drastically reduced after BSW addition from 2.27 to 2.19, 2.11, 2.05, and 1.86% for T1, T2, T3, and T4 BSW treatments, correspondingly. This might be due to the addition of BSW with no OM, which reduced OM proportion in the experimental soil. Similarly, the redox potential of soil was also significantly ($p < 0.05$) reduced after the BSW amendment, i.e., from 318.84 mV in control to 89 mV in T4. The reduction in redox potential was less in the T1 treatment (9.37%) while it reached maximum in the T4 treatment (256.75%). Herein, the mixing of BSW could result in the release of certain reducing chemical species in the soil, which could deplete the oxygen levels, thereby reducing redox potential [33].

Table 1. Effect of battery scrap waste (BSW) on physicochemical and heavy metal properties of experimental soil.

Properties	Experimental Treatments				
	Control	T1: 1% BSW	T2: 2% BSW	T3: 3% BSW	T4: 4% BSW
pH	7.23 ± 0.04 [a]	7.55 ± 0.03 [b]	7.63 ± 0.07 [b]	7.91 ± 0.05 [c]	8.35 ± 0.10 [d]
CEC (cmol·kg^{-1})	8.18 ± 0.09 [a]	8.60 ± 0.12 [b]	8.79 ± 0.06 [b]	8.91 ± 0.09 [bc]	9.04 ± 0.03 [c]
Organic matter (%)	2.27 ± 0.03 [a]	2.19 ± 0.02 [b]	2.11 ± 0.04 [c]	2.05 ± 0.03 [c]	1.86 ± 0.06 [d]
Redox potential (mV)	318.84 ± 4.10 [a]	291.50 ± 9.81 [b]	210.73 ± 6.30 [c]	163.15 ± 10.21 [d]	89.37 ± 7.58 [e]
Clay (%)	19.20 ± 1.80	-	-	-	-
Silt (%)	42.10 ± 2.90	-	-	-	-
Sand (%)	38.60 ± 1.40	-	-	-	-
Total Cd (mg·kg^{-1}) *	0.04 ± 0.01 [a]	40.48 ± 0.52 [b]	75.20 ± 2.06 [c]	117.36 ± 7.10 [d]	152.02 ± 4.75 [e]
Total Pb (mg·kg^{-1}) *	0.06 ± 0.01 [a]	110.06 ± 6.84 [b]	217.91 ± 10.38 [c]	340.40 ± 15.24	470.15 ± 12.62 [e]
Bioavailable Cd (mg·kg^{-1})	0.00 ± 0.00 [a]	11.26 ± 0.30 [b]	24.50 ± 1.05 [c]	38.17 ± 2.18 [d]	49.40 ± 3.85 [d]
Bioavailable Pb (mg·kg^{-1})	0.01 ± 0.00 [a]	32.50 ± 0.85 [b]	84.21 ± 5.30 [c]	107.82 ± 9.46 [d]	152.24 ± 14.67 [e]

Values are mean ± SD of three replicates (n = 3); CEC: cation exchange capacity; the same letters (a–e) indicate no significant difference among treatment groups at $p < 0.05$; *: threshold value (Cd: 0.80 mg·kg^{-1} and Pb: 85 mg·kg^{-1}) in the soil as recommended by WHO (1996).

On the other hand, the total and bioavailable concentrations of selected HMs, i.e., Cd and Pb, were significantly ($p < 0.05$) increased after BSW treatment. BSW are generally rich in metallic and electrolyte wastes, which could be the source of Cd and Pb. Cd and Pb metals are primarily cast-off in the manufacturing of Ni-Cd and Pb-acid batteries, which are commonly used for wireless communication devices, while Pb-acid batteries are used in automobiles, solar plates, inverters, power backup systems, etc. The levels of Cd and Pb were considerably higher in BSW-treated soil as suggested by the threshold limits of the WHO [44]. The impact of different types of battery waste has been previously outlined by other authors [45–47]. Among them, Chowdhury et al. [45], investigated the impact of Pb-acid battery waste on the soil properties in a residential community in Bangladesh. They reported that Pb concentration in soil ranged between 94 and 119,000 mg·kg^{-1} against the threshold value of 80.00 mg·kg^{-1}. Similarly, Ogundiran et al. [46] also studied the soil properties of abandoned Pb-acid battery dumpsite and found that the concentration of Pb and Cd reached >25,900 and 3.3 mg·kg^{-1} maximally up to a depth of 0–15 cm. Another study by Orjiakor et al. [47] found that automobile batter waste contamination significantly ($p < 0.01$) induced the soil physicochemical properties and elevated levels of Pb, Zn, Cd, Cr, and Cu HMs compared with those in non-affected soils. Therefore, the results of these reports are in accordance with those obtained in the present study.

3.2. Cd and Pb Removal by Coco Grass (*C. rotundus*)

In this study, *C. rotundus* demonstrated a significant ability to remove Cd and Pb contents from BSW-contaminated soils. In particular, the contents of Cd and Pb significantly ($p < 0.05$) declined after 45 days of phytoextraction experiments (Table 2). Although the contents of both Cd and Pb removed were higher in a higher dose of BSW mixing, the removal efficiency was reported to decline with an increase in BSW dose from 1 to 4%. The values of removal efficiency for control, T1, T2, T3, and T4 treatments accounted for 75, 35, 44, 31, and 26% for Cd, while the values for Pb were 83, 48, 40, 29, and 22%, respectively. The average Cd and Pb removal rates were found to be increasing with an increase in BSW dose. However, removal rates of both Cd and Pb were more identical in

T3 and T4 treatments, which indicates that the plant may have reached saturation level for BSW loading.

Table 2. Changes in the soil Cd and Pb concentration before and after phytoextraction using coco grass (*C. rotundus*) grown in soils amended with battery scrap waste (BSW).

Heavy Metal	Variable	Experimental Treatments				
		Control	T1: 1% BSW	T2: 2% BSW	T3: 3% BSW	T4: 4% BSW
Cd	Initial concentration (mg·kg $^{-1}$)	0.04 ± 0.01 ^{a}	40.48 ± 0.52 ^{a}	75.20 ± 2.06 ^{a}	117.36 ± 7.10 ^{a}	152.02 ± 4.75 ^{a}
	Final concentration (mg·kg $^{-1}$)	0.01 ± 0.00 ^{b}	25.98 ± 1.75 ^{b}	42.08 ± 3.63 ^{b}	80.40 ± 4.60 ^{b}	112.30 ± 6.14 ^{b}
	Removal efficiency (%)	75	35	44	31	26
	Removal rate (mg·kg $^{-1}$day $^{-1}$)	0.001	0.242	0.552	0.616	0.662
Pb	Initial concentration (mg·kg $^{-1}$)	0.06 ± 0.01 ^{a}	110.06 ± 6.84 ^{a}	217.91 ± 10.38 ^{a}	340.40 ± 15.24 ^{a}	470.15 ± 12.62 ^{a}
	Final concentration (mg·kg $^{-1}$)	0.01 ± 0.00 ^{b}	56.26 ± 2.70 ^{b}	130.50 ± 7.41 ^{b}	240.00 ± 10.65 ^{b}	365.09 ± 18.34 ^{b}
	Removal efficiency (%)	83	48	40	29	22
	Removal rate (mg·kg $^{-1}$day $^{-1}$)	0.001	0.897	1.457	1.673	1.751

Values are mean ± SD of three replicates (n = 3); the same letters (a, b) indicate no significant difference between initial and final concentration values at $p < 0.05$.

Recent studies have demonstrated the capability of *C. rotundus* to eradicate different hazardous heavy metals [48]. Garba et al. [49] reported that *C. rotundus* can tolerate Cd, Pb, Zn, and Ni stress of up to 1000 mg·kg $^{-1}$. Another study by Bordoloi and Basumatary [50] showed that *C. rotundus* significantly removed Pb (43.8%), Mn (27%), and Cd (31%) metals from soils contaminated with crude soil. Similarly, Jahan-Nejati et al. [51] studied the Cu removal performance of *C. rotundus* under different mixing rates (0–300 mg·kg $^{-1}$) in pod soil. They observed the quadratic model showing good fitness and optimum rate for Cu removal over time. Likewise, Tripathy et al. [52] studied the impact of pulp and paper mill effluent on Hg and Cd accumulation in *C. rotundus*. They reported that *C. rotundus* had a high potential to remove Hg and Cd from pulp and paper mill effluent-contaminated sites. These reports outlined that *C. rotundus* is considered a good phytoextraction candidate for the removal of hazardous HMs from contaminated sites [50], which is also supported by the results obtained in the current study. Therefore, our study also confirmed that Cd and Pb were efficiently removed by *C. rotundus* from BSW-mixed soil.

3.3. Cd and Pb Bioaccumulation Efficiency of Coco Grass (C. rotundus)

Phytoremediation plants have evolved to grow in hostile soil environments. Such plants can lower HM stress in their rhizosphere by accumulating them in their vegetative parts, thereby contributing to soil decontamination [53]. This feature makes them ideal candidates for soil remediation using green technology. In the present investigation, it was observed that *C. rotundus* significantly ($p < 0.05$) accumulated the contents of Cd and Pb from BSW-contaminated soils (Table 3). However, root parts of *C. rotundus* showed higher Cd and Pb bioaccumulation compared with the shoot and whole plant body. The overall order of Cd and Pb bioaccumulation by *C. rotundus* was observed as root > whole plant > shoot. When compared with the control treatment, the levels of Cd and Pb in *C. rotundus* tissues increased with an increase in the BSW mixing dose. However, there was a much less significant difference between the T3 and T4 treatments, which might be because the plant already achieved its saturation point in the 3% BSW treatment. The highest concentrations of Cd and Pb were encountered in root parts of *C. rotundus* for T4 treatment, i.e., 38.81 and 109.06 mg·kg $^{-1}$, respectively.

Table 3. Bioaccumulation and translocation of Cd and Pb by phytoextraction using coco grass (*C. rotundus*) grown in soils amended with battery scrap waste (BSW).

Heavy Metals	Experimental Treatments	Concentration ($mg \cdot kg^{-1}$ dwt.)			Bioaccumulation Factor (Soil → Plant)			Translocation Factor (Root → Shoot)
		Shoot	Root	Whole Plant	Shoot	Root	Whole Plant	
Cd	Control	0.01 ± 0.00^{a}	0.03 ± 0.01^{a}	0.02 ± 0.01^{a}	0.25	0.75	0.50	0.33
	T1: 1% BSW	4.90 ± 0.13^{b}	12.50 ± 0.28^{b}	9.67 ± 0.19^{b}	0.12	0.31	0.24	0.39
	T2: 2% BSW	13.04 ± 0.50^{c}	30.10 ± 1.07^{c}	22.08 ± 0.65^{c}	0.17	0.40	0.29	0.43
	T3: 3% BSW	17.62 ± 1.17^{d}	37.90 ± 2.24^{d}	24.64 ± 1.72^{cd}	0.15	0.32	0.21	0.46
	T4: 4% BSW	21.30 ± 0.92^{e}	38.81 ± 1.59^{d}	27.43 ± 1.24^{d}	0.14	0.26	0.18	0.55
Pb	Control	0.01 ± 0.01^{a}	0.04 ± 0.01^{a}	0.03 ± 0.01^{a}	0.17	0.67	0.50	0.25
	T1: 1% BSW	7.13 ± 0.20^{b}	30.85 ± 1.34^{b}	21.62 ± 0.90^{b}	0.06	0.28	0.20	0.23
	T2: 2% BSW	16.04 ± 0.61^{c}	87.40 ± 7.95^{c}	55.17 ± 4.34^{c}	0.07	0.40	0.25	0.18
	T3: 3% BSW	19.32 ± 0.49^{d}	103.27 ± 4.01^{d}	69.01 ± 9.42^{cd}	0.06	0.30	0.20	0.19
	T4: 4% BSW	22.65 ± 1.80^{de}	109.06 ± 6.27^{d}	76.10 ± 6.05^{d}	0.05	0.23	0.16	0.21

Values are mean ± SD of three replicates ($n = 3$); the same letters (a–e) indicate no significant difference among treatment groups at $p < 0.05$.

Long-term monitoring of BAF in polluted soils is helpful in understanding the effectiveness of phytoremediation over time and setting up regulatory compliances for the same. BAF also helps in determining the frequency of HMs migration to upper trophic levels within the food chain. The highest BAF values were reported for root parts, which were subjected to the highest bioaccumulation amount. However, BAF decreased with an increase in BSW dose, which was largely associated with concentration-induced toxicity in soil medium. Similarly, the Tf analysis showed that the migration of Cd and Pb from root to shoot parts of *C. rotundus* was significantly associated with the BSW mixing rate. The Tf values of Cd and Pb increased with an increase in their bioavailable concertation in the soil. Herein, the highest Tf values reported for Cd and Pb were 0.55 and 0.25, respectively. Nevertheless, the Tf values for Cd were positively associated with BSW dose while those for Pb had negative associations. The Pearson correlation matrix given in Figure 3 shows that selected soil parameters (pH, CEC, OM, RP, and HMs) were significantly ($p < 0.05$) correlated with the HMs bioaccumulation by *C. rotundus*. In general, pH was negatively correlated with OM ($r = -0.99$) and RP ($r = -0.97$) of experimental soil. However, the BSW mixing rate showed a significant negative influence on soil OM and RP. The contents of HMs in both soil and *C. rotundus* were negatively correlated. Herein, all other soil parameters showed a significant positive correlation with soil and plant HM contents.

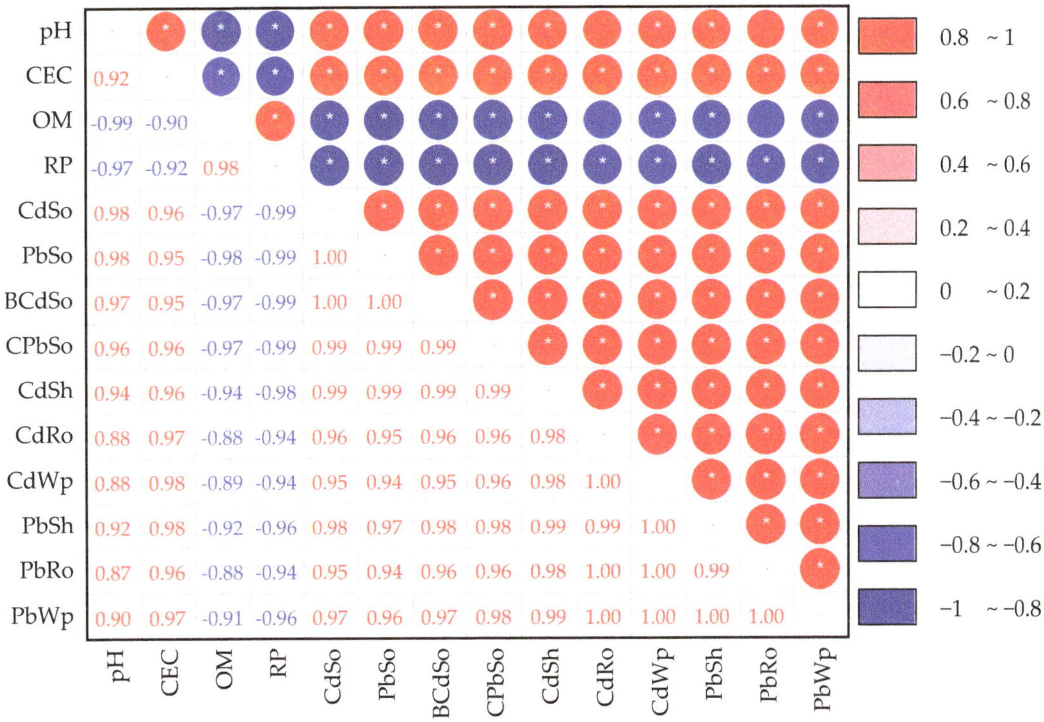

Figure 3. Pearson correlation matrix for interaction between soil properties and heavy metal uptake by coco grass (*C. rotundus*) grown in soils amended with battery scrap waste (BSW) (CEC: cation exchange capacity; OM: organic matter; RP: redox potential; CdSo: Cd in soil; PbSo: Pb in soil; CdSh: Cd in the shoot; CdRo: Cd in the root CdWp: Cd in the whole plant; PbSh: Pb in the shoot; PbRo: Pb in the root; and PbWp: Pb in the whole plant; *: significant at $p < 0.05$).

Bioaccumulation of several HMs by *C. rotundus* has been previously reported in several studies [10,54–56]. In particular, Khandare et al. [54] investigated the bioaccumulation and translocation of fluoride (F) by *C. rotundus*. They found that *C. rotundus* showed good potential for accumulating F contents within leaf and root parts where the highest BAF and Tf were 1.07 and 0.52, respectively. Halder et al. [55] also found that *C. rotundus* accumulated >350 mg·kg^{-1} of Pb in roots while only <50 mg·kg^{-1} in shoot parts accounting for BAF values of >20, respectively. Moreover, Nwaichi et al. [32] reported that *C. rotundus* accumulated significant contents of Pb (0.18), Cr (0.60), Cd (0.01), and Ni (0.1) from crude oil-contaminated soils. Similarly, Khan [56] observed HMs bioaccumulation by *C. rotundus* growing near the riparian zone of Lahat, Malaysia. ICP-MS analysis revealed that root parts of *C. rotundus* had the highest HM concentration with maximum BAF and Tf values, which is in line with those observed in the current study.

3.4. *Effect of Battery Waste on Growth, Biochemical, and Enzyme Response of* C. rotundus

Since HM stress may result in inefficient growth and physiological mechanisms, it is essential to monitor the biochemical response to ensure the suitability of plants in phytoextraction. Table 4 shows the effect of BSW amendment on the growth, biochemical, and enzyme response of *C. rotundus* in this study. It was observed that BSW application significantly ($p < 0.05$) reduced the plant's growth and biochemical parameters. Specifically, control and T1 showed a non-significant ($p > 0.05$) difference for the change in plant height and weight, and root length; however, these were significantly ($p < 0.05$) higher in the rest of the treatments, i.e., T2, T3, and T4. The maximum average plant height (34.27 cm), fresh weight (6.35 g), dry weight (1.80 g), root length (8.56 cm), and relative growth rate (0.14 g·day^{-1}) were found in the control treatment whereas the minimum was observed in the T4.

Table 4. Impact of Cd and Pb on plant growth, biochemical, and enzyme activity responses of coco grass (*C. rotundus*) grown in soils amended with battery scrap waste (BSW).

Parameters	Experimental Treatments				
	Control	**T1: 1% BSW**	**T2: 2% BSW**	**T3: 3% BSW**	**T4: 4% BSW**
Plant height (cm)	34.27 ± 2.50 ^{c}	32.30 ± 1.74 ^{c}	27.93 ± 1.25 ^{b}	24.68 ± 2.40 ^{ab}	21.01 ± 1.18 ^{a}
Fresh weight (g·plant^{-1})	6.35 ± 0.04 ^{e}	5.84 ± 0.09 ^{d}	4.16 ± 0.13 ^{c}	3.95 ± 0.05 ^{b}	3.60 ± 0.10 ^{a}
Dry weight (g·plant^{-1})	1.80 ± 0.04 ^{de}	1.71 ± 0.07 ^{d}	1.40 ± 0.06 ^{c}	1.16 ± 0.12 ^{ab}	1.09 ± 0.05 ^{a}
Root length (g)	8.56 ± 0.12 ^{de}	8.20 ± 0.15 ^{d}	7.58 ± 0.09 ^{c}	7.10 ± 0.22 ^{b}	6.74 ± 0.08 ^{a}
Relative growth rate (g·day^{-1})	0.14	0.13	0.09	0.09	0.08
Chlorophyll content (mg·g^{-1} fwt.)	2.42 ± 0.04 ^{e}	2.31 ± 0.02 ^{d}	2.05 ± 0.03 ^{c}	1.92 ± 0.06 ^{b}	1.80 ± 0.04 ^{a}
Carotenoids (mg·g^{-1} fwt.)	3.78 ± 0.05 ^{e}	3.60 ± 0.09 ^{d}	3.11 ± 0.10 ^{bc}	2.80 ± 0.13 ^{b}	2.56 ± 0.07 ^{a}
Tuber carbohydrates (%)	24.98 ± 1.18 ^{cd}	22.10 ± 1.70 ^{bc}	19.66 ± 0.91 ^{b}	16.05 ± 1.58 ^{a}	15.93 ± 2.20 ^{a}
Superoxide dismutase (U·mg^{-1} P)	2.60 ± 0.08 ^{a}	3.51 ± 0.14 ^{b}	5.73 ± 0.32 ^{c}	6.04 ± 0.21 ^{cd}	8.25 ± 0.74 ^{e}
Catalase (μmol·min^{-1}mg^{-1} P)	1.70 ± 0.09 ^{a}	2.25 ± 0.13 ^{b}	3.19 ± 0.06 ^{c}	4.01 ± 0.02 ^{d}	4.10 ± 0.04 ^{e}
Ascorbate peroxidase (mM·mg^{-1} P)	3.04 ± 0.02 ^{a}	5.20 ± 0.10 ^{b}	7.58 ± 0.07 ^{c}	11.82 ± 0.40 ^{d}	12.09 ± 1.15 ^{d}

Values are mean ± SD of three replicates (n = 3); the same letters (a–e) indicate no significant difference among treatment groups at $p < 0.05$.

A sharp decline in growth parameters of *C. rotundus* might be due to the high Cd and Pb toxicity of soil that came from BSW. Moreover, the biochemical parameters of *C. rotundus*, such as chlorophyll content (2.42 mg·g^{-1}), carotenoids (3.78 mg·g^{-1}), and tuber carbohydrate (24.98%), were also reduced in all BSW treatments. As an important cell constituent, chlorophyll biosynthesis is negatively affected by the presence of hazardous HMs [57]. Similarly, carotenoids play an important role as photoprotectors and antioxidants in plant cells [58]. Reduction in carotenoids may result in several cellular deficiencies, including inefficient cell defense and lowering of photosynthesis rates [59]. The contents of carbohydrates were also reduced from 24.98 to 1593, which is an indication of inefficient plant growth and failed photosynthesis. This is supported by the findings of present experiments, which depict a nearly two-fold reduction in both chlorophyll and carotenoids of *C. rotundus*.

SOD acts as an important enzyme to combat oxidative defense in plant systems by the dismutation of superoxide radicals [60]. SOD levels may be elevated if plants are exposed to oxidative conditions, which helps in converting the harmful radicals to H_2O_2 [61]. BSW amendment in the current study could have resulted in an increased number of reactive oxidizing species, which is detrimental to plants. For this, SOD levels were inclined from 2.60 to 8.25 U·mg^{-1} P (i.e., 217.30%) in control to T4 treatment, indicating that *C. rotundus* showed the ability to neutralize the oxidative stress caused by Cd and Pb. Similarly, CAT works after SOD by converting the H_2O_2 molecules to H_2O and O_2 molecules, thus accomplishing a complete defense against oxidative stress [62]. Excess H_2O_2 accumulation in plant cells could result in significant cell injuries. The elevated H_2O_2 levels are neutralized by CAT activity, which was observed in the current study where CAT levels significantly increased from 1.70 to 4.10 µmol·min^{-1}mg^{-1} P with an increase in BSW treatment dose. Similarly, ascorbate peroxidase also breaks down excess H_2O_2 by utilizing ascorbate (vitamin C) molecules [63]. The levels of ascorbate peroxidase were significantly increased from 2.04 to 122.09 mM·mg^{-1} P. However, the magnitude of these specific responses to upregulate stress caused by Cd and Pb may vary according to the BSW mixing rate, exposure, duration, and plant species. The Pearson correlation matrix in Figure 4 shows that soil properties had a significant impact on the growth, biochemical, and enzymatic response of *C. rotundus*.

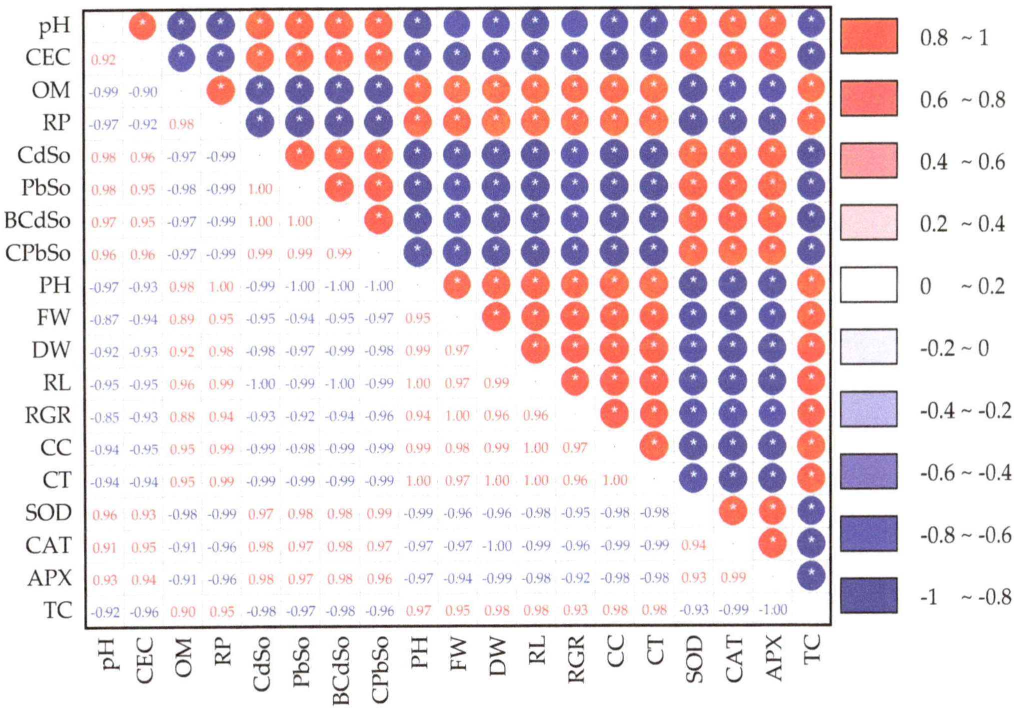

Figure 4. Pearson correlation matrix for interaction between soil properties, growth, and the enzymatic response of coco grass (*C. rotundus*) grown in soils amended with battery scrap waste (BSW) (CEC: cation exchange capacity; OM: organic matter; RP: redox potential; CdSo: Cd in soil; PbSo: Pb in soil; PH: plant height; FW: fresh weight; DW: dry weight; RL: root length; RGR: relative growth rate; CC: chlorophyll content; CT: carotenoids; SOD: superoxide dismutase; CAT: catalase; APX: ascorbate peroxidase; and TC: tuber carbohydrates; *: significant at $p < 0.05$).

The impact of hazardous HMs on plant growth, and biochemical, and enzymatic responses has been widely investigated by several authors [33,64]. Kriti et al. [33] studied the impact of Ni and Cd-rich battery waste on the physiological and enzyme response of two plants: vetiver (*Chrysopogon zizanioides* L.) and lemongrass (*Cymbopogon citratus* DC. Stapf) and found that Ni and Cd levels in the soil had significant detrimental effects on SOD, chlorophyll, carotenoid, and ascorbate peroxidase of both plants. Similarly, Adejumo et al. [64] reported that the bachelor's button (*Gomphrena celosoides* Mart.) plant showed a significant decline in soils contaminated with Pb and Cr from battery waste dumpsites in Nigeria. Overall, the results of the present study show that high Cd and Pb contents in soil could bring harmful effects on the growth and biochemical and enzymatic response of *C. rotundus* grown in BSW-contaminated soil.

3.5. Predictive Models for Cd and Pb Uptake by C. rotundus

The uptake of HMs by plants can be influenced by several factors, such as soil properties (pH, OM, HM's availability), type of plant, and environmental conditions. MLR and ANN are widely used to construct predictive relationships between these parameters and the amount of HMs absorbed by phytoremediation plants [65]. In this study, it was observed that both MLR and ANN models showed good fitness to predict the uptake of Cd and Pb by *C. rotundus* tissues. Table 5 shows the developed MLR- and ANN-based prediction models and their validation parameters. For MLR, the models showed good coefficient of determination (R^2) values ranging from 0.98 to 0.99. Herein, pH showed a negative coefficient for both HMs uptake by shoot, root, and whole plant parts of *C. rotundus*, while other independent variables (CEC, OM, and HMs) had positive coefficient values. The ME values of MLR models ranged between 0.98 and 0.99, while RMSE was recorded >5.72, indicating that the developed equations can predict HMs uptake with high efficiency and low error. On the other hand, ANN showed slightly better R^2 values (0.99) for all models. The ME and RMSE values were recorded as 0.99 and >0.84, respectively, which indicates better goodness of fit. As depicted in Figure 5, the experimental vs. model predicted scatter plot clearly shows that responses predicted by ANN models were highly close to experimental values compared with those of MLR models. This supports the fact of low prediction error.

However, MLR requires manual feature extraction in which the selection of suitable input parameters is decided based on their overall influence on the model. This problem is efficiently addressed by an ANN, which can automatically identify relevant variables from the input dataset, thereby reducing the need for manual feature extraction. Moreover, ANN models are known to have better prediction accuracy if trained using a well-curated and large number of input data points [66]. ANN algorithms can capture nonlinear relationships between input and output variables, which is generally difficult for other models, such as MLR. Although MLR is a traditional predictive technique that creates a linear function between input and output response, it is still useful for constructing mathematical models for simple and small-scale experiments and it provides valuable insight into HM uptake predictions [39].

Recent studies have employed linear and nonlinear models for the prediction of HMs uptake by plants under different phytoremediation systems [67,68]. However, only a few employed prediction modeling for BSW-based phytoremediation systems [25,33,69]. Among them, Kriti et al. [33] developed MLR and nonlinear sigmoid models for the prediction of Ni and Cd by *C. zizanioides* and *C. citratus* plants grown in soils mixed with battery electrolyte waste. They reported that MLR models were more efficient for Ni and Cd prediction compared with sigmoid models as indicated by experimental vs. predicted, R^2 (>0.88), and Nash–Sutcliffe ME (>0.87) values. Similarly, Priyadarshini et al. [25] created MLR and machine learning-based prediction models for HM's recovery from spent Zn-Mn batteries. They also confirmed that machine-learning models had better fitness compared with MLR. In their study, Bajestani et al. [69] evaluated the application of quadric regression modeling (response surface method) for optimization of Ni, Cd, and Co bio-recovery from spent household battery waste using *Acidithiobacillus ferrooxidans*. Thus, the combination of both MLR and ANN models could provide complementary strengths to improved prediction of HMs uptake by *C. rotundus*.

Table 5. Comparative assessment of MLR and ANN predictive models for Cd and Pb uptake by coco grass (*C. rotundus*) grown in soils amended with battery scrap waste (BSW).

Model Type	Heavy Metals	Plant Parts	Model Equation	R^2	ME	RMSE
MLR	Cd	Shoot	$40.82 - 8.72 \times pH + 1.88 \times CEC + 2.89 \times OM + 0.20 \times HM_{So}$	0.99	0.99	0.67
		Root	$93.48 - 36.00 \times pH + 16.51 \times CEC + 14.05 \times OM + 0.46 \times HM_{So}$	0.99	0.99	1.34
		Whole Plant	$20.14 - 22.24 \times pH + 18.35 \times CEC - 4.06 \times OM + 0.22 \times HM_{So}$	0.98	0.98	1.43
	Pb	Shoot	$4.39 - 14.61 \times pH + 10.70 \times CEC + 6.20 \times OM + 0.06 \times HM_{So}$	0.98	0.98	1.07
		Root	$296.99 - 129.12 \times pH + 55.68 \times CEC + 80.29 \times OM + 0.50 \times HM_{So}$	0.98	0.98	5.72
		Whole Plant	$43.87 - 63.20 \times pH + 30.51 \times CEC + 72.24 \times OM + 0.31 \times HM_{So}$	0.98	0.98	3.21
ANN	Cd	Shoot	-	0.99	0.99	0.23
		Root	-	0.99	0.99	0.80
		Whole Plant	-	0.99	0.99	0.45
	Pb	Shoot	-	0.99	0.99	0.30
		Root	-	0.99	0.99	0.34
		Whole Plant	-	0.99	0.99	0.84

MLR: multiple linear regression; -: not applicable; ANN: artificial neural network; ME: model efficiency; RMSE: root mean square error.

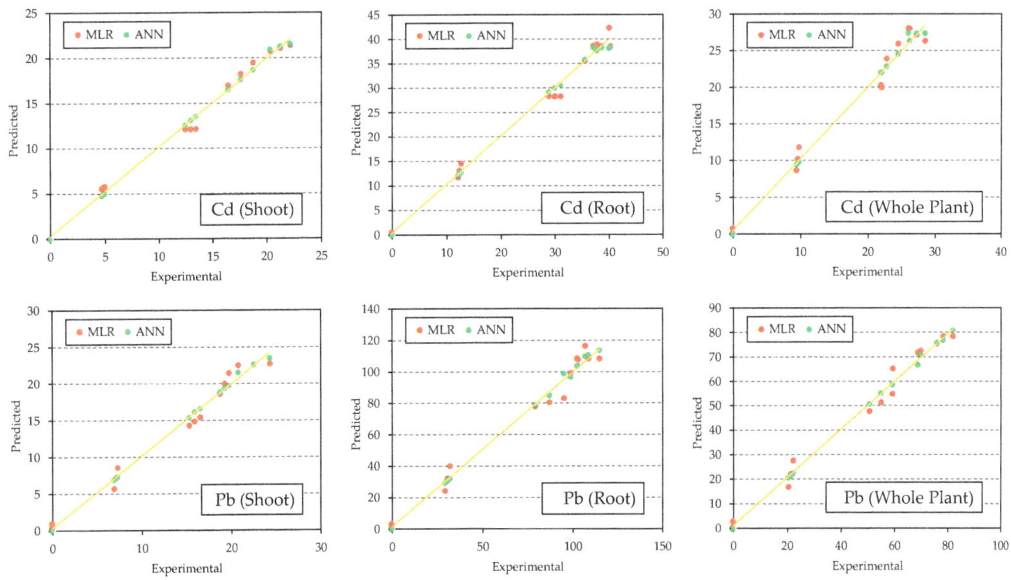

Figure 5. Experimental vs. model predicted (MLR: multiple linear regression; ANN: artificial neural network) values for Cd and Pb uptake by coco grass (*C. rotundus*) grown in soils amended with battery scrap waste (BSW).

4. Conclusions

The results revealed that BSW mixing in soil significantly elevated the levels of Cd and Pb ($p < 0.05$), which were maximally taken by *C. rotundus* into root parts. The excess uptake of Cd and Pb by *C. rotundus* resulted in reduced growth and biochemical parameters while inducing the levels of stress enzymes. The BAF and Tf values also indicated that *C. rotundus* was a Cd- and Pb-hyperaccumulator plant that can be used to clean up BSW-contaminated sites. Furthermore, prediction modeling results using MLR and ANN showed that both approaches were efficient for effective HMs uptake by *C. rotundus*. However, ANN models gave better results when tested using different validation tools. Overall, this study concluded that *C. rotundus* can be used as a potential phytoremediation plant for Cd and Pb phytoextraction from BSW-contaminated soils. Further studies on the bioaccumulation of other toxic HMs and their physiological and microscopic effects on *C. rotundus* with post-characterization of harvested biomass are highly recommended.

Author Contributions: Conceptualization, A.A.A.-H., M.A.T., P.K. (Pankaj Kumar) and E.M.E.; Methodology, A.A.A.-H., I.Š., V.K. and P.K. (Pankaj Kumar); Software, I.Š. and P.K. (Pankaj Kumar); Validation, A.A.A.-H., M.A.T., I.Š., M.G., B.A., K.S.C., P.K. (Piyush Kumar), V.K. and E.M.E.; Formal analysis, M.G., P.K. (Piyush Kumar) and P.K. (Pankaj Kumar); Investigation, M.G., P.K. (Piyush Kumar) and P.K. (Pankaj Kumar); Resources, V.K.; Data curation, I.Š., B.A. and K.S.C.; Writing—original draft, P.K. (Pankaj Kumar); Writing—review & editing, M.A.T., I.Š., M.G., B.A., K.S.C., P.K. (Piyush Kumar), V.K. and E.M.E.; Visualization, B.A. and K.S.C.; Supervision, E.M.E.; Project administration, A.A.A.-H., M.A.T. and E.M.E.; Funding acquisition, A.A.A.-H. and M.A.T. All authors have read and agreed to the published version of the manuscript.

Funding: This research was funded by Princess Nourah bint Abdulrahman University Researchers Supporting Project number (PNURSP2023R93), Princess Nourah bint Abdulrahman University, Riyadh, Saudi Arabia. This research was funded by the Deanship of Scientific Research at King Khalid University, Abha, Saudi Arabia (grant number RGP.2/220/44).

Institutional Review Board Statement: Not applicable.

Data Availability Statement: Not applicable.

Acknowledgments: The authors express their gratitude to Princess Nourah bint Abdulrahman University Researchers Supporting Project number (PNURSP2023R93), Princess Nourah bint Abdulrahman University, Riyadh, Saudi Arabia. The authors extend their appreciation to the Deanship of Scientific Research at King Khalid University, Abha, Saudi Arabia, for funding this work through the Research Group Project under grant number RGP.2/220/44.

Conflicts of Interest: The authors declare no conflict of interest.

References

1. Schmidt-Rohr, K. How Batteries Store and Release Energy: Explaining Basic Electrochemistry. *J. Chem. Educ.* **2018**, *95*, 1801–1810. [CrossRef]
2. Cho, J.; Jeong, S.; Kim, Y. Commercial and Research Battery Technologies for Electrical Energy Storage Applications. *Prog. Energy Combust. Sci.* **2015**, *48*, 84–101. [CrossRef]
3. Grand View Research. *Battery Market Size, Share & Trends Analysis Report By Product (Lead Acid, Li-Ion, Nickle Metal Hydride, Ni-Cd)*; Grand View Research: San Francisco, CA, USA, 2020.
4. Kala, S.; Mishra, A. Battery Recycling Opportunity and Challenges in India. *Mater. Today Proc.* **2021**, *46*, 1543–1556. [CrossRef]
5. Dasila, H.; Joshi, D.; Verma, S.; Maithani, D.; Rawat, S.K.; Kumar, A.; Suyal, N.; Kumar, N.; Suyal, D.C. Hazardous Waste: Impact and Disposal Strategies. In *Advanced Microbial Techniques in Agriculture, Environment, and Health Management*; Elsevier: Amsterdam, The Netherlands, 2023; pp. 153–166. ISBN 9780323916431.
6. Rarotra, S.; Sahu, S.; Kumar, P.; Kim, K.; Tsang, Y.F.; Kumar, V.; Kumar, P.; Srinivasan, M.; Veksha, A.; Lisak, G. Progress and Challenges on Battery Waste Management: A Critical Review. *ChemistrySelect* **2020**, *5*, 6182–6193. [CrossRef]
7. Zeng, X.; Li, J. Spent Rechargeable Lithium Batteries in E-Waste: Composition and Its Implications. *Front. Environ. Sci. Eng.* **2014**, *8*, 792–796. [CrossRef]
8. Sivaramanan, S. E-Waste Management, Disposal and Its Impacts on the Environment. *Univers. J. Environ. Res. Technol.* **2013**, *3*, 531–537.
9. Tutic, A.; Novakovic, S.; Lutovac, M.; Biocanin, R.; Ketin, S.; Omerovic, N. The Heavy Metals in Agrosystems and Impact on Health and Quality of Life. *Open Access Maced. J. Med. Sci.* **2015**, *3*, 345–355. [CrossRef]

10. Singh, R.; Gautam, N.; Mishra, A.; Gupta, R. Heavy Metals and Living Systems: An Overview. *Indian J. Pharmacol.* **2011**, *43*, 246. [CrossRef]
11. Nordberg, G.F.; Fowler, B.A. *Risk Assessment for Human Metal Exposures*; Elsevier: Amsterdam, The Netherlands, 2019; ISBN 9780128042274.
12. Abdel-Rahman, G. Heavy Metals, Definition, Sources of Food Contamination, Incidence, Impacts and Remediation: A Literature Review with Recent Updates. *Egypt. J. Chem.* **2021**, *65*, 419–437. [CrossRef]
13. Collin, S.; Baskar, A.; Geevarghese, D.M.; Ali, M.N.V.S.; Bahubali, P.; Choudhary, R.; Lvov, V.; Tovar, G.I.; Senatov, F.; Koppala, S.; et al. Bioaccumulation of Lead (Pb) and Its Effects in Plants: A Review. *J. Hazard. Mater. Lett.* **2022**, *3*, 100064. [CrossRef]
14. Asare, M.O.; Száková, J.; Tlustoš, P. The Fate of Secondary Metabolites in Plants Growing on Cd-, As-, and Pb-Contaminated Soils—A Comprehensive Review. *Environ. Sci. Pollut. Res.* **2022**, *30*, 11378–11398. [CrossRef] [PubMed]
15. Kumar, S.; Rahman, M.A.; Islam, M.R.; Hashem, M.A.; Rahman, M.M. Lead and Other Elements-Based Pollution in Soil, Crops and Water near a Lead-Acid Battery Recycling Factory in Bangladesh. *Chemosphere* **2022**, *290*, 133288. [CrossRef] [PubMed]
16. Tauqeer, H.M.; Basharat, Z.; Adnan Ramzani, P.M.; Farhad, M.; Lewińska, K.; Turan, V.; Karczewska, A.; Khan, S.A.; Faran, G.; Iqbal, M. Aspergillus Niger-Mediated Release of Phosphates from Fish Bone Char Reduces Pb Phytoavailability in Pb-Acid Batteries Polluted Soil, and Accumulation in Fenugreek. *Environ. Pollut.* **2022**, *313*, 120064. [CrossRef] [PubMed]
17. Saxena, G.; Purchase, D.; Mulla, S.I.; Saratale, G.D.; Bharagava, R.N. Phytoremediation of Heavy Metal-Contaminated Sites: Eco-Environmental Concerns, Field Studies, Sustainability Issues, and Future Prospects. *Rev. Environ. Contam. Toxicol.* **2019**, *249*, 71–131.
18. Henschel, J.; Mense, M.; Harte, P.; Diehl, M.; Buchmann, J.; Kux, F.; Schlatt, L.; Karst, U.; Hensel, A.; Winter, M.; et al. Phytoremediation of Soil Contaminated with Lithium Ion Battery Active Materials—A Proof-of-Concept Study. *Recycling* **2020**, *5*, 26. [CrossRef]
19. Justin, M.Z.; Pajk, N.; Zupanc, V.; Zupančič, M. Phytoremediation of Landfill Leachate and Compost Wastewater by Irrigation of Populus and Salix: Biomass and Growth Response. *Waste Manag.* **2010**, *30*, 1032–1042. [CrossRef]
20. Chehregani, A.; Noori, M.; Yazdi, H.L. Phytoremediation of Heavy-Metal-Polluted Soils: Screening for New Accumulator Plants in Angouran Mine (Iran) and Evaluation of Removal Ability. *Ecotoxicol. Environ. Saf.* **2009**, *72*, 1349–1353. [CrossRef]
21. Siyar, R.; Doulati Ardejani, F.; Norouzi, P.; Maghsoudy, S.; Yavarzadeh, M.; Taherdangkoo, R.; Butscher, C. Phytoremediation Potential of Native Hyperaccumulator Plants Growing on Heavy Metal-Contaminated Soil of Khatunabad Copper Smelter and Refinery, Iran. *Water* **2022**, *14*, 3597. [CrossRef]
22. Manoj, S.R.; Karthik, C.; Kadirvelu, K.; Arulselvi, P.I.; Shanmugasundaram, T.; Bruno, B.; Rajkumar, M. Understanding the Molecular Mechanisms for the Enhanced Phytoremediation of Heavy Metals through Plant Growth Promoting Rhizobacteria: A Review. *J. Environ. Manag.* **2020**, *254*, 109779. [CrossRef]
23. Wang, Y.; Li, S.; Wang, X.; Xu, J.; Li, T.; Zhu, J.; Yang, R.; Wang, J.; Chang, M.; Wang, L. Biochelator Assisted Phytoremediation for Cadmium (Cd) Pollution in Paddy Field. *Sustainability* **2021**, *13*, 12170. [CrossRef]
24. Kumar, S.; Dube, K.K.; Rai, J.P.N. *Mathematical Model for Phytoremediation of Pulp and Paper Industry Wastewater*; CSIR: New Delhi, India, 2005; pp. 717–721.
25. Priyadarshini, J.; Elangovan, M.; Mahdal, M.; Jayasudha, M. Machine-Learning-Assisted Prediction of Maximum Metal Recovery from Spent Zinc–Manganese Batteries. *Processes* **2022**, *10*, 1034. [CrossRef]
26. Peerzada, A.M. Biology, Agricultural Impact, and Management of *Cyperus rotundus* L.: The World's Most Tenacious Weed. *Acta Physiol. Plant* **2017**, *39*, 270. [CrossRef]
27. Baloch, A.H. The Biology of Balochistani Weed: *Cyperus rotundus* Linnaeus. A Review. *Pure Appl. Biol.* **2015**, *4*, 171–180. [CrossRef]
28. Stoller, E.W.; Sweet, R.D. Biology and Life Cycle of Purple and Yellow Nutsedges (*Cyperus rotundus* and *C. Esculentus*). *Weed Technol.* **1987**, *1*, 66–73. [CrossRef]
29. Peerzada, A.M.; Ali, H.H.; Naeem, M.; Latif, M.; Bukhari, A.H.; Tanveer, A. *Cyperus rotundus* L.: Traditional Uses, Phytochemistry, and Pharmacological Activities. *J. Ethnopharmacol.* **2015**, *174*, 540–560. [CrossRef] [PubMed]
30. Jain, P.K.; Das, D. Ethnopharmacological Study of *Cyperus rotundus* a Herb Used By Tribal Community As a Traditional Medicine for Treating Various Diseases. *Innov. J. Ayruvedic Sci.* **2016**, *4*, 4–6.
31. Basumatary, B.; Saikia, R.; Bordoloi, S. Phytoremediation of Crude Oil Contaminated Soil Using Nut Grass, *Cyperus rotundus*. *J. Environ. Biol.* **2012**, *33*, 891–896.
32. Nwaichi, E.O.; Chukwuere, C.O.; Abosi, P.J.; Onukwuru, G.I. Phytoremediation Of Crude Oil Impacted Soil Using Purple Nutsedge. *J. Appl. Sci. Environ. Manag.* **2021**, *25*, 475–479. [CrossRef]
33. Kriti; Basant, N.; Singh, J.; Kumari, B.; Sinam, G.; Gautam, A.; Singh, G.; Swapnil; Mishra, K.; Mallick, S. Nickel and Cadmium Phytoextraction Efficiencies of Vetiver and Lemongrass Grown on Ni–Cd Battery Waste Contaminated Soil: A Comparative Study of Linear and Nonlinear Models. *J. Environ. Manag.* **2021**, *295*, 113144. [CrossRef]
34. Walkley, A.; Black, I.A. An Examination of the Degtjareff Method for Determining Soil Organic Matter, and a Proposed Modification of the Chromic Acid Titration Method. *Soil Sci.* **1934**, *37*, 29–38. [CrossRef]
35. Bharti, M.; Kamboj, N.; Kamboj, V.; Bisht, A.; Kumar, A. Dynamics of Soil Cationic Micronutrients in Different Land Use Systems in Lower Shiwalik Region of Uttarakhand, India. In *Environmental Pollution and Natural Resource Management*; Springer Proceedings in Earth and Environmental Sciences Series; Springer: Cham, Switzerland, 2022; pp. 185–199.

36. Fathabad, A.E.; Shariatifar, N.; Moazzen, M.; Nazmara, S.; Fakhri, Y.; Alimohammadi, M.; Azari, A.; Mousavi Khaneghah, A. Determination of Heavy Metal Content of Processed Fruit Products from Tehran's Market Using ICP-OES: A Risk Assessment Study. *Food Chem. Toxicol.* **2018**, *115*, 436–446. [CrossRef] [PubMed]
37. Kanwal, A.; Farhan, M.; Sharif, F.; Hayyat, M.U.; Shahzad, L.; Ghafoor, G.Z. Effect of Industrial Wastewater on Wheat Germination, Growth, Yield, Nutrients and Bioaccumulation of Lead. *Sci. Rep.* **2020**, *10*, 11361. [CrossRef] [PubMed]
38. Ismail, N.I.; Abdullah SR, S.; Idris, M.; Hasan, H.A.; Halmi MI, E.; Al Sbani, N.H.; Jehawi, O.H. Simultaneous Bioaccumulation and Translocation of Iron and Aluminium from Mining Wastewater by Scirpus Grossus. *Desalination Water Treat.* **2019**, *163*, 133–142. [CrossRef]
39. Eid, E.M.; Shaltout, K.H.; Alamri, S.A.M.; Sewelam, N.A.; Galal, T.M.; Brima, E.I. Prediction Models for Evaluating Heavy Metal Uptake by Pisum Sativum L. in Soil Amended with Sewage Sludge. *J. Environ. Sci. Health Part A* **2020**, *55*, 151–160. [CrossRef]
40. Galal, T.M.; Eid, E.M.; Dakhil, M.A.; Hassan, L.M. Bioaccumulation and Rhizofiltration Potential of *Pistia stratiotes* L. for Mitigating Water Pollution in the Egyptian Wetlands. *Int. J. Phytoremediat.* **2018**, *20*, 440–447. [CrossRef] [PubMed]
41. Sen, S.; Nandi, S.; Dutta, S. Application of RSM and ANN for Optimization and Modeling of Biosorption of Chromium(VI) Using Cyanobacterial Biomass. *Appl. Water Sci.* **2018**, *8*, 148. [CrossRef]
42. AL-Huqail, A.A.; Kumar, P.; Abou Fayssal, S.; Adelodun, B.; Širić, I.; Goala, M.; Choi, K.S.; Taher, M.A.; El-Kholy, A.S.; Eid, E.M. Sustainable Use of Sewage Sludge for Marigold (*Tagetes erecta* L.) Cultivation: Experimental and Predictive Modeling Studies on Heavy Metal Accumulation. *Horticulturae* **2023**, *9*, 447. [CrossRef]
43. Kyzioł, J. Effect of Physical Properties and Cation Exchange Capacity on Sorption of Heavy Metals onto Peats. *Pol. J. Environ. Stud.* **2002**, *11*, 713–718.
44. World Health Organization. *WHO Permissible Limits of Heavy Metals in Soil and Plants*; World Health Organization: Geneva, Switzerland, 1996.
45. Chowdhury, K.I.A.; Nurunnahar, S.; Kabir, M.L.; Islam, M.T.; Baker, M.; Islam, M.S.; Rahman, M.; Hasan, M.A.; Sikder, A.; Kwong, L.H.; et al. Child Lead Exposure near Abandoned Lead Acid Battery Recycling Sites in a Residential Community in Bangladesh: Risk Factors and the Impact of Soil Remediation on Blood Lead Levels. *Environ. Res.* **2021**, *194*, 110689. [CrossRef]
46. Ogundiran, M.B.; Osibanjo, O. Mobility and Speciation of Heavy Metals in Soils Impacted by Hazardous Waste. *Chem. Speciat. Bioavailab.* **2009**, *21*, 59–69. [CrossRef]
47. Orjiakor, P.; Atuanya, E. Effects of Automobile Battery Wastes on Physicochemical Properties of Soil in Benin City, Edo State. *Glob. J. Pure Appl. Sci.* **2015**, *21*, 129. [CrossRef]
48. Ariyachandra, S.P.; Alwis, I.S.; Wimalasiri, E.M. Phytoremediation Potential of Heavy Metals by *Cyperus rotundus*. *Rev. Agric. Sci.* **2023**, *11*, 20–35. [CrossRef]
49. Garba, S.T.; Gudusu, M.; Inuwa, L.B. Accumulation Ability of the Native Grass Species, *Cyperus rotundus* for the Heavy Metals; Zinc (Zn), Cadmium (Cd), Nickel (Ni) and Lead (Pb). *Int. Res. J. Pure Appl. Chem.* **2018**, *17*, 1–15. [CrossRef]
50. Bordoloi, S.; Basumatary, B. A Study on Degradation of Heavy Metals in Crude Oil-Contaminated Soil Using *Cyperus rotundus*. In *Phytoremediation*; Springer International Publishing: Cham, Switzerland, 2016; pp. 53–60; ISBN 9783319418117.
51. Jahan-Nejati, S.; Jowkar-Tangkarami, M.; Taei-Semiromi, J. *Cyperus rotundus*: A Safe Forage or Hyper Phytostabilizer Species in Copper Contaminated Soils. *Int. J. Phytoremediat.* **2021**, *23*, 1212–1221. [CrossRef]
52. Tripathy, A.P.; Dixit, P.K.; Panigrahi, A.K. Impact of Effluent of Pulp & Paper Industry on the Flora of River Basin at Jaykaypur, Odisha, India and Its Ecological Implications. *Environ. Res.* **2022**, *204*, 111769. [CrossRef] [PubMed]
53. DalCorso, G.; Fasani, E.; Manara, A.; Visioli, G.; Furini, A. Heavy Metal Pollutions: State of the Art and Innovation in Phytoremediation. *Int. J. Mol. Sci.* **2019**, *20*, 3412. [CrossRef] [PubMed]
54. Khandare, R.V.; Watharkar, A.D.; Pawar, P.K.; Jagtap, A.A.; Desai, N.S. Hydrophytic Plants Canna Indica, Epipremnum Aureum, Cyperus Alternifolius and *Cyperus rotundus* for Phytoremediation of Fluoride from Water. *Environ. Technol. Innov.* **2021**, *21*, 101234. [CrossRef]
55. Halder, L.; Arce, L.D.; Yllano, O.B. Bioaccumulation and Bioconcentration of Pb in the Tissues of Eight Weed Species. *J. Int. Sch. Conf.* **2016**, *1*, 14–18.
56. Khan, A.M. Accumulation, Uptake and Bioavailability of Rare Earth Elements (REEs) in Soil Grown Plants from Ex-Mining Area in Perak, Malaysia. *Appl. Ecol. Environ. Res.* **2017**, *15*, 117–133. [CrossRef]
57. Shakya, K.; Chettri, M.K.; Sawidis, T. Impact of Heavy Metals (Copper, Zinc, and Lead) on the Chlorophyll Content of Some Mosses. *Arch. Environ. Contam. Toxicol.* **2008**, *54*, 412–421. [CrossRef]
58. Zeeshan, M.; Ahmad, W.; Hussain, F.; Ahamd, W.; Numan, M.; Shah, M.; Ahmad, I. Phytostabalization of the Heavy Metals in the Soil with Biochar Applications, the Impact on Chlorophyll, Carotene, Soil Fertility and Tomato Crop Yield. *J. Clean. Prod.* **2020**, *255*, 120318. [CrossRef]
59. Rai, R.; Agrawal, M.; Agrawal, S.B. Impact of Heavy Metals on Physiological Processes of Plants: With Special Reference to Photosynthetic System. In *Plant Responses to Xenobiotics*; Springer: Singapore, 2016; pp. 127–140. ISBN 9789811028601.
60. Rajput, V.D.; Harish; Singh, R.K.; Verma, K.K.; Sharma, L.; Quiroz-Figueroa, F.R.; Meena, M.; Gour, V.S.; Minkina, T.; Sushkova, S. Recent Developments in Enzymatic Antioxidant Defence Mechanism in Plants with Special Reference to Abiotic Stress. *Biology* **2021**, *10*, 267. [CrossRef] [PubMed]

61. Dudonné, S.; Vitrac, X.; Coutiére, P.; Woillez, M.; Mérillon, J.M. Comparative Study of Antioxidant Properties and Total Phenolic Content of 30 Plant Extracts of Industrial Interest Using DPPH, ABTS, FRAP, SOD, and ORAC Assays. *J. Agric. Food Chem.* **2009**, *57*, 1768–1774. [CrossRef] [PubMed]
62. Ighodaro, O.M.; Akinloye, O.A. First Line Defence Antioxidants-Superoxide Dismutase (SOD), Catalase (CAT) and Glutathione Peroxidase (GPX): Their Fundamental Role in the Entire Antioxidant Defence Grid. *Alex. J. Med.* **2018**, *54*, 287–293. [CrossRef]
63. Jena, A.B.; Samal, R.R.; Bhol, N.K.; Duttaroy, A.K. Cellular Red-Ox System in Health and Disease: The Latest Update. *Biomed. Pharmacother.* **2023**, *162*, 114606. [CrossRef]
64. Adejumo, S.A.; Tiwari, S.; Thul, S.; Sarangi, B.K. Evaluation of Lead and Chromium Tolerance and Accumulation Level in Gomphrena Celosoides: A Novel Metal Accumulator from Lead Acid Battery Waste Contaminated Site in Nigeria. *Int. J. Phytoremediat.* **2019**, *21*, 1341–1355. [CrossRef]
65. Zhao, B.; Zhu, W.; Hao, S.; Hua, M.; Liao, Q.; Jing, Y.; Liu, L.; Gu, X. Prediction Heavy Metals Accumulation Risk in Rice Using Machine Learning and Mapping Pollution Risk. *J. Hazard. Mater.* **2023**, *448*, 130879. [CrossRef]
66. Kumar, P.; Kumar, V.; Singh, J.; Kumar, P. Electrokinetic Assisted Anaerobic Digestion of Spent Mushroom Substrate Supplemented with Sugar Mill Wastewater for Enhanced Biogas Production. *Renew. Energy* **2021**, *179*, 418–426. [CrossRef]
67. Chen, P.; Zhang, H.-M.; Yao, B.-M.; Chen, S.-C.; Sun, G.-X.; Zhu, Y.-G. Bioavailable Arsenic and Amorphous Iron Oxides Provide Reliable Predictions for Arsenic Transfer in Soil-Wheat System. *J. Hazard. Mater.* **2020**, *383*, 121160. [CrossRef]
68. Chiou, W.-Y.; Hsu, F.-C. Copper Toxicity and Prediction Models of Copper Content in Leafy Vegetables. *Sustainability* **2019**, *11*, 6215. [CrossRef]
69. Ijadi Bajestani, M.; Mousavi, S.M.; Shojaosadati, S.A. Bioleaching of Heavy Metals from Spent Household Batteries Using Acidithiobacillus Ferrooxidans: Statistical Evaluation and Optimization. *Sep. Purif. Technol.* **2014**, *132*, 309–316. [CrossRef]

agriculture

Article

Evaluation of the Dynamic Tube Method for Measuring Ammonia Emissions after Liquid Manure Application

Martin ten Huf * and Hans-Werner Olfs

Faculty of Agricultural Sciences and Landscape Architecture, University of Applied Sciences Osnabrück, 49090 Osnabrück, Germany; h-w.olfs@hs-osnabrueck.de
* Correspondence: martin.ten-huf@hs-osnabrueck.de

Abstract: Easy and inexpensive methods for measuring ammonia emissions in multi-plot field trials allow the comparison of several treatments with liquid manure application. One approach that might be suitable under these conditions is the dynamic tube method (DTM). Applying the DTM, a mobile chamber system is placed on the soil surface, and the air volume within is exchanged at a constant rate for approx. 90 s. with an automated pump. This procedure is assumed to achieve an equilibrium ammonia concentration within the system. Subsequently, a measurement is performed using an ammonia-sensitive detector tube. Ammonia fluxes are calculated based on an empirical model that also takes into account the background ammonia concentration measured on unfertilized control plots. Between measurements on different plots, the chamber system is flushed with ambient air and cleaned with paper towels to minimize contamination with ammonia. The aim of this study was to determine important prerequisites and boundary conditions for the application of the DTM. We conducted a laboratory experiment to test if the ammonia concentration remains stable while performing a measurement. Furthermore, we investigated the cleaning procedure and the effect of potential ammonia carryover on cumulated emissions under field conditions following liquid manure application. The laboratory experiment indicated that the premeasurement phase to ensure a constant ammonia concentration is not sufficient. The concentration only stabilized after performing more than 100 pump strokes, with 20 pump strokes (lasting approximately 90 s) being the recommendation. However, the duration of performing a measurement can vary substantially, and linear conversion accounts for those differences, so a stable concentration is mandatory. Further experiments showed that the cleaning procedure is not sufficient under field conditions. Thirty minutes after performing measurements on high emitting plots, which resulted in an ammonia concentration of approx. 10 ppm in the chamber, we detected a residual concentration of 2 ppm. This contamination may affect measurements on plots with liquid manure application as well as on untreated control plots. In a field experiment with trailing hose application of liquid manure, we subsequently demonstrated that the calculation of cumulative ammonia emissions can vary by a factor of three, depending on the degree of chamber system contamination when measuring control plots. When the ammonia background values were determined by an uncontaminated chamber system that was used to measure only control plots, cumulative ammonia emissions were approximately 9 kg NH_3-N ha^{-1}. However, when ammonia background values were determined using the contaminated chamber system that was also used to measure on plots with liquid manure application, the calculation of cumulative ammonia losses indicated approximately 3 kg NH_3-N ha^{-1}. Based on these results, it can be concluded that a new empirical DTM calibration is needed for multi-plot field experiments with high-emitting treatments.

Keywords: ammonia background concentration; chamber system contamination; multi-plot field trials

1. Introduction

Nitrogen contained in liquid manure is an essential plant nutrient [1]. However, ammonium (NH_4^+) can be easily converted to gaseous ammonia (NH_3) following liquid manure application [2]. The emitted NH_3 is either deposited locally or transported over long distances, where it increases airborne deposition of reactive nitrogen [1,3,4]. Besides detrimental effects on human health [5,6] and nonagrarian ecosystems [7], the nitrogen use efficiency of the applied liquid manure is also reduced [8]. Thus, abatement of NH_3 emissions following liquid manure application is a priority for many countries worldwide [9,10]. For example, the European Union defined limits for maximum NH_3 emissions [11], forcing member states to reduce emissions. Hence, optimized techniques to apply liquid manures are mandatory. However, to assess the effects of different application techniques, NH_3 emissions need to be quantified in multi-plot field trials [12]. Standard micrometeorological methods such as the integrated horizontal flux method [13,14] or wind tunnels [15] for quantifying NH_3 emissions require large experimental plots, expensive equipment, or an in-field electric power supply [12]. Thus, applying standard methods in multi-plot field trials is hardly possible [16].

An alternative, which might be feasible for multi-plot field trials, is the dynamic tube method (DTM). Originally the DTM was developed for evaluating CO_2 formation in soils [17], and later, it was adapted for assessing NH_3 fluxes following nitrogen fertilization [16,18]. This original setup did not consider wind speed effects on NH_3 emissions. Therefore, DTM NH_3 "raw fluxes" were calibrated with the so-called "Integrated Horizontal Flux" method, taking into consideration the wind speed at the time of measurement [12,19,20]. Nowadays, the wind speed corrected DTM is used in combination with so-called "passive samplers" (i.e., plastic bottles with small openings for air exchange that are filled with an acid solution [21,22]). These acid traps are placed in each plot of a field experiment to collect NH_3 to assess relative differences between plots, while the wind speed corrected DTM is used on a few selected plots to scale those relative differences [12]. This approach was used in a wide range of experiments in Germany [4,23–27] and Denmark [28].

For DTM measurements, a stainless steel chamber system, including four individual circular chambers with 11.5 cm diameter and small openings for air exchange, is placed on the soil surface where N fertilizer was applied [12,19]. Tubings connect the chambers with an automated pump. Before starting the measurement, 20 pump strokes are performed in 90 s to flush the volume of the chamber system while already placed on the soil surface where the measurement is planned. This procedure is assumed to ensure a "steady" NH_3 concentration when the measurement is carried out. Subsequently, an NH_3 detector tube is inserted, which displays the NH_3 concentration after a defined number of pump strokes by the color change in a reactant, reaching one of the calibration marks on the transparent glass tube. If no mark is reached after performing the specified number of pump strokes, the measurement is continued. Depending on the expected NH_3 concentration, different detector tubes with a varying default number of strokes are used. Different numbers of pump strokes are taken into consideration by linear conversion to the default number of strokes of the respective detector tube [12].

Between measurements, the chamber system is cleaned with paper towels and flushed with ambient air to minimize contamination with NH_3 [12]. When measurements are conducted under field conditions, the NH_3 concentration of unfertilized plots is considered as background and subtracted from the concentration measured in fertilized plots [12,16]. This background-adjusted NH_3 concentration is then used to calculate NH_3-N raw fluxes. Subsequently, the raw flux is adjusted for the wind speed at the time of measurement by an empirical formula since the wind would affect actual in-field NH_3 emissions [29,30], but the chamber system inhibits those effects [19]. Two empirical formulas were developed for different canopy heights [19]. Finally, NH_3 emissions are cumulated by linear interpolation of wind speed-adjusted fluxes between measurements.

The overall aim of this study was to propose an improved approach to use the DTM. Therefore, we formulated several subordinate objectives. The first was to determine whether the NH_3 concentration remains stable during the actual measurement period to allow the comparison of deviating numbers of pump strokes and/or the use of different detector tubes. The second objective was to evaluate the on-site cleaning procedure with paper towels and the flushing of the chamber system with ambient air. Possible NH_3 carryover could affect both the measured NH_3 concentration in the treated plots and in the unfertilized control, which is considered as background. Finally, we examined how different ways of accounting for background NH_3 levels affect the calculated values for cumulative NH_3 emissions.

2. Materials and Methods

2.1. Constancy of the NH_3 Concentration within the Chamber System

In a laboratory experiment (Figure 1), we filled four squared aluminum trays (12 × 12 cm) with 25 mL pig slurry and 25 mL distilled water so that the bottom of those trays was completely covered. The slurry NH_4-N content was 1.4 kg m^3, the pH was 8.4, and it contained 1.3% dry matter. The temperature in the laboratory was set to 13.5 °C. Sixty minutes after preparing the trays, the background NH_3 concentration within the laboratory was checked using a 0.25a detector tube (Drägerwerk AG, Lübeck, Germany; Table 1) with a detection range of 0.25–3 ppm. Subsequently, each of the four chambers of the DTM measuring system was placed into one of the four aluminum trays, a 5a detector tube (Drägerwerk AG, Lübeck, Germany; Table 1) was inserted, and a measurement was started (Figure 1). After performing the default number of strokes (in case no calibration mark was reached, the measurement was continued), the NH_3 concentration was noted, a new 5a detector tube was inserted, and the next measurement was started immediately. Overall, ten consecutive measurements were carried out, and the NH_3 concentration, as well as the required pump strokes of each measurement, were noted.

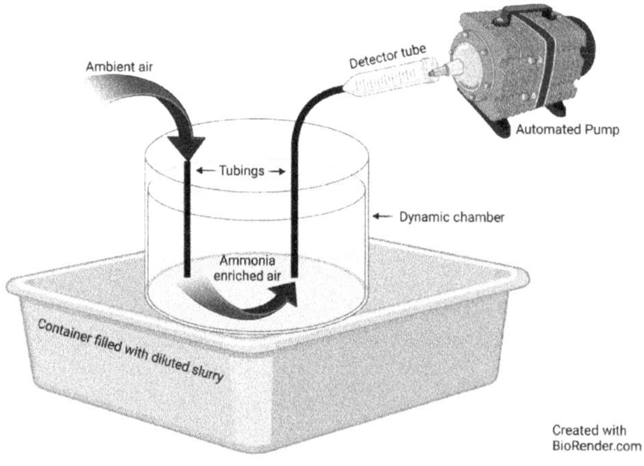

Figure 1. Schematic illustration of the laboratory experiment (only one of the four chambers within the four slurry-filled containers is displayed).

Table 1. Ammonia detector tubes (Drägerwerk AG, Lübeck, Germany).

Detector Tube	Detection Range (ppm)	Default Stroke Number	Time per Stroke (s)
0.25a	0.25–3	10	4.5
2a	2–30	5	6.5
5a	5–70	10	4.5

ppm = parts per million, s = second.

2.2. Assessment of the In-Field Chamber System Cleaning Procedure

On 23 June 2020, we conducted a grassland experiment on a farm close to Osnabrück in northwest Germany. The experiment included four different slurry application techniques and unfertilized control plots in a randomized block design with four replicates. More detailed information on that grassland experiment (site Osnabrück 2020) is published in Nyameasem et al. [24]. Immediately after slurry application, the chamber system was placed on the slurry-covered soil surface, and a DTM measurement was performed. After each measurement, the chamber system was thoroughly cleaned with paper towels and flushed with ambient air as described by Pacholski [12] and subsequently moved to another plot to carry out the next measurement. Within the next 100 min, all 16 plots with slurry application were measured once. The final measurement in this series indicated an NH_3 concentration of 10 ppm within the chamber system. Thirty minutes after measuring the last plot with slurry application, we started a series of eight consecutive measurements on unfertilized control plots using this contaminated chamber system (CCS). The control plots were measured in blockwise order, and we performed two measurement cycles so that each of the four control plots was measured twice. Immediately before and directly after these eight measurements, a so-called "uncontaminated chamber system" (UCC; i.e., not used for previous measurements on fertilized plots) was also used to measure the NH_3 concentrations in unfertilized control plots.

2.3. Winter Wheat Field Trial to Estimate the Influence of Ubiquitous NH_3 Concentration and Chamber System Contamination

Further tests were conducted in a winter wheat (*Triticum aestivum* L.) field trial carried out close to Osnabrück in Lower Saxony between 28 April and 2 May 2020 on a loamy soil (41% sand, 51% silt, and 8% clay; pH 6.1). During this 5-day period, air temperature (11.1 °C on average at 1 m height), wind speed (1.5 m s^{-1} on average at 2 m height), and rainfall (total 12.8 mm) were detected by a nearby weather station. Slurry and acidified slurry were applied by trailing hose technique using a custom-made slurry spreader based on an applicator system from Samson Agro A/S (Viborg, Denmark). Distance between slurry bands was 25 cm. Overall, 26.6 m^3 slurry ha^{-1} with an NH_4-N content of 2.3 kg m^3 was applied, leading to an NH_4-N application rate of 61 kg ha^{-1}. The slurry dry matter content was 8.0%, and the slurry pH was 7.8 for the nontreated slurry and 6.1 for the acidified slurry. Acidification was performed by adding sulfuric acid to a 1-m^3 tank filled with slurry until the target pH was reached while the substrate was thoroughly mixed. Additional control plots without N fertilization were also implemented to assess the background NH_3 concentration. All treatments were setup in a randomized block design with four replicates. The plot size was 9 × 9 m, and additional unfertilized interspaces of 9 m minimized cross-contamination by NH_3 drift between individual plots.

DTM measurements started immediately after slurry application in each individual plot using an NH_3-sensitive detector tube adapted to the expected concentration (Table 1). For treatments with slurry application, the chamber system was centered on the slurry bands, covering 11.5 cm of the 25 cm distance between two slurry bands. Therefore, we adjusted the raw flux calculation (Equation (1)) by applying the factor 0.46. To calculate NH_3 emission rates, we applied the formula for wind speed correction developed by Pacholski [12] for low canopy height (Equation (2)). On the first day, we performed five measurements per plot; on the second day, we carried out two measurement cycles and during the remainder of the 5-day measurement period, only one measurement cycle was carried out each day. We used two separate chamber systems, i.e., one chamber system was used on all plots ("contaminated chamber system" [CCS]), whereas the second chamber system was used exclusively on unfertilized control plots ("uncontaminated chamber system" [UCS]).

Equation (1)—raw flux calculation according to Pacholski et al. [19]:

$$F_{Ng} = vol. * conc * \frac{1013(\text{hPa})}{pact} * \left(\frac{696.1\ (\frac{\text{mg}}{\text{L}}) * 298.15\text{K}}{(273.15 + Tact)\text{K}} \right) * 10^{-6} * \frac{14\ (\frac{\text{g}}{\text{mol}})}{17\ (\frac{\text{g}}{\text{mol}})} * \frac{10000(\text{cm}^2/\text{m}^2)}{415(\text{cm}^2)} * \frac{3600(\frac{\text{s}}{\text{h}})}{dur(\text{s})} \quad (1)$$

where F_{Ng} = ammonia flux (mg N m^{-2} h^{-1}), *vol.* [1] = volume of air passed through the system (l), *conc* [1] = detector tube reading minus background concentration (ppm). The background concentration is defined as the detector tube reading on unfertilized control plots. Measurements in treated and control plots should be closely related in space and time; *pact* = actual air pressure (hPa), *Tact* = air temperature at measurement (°C), 14/17 = conversion from NH_3 to N by their atomic weight, 415 = area covered by chambers (cm^2), *dur* [1] = duration of measurement (s), and [1] = by linear conversion related to the default stroke number of the detector tube used.

Equation (2)—wind speed correction for low canopy height according to *Pacholski* et al. [19]:

$$\ln(absolute\ flux) = 0.444 * \ln(raw\ flux) + 0.59 * \ln(V_{wind}) \quad (2)$$

where *absolute flux* = ammonia flux (kg N ha^{-1} h^{-1}), *raw flux* = ammonia flux (kg N ha^{-1} h^{-1}) calculated by Equation (1), and V_{wind} = wind speed (m s^{-1}) at 2 m height during the measurement.

2.4. Data Analyses

For assessing the constancy of the NH_3 concentration within the chamber system, we plotted the cumulative number of strokes on the x-axis and the measured NH_3 concentration on the y-axis. For calculating the regression function, we used natural logarithmic transformation. Subsequently, we calculated the coefficient of determination (R^2) and the significance of the slope using IBM SPSS statistics 28.

To evaluate the in-field chamber system cleaning procedure, we plotted the time after measuring on high emission plots on the x-axis and the measured NH_3 concentration on the y-axis. For calculating the regression function, we used exponential transformation, and subsequently, the R^2 value and significance of the slope were calculated using IBM SPSS statistics 28.

The influence of ubiquitous NH_3 concentration and chamber system contamination was evaluated by plotting the hours after slurry application on the x-axis and treatment mean NH_3 concentrations on the y-axis. For the unfertilized control, we present treatment mean values measured by the UCS as well as treatment mean values measured by the CCS. We used logarithmic transformation of the x-axis since the majority of emissions occurred within the first eight hours of the five-day experiment.

Furthermore, we calculated the cumulated NH_3 emissions based on three different approaches:

1. No background subtraction regarding the raw flux calculation;
2. Background subtraction based on UCS data to estimate the influence of the ubiquitous NH_3 concentration;
3. Background subtraction based on CCS data.

Subsequent to raw flux calculation (Equation (1)), the formula for wind speed correction (Equation (2)) was applied, and emissions were cumulated by linear interpolation for all three approaches.

3. Results

3.1. Constancy of the NH_3 Concentration within the Chamber System

Prior to the start of the experiment, the background NH_3 concentration in the laboratory was 0.25 ppm. After performing 16 strokes with a 5a detector tube inserted while the chamber system was placed on the aluminum trays filled with slurry, the measured NH_3 concentration was 3.1 ppm (Figure 2). For 20 strokes, corresponding to the premeasurement phase to establish a constant NH_3 concentration in the chamber system [12], we calculated a concentration of 6.3 ppm. After 30 strokes (i.e., detector tube reading on the field), the NH_3 concertation increased to 10.2 ppm. Thereafter, the concentration in the chamber system continued to increase until a concentration of 23.1 ppm was reached at 111 strokes. No further increase in concentration was detected thereafter.

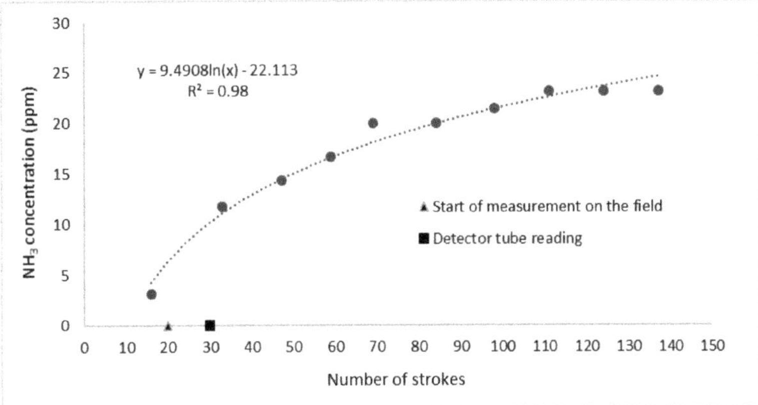

Figure 2. Increase in the NH_3 concentration in the chamber system following ongoing measurements on a surface covered with pig slurry. The slope of the equation is highly significant ($p < 0.001$). R^2 = coefficient of determination, ln = natural logarithm.

3.2. *Assessment of the In-Field Chamber System Cleaning Procedure*

The first measurement on the unfertilized control indicated an NH_3 concentration of 2 ppm, while for the second measurement, the concentration decreased to 1.33 ppm (Figure 3). Thereafter, the decrease in the NH_3 concentration in the chamber system slowed down and leveled off at the seventh consecutive measurement (0.5 ppm). Approximately one hour after measuring the last plot with slurry application, we performed the eighth measurement on the unfertilized control, where the NH_3 concentration was still 0.5 ppm. No NH_3 was detected during simultaneous measurements with a chamber system not used for previous measurements on plots with slurry application (Figure 3).

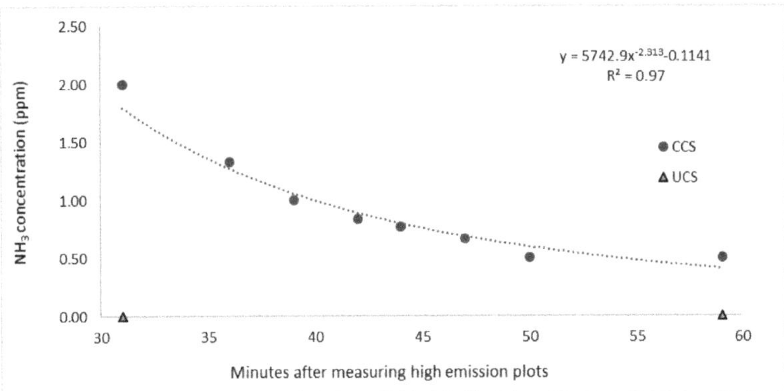

Figure 3. Decrease in the NH_3 concentration within a contaminated chamber system induced by consecutive measurements on unfertilized plots. The slope of the equation is highly significant ($p < 0.001$); R^2 = coefficient of determination, CCS = contaminated chamber system, and UCS = uncontaminated chamber system.

3.3. *Influence of Ubiquitous NH_3 Concentration and Chamber System Contamination*

Figure 4 shows the NH_3 concentration within the chamber system for two treatments with slurry application and for unfertilized control plots in a winter wheat field experiment. Two separate chamber systems were used for the measurements on the control plots. One chamber system was exclusively used for control plots (UCS), whereas the other was also used for plots with slurry application (CCS). Within the first four hours after application, the NH_3 concentration measured for trailing hose application of slurry increased to approximately 11.2 ppm, while for trailing hose application of acidified slurry, the NH_3 concentration reached only 5.3 ppm (Figure 4). Four hours after the start of the experiment, the CCS indicated an NH_3 concentration of 3.9 ppm for the control plots, while the UCS did not indicate any NH_3. Within eight hours after application, the NH_3 concentration in the trailing hose treatment without acidification decreased to approximately 2 ppm. For the trailing hose treatment with acidification and the control treatment measured by the CCS, the NH_3 concentration was approximately 1.5 ppm. The NH_3 concentration in the control measured by the UCS increased to 0.5 ppm. Thereafter, the NH_3 concentration gradually decreased until no NH_3 was measured 72 h after the slurry application started. Plots with slurry application and control plots measured with the CCS exhibited a similar pattern for the period 24 to 96 h after application: 24 h after application, the concentration was approximately 0.5 ppm, and thereafter, it declined to 0.1 ppm 96 h after application.

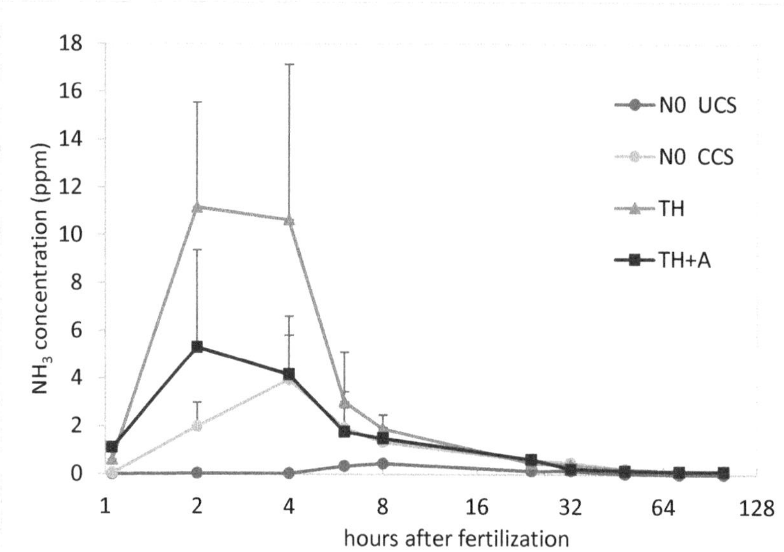

Figure 4. Mean NH_3 concentrations within the chamber system for different treatments and unfertilized control plots following slurry application in a winter wheat field experiment. Logarithmic transformation of the x-axis was used to highlight the first day after application when the NH_3 concentration was highest. Error bars represent the standard deviation (to minimize overlapping, only the positive standard deviation is shown). N0 UCS = control plots without slurry application measured with an uncontaminated chamber system exclusively used for measuring control plots, N0 CCS = control plots without slurry application measured with a contaminated chamber system also used for measuring plots with slurry application, TH = trailing hose application of slurry, and TH+A = trailing hose application of acidified slurry.

Figure 5 shows cumulated NH_3 emissions for two treatments with slurry application for the 5-day measuring period in the winter wheat field experiment. We used three different approaches for calculating cumulative NH_3 emissions: (1) no background subtraction,

(2) background subtraction based on UCS data, and (3) background subtraction based on CCS data. When applying no background subtraction, NH_3-N emissions sum up to 9.7 kg ha^{-1} following slurry application by trailing hose. Background subtraction based on UCS data resulted in NH_3-N emissions of 8.8 kg ha^{-1}, and when background subtraction was based on CCS data, we calculated NH_3-N emissions of only 3.0 kg ha^{-1}. Slightly lower emissions were found for the trailing hose treatment with slurry acidification for these three approaches. The standard deviation of the four replicates for the approaches "no background subtraction" and "background subtraction based on UCS data" overlaps for both treatments with slurry application. However, when looking at individual plots, "no background subtraction" always leads to higher NH_3-N emissions than "background subtraction based on UCS data".

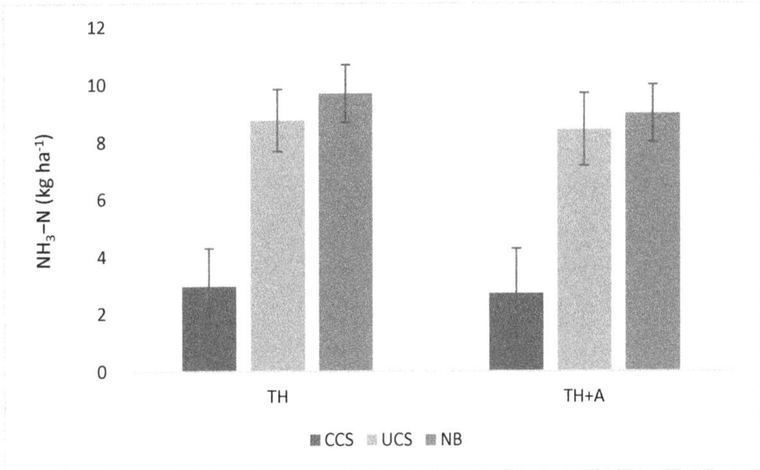

Figure 5. Cumulated NH_3 emissions for three approaches to subtract background values for calculating NH_3 emissions following two slurry application techniques in a winter wheat field experiment. The error bars represent the standard deviation (n = 4). TH = trailing hose application of slurry, TH+A = trailing hose application of acidified slurry, CCS = background subtraction based on data obtained from a contaminated chamber system also used for measuring plots with slurry application, UCS = background subtraction based on data obtained from an uncontaminated chamber exclusively used for measuring unfertilized control plots, and NB = no background subtraction applied.

4. Discussion

4.1. Constancy of the NH_3 Concentration within the Chamber System

For the laboratory test, the relevant experimental conditions (i.e., temperature and NH_3 concentration) were chosen to be comparable to conditions typical for a field experiment on the first day after slurry application. Obviously, the NH_3 concentration in the chamber system is not stable after performing 20 pump strokes (Figure 2). Different detector tubes with varying standard stroke numbers can be used for the DTM, and if no calibration mark is reached after performing the standard stroke number of the respective detector tube, the measurement should be extended [12]. However, that means that linear conversion between the alternating number of pump strokes as required for Equation (1) [16,19] is problematic since the NH_3 concentration within the chamber system is not stable. Furthermore, changing the number of initial pump strokes prior to the actual measurement would affect raw flux calculation and the subsequent result of the empirical wind speed correction. Therefore, the "raw flux" [16,19] cannot be regarded as a flux and should be viewed as an empirical factor based on the NH_3 concentration within the chamber system.

4.2. Assessment of the In-Field Chamber System Cleaning Procedure

Our data (Figure 3) revealed that the in-field cleaning procedure is not sufficient to reduce chamber system contamination. Cleaning by using paper towels removes dirt sticking to the chamber system but might not affect NH_3 retained at chamber walls or tubings or solved in condensed water anywhere in the system. Flushing by ambient air addresses this problem, but it might take too long to reduce contamination under field conditions. Thirty minutes after measuring on high emission plots, the NH_3 concentration in the CCS was still as high as 2 ppm, while simultaneous measurements with a UCS did not indicate any NH_3. Even one hour after measuring on high emission plots, the concentration was still 0.5 ppm. In that period, eight consecutive measurements were performed so that the chamber system was constantly flushed with ambient air. The inadequacy of the cleaning procedure for the chamber system under field conditions means that the calculation of NH_3 emissions is affected by ubiquitous NH_3 but also by chamber system contamination.

4.3. Influence of Ubiquitous NH_3 Concentration and Chamber System Contamination

In our multi-plot field trial, the NH_3 background in the control plots was much more influenced by chamber system contamination than by ubiquitous NH_3. Measurements with the UCS did not result in NH_3 concentrations above 0.5 ppm, whereas for the CCS concentrations, up to 3.9 ppm in unfertilized control plots on the day of slurry application were detected (Figure 4). Obviously, in fertilized plots, the chamber system was contaminated with similar amounts of residual NH_3 from previous measurements. Moreover, the background value within the chamber system might vary depending on the emissions in the previously measured plot. This might explain the high standard deviation on the day of slurry application (Figure 4).

Furthermore, the so-called background NH_3 concentration required for the calculations according to Equation (1) [12,16,19] is not exactly specified. We showed that cumulated NH_3 emissions differ depending on its definition (Figure 5). When background subtraction was solely based on the ubiquitous NH_3 concentration measured by a UCS, cumulated NH_3 emissions were approximately threefold higher compared to calculations using background values measured with a CCS. Furthermore, it needs to be considered that the NH_3 background of the chamber system and actual NH_3 emissions caused by fertilization might not be additive (i.e., the NH_3 concentration measured in a treated plot increases by the amount of the NH_3 background concentration). The laboratory experiment (Figure 2) revealed that after 30 strokes, corresponding to the timing of reading the detector tube in the field, the NH_3 concentration was 10.2 ppm. The measured background was 0.25 ppm leading to a background-adjusted concentration according to the recommendations from Pacholski et al. [19] of 9.95 ppm. However, when considering the NH_3 concentration after 30 strokes (10.2 ppm) as background for the measurement after 60 strokes (16.7 ppm), the value of the background adjusted NH_3 concentration would only be 6.5 ppm. This illustrates that, at least for very high chamber system NH_3 concentrations, background and emissions caused by fertilization are not additive. Thus, simply subtracting background values is problematic.

It has to be kept in mind that DTM raw flux calibration was performed in only two fertilized plots and one unfertilized area [19], so the relevance of chamber system contamination was probably much lower than in our multi-plot field experiment (Figure 4). Currently, the DTM is frequently used to quantify relative differences in NH_3 emission between treatments (e.g., different techniques for slurry application) as determined by acid traps placed in the center of each plot [12,21]. For this purpose, DTM measurements are only performed in one or a few selected plots so that the conditions are closer to the initial calibration setup. When applied with great care regarding chamber system contamination, this approach might still be the best alternative for estimating NH_3 emissions in multi-plot field trials. To reduce the risk of nonrepresentative background values due to chamber system contamination, we suggest using at least two chamber systems. One chamber system is used for treated plots as well as for control plots, whereas another chamber system measures exclusively on control plots to determine the ubiquitous NH_3 concentration. Subsequently, the results of both chamber systems are compared. If a discrepancy between the chamber systems occurs regarding the measurement on the control plots, the chamber system used for treated plots should be thoroughly cleaned, or another chamber system should be used.

5. Conclusions

Our objective was to test important prerequisites and boundary conditions for the application of the DTM technique in order to propose an improved approach. We showed that (1) the NH_3 concentration in the chamber is not stable and that (2) the cleaning procedure with paper towels and flushing with ambient air is not sufficient under field conditions. Therefore, NH_3 carryover affects the measurements in both the treated plots and the unfertilized control plots. This affects the calculation of NH_3 emissions.

Ultimately, the DTM approach requires a new empirical calibration that considers these problems. Instead of using analog detector tubes with calibration marks, the use of a digital NH_3 detector would allow for performing measurements after exchanging a specified air volume in a specified time. Furthermore, we advise using separate chamber systems for each individual plot so that cross-contamination is prevented. Overall, the new empirical model should be based on the factors (1) NH_3 concentration measured in the fertilized plot, (2) NH_3 concentration measured in a control plot using the CCS as well as (3) a UCS and (4) wind speed. For the current empirical model, the wind speed at the time of measurement is used. However, the wind speed at the time of measurement is then used to extrapolate emissions until the next measurement is performed. Therefore, we suggest including the average wind speed between the two measurements in the new model.

Moreover, it has to be considered that contamination might not only affect DTM measurements. It might be a common problem for all chamber systems and wind tunnels since NH_3 is highly reactive and readily soluble in water, which makes it likely to be retained anywhere in those systems [31]. Therefore, we suggest that all systems currently used to estimate NH_3 emissions should be tested for contamination. Otherwise, it needs to be considered that NH_3 emissions, estimated by those systems, are not only caused by the effect(s) of interest (e.g., N fertilization) but also by the system's specific NH_3 release after it was initially exposed to an NH_3 source.

Author Contributions: Conceptualization, M.t.H. and H.-W.O.; methodology, M.t.H.; software, M.t.H.; validation, H.-W.O.; formal analysis, M.t.H.; investigation, M.t.H.; resources, H.-W.O.; data curation, M.t.H.; writing—original draft preparation, M.t.H.; writing—review and editing, H.-W.O.; visualization, M.t.H.; supervision, H.-W.O.; project administration, H.-W.O.; funding acquisition, H.-W.O. All authors have read and agreed to the published version of the manuscript.

Funding: This research was funded by the German Federal Ministry of Food and Agriculture (BMEL; grant number 281B300916) within the project "Mitigation of ammonia and greenhouse gas emission and improving nitrogen use efficiency by innovative slurry and digestate application techniques for growing crops" based on a decision of the Parliament of the Federal Republic of Germany via the Federal Office for Agriculture and Food (BLE).

Institutional Review Board Statement: Not applicable.

Data Availability Statement: The data presented in this study are available on request from the corresponding author.

Acknowledgments: We would like to thank everybody who helped us with the field experiments and laboratory analysis, and we acknowledge Samson Agro A/S for their support during the development of the experimental device for applying the liquid manure.

Conflicts of Interest: The authors declare no conflict of interest. The funders had no role in the design of the study; in the collection, analyses, or interpretation of data; in the writing of the manuscript; or in the decision to publish the results.

References

1. Sutton, M.A.; Oenema, O.; Erisman, J.W.; Leip, A.; van Grinsven, H.; Winiwarter, W. Too Much of a Good Thing. *Nature* **2011**, *472*, 159–161. [CrossRef] [PubMed]
2. Sommer, S.G.; Génermont, S.; Cellier, P.; Hutchings, N.J.; Olesen, J.E.; Morvan, T. Processes Controlling Ammonia Emission from Livestock Slurry in the Field. *Europ. J. Agron.* **2003**, *19*, 465–486. [CrossRef]
3. Asman, W.A.H.; Sutton, M.A.; Schjorring, J.K. Ammonia: Emission, Atmospheric Transport and Deposition. *New Phytol.* **1998**, *139*, 27–48. [CrossRef]
4. Ni, K.; Köster, J.R.; Seidel, A.; Pacholski, A. Field Measurement of Ammonia Emissions After Nitrogen Fertilization—A Comparison between Micrometeorological and Chamber Methods. *Eur. J. Agron.* **2015**, *71*, 115–122. [CrossRef]
5. Lelieveld, J.; Evans, J.S.; Fnais, M.; Giannadaki, D.; Pozzer, A. The Contribution of Outdoor Air Pollution Sources to Premature Mortality on a Global Scale. *Nature* **2015**, *525*, 367–371. [CrossRef]
6. Van Damme, M.; Clarisse, L.; Whitburn, S.; Hadji-Lazaro, J.; Hurtmans, D.; Clerbaux, C.; Coheur, P.-F. Industrial and Agricultural Ammonia Point Sources Exposed. *Nature* **2018**, *564*, 99–103. [CrossRef]
7. Emmerling, C.; Krein, A.; Junk, J. Meta-Analysis of Strategies to Reduce NH_3 Emissions from Slurries in European Agriculture and Consequences for Greenhouse Gas Emissions. *Agronomy* **2020**, *10*, 1633. [CrossRef]
8. Liu, L.; Xu, W.; Lu, X.; Zhong, B.; Guo, Y.; Lu, X.; Zhao, Y.; He, W.; Wang, S.; Zhang, X.; et al. Exploring Global Changes in Agricultural Ammonia Emissions and Their Contribution to Nitrogen Deposition Since 1980. *Proc. Natl. Acad. Sci. USA* **2022**, *119*, e2121998119. [CrossRef]
9. Webb, J.; Menzi, H.; Pain, B.F.; Misselbrook, T.H.; Dämmgen, U.; Hendriks, H.; Döhler, H. Managing Ammonia Emissions from Livestock Production in Europe. *Environ. Pollut.* **2005**, *135*, 399–406. [CrossRef]
10. Webb, J.; Pain, B.; Bittman, S.; Morgan, J. The Impacts of Manure Application Methods on Emissions of Ammonia, Nitrous Oxide and on Crop Response—A Review. *Agr. Ecosyst. Environ.* **2010**, *137*, 39–46. [CrossRef]
11. European Environment Agency. *Air Pollution: National Emission Reduction*; European Environment Agency: Copenhagen, Denmark, 2016.
12. Pacholski, A. Calibrated Passive Sampling—Multi-Plot Field Measurements of NH_3 Emissions with a Combination of Dynamic Tube Method and Passive Samplers. *J. Vis. Exp.* **2016**, *109*, e53273. [CrossRef]
13. Leuning, R.; Freney, J.R.; Denmead, O.T.; Simpson, J.R. A Sampler for Measuring Atmospheric Ammonia Flux. *Atmos. Environ. (1967)* **1985**, *19*, 1117–1124. [CrossRef]
14. Sherlock, R.R.; Freney, J.R.; Smith, N.P.; Cameron, K.C. Evaluation of a Sampler for Assessing Ammonia Losses from Fertilized Fields. *Fertil. Res.* **1989**, *21*, 61–66. [CrossRef]
15. Sommer, S.G.; Misselbrook, T.H. A Review of Ammonia Emission Measured Using Wind Tunnels Compared with Micrometeorological Techniques. *Soil Use Manag.* **2016**, *32*, 101–108. [CrossRef]
16. Roelcke, M.; Li, S.X.; Tian, X.H.; Gao, Y.J.; Richter, J. In Situ Comparisons of Ammonia Volatilization from N Fertilizers in Chinese Loess Soils. *Nutr. Cycl. Agroecosyst.* **2002**, *62*, 73–88. [CrossRef]
17. Richter, J. Zur Methodik des Bodengashaushaltes. II. Ergebnisse und Diskussion. *Z. Pflanzenernaehr. Bodenk.* **1972**, *132*, 220–239. [CrossRef]
18. Roelcke, M.A.P. Die Ammoniak-Volatilisation nach Ausbringung von Mineraldünger-Stickstoff in Carbonatreichen Chinesischen Löß-Ackerböden. Ph.D. Thesis, Braunschweig Technical University, Braunschweig, Germany, 1994.
19. Pacholski, A.; Cai, G.; Nieder, R.; Richter, J.; Fan, X.; Zhu, Z.; Roelcke, M. Calibration of a Simple Method for Determining Ammonia Volatilization in the Field—Comparative Measurements in Henan Province, China. *Nutr. Cycl. Agroecosyst.* **2006**, *74*, 259–273. [CrossRef]
20. Pacholski, A.; Cai, G.-X.; Fan, X.-H.; Ding, H.; Chen, D.; Nieder, R.; Roelcke, M. Comparison of Different Methods for the Measurement of Ammonia Volatilization After Urea Application in Henan Province, China. *Z. Pflanzenernähr. Bodenk.* **2008**, *171*, 361–369. [CrossRef]
21. Gericke, D.; Pacholski, A.; Kage, H. Measurement of Ammonia Emissions in Multi-Plot Field Experiments. *Biosyst. Eng.* **2011**, *108*, 164–173. [CrossRef]
22. Vandré, R.; Kaupenjohann, M. In Situ Measurement of Ammonia Emissions from Organic Fertilizers in Plot Experiments. *Soil Sci. Soc. Am. J.* **1998**, *62*, 467–473. [CrossRef]
23. Ni, K.; Pacholski, A.; Kage, H. Ammonia Volatilization after Application of Urea to Winter Wheat Over 3 Years Affected by Novel Urease and Nitrification Inhibitors. *Agric. Ecosyst. Environ.* **2014**, *197*, 184–194. [CrossRef]
24. Nyameasem, J.K.; Zutz, M.; Kluß, C.; Ten Huf, M.; Essich, C.; Buchen-Tschiskale, C.; Ruser, R.; Flessa, H.; Olfs, H.-W.; Taube, F.; et al. Impact of Cattle Slurry Application Methods on Ammonia Losses and Grassland Nitrogen Use Efficiency. *Environ. Pollut.* **2022**, *315*, 120302. [CrossRef] [PubMed]
25. Quakernack, R.; Pacholski, A.; Techow, A.; Herrmann, A.; Taube, F.; Kage, H. Ammonia Volatilization and Yield Response of Energy Crops After Fertilization with Biogas Residues in a Coastal Marsh of Northern Germany. *Agric. Ecosyst. Environ.* **2012**, *160*, 66–74. [CrossRef]
26. Seidel, A.; Pacholski, A.; Nyord, T.; Vestergaard, A.; Pahlmann, I.; Herrmann, A.; Kage, H. Effects of Acidification and Injection of Pasture Applied Cattle Slurry on Ammonia Losses, N2O Emissions and Crop N Uptake. *Agric. Ecosyst. Environ.* **2017**, *247*, 23–32. [CrossRef]

27. Ten Huf, M.; Reinsch, T.; Zutz, M.; Essich, C.; Ruser, R.; Buchen-Tschiskale, C.; Flessa, H.; Olfs, H.-W. Effects of Liquid Manure Application Techniques on Ammonia Emission and Winter Wheat Yield. *Agronomy* **2023**, *13*, 472. [CrossRef]
28. Wagner, C.; Nyord, T.; Vestergaard, A.V.; Hafner, S.D.; Pacholski, A.S. Acidification Effects on In Situ Ammonia Emissions and Cereal Yields Depending on Slurry Type and Application Method. *Agriculture* **2021**, *11*, 1053. [CrossRef]
29. Freney, J.R.; Simpson, J.R.; Denmead, O.T. Volatilization of Ammonia. In *Gaseous Loss of Nitrogen from Plant-Soil Systems*; Freney, J.R., Simpson, J.R., Eds.; Springer: Dordrecht, The Netherlands, 1983; pp. 1–32. ISBN 978-90-481-8276-3.
30. Sommer, S. Modelling Ammonia Volatilization from Animal Slurry Applied with Trail Hoses to Cereals. *Atmos. Environ.* **2000**, *34*, 2361–2372. [CrossRef]
31. Denmead, O.T. Micrometeorological Methods for Measuring Gaseous Losses of Nitrogen in the Field. In *Gaseous Loss of Nitrogen from Plant-Soil Systems*; Freney, J.R., Simpson, J.R., Eds.; Springer: Dordrecht, The Netherlands, 1983; pp. 133–157. ISBN 978-90-481-8276-3.

agriculture

MDPI

Article

Does Livelihood Determine Attitude? The Impact of Farmers' Livelihood Capital on the Performance of Agricultural Non-Point Source Pollution Management: An Empirical Investigation in Yilong Lake Basin, China

Ning Zhou [1], Fanglei Zhong [2,*] and Yanjie Yin [2]

[1] School of Humanities and Social Sciences, Beijing Institute of Technology, Beijing 100081, China; 7520200111@bit.edu.cn

[2] School of Economics, Minzu University of China, Beijing 100081, China

* Correspondence: zfl@muc.edu.cn

Abstract: Agricultural non-point source pollution is intricately connected to the rural population's production and lifestyle. The heterogeneous composition of livelihood capital results in varied livelihood types, influencing the farmers' attitudes and perceptions of the treatment projects. This ultimately causes discrepancies in the farmers' evaluation of agricultural non-point source pollution control. In this study, a participatory evaluation method was employed to evaluate the performance of agricultural non-point source pollution control projects in the Yilong Lake Watershed of Yunnan Province and analyze the underlying reasons for the differing performance evaluations. The findings revealed that pure farmers' performance evaluation value for agricultural non-point source pollution control projects in Yilong Lake Basin was 0.4811 (with the full mark being 1), with a general evaluation grade. Part-time business households had a performance evaluation value of 0.5969, also with a general evaluation grade, while non-farmers had a performance evaluation value of 0.7057, with a good evaluation grade. The performance evaluation value ranked from highest to lowest is non-farmer > part-time farmer > pure farmer. The main factor affecting the variation in farmers' performance evaluation is the key index of different livelihood capital. If pollution control projects can promote the adjustment of farmers' livelihood capital types, it can enhance not only the performance evaluation degree of farmers, but also the sustainability of farmers' livelihoods and increase their adaptability to livelihood risks. Therefore, pollution control projects should consider farmers' livelihood capital types and be implemented accurately to improve farmers' satisfaction and sustainability.

Keywords: livelihood capital; livelihood type; agricultural non-point source pollution control; governance project performance evaluation

1. Introduction

At present, agricultural non-point source pollution has emerged as the principal contributor to water pollution in China [1]. As per the "Second National Pollution Source Census Report" in 2017, agricultural sources discharged 10.67 million tons of chemical oxygen demand (COD), accounting for 49.77% of the total COD emissions in the nation. Furthermore, 216,200 tons of ammonia nitrogen (NH3-N) were emitted, making up 22.44% of the nationwide NH3-N emissions. Additionally, agricultural sources released 1.4149 million tons of total nitrogen (TN), which accounts for 46.52% of the total TN emissions in water pollutants, and 212,000 tons of total phosphorus (TP), constituting 67.22% of the entire TP emissions in water pollutants. The agricultural non-point source pollution perilously threatens China's food security, food safety, and the sustainable development of society [2]. The Chinese government holds great concern regarding agricultural non-point source pollution. In October 2022, the report of the 20th National Congress of the Communist Party of China emphasized "enhancing soil pollution control at its source and treating new pollutants" and subsequently initiated a series of measures and policies for governance. For instance, in the "14th Five-Year Plan" for National Economic and Social Development, specific implementation objectives and strategies have been proposed for the administration of agricultural non-point source pollution: "Profoundly execute the reduction of pesticides and fertilizers, curb agricultural film pollution, enhance the utilization rate of agricultural film recycling, encourage comprehensive utilization of straw, and promote the resource utilization of livestock and poultry manure".

Research on the management of agricultural non-point source pollution was initiated in the 1970s in Europe and America, with a focus on the natural mechanisms of non-point source pollution. From the perspective of pollution generation mechanisms, various legal, economic, and technological measures have been implemented to prevent and control pollution, along with corresponding strategies [3,4]. However, this engineering-centric approach may lead to merely treating the symptoms instead of addressing the root causes. Therefore, it is essential to evaluate the effectiveness of agricultural non-point source pollution control to continuously improve the measures and policies. Currently, most evaluations concentrate on engineering performance indicators [5,6], with little consideration of multi-stakeholder perspectives. As a matter of fact, since agricultural non-point source pollution is closely linked to the production and livelihood behavior of farmers and other stakeholders, it is crucial to incorporate more multi-stakeholder evaluation perspectives to systematically assess the effectiveness of agricultural non-point source pollution control. The overall evaluation of farmers regarding the effectiveness of agricultural non-point source pollution control not only provides the most intuitive evaluation result but also reflects their recognition of pollution control, which is strongly related to their willingness to implement environmentally friendly behavior. Agricultural non-point source pollution control has the most significant impact on farmers, and government-led pollution control efforts cannot succeed without farmers' participation, cooperation, and support. Therefore, strengthening the evaluation of the effectiveness of agricultural non-point source pollution control from the perspective of farmers and studying the influencing factors will help comprehensively evaluate pollution control effectiveness and understand the factors influencing farmers' production, management, consumption behavior, and adoption of modern technology. This, in turn, will enable the implementation of a series of policy measures, such as taxation, fiscal compensation, and education incentives, to guide farmers' behavior, encourage voluntary adoption of environmentally friendly technology, and achieve the goal of controlling agricultural non-point source pollution [7–9].

Currently, domestic and foreign scholars' debates on farmers' behavior mainly focus on whether farmers are fully rational economic actors and different understandings of farmers' rationality will directly affect the analytical perspective and results of farmers' behavior. Herbert Simon argued that "rationality refers to the behavior that achieves the given goals within the given conditions and constraints" [10]. Based on Simon's bounded rationality hypothesis, this paper believes that although farmers subjectively base their behavioral decisions on profit or utility maximization, their rational behavior in decision-making exhibits certain differences from fully rational behavior due to the influence of multiple factors such as individual abilities, experience, knowledge structures, preferences, and resource endowments. Farmers' decisions regarding production are intricately linked with their strategies for sustaining their livelihoods, and their livelihood capital plays a crucial role in shaping these strategies. Livelihood capital not only reflects farmers' basic livelihood status but also has a significant impact on their investment decisions, production management, human-land relationships, and behavioral intentions [11–13]. The strategies that farmers adopt to sustain their livelihoods determine the nature and intensity of their impact on the local ecological environment, with profound implications for the ecological environment [14,15]. In recent years, research on farmers' livelihoods has become a hot topic in academia, covering various aspects such as poverty alleviation, ecological migration, ecological construction, tourism, cultural change, and environmental change. The academic community's research on the relationship between policy systems, environmental change, land use, and farmers' livelihoods has been relatively in-depth. For example, Sumac Elisa Cárdenas Oleas (2015) analyzed the impact and role of the government's support for the quinoa industry on the sustainable livelihood development of poor rural populations under the background of policy formulation and implementation [16]; Bhandari et al. (2013) used the sustainable livelihood approach to analyze the degree of influence of household and community assets from a micro perspective on farmers' livelihood transformation strategies [17]; Milan et al. (2014) conducted empirical research on the relationship between climate change, environmental change, and mountain migration, studying the differences in livelihood and population mobility patterns in households at different altitudes [18]. However, research on pollution control from the perspective of farmers' livelihoods remains limited. While some scholars have explored the relationship between pollution control and farmers' livelihoods [19], they have only examined the relationship between these two factors [20,21]. Other scholars have focused on the current situation and challenges of environmental governance from the perspective of social capital, exploring the role of social capital in environmental governance and making suggestions and recommendations [22]. However, these studies only examine the external environment and the behavioral control imposed by policies and institutions, without establishing the internal mechanisms through which livelihood capital influences livelihood behavior and, consequently, farmers' evaluations of pollution control effectiveness.

Yunnan Province boasts abundant water resources and numerous natural lakes, serving both as a crucial ecological barrier and water source protection zone in China and as a relatively vulnerable area in terms of ecology and economic development. The nine basin areas of Yunnan Province's major plateau lakes are the most densely populated, with the most frequent human activities and the most developed economy. Reinforcing the protection of these nine major plateau lake basins plays a pivotal role in Yunnan Province's development [23]. As one of the nine major plateau lakes in Yunnan Province, Yilong Lake has long maintained its water quality between Class V and inferior Class V, severely affecting the lake's ecological service function and the ecological safety of local residents [24]. The agricultural non-point source pollution control project is a critical undertaking for environmental improvement in the Yilong Lake basin, with its governance performance and effects on local farmers' livelihoods under wide-ranging scrutiny from various sectors of society.

Therefore, in summary, this study focuses on evaluating the performance of governance projects based on different types of livelihood capital among farmers, analyzes the effects of farmers' livelihood capital on governance performance evaluation, constructs an analytical framework for the relationship between livelihood capital and governance performance, and provides empirical evidence from the agricultural non-point source pollution control project in Yilong Lake, Yunnan Province to comprehend the behavior and willingness of local farmers to participate in governance. This study will furnish crucial reference value to the government to introduce targeted policies and counter the detrimental impacts of farmers' production activities on the environment. Additionally, it will provide a novel research perspective for evaluating the project's performance in pollution control, along with offering guidance for environmental pollution control and the sustainable development of the basin economy.

2. Materials and Methods

2.1. Overview of the Study Area

Yilong Lake is situated in Shiping County, Honghe Hani and Yi Autonomous Prefecture, Yunnan Province, and serves as the source of the Pearl River, the third longest river in China and the largest river system in southern China. Being the only lake in Honghe Prefecture with an area exceeding 30 square kilometers, it is revered as the "mother lake" of Honghe Prefecture. Furthermore, it is also the southernmost plateau freshwater lake in

China, thereby underscoring the strategic importance of ensuring the ecological security and sustainable development of the economic society in the Yilong Lake basin, not just for Shiping County, but also for the southern Yunnan region.

The Yilong Lake basin serves as the hub of economic activity in Yunnan Province's Shiping County, boasting a high intensity of agricultural development. However, agricultural non-point source pollution has become the primary cause of water pollution in the lake [25]. Farmers' use of unreasonable farming methods has resulted in the daunting task of reducing pollution. The high-polluting agricultural cultivation techniques are widespread in lakeside, dam, and semi-mountainous areas. The cultivation of economically valuable crops, such as high-profit vegetables and fruits, which contribute to increased pollution, is continually expanding. Crop rotation and replanting indices are escalating, while the usage of pesticides and fertilizers is increasing each year. Rural living and the pollution generated by free-range livestock and poultry contribute to 51.4%, 45.1%, 49.0%, and 70.4% of the total pollution load into Yilong Lake as per statistics, thereby posing challenges to agricultural and rural pollution control [26].

Presently, the government of Shiping County has initiated the Agricultural Non-Point Source Pollution Control Project within the Yilong Lake Basin. This project aims to curb non-point source pollution in farmlands on the north bank of Yilong Lake, along the urban river, and facilitate comprehensive improvements in large-scale livestock and poultry farming along the lake and river. It also involves construction efforts that strengthen agricultural environmental protection measures, such as the restoration of ponds to the lake, soil testing, and formula fertilization, farmland land circulation, and reduced use and control of pesticides and fertilizers. The project will mainly affect Yilong Town and BaXin Town, covering 12 communities and village committees, including Dong Cheng Community, Dashui, Darui City, Dou Di Wan, Gaojiawan, Maohe Village Committee, and Haixi, Wangjiachong, Old Street, Bailang, Baxin, and Xinjie Village Committees. They have jurisdiction over a total of 80 natural villages. However, due to the wide distribution of agricultural non-point source pollution in the Yilong Lake Basin and the project being in its initial stages, the supervisory management mechanism remains imperfect. Moreover, since governance policies and project implementation significantly impact farmers' livelihoods, which are tied to their fundamental interests, they have exhibited strong resistance to the implementation of agricultural non-point source pollution control projects. This has resulted in slow progress, long cycles, and significant difficulty in control. This article primarily investigates villages along the lake in Yilong Town and BaXin Town (see Figure 1).

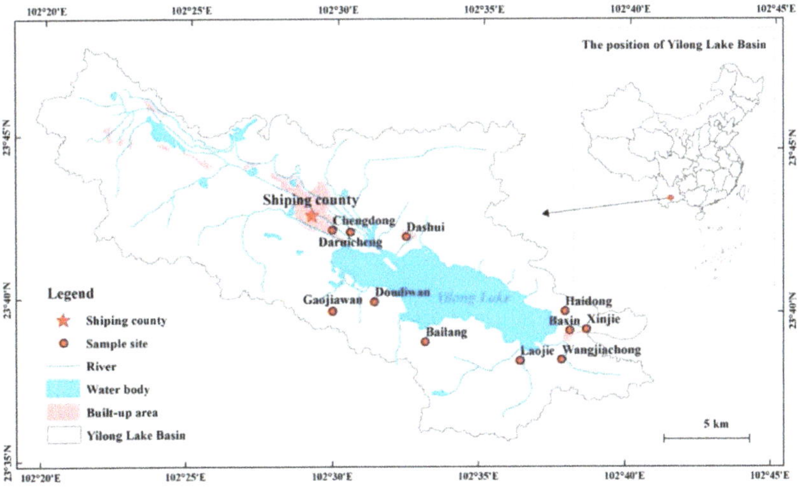

Figure 1. Overview of the study area and the locations of the Yilong Lake.

2.2. Data Collection

Using random sampling and Participatory Rural Appraisal (PRA), the research team conducted a field investigation on the management of agricultural non-point source pollution in the Yilong Lake Basin from July to September 2019. To facilitate the collection and processing of research data and ensure data validity, prior to conducting the questionnaire and interview with farmers, the grassroots staff of each village committee assisted in classifying farmers in the target natural villages into pure farmers, part-time farmers, and non-farmers according to the classification criteria for farmer livelihoods in this article. After classification, a random sample of the three types of farmers was obtained for investigation.

The research team performed field research on agricultural non-point source pollution control in the Yilong Lake Basin from July to September 2019, utilizing a random sampling method and Participatory Rural Appraisal (PRA). To ensure data validity and simplify the collection and processing of research data, the team collaborated with grassroots staff from various village committees to classify households based on the criteria provided in this article. The households were categorized into three types: pure farming households, part-time farming households, and non-farming households. Following this, questionnaire surveys and interviews were conducted with random samples from each of these household types.

The research team initially focused on 11 natural villages along the north bank, southwest bank, and west bank of the Yilong Lake in the jurisdiction of Yilong Town, which are close to Shiping County. These villages included the City East Community, Da Shui He, Da Rui Cheng, Dou Di Wan, Gao Jia Wan, and Mao He Village Committee, with a total sample size of 99 households classified into pure farming households, part-time farming households, and non-farming households. Later, the research covered 11 natural villages along the east bank, south bank, and southeast bank of Yilong Lake in the jurisdiction of Ba Xin Town, including Hai Dong, Wang Jia Chong, Lao Jie, Bai Lang, Ba Xin, and Xin Jie Village Committee, with a sample size of 95 households classified into pure farming households, part-time farming households, and non-farming households. Overall, the research collected 194 random samples, with 75 pure farming households, 63 part-time farming households, and 56 non-farming households.

During the course of our research, we conducted semi-structured interviews with households randomly sampled from the population. Each interview lasted between 20 and 40 min and was designed to cover a range of topics. These included the purpose of the questionnaire, the rationale behind the design of the response options, basic information about the households, their livelihood capital situation, their attitudes towards environmental governance, their evaluation of specific indicators related to agricultural non-point source pollution control, the specific factors affecting their livelihoods with regard to agricultural non-point source pollution control, and their opinions and suggestions on the topic. The five-level Likert scale, i.e., "very dissatisfied, dissatisfied, average, satisfied, very satisfied", was used to evaluate farmers' satisfaction with pollution control.

2.3. Research Methods

2.3.1. Method for Classifying Livelihood Types of Farm Households

Drawing on relevant research findings on household types [27,28] and the specific context of the Yilong Lake basin, this investigation categorized farming households' livelihood types based on the proportion of non-agricultural income in their total household income. Pure farming households were defined as those whose non-agricultural income accounted for less than 30% of their total household income; part-time farming households were those with non-agricultural income accounting for 30–90% of total household income; and non-farming households were those with non-agricultural income accounting for more than 90% of total household income. Among the sampled households, there were 75 pure farming households, representing 38.66% of the total surveyed, 63 part-time farming households, accounting for 32.47%, and 56 non-farming households, accounting for 28.87% of the total surveyed.

2.3.2. Indicators and Measures for Evaluating the Performance of Pollution Control Projects

Agricultural non-point source pollution has a strong connection with farm households, and its control effectiveness can be assessed through feedback from such households. Hence, this study focuses on evaluating and measuring performance indicators based on farm households' perceptions. Taking into account the specific circumstances of agricultural non-point source pollution control projects in the Yilong Lake basin, the performance evaluation covers four aspects: "observable pollution control effects, the influence of control projects on farm household livelihoods, participation of farm households, and satisfaction of farm households" (as depicted in Table 1).

Table 1. Performance evaluation indexes and weights of agricultural non-point source pollution control projects in the Yilong Lake Basin.

Level Indicators	The Weight	The Secondary Indicators	The Weight
Impacts of governance projects on farmers S1	0.4	impacts on actual labor T_1	0.3
		Influence on cultivated land T_2	0.3
		Influence on farmland fertility and crop yield T_3	0.3
		Improvement of ecological awareness T_4	0.1
Perceived governance effect S2	0.3	Improvement of human settlement environment T_5	0.15
		Improvement of water quality of Yilong Lake T_6	0.15
		Reduction in the application of chemical fertilizers and pesticides T_7	0.4
		Improvement of farmland environment T_8	0.3
Household participation S3	0.15	participation in decision-making T_9	0.33
		Participation in planning T_{10}	0.33
		Participate in the management and maintenance T_{11}	0.34
Farmer satisfaction S4	0.15	Incentive policy and compensation T_{13}	0.4
		Project construction planning and completion rate T_{14}	0.2
		Post-project management and maintenance T_{15}	0.2
		Environmental protection publicity, education, and training T_{16}	0.2

1. Specific indicators decomposition;

The impact of the pollution control project on farmers (S1) is assessed based on four dimensions: the effect on actual labor force (T1), cultivated land area (T2), soil fertility and crop yield (T3), and improvement of ecological awareness (T4).

The "perceptible effect of pollution control" (S2) is measured by four sub-indicators: improvement of the living environment (T5), the water quality of Yilong Lake (T6), reduction of fertilizer and pesticide use (T7), and improvement of the agricultural environment (T8). This aspect is a vital indicator for evaluating the environmental management performance of Yilong Lake, and the evaluation indicators in this paper are based on the farmers' subjective perceptions.

Farmers' participation (S3) in the project plays a crucial role in enhancing governance ability and environmental awareness among the public, as well as facilitating collaboration and communication among different stakeholders. The sub-indicators of farmers' participation mainly include participation in decision-making (T9), planning (T10), and management and maintenance (T11).

Farmers' satisfaction (S4) is an important indicator for measuring the project's performance. As beneficiaries and participants in the governance process, farmers' satisfaction is mainly decomposed into policy incentives and compensation (T12), project planning and completion rate (T13), post-construction management and maintenance of the project (T14), and environmental publicity, education, and training (T15).

2. Determination of Weight Values;

This study adopts the Analytic Hierarchy Process (AHP) to assign weight values to the evaluation indicators through a combination of qualitative and quantitative methods based on expert ratings [29]. The AHP analysis method relies mainly on expert knowledge and experience to ensure a scientific and accurate reasoning process. It is a standardized decision-making method that has been widely used in various fields [30]. A panel of 10 experts in public management and agricultural pollution source control, as suggested by relevant research results [31], was invited to form an expert system. They were requested to fill out the judgment matrix for the first-level indicators in Table 1, following the scale instructions, to obtain the judgment matrix. To eliminate systematic errors, 10 farmers were randomly chosen, excluding those sampled in the survey. The judgment matrices for the relevant indicators were completed according to the scale table, as displayed in Table 2.

Table 2. Scale description table.

Scale Value	Description
1	Both indicators are equally important
3	One indicator is more important than the other
5	One indicator is more important than the other
7	One indicator is more strongly important than the other
9	One indicator is more important than the other
2, 4, 6, 8	Judge the median value

Based on the aforementioned process, a judgment matrix table was obtained, which was inputted into the YAAHP (Yet Another AHP) software, resulting in the judgment matrices and weight values for each indicator. To eliminate the bias of individual perception of indicator weights, the average value of the judgment matrix weight from ten decision matrices was taken as the final weight. The calculation results are shown in Table 1.

3. Calculation of farm household performance evaluation values.

The performance evaluation criteria were assessed based on the weighted average value of each indicator. The performance value of each evaluation indicator at each level was obtained by multiplying the weight value with the average value of the corresponding indicator. The performance evaluation values of the second-level indicators were then totaled and multiplied by their corresponding weight values of the first-level indicators. This resulted in the performance evaluation value of the first-level evaluation indicator. The total performance evaluation value of the farm household for a specific livelihood type was determined by summing up the performance evaluation values of each first-level indicator.

2.3.3. Livelihood Capital Evaluation Indicators and Measurement for Farmers

1. Measurement indicators and weights of livelihood capital;

The study first designed measurement indicators suitable for assessing the livelihood capital of farmers in the Yilong Lake Basin, based on the indicator system of livelihood capital quantification research results [32,33]. The livelihood capital of farmers was evaluated based on the sustainable livelihood framework proposed by the UK Department for International Development (DFID), which includes human capital, natural capital, physical capital, financial capital, and social capital [34,35]. Five livelihood capital measurement indicators were adopted for this study, as shown in Table 3.

Table 3. The index, value, and weight of livelihood capital of farmers in the Yilong Lake watershed.

Livelihood Capital Measures	The Secondary Indicators	The Assignment	Pure Farmers	Part-Time Farmers	Non-Farmers	The Weight
human capital	The overall labor capacity of the family V1	the labor force of more than 6 people is 5, 5–6, 4, 3–4, 3, 1–2, 2, 0, 1.	2.8	3.8	2.3	0.568
	The education level of the adult labor force is V2	5 for junior college or above, 4 for senior high school or technical secondary school, 3 for junior high school, 2 for primary school, and 1 for illiteracy.	3.1	3.9	3.6	0.432
natural capital	Arable land area W1	household per capital arable land area	0.89 Mu/person	0.35 Mu/person	0.08 Mu/person	0.585
	The cultivated land quality W2	cultivated land quality is 5, 4, 3, 2 and 1	4.1	2.9	1.8	0.415
material capital	Per capital housing area X1	5 for more than 45 m^2; 35–45 m^2 is 4; 25 to 35 m^2 for 3; 15 to 25 m^2 is 2; Less than 15 m^2 is 1.	3.4	3.6	3.5	0.252
	Housing structure X2	steel concrete building 5, brick concrete building 4; Brick and wood house 3; Civil Building 2; Vegetation house is 1 (per capital housing area and housing structure are two indicators quantified)	4.1	4.2	4.0	0.252
	Household fixed assets X3	The proportion of the number of consumer durable, household assets, types of agricultural facilities and equipment, livestock, and poultry owned by farmers in the total number of categories listed in the questionnaire (transportation, household appliances, furniture; Water pumps, weeding machines, harvesting machines, greenhouses, and other agricultural facilities and equipment)	67%	78%	61%	0.496
financial capital	Credit capacity Y1	The likelihood of obtaining a loan. It is 5, it is bigger than 4, it is usually 3, it is smaller than 2, and it is not 1.	3.2	3.9	4.6	0.348
	Cash income Y2	Per capital annual cash income	1.1 Ten thousand	1.4 Ten thousand	1.5 Ten thousand	0.416
	Government subsidy Y3	Government rent subsidy is 2500 yuan/mu per year	2225 yuan	875 yuan	200 yuan	0.236

Table 3. *Cont.*

Livelihood Capital Measures	The Secondary Indicators	The Assignment	Pure Farmers	Part-Time Farmers	Non-Farmers	The Weight
Social capital	Risk assistance ability Z1	The possibility that relatives and friends will offer assistance to you when you are in trouble. 5 for sure, 4 for larger, 3 for general, 2 for smaller, and not 1.	2.4	3.5	4.6	0.275
	In social network Z2	family members, the number of people who had worked as township or village cadres, public servants, technicians, teachers, doctors, enterprise and public institution workers, soldiers, and other professions: 5 for more than 4, 4 for 3, 2 for 3, 1 for 2, and 1 for none	1.7	2.4	3.6	0.241
	Participation in community organizations Z3	5 represents every-time participation, 4 represents frequent participation, 3 represents average participation, 2 represents less frequent participation, and 1 represents little participation.	2.3	3.9	3.4	0.234
	Social trust Z4	The degree of trust of villagers, relatives and friends, and village cadres is 5, 4, 3, 2, and 1	4.1	3.4	2.7	0.250

Next, the weight of each indicator is determined. A panel of 13 experts, including grassroots personnel involved in the environmental management of Yilong Lake and those familiar with the livelihood status of farmers in the basin, was invited to form an expert system. The AHP method was used for data processing, and the weight calculation steps were the same as the weight method for evaluating the performance of agricultural pollution control projects (as shown in Table 2). The weight values for each indicator were obtained (as shown in Table 3). The weight determination results indicated that overall household labor capacity had the highest weight for human capital, at 0.568. For natural capital, the weight of per capital cultivated land area was higher, at 0.585. For physical capital, the weight of per capital fixed asset value of the household was higher, at 0.496. For financial capital, the weight of cash income was higher, at 0.416. For social capital, the weight of risk-assistance capacity was relatively higher at 0.275. Overall, the study found that in the Yilong Lake Basin, overall household labor capacity, cultivated land area, asset value, cash income, and risk-assistance capacity have a greater impact on farmers' livelihood capital. Analyzing the composition of farmers' livelihood capital in the context of environmental governance can provide insights into their resilience.

2. Livelihood capital measurement.

The field survey data collected from Yilong Lake was standardized using the deviation standardization method, which involved applying the following formula to the data from different dimensions:

$$x^* = \frac{x - v_{min}}{v_{max} - v_{min}} \quad (1)$$

where x^* represents the data after processing, x represents the data before processing, and the standardized value is obtained after calculation. Through the weight value of each index and the standardized value of data, the livelihood capital value of farmers can be obtained, and the formula is as follows:

$$W = \sum_{i=0}^{n} P_i Q_i \tag{2}$$

where P_i is the weight of the evaluation index and Q_i is the standardized value of this index.

2.3.4. Analysis Model of the Impact of Livelihood Capital on Performance Evaluation

The least squares regression model is used to analyze the impact of explanatory variables on the dependent variable, which in this study is the relationship between the livelihood capital of farmers and the performance evaluation of different livelihood types. The method aims to minimize the sum of squared errors between predicted values and actual observations in the regression empirical model [36,37]. Therefore, this study uses the least squares method (OLS) to perform linear regression on the relationship between farmers' livelihood capital and performance evaluation, in order to examine the impact of various components of livelihood capital on the performance evaluation of different types of livelihoods. The regression model is constructed using Eviews software, and the specific formula is as follows: The regression model is constructed using E-views software, and the specific formula is as follows:

$$\begin{gathered}\mathrm{Per}_n = \beta_0 + \beta_1 V_1 + \beta_2 V_2 + \beta_3 W_1 + \beta_4 W_2 + \\ \beta_5 X_1 + \beta_6 X_2 + \beta_7 X_3 + \beta_8 Y_1 + \beta_9 Y_2 + \beta_{10} Y_3 + \beta_{11} Z_1 + \beta_{12} Z_2 + \beta_{13} Z_3 + \beta_{14} Z_4 + c\end{gathered} \tag{3}$$

In Formula (3), subscript n represents different peasant households; C is a constant. The specific research roadmap is shown below (see Figure 2):

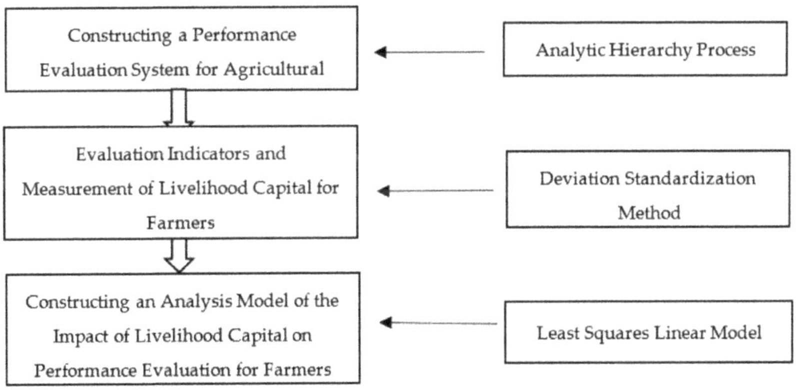

Figure 2. Flow chart of specific research roadmap.

3. Results

3.1. Farmers' Performance Evaluation of Farmland Non-Point Source Pollution Control Projects

Based on the aforementioned method, the survey results were computed, and the performance evaluation values of farmland non-point source pollution control were derived for three types of households: pure farming households, part-time farming households, and non-farming households. The results are presented in Table 4.

Table 4. Performance evaluation of non-point source pollution projects by farmers with different livelihood types.

Level 1 Indicators (Weight)	Secondary Indicators (Weight)	Pure Farmers	Part-Time Farmers	Non-Farmers
S1 (0.4)	T1 (0.3)	2.267	3.222	3.946
	T2 (0.3)	1.747	2.857	3.786
	T3 (0.3)	2.347	3.222	3.821
	T4 (0.1)	2.600	3.238	3.982
S2 (0.3)	T5 (0.15)	3.267	3.571	3.768
	T6 (0.15)	3.320	3.397	3.357
	T7 (0.4)	2.600	3.413	3.911
	T8 (0.3)	3.053	3.222	3.714
S3 (0.15)	T9 (0.33)	1.160	1.349	1.750
	T10 (0.33)	1.107	1.460	1.571
	T11 (0.34)	2.180	2.683	2.446
S4 (0.15)	T12 (0.4)	2.227	2.270	3.589
	T13 (0.2)	3.893	3.921	4.107
	T14 (0.2)	3.440	3.619	3.839
	T15 (0.2)	2.600	2.921	3.857

The maximum value of agricultural non-point source pollution project performance evaluation is 1, and the minimum value is 0. This evaluation index ranges from 0.0000 to 1.0000 and is divided into five levels, ranging from level 1 to level 5. Level 1 corresponds to an evaluation index range of 0.0000 to 0.2000, while level 5 corresponds to an evaluation index range of 0.8000 to 1.0000. According to this level division, the meaning of the evaluation index in different ranges are as follows: level 1 is very poor; level 2 is poor; level 3 is general; level 4 is good; and level 5 is very good. To simply and intuitively reflect the final results of environmental governance, the performance evaluation index grading standards of the Yilong Lake farmland non-point source pollution control project are shown in Table 5:

Table 5. The grading standard of Yilong Lake environmental governance project performance.

Evaluation Index	0.0000–0.2000	0.2001–0.4000	0.4001–0.6000	0.6001–0.8000	0.8000–1.0000
Level	1	2	3	4	5
Meaning	very poor	poor	general	good	very good

Based on the evaluation index system presented in Table 5, the performance grade of farmers with different livelihood types has shown in Table 6:

Table 6. The performance evaluation value and performance grade of farmers with different livelihood types.

Level 1 Index	Pure Farmers		Part-Time Farmers		Non-Farmers	
	Value Rating	Performance Rating	Value Rating	Performance Rating	Value Rating	Performance Rating
S1	0.4337	general	0.6882	good	0.7728	good
S2	0.5888	general	0.6754	good	0.7495	good
S3	0.2979	poor	0.3678	poor	0.3855	poor
S4	0.5755	general	0.6000	good	0.7592	good
Total performance	0.4811	general	0.5969	general	0.7057	good

Table 6 indicates variations in the assessment of the agricultural non-point source pollution control performance among different types of households. Pure farming households view the total performance of agricultural non-point source pollution control as 0.4811, with a performance level of "general"; part-time farming households rate the total performance value as 0.5969, with a performance level of "general"; non-farming households assess the total performance value as 0.7057, with a performance level of "good".

Livelihood type emerges as a vital determinant influencing farmers' evaluation of non-point source pollution control performance. Pollution control projects that have a greater impact on farmers' livelihoods receive significantly lower evaluation results than those with less impact on livelihoods, but all show relatively high recognition and satisfaction with the ecological environment improvement performance of pollution control. In terms of the S1 index, which reflects the effect of control measures on farmers, the average evaluation values of pure farming households for labor force impact, cultivated land area, farmland fertility, yield, and household income are all lower than those of part-time farming households and non-farming households. From the S2 index's perspective, which reflects the perceivable effect of control measures, pure farming households receive the lowest evaluation value, followed by part-time farming households, while non-farming households have the highest evaluation value. As for the S4 index, which mirrors farmers' satisfaction, the assessment values of the three types of livelihood households follow the order of pure farming households < part-time farming households < non-farming households.

The agricultural non-point source pollution control projects in the Yilong Lake Basin are closely intertwined with agricultural production activities. Pure farming households, due to their larger cultivated land areas, are more directly affected in the short term compared to part-time farming households and non-farming households. For instance, the implementation of wastewater recycling and soil testing, and formula fertilization projects, which are directly related to the amount of fertilizer used and crop yield, can result in increased costs of water-saving irrigation and impact the fertility of farmland. Such policies can have the greatest impact on the interests of pure farming households, thereby affecting their willingness to participate and subsequently influencing their evaluation of the project's control performance. From the perspective of farmers' participation, the evaluation values of the three types of farming households are in the order of pure farming households < part-time farming households < non-farming households. The participation degree of farmers depends on their education level and economic interests. Education can enhance their environmental awareness and willingness to participate in environmental governance, which ultimately depends on the benefits and losses to oneself.

Overall, indicators that are closely related to farmers' livelihoods make them more concerned about short-term direct benefits and losses. If the benefits are high, their evaluation will be high, and if the losses are significant, their evaluation will be low. Therefore, it is essential to consider compensation mechanisms for control projects while also strengthening promotion and education on farmers' long-term ecological benefits awareness.

3.2. Impact of Livelihood Capital on Performance Evaluation of Governance Projects of Different Livelihood Types of Peasant Households

This article employs the least squares method to establish a regression model for investigating the correlation between livelihood capital indicators and the performance evaluation of agricultural non-point source pollution control projects among the three types of households. The outcomes are presented in Table 7. It is evident that households with distinct livelihood types exhibit dissimilar evaluations regarding the effectiveness of agricultural non-point source pollution control projects in the Yilong Lake basin. The marked disparities in their livelihood capital configuration are likely the primary cause. As the households are influenced by different livelihood capital indicator factors, their evaluations of project performance vary accordingly.

Table 7. The influence of livelihood capital on the project performance evaluation of farmers with different livelihood types.

Livelihood Capital Index		Performance Evaluation Value		
Level 1 indicators	Secondary indicators	Pure farmers	Part-time farmers	Non-farmers
Human capital	V1	−0.044 *** (−2.71)	−0.163 *** (−5.89)	−0.007 (−0.63)
	V2	0.001 (0.07)	0.111 *** (3.28)	0.014 * (2.03)
Natural capital	W1	−1.085 *** (−2.12)	−0.050 * (−1.00)	−0.097 (−0.40)
	W2	−0.063 *** (−3.40)	0.002 (0.13)	0.005 (0.41)
Physical capital	X1	−0.006 (−0.81)	0.002 (0.12)	−0.001 (−0.07)
	X2	−0.013 (−1.87)	0.074 (2.47)	0.009 (0.67)
	X3	0.093 * (0.30)	0.337 *** (0.48)	−0.566 (−1.21)
Financial capital	Y1	−0.006 (−0.81)	−0.010 (−0.78)	0.011 *** (0.51)
	Y2	0.437 ** (4.46)	0.166 * (2.69)	0.480 *** (5.69)
	Y3	1.609 *** (3.01)	−0.370 (−0.81)	−0.647 (−1.06)
Social capital	Z1	−0.003 (−0.20)	−0.004 (−0.30)	0.070 (3.06)
	Z2	0.016 (1.88)	0.004 (0.29)	0.096 *** (3.65)
	Z3	0.002 (0.17)	0.005 *** (0.14)	0.009 (0.77)
	Z4	0.072 *** (3.27)	−0.003 (−0.23)	−0.003 (−0.26)
	_cons	1.821 *** (6.05)	3.226 *** (5.69)	2.735 *** (9.99)

Note: ***, **, and * represent significance at the 1%, 5%, and 10% levels, respectively, and the brackets represent the T statistic.

3.2.1. The Influence of Livelihood Capital of Pure Farming Households on the Evaluation of Governance Project Performance

Within the human capital of households solely dedicated to farming, the V1 index measuring the adult labor force within the family has a substantial negative influence on the assessment of project performance. Specifically, for each incremental unit of family adult labor, the evaluation of governance project performance in pure farming households diminishes by 0.044. This denotes that the greater the labor force within pure farming households, the more significant the impact of agricultural source pollution control projects upon them. This observation is confirmed by the T1 index, revealing that the evaluation of governance projects' impact on the actual labor force in pure farming households is the lowest among the three household types. The source pollution control projects enacted within the Yilong Lake Basin, including the reversion of farmland to forests, wetlands, and other endeavors, carry the greatest influence on the labor force of farming households. Given that pure farming households solely rely on income from farming, they are bound to experience income reduction in the short term if the cultivated land area significantly decreases and the labor force becomes idle. Thus, the more labor force present in pure farming households, the more probable they are to hold negative appraisals of governance projects and their effects.

Regarding natural capital, the W1 indicator measuring the cultivated land area has a pronounced negative impact on the evaluation of project performance for pure farming households. With every unit increase in the cultivated land area, the project performance evaluation declines by 1.085. The mean cultivated land area among pure farming households in the sample is 0.89 mu, the highest among the three household types. Agricultural source pollution control projects directly or indirectly affect cultivated land, with pure farming households being most significantly impacted. In the project performance evaluation system that pertains to cultivated land indicators, pure farming households are consistently given the lowest evaluations. The W2 indicator measuring the quality of cultivated land similarly has a negative impact on the evaluation of project performance for pure farming households. The soil in the Yilong Lake basin is relatively fertile, and it is well-suited for the cultivation of various crops such as vegetables, soybeans, wheat, and corn. The quality of cultivated land is directly proportional to the crop yield and economic benefits. Therefore, the better the quality of cultivated land, the more substantial the impact of the control project on the households, resulting in a lower evaluation. Pure farming households have the largest cultivated land area and the highest proportion of good-quality cultivated land. As a result, their evaluation is significantly and negatively influenced by cultivated land quality. This also explains why the evaluation of pure farming households is the lowest in indicators such as the impact of control projects on farmland fertility, the use of fertilizers and pesticides, and the improvement of the farmland environment.

In the realm of financial capital, subsidies from the government towards the Y3 index have a noteworthy and affirmative influence on the evaluation of the pure farmers' performance in project governance. The term "government subsidies" denotes compensation payments made to farmers for various matters, including land expropriation and rent. With each unit increase in government compensation, the assessment of the pure farmers' performance ascends by 1.609. Policies such as the restoration of farmland to forests and lakes, the restoration of wetlands, and the restoration of ponds to lakes offer a higher level of compensation and rent, resulting in greater satisfaction among farmers, and leading to enhanced performance evaluations. To pure farmers, farmland represents their livelihood capital and natural capital, yet it also connotes their significant reliance on farmland, limited ability to transition to other livelihoods, and therefore high expectations for compensation. The pure farmers demonstrate particularly low evaluations of incentive policies and compensation T12 indices, reflecting their dissatisfaction with compensation payments. Agricultural non-point source pollution control projects entail land recapture and leasing matters, and the fairness of compensation pricing and timely payment of compensation payments are closely linked to farmers' economic interests. Presently, the compensation payment for the restoration of farmland to lakes is 5000 yuan per mu, payable as a lump sum, while the rent for the restoration of wetlands is 2500 yuan per year, payable every three years. This indicates that after the land is recaptured; if farmers do not expeditiously transition to other livelihoods, they will solely receive compensation payments and have no income. Projects such as the restoration of ponds to lakes and the restoration of farmland to forests do not offer any compensation payments. Of course, most of these farmlands were illegally occupied by farmers, and while farmers can understand the recapture of land, they have difficulty accepting the absence of compensation for crops grown on the land or fish raised in the ponds. Based on current market prices, farmers' annual income from farming is around 10,000 yuan, and the compensation payments and rents for the control projects fall far below farmers' expectations. For pure farmers, who are constrained by their livelihood capital, the unfavorable impact of government subsidies Y3 on performance evaluation explains why they have a low evaluation of incentive policies and compensation T12 indices.

The social trust level Z4 indicator within the social capital of pure rural households exhibits a significant positive impact on the evaluation of project governance performance in financial capital. With each one-unit increase in social trust level, the performance evaluation value of pure rural households increases by 0.072. The higher the social trust level of pure rural households, the more robust their willingness to engage in governance, and the better their evaluation of governance performance. Interviews with pure rural households reveal that they hold a relatively favorable view of the effectiveness of environmental governance and the government's work in governance, believing that the implementation of environmental governance has greatly helped improve the local ecological environment. Despite their dissatisfaction with the amount and promptness of compensation, and their concerns about the duration of land requisition, most rural households still respond positively to policies. They dutifully submit their land within the prescribed time and express their support for pollution control projects.

3.2.2. Impact of Livelihood Capital of Part-Time Farmers on Governance Project Performance Evaluation

Within households engaged in part-time farming, the human capital indicator of V1, representing family adult labor, has a notable detrimental impact on the performance evaluation of governance projects. Specifically, each increase of one unit in family adult labor results in a decrease of 0.163 in performance evaluation. As part-time farmers lack an advantage in natural capital, family labor capacity is a crucial form of livelihood capital. Therefore, they must acquire not only agricultural production knowledge and technology but also other industry-specific livelihood skills. Governance projects place an increased urgency on the transition of family labor force and skills from non-agricultural to non-farming industries, which poses difficulties in adaptation and transition, testing the livelihood capacity of part-time farmers. Hence, the T1 indicator of labor impact on governance project performance evaluation is relatively low for this group. In contrast, the V2 indicator of adult labor education level has a significantly positive impact on governance project performance evaluation, increasing by 0.111 for each unit increase in the adult labor education level of households engaged in part-time farming. This finding indicates that the higher the education level of family labor, the more positively they participate in and evaluate governance work.

Concerning material capital, the X3 indicator of the quantity of family fixed assets has a substantial positive impact on governance project performance evaluation, increasing by 0.337 for each unit increase in family fixed asset quantity. Owning agricultural equipment and facilities, such as water pumps, weed cutters, harvesters, greenhouses, and more vehicles, enhances the efficiency of part-time farmers' work, strengthens their ability to withstand risks, and enables them to carry out meticulous farming.

Concerning social capital, the Z3 indicator of participation in community organizations has a significantly positive impact on governance project performance evaluation, increasing by 0.005 for each unit increase in participation in community activities. Field surveys reveal that among the three types of farmers' livelihoods, part-time farmers exhibit the highest proportion of participation in community activities, the strongest awareness of environmental protection, and the greatest support for collective work and public service. Part-time farmers achieve a higher evaluation than pure farmers and non-farmers on the T11 indicator of farmer participation in management and maintenance, indicating that more part-time farmers participate in the management and post-maintenance of environmental governance projects, and they evaluate their participation in this work more highly.

3.2.3. Impact of Non-Farm Livelihood Capital on Governance Project Performance Evaluation

The indicator of credit capacity Y1 in non-agricultural financial capital has a substantial positive impact on the evaluation of governance projects. Each increase in the credit capacity of non-agricultural households results in a performance evaluation value increase of 0.011. The livelihoods of non-agricultural households in the Yilong Lake Basin mainly involve tertiary industries such as accommodation, catering, and retail. Non-agricultural households with robust credit capacity possess stronger livelihood capabilities and are better equipped to tackle the impacts of governance projects.

The cash income Y2 indicator also has a notable positive effect on the performance evaluation of governance projects. For every unit increase in non-agricultural households' cash income, the performance evaluation value increases by 0.48. Non-agricultural households with higher cash income possess stronger livelihood capabilities and are less impacted by governance measures.

The case of non-agricultural households in Mao Ha Village serves as a prime example to demonstrate this relationship. In this village, a significant proportion of non-agricultural households were previously part-time farmers who shifted to non-farming livelihoods due to the implementation of policies such as the ban on fish farming in cages, the prohibition of "restaurants on boats", and the ban on fishing during the non-fishing season. To aid these fishermen in transitioning to reasonable livelihood paths, the government established a new boat restaurant business in the ponds of the tourist center on the west bank of Yilong Lake. This fish pond is separated from Yilong Lake by a circumferential road, which not only preserves the tourist attraction of boat restaurants in Yilong Lake but also addresses the water pollution issue caused by traditional boat restaurants while also taking into account farmers' livelihood concerns.

Several projects of Yilong Lake Tourist Fishing Boat Company are collectively invested in by Mao Ha Village residents. With government support and assistance, these villagers raised investment funds through borrowing, which are managed and operated by the company. Shareholders of the boat restaurant project receive dividends of approximately 40,000–50,000 yuan per year. Villagers who wish to become employees of the fishing boat company can also receive a monthly income. Thus, the villagers of Maoha Village are particularly content and supportive of the government's environmental governance, and their performance evaluation of the source pollution control project is higher than the average evaluation of other villages along the lake.

In non-agricultural social capital, the social network Z2 indicator has a positive and significant effect on the performance evaluation value of governance projects. Each increase in non-agricultural households' social network results in a performance evaluation increase of 0.096. Survey data reveals that the number of people employed in government agencies, public institutions, and village-level units in non-agricultural households is significantly higher than in part-time and pure farming households. The stronger the social network of non-agricultural households, the greater the promotional efforts of enterprise and institution family members in promoting environmental protection to farmers. This is particularly typical of the villagers in Xiaoshui, Dashui, Longjing, Shaqiao, and Renshou. In the policies of returning farmland to lakes and wetlands, even if the villagers are dissatisfied with the compensation, these socially networked farmers are the first to respond to the policy by surrendering their land. These non-agricultural households originally had very little cultivated land, and after surrendering their cultivated land, they became landless farmers. However, they responded to the governance policy in a timely manner under the persuasion of family members working in government and enterprise departments.

4. Discussion

This paper takes the agricultural non-point source pollution control project in the Yilong Lake basin as the background, and through a comparison and summary of the characteristics of different types of farmers' livelihood capital, analyzes the effect of different types of farmers' livelihood capital on the performance evaluation of agricultural non-point source pollution control projects.

Firstly, the agricultural non-point source pollution control project requires coordination among various stakeholders, with farmers being the most fundamental participating party. Their level of identification and participation is a crucial aspect of the long-term sustainability of pollution control [38]. Currently, the performance evaluation of pollution control focuses on natural engineering indicators, lacking subjective evaluation of the project's performance by farmers [39–41]. Finding measurable indicators from the perspective of farmers and designing performance evaluation indicators based on the research objectives is an innovation and breakthrough from the previous focus only on environmental, social, and economic performance indicators. Comparing farmers' livelihoods and environmental governance effects and analyzing the impact of farmers on environmental governance performance can solve the problem of academic emphasis on result performance in environmental governance project performance evaluation and provide a new perspective for agricultural non-point source pollution control and government performance evaluation. Surveying farmers' subjective evaluation and analyzing influencing factors from the perspective of farmers' livelihood capital framework can undoubtedly further propose policies for improving governance projects, promoting farmers' recognition and participation in governance projects at the grassroots level, and greatly benefiting the successful implementation of projects [42]. However, this study uses cross-sectional data and lacks longitudinal comparative data, which fails to comprehensively reflect the influencing situation. The research project involves environmental protection issues, and many data are sensitive, which affects the comprehensiveness and accessibility of the data to a certain extent. There is still room for improvement in the scientific rationality and comprehensive synthesis of indicator selection and application methods.

Secondly, the livelihood capital of farmers in the Yilong Lake basin is generally low. The adaptation effect of farmers to changes in the living environment after non-point source pollution control is not ideal, and the sustainable development of their livelihoods is greatly restricted, mainly manifested in the insufficient buffering capacity of their livelihoods, poor stability, and low productivity [43]. Changing farmers' livelihood strategies is an effective way to improve the performance evaluation of agricultural non-point source pollution control projects. In fact, in the performance evaluation of the Yilong Lake basin control project by the three types of farmers' capital, pure farmers, dual farmers, and non-farmers, their dependence on the ecological environment quality is gradually decreasing [44,45], and their evaluation of the project's performance is gradually increasing. Therefore, to improve the performance evaluation, it is necessary to start from a deeper perspective. This paper believes that changing livelihood types is an important way to improve performance evaluation, and the primary way is to increase the proportion of non-farmers. Since non-farmers have a relatively small dependence on and impact on the ecological environment, non-farm livelihoods reduce environmental damage and consumption while creating sufficient employment opportunities and sources of income [46,47]. Therefore, promoting farmers to shift from agriculture to non-farm activities and increasing the degree of diversification is a new opportunity chain for non-point source pollution control. However, this process is bound to be complex and slow. A more feasible method is to gradually shift pure farmers to dual farmers or non-farmers, encourage local residents to change their livelihood strategies or achieve livelihood diversification, and enable more pure farmers to achieve diversion through labor transfer or industrial upgrading, thereby promoting the overall improvement of project performance evaluation.

Thirdly, optimizing the structure of livelihood capital can improve the performance evaluation of agricultural non-point source pollution control projects. After the discussion above, it is worth focusing on how to effectively change farmers' livelihood types. Since the differences in livelihood capital help explain environmental differences, non-farm activities promote the diversification of farmers' livelihoods, enhance their autonomy, and reduce their dependence on the environment and resources [48]. Therefore, changing the combination of livelihood capital and increasing the proportion of non-farm livelihoods is an important way to fundamentally improve performance by changing livelihood types. Currently, the diversification of non-farm livelihoods is the main trend of China's livelihood strategy transformation [49]. With the government's promotion, it is easier to successfully transform the types of farmers' income by increasing the proportion of non-farm capital. Therefore, it is necessary to adopt targeted strategies to adjust the composition of farmers' livelihood capital to achieve livelihood diversification and transform their current situation.

(1) For pure farmers, overall labor capacity, land leasing, family asset value, cash income, and social network are significantly positively correlated with the performance evalu-

ation of agricultural non-point source pollution control projects. This indicates that the production and life of pure farmers have a direct dependence on the resources and environment in the basin, which indirectly reflects that the performance evaluation of pure farmers is largely determined by their existing abilities and capital. Therefore, to improve the performance evaluation of agricultural non-point source pollution control projects, it is necessary to first increase the cash subsidy for pure farmers. The government should provide various supports, carry out vocational skills training, education, and cultural training tailored to local conditions to improve the overall labor capacity of pure farmers; promote land leasing through measures such as providing small loans and government transaction platforms, strengthen social networks, and increase the proportion of human capital, financial capital, and social capital in the total livelihood capital, promote the transition from farmers to non-farmers, and fundamentally improve the ability to control and respond to agricultural non-point source pollution.

(2) For part-time farmers, overall labor capacity, property value, and risk assistance capacity are significantly positively correlated with the performance evaluation of agricultural non-point source pollution control projects. Compared to pure farmers and non-farmers, part-time farmers' work is more complex and they face more risks and difficulties. Therefore, the overall labor capacity requirements of part-time farmers are the highest among the three livelihood types. They not only require professional agricultural production capabilities similar to pure farmers, but also need high non-agricultural sensitivity similar to non-farmers. The complex livelihood type makes the risk assistance capacity requirements of part-time farmers fall between pure farmers and non-farmers. Therefore, it is very important to provide the necessary training for part-time farmers to improve the performance evaluation of the project. In addition to organizing experts to regularly train part-time farmers in agriculture, animal husbandry, and other knowledge, attention should also be paid to helping solve other problems encountered by part-time farmers in production and life. Both agricultural labor skills and non-agricultural development capabilities should be considered to meet both skill requirements. Furthermore, part-time farmers have a higher demand for risk avoidance, and the government should create conditions to play the role of agricultural insurance in protecting farmers, appropriately increasing premium subsidies so that planting and breeding industries are protected from natural disasters and accidents, and reducing economic losses caused by them.

(3) For non-farmers, the indicators of the education level of the adult labor force, property value, credit capacity, and risk assistance capacity are significantly positively correlated with the performance evaluation of agricultural non-point source pollution control projects. Since the income of non-farmers does not depend on the use of agricultural resources, a feasible strategy is to reduce their compensation for agricultural non-point source pollution while providing convenient conditions for their production and operation, such as lowering their loan threshold, increasing their loan amount, reducing taxes and fees, providing technical support, and helping to reduce investment risks.

5. Conclusions

The livelihood type is an important factor affecting the performance of agricultural non-point source pollution control. Farmers of different livelihood types have different evaluations of the performance of control projects, which is closely related to the composition of their livelihood capital. Livelihood capital reflects the livelihood status of farmers, directly affecting their behavior willingness, in turn, affecting their livelihood type, thereby affecting their evaluation of the performance of agricultural non-point source pollution control. Therefore, this article constructs a framework for evaluating the performance of control projects based on livelihood capital and different livelihood types of farmers to understand their behavior and willingness to participate in control projects, the following conclusions can be drawn:

(1) According to the actual livelihood status of farmers in the Yilong Lake basin, this article divides them into three types: pure farmers, part-time farmers, and non-farmers, to evaluate the performance of agricultural non-point source pollution control projects. The results show that the performance evaluation of pure farmers on the control project is 0.4811, and the evaluation level is general; the performance evaluation of part-time farmers on the control project is 0.5969, and the evaluation level is general; the performance evaluation of non-farmers on the control project is 0.7057, and the evaluation level is good. It can be seen that there are differences in the performance evaluation of farmers of different livelihood types, with the performance evaluation values ranked from high to low as non-agricultural households > part-time agricultural households > pure agricultural households. The performance results calculated by this method are consistent with our field research. This suggests that the performance evaluation system for environmental pollution control projects based on farmers' livelihoods is scientific and reasonable.
(2) Using livelihood capital as a framework, it was discovered that the indicator of household labor capacity in human capital has a significant negative impact on the performance evaluation of pure farmers and part-time farmers; the education level of adult labor has a significant positive impact on the performance evaluation of part-time farmers and non-farmers. In natural capital, the indicator of cultivated land area has a significant negative impact on the performance evaluation of pure farmers and a negative impact on the performance evaluation of part-time farmers; the cultivated land quality indicator has a significant negative impact on the performance evaluation of pure farmers. In material capital, the indicator of fixed assets of the household has a significant positive impact on the performance evaluation of part-time farmers. In financial capital, the credit capacity indicator has a significant positive impact on the performance evaluation of non-farmers; the cash income indicator has a significant positive impact on the performance evaluation of non-farmers and a positive impact on the performance evaluation of pure farmers and part-time farmers. The government subsidy indicator has a significant positive impact on the performance evaluation of pure farmers; the social network indicator in social capital has a significant positive impact on the performance evaluation of non-farmers; the participation in community organization activities indicator has a significant positive impact on the performance evaluation of part-time farmers; and social trust has a significant positive impact on the performance evaluation of pure farmers. Based on the above conclusions, it can be seen that the deep reasons for the differences in performance evaluation values lie in the different compositions of livelihood capital among farmers of different livelihood types. Key livelihood indicators that have a significant impact on performance evaluation values influence farmers' attitudes and behaviors towards agricultural non-point source pollution control. Therefore, by adjusting the influential factors of significant livelihood capital indicators and targeting the changes in the livelihood capital capacity of farmers of different livelihood types, we can improve farmers' sustainable livelihood and risk avoidance capacities, thereby changing their evaluation values of the performance of agricultural non-point source pollution control.

6. Policy Implications

Based on the discussion and conclusion, the following recommendations can be made for agricultural non-point source pollution control projects:

(1) Strengthen the diversification of governance projects and compensation standards and methods from the perspective of farmers' needs. Different compensation policies should be implemented based on the livelihood composition of farmers. For pure farmers, who have the largest cultivated land area and suffer the greatest impact

from governance, the government should make corresponding adjustments to the actual situation and provide additional subsidies based on market demand. For part-time farmers, the government should establish a platform for capacity building, provide employment guidance and labor skills training, broaden the employment channels, and improve their labor mobility. For non-farmers, their compensation can be appropriately reduced, while facilitating their non-agricultural production and management, such as lowering their loan thresholds, increasing loan amounts, reducing taxes and fees, providing technical support, and helping to reduce investment risks to ensure their sustainable livelihoods.

(2) Enhance policy transparency and information symmetry to improve farmers' participation. Farmers' participation in pollution control is evaluated as poor for all three livelihood types, which has a negative impact on their comprehensive performance evaluation. A new model that combines government control measures, market incentives, and farmer participation mechanisms is needed to fundamentally solve the problem. The transparency of governance policy formulation should be enhanced, and farmers' participation in policy formulation and planning should be increased to construct a new model that combines government control and farmer participation mechanisms.

(3) Promote farmers to change their livelihood strategies, optimize their livelihood capital structure, and improve their risk resistance. Encouraging farmers to transition from agriculture to non-farm activities is an opportunity to increase the degree of part-time farming and change the opportunity chain in non-point source pollution control. Gradually shifting from pure farmers to part-time farmers or non-farmers encourages local residents to change their livelihood strategies or achieve diversification, thereby enabling more pure farmers to transfer their labor or upgrade their industries to achieve diversion. Non-farm activities promote the diversification of farmers' livelihoods, enhance their dominant position, reduce their dependence on the environment and resources, and form environmentally friendly behaviors to solve agricultural non-point source pollution problems at the source.

Author Contributions: Conceptualization, N.Z. and F.Z.; data curation, N.Z.; investigation, F.Z.; formal analysis, N.Z. and F.Z.; methodology, F.Z.; writing—original draft preparation, N.Z., F.Z. and Y.Y.; writing—review and editing, F.Z. and Y.Y.; supervision, F.Z. All authors have read and agreed to the published version of the manuscript.

Funding: This work was supported by the National Key R&D Program of China under Grant 2022YFC3800705; the Youth Innovation Promotion Association of the Chinese Academy of Sciences under Grant 2020424, and the National Natural Science Foundation of China under Grants 41801208.

Institutional Review Board Statement: Not applicable.

Data Availability Statement: Relevant data can be obtained by contacting the corresponding author with reasonable reasons.

Conflicts of Interest: The authors declare no conflict of interest.

References

1. Hu, Y.; Lin, Y.; Jin, S. Agricultural non-point source pollution situation and policy orientation during the "14th Five-Year Plan" period: A comparative analysis based on two national pollution source censuses. *Environ. Prot.* **2021**, *49*, 31–36.
2. Jin, S.; Xing, X. Trend analysis, policy review and countermeasure suggestions for agricultural non-point source pollution. *Chin. Agric. Sci.* **2018**, *51*, 593–600.
3. Lin, H.; Liu, M.; Hong, C.; Li, W.; Min, Q. Evaluation of the policy system for prevention and control of agricultural non-point source pollution in China. *Environ. Pollut. Prev.* **2015**, *37*, 90–95.
4. Wang, Y.; Wang, H.; Zheng, Y.; Sun, X. Research progress on methods and control technologies for agricultural non-point source pollution. *Chin. J. Agric. Resour. Reg. Plan.* **2021**, *42*, 25–33.
5. Ryu, H.-D.; Kim, M.-S.; Chung, E.G.; Baek, U.-I.; Kim, S.-J.; Kim, D.-W.; Kim, Y.S.; Lee, J.-K. Assessment and identification of nitrogen pollution sources in the Cheongmi River with intensive livestock farming areas, Korea. *Environ. Sci. Pollut. Res.* **2018**, *25*, 13499–13510. [CrossRef] [PubMed]

6. Cheng, L.; Lin, Q.; Li, X.; Wen, C. Performance Evaluation and Influencing Factors Analysis of "San Sheng" Environmental Governance in Chinese Rural Areas. *Stat. Decis.* **2022**, *38*, 88–92.
7. Shi, K.; Shang, J.; Yang, G. Review and Improvement Strategy of Agricultural Non-point Source Pollution Prevention and Control Policies from the Perspective of Farmers. *Issues Agric. Econ.* **2020**, *3*, 136–142.
8. He, K.; Zhang, J.; Zhang, L.; Wu, X. Interpersonal Trust, Institutional Trust and Farmers' Willingness to Participate in Environmental Governance: Taking Agricultural Waste Resource Utilization as an Example. *Manag. World* **2015**, *5*, 75–88.
9. Fan, S.; Zhang, T.; Li, M. The credibility and bargaining during the process of policy implementation—A case study of China's prohibition of open burning of crop straw policy. *J. Chin. Gov.* **2020**, *6*, 1–24. [CrossRef]
10. Simon, H.A. Rational decision making in business organizations. *Am. Econ. Rev.* **1979**, *69*, 493–513.
11. Zhao, X. Impacts of livelihood patterns on household energy consumption patterns: A case study of Gannan Plateau. *Acta Ecol. Sin.* **2015**, *35*, 1610–1619.
12. Zhang, L.; Zhang, Y.; Yan, J.; Wu, Y. Study on the relationship between land use and livelihoods in mountainous areas of the Eastern Tibetan Plateau. *Acta Geogr. Sin.* **2008**, *63*, 377–385.
13. Zhang, F.; Zhao, X. A review on the ecological effects of peasant household livelihood transition in China. *Acta Ecol. Sin.* **2015**, *35*, 3157–3164.
14. Andati, P.; Majiwa, E.; Ngigi, M.; Mbeche, R.; Ateka, J. Determinants of adoption of climate smart agricultural technologies among potato farmers in Kenya: Does entrepreneurial orientation play a role? *Sustain. Technol. Entrep.* **2022**, *1*, 100017. [CrossRef]
15. Liu, W.; Zhou, W.; Lu, L. An innovative digitization evaluation scheme for Spatio-temporal coordination relationship between multiple knowledge driven rural economic development and agricultural ecological environment—Coupling coordination model analysis based on Guangxi. *J. Innov. Knowl.* **2022**, *7*, 100208. [CrossRef]
16. Cárdenas Oleas, S.E. Case Study: A Sustainable Livelihoods Analysis of a Non-Governmental Organization Working with Quinoa Producers in the Province of Chimborazo, Ecuador. Master's Thesis, Iowa State University, Ames, IA, USA, 1 January 2015.
17. Bhandari, P.B. Rural livelihood change? Household capital, community resources and livelihood transition. *J. Rural. Stud.* **2013**, *32*, 126–136. [CrossRef]
18. Milan, A.; Ho, R. Livelihood and migration patterns at different altitudes in the Central Highlands of Peru. *Clim. Dev.* **2014**, *6*, 69–76. [CrossRef]
19. Xu, L.; Zhou, Z.; Du, J. An Evolutionary Game Model for the Multi-Agent Co-Governance of Agricultural Non-Point Source Pollution Control under Intensive Management Pattern in China. *Int. J. Environ. Res. Public Health* **2020**, *17*, 2472. [CrossRef]
20. Oumer, A.M.; Hjortsø, C.N.; De Neergaard, A. Understanding the relationship between livelihood strategy and soil management: Empirical insights from the central highlands of Ethiopia. *Food Secur.* **2013**, *5*, 143–156. [CrossRef]
21. Mumuni, E.; Oladele, O.I. Access to Livelihood Capitals and Propensity for Entrepreneurship amongst Rice Farmers in Ghana. *Agric. Food Secur.* **2016**, *5*, 1–11. [CrossRef]
22. Du, Y.; Liu, P.; Bao, C.; Su, S. Research on rural environmental governance from the perspective of social capital—A case study of farming pollution in J Village in underdeveloped area. *J. Public Manag.* **2016**, *13*, 101–111.
23. Kong, Y.; Yu, Y.; Su, B. Current management system and suggestions for improvement of lake basins in nine plateau regions of Yunnan. *Chin. J. Hydroecol.* **2018**, *39*, 67–75.
24. Liu, W.; Ning, P.; Sun, X.; Ma, Q. The Evolution Tendency of Water Quality and Aquatic System in Lake Yilong. *J. Environ. Sci. Technol.* **2018**, *41*, 281–287.
25. Wu, H.; Chang, F.; Zhang, H.; Duan, L.; Zhang, X.; Peng, W.; Zhang, Y.; Liu, Q.; Liu, F. Changes of stable isotopes of organic carbon and nitrogen in Lake sediments from Yilong Lake and their environmental implications. *Chin. J. Ecol.* **2020**, *39*, 2478–2487.
26. Zou, R.; Su, H.; Yu, Y.; Wang, J.; Ye, R.; Liu, Y. Research on precise pollution control decision based on water quality objective in Yilong Lake Basin. *Acta Sci. Nat. Univ. Sin.* **2018**, *54*, 426–434.
27. Li, Y.; Wang, Y.; Luo, J. Rural land transfer, livelihood strategy and household income: Based on the analysis of six provinces and cities in Western China. *Rural. Econ.* **2020**, *9*, 51–58.
28. Wu, K.; Liu, Q.; Zhang, J.; Yang, X. Under drought stress of minqin oasis farmers livelihood vulnerability and adaptation patterns. *J. Econ. Geogr.* **2019**, *33*, 157–167.
29. Deng, X.; Li, J.; Zeng, H.; Chen, J.; Zhao, J. Analysis and Application of AHP Weight Calculation Method. *Math. Pract. Underst.* **2012**, *42*, 93–100.
30. Xu, J. *Metrological Geography*; Higher Education Press: Beijing, China, 2006; pp. 226–253.
31. Fan, S.; Xu, J.; Chen, Y. Research on performance evaluation of desertification control policy based on public value. *Arid. Land Geogr.* **2013**, *36*, 905–987.
32. Scoones, I. IDS Working Paper 72. In *Sustainable Rural Livelihoods: A Framework for Analysis*; Institute of Development Studies: Brighton, UK, 1998.
33. Li, J.; Li, Y.; Tai, X.; Li, C. Analysis on the Livelihood of rural households in the poverty-stricken mountainous areas of western China under the framework of Sustainable Livelihood Analysis. *China Rural. Surv.* **2009**, *5*, 29–38.
34. Ashley, C.; Carney, D. *Sustainable Livelihoods: Lessons from Early Experience*; Department for International Development: London, UK, 1999.
35. DFID. *DFID Sustainable Livelihood Guidance Sheets*; Department for International Development: London, UK, 1999.
36. Tong, D.; Luo, Z.; Feng, C. State policy of poverty reduction effect and space heterogeneity. *J. Econ. Geogr.* **2021**, *9*, 176–184.

37. Wang, W.; Zhou, L.; Sun, Y.; Chen, Y.; Yang, G. Factors influencing life satisfaction of farmers in Yanchi County, Ningxia under the background of grazing ban policy. *Acta Ecol. Sin.* **2021**, *41*, 9282–9291.
38. Wan, W.; Han, Y.; Wu, H.; Liu, F.; Liu, Z. Application of the source–sink landscape method in the evaluation of agricultural non-point source pollution: First estimation of an orchard-dominated area in China. *Agric. Water Manag.* **2021**, *252*, 106910. [CrossRef]
39. Camacho-Cuena, E.; Requate, T. The regulation of non-point source pollution and risk preferences: An experimental approach. *Ecol. Econ.* **2012**, *73*, 179–187. [CrossRef]
40. Zhan, X.Y.; Zhang, A.P.; Zhang, Q.W. Controlling agricultural non-point source pollution: Thinking and practice in the era of agricultural green high-quality development. *Nongye Gongcheng Xuebao/Trans. Chin. Soc. Agric. Eng.* **2020**, *36*, 1–7.
41. Plunge, S.; Gudas, M.; Povilaitis, A. Effectiveness of best management practices for non-point source agricultural water pollution control with changing climate—Lithuania's case. *Agric. Water Manag.* **2022**, *267*, 107635. [CrossRef]
42. De Leo, S.; Di Fonzo, A.; Giuca, S.; Gaito, M.; Bonati, G. Economic Implications for Farmers in Adopting Climate Adaptation Measures in Italian Agriculture. *Land* **2023**, *12*, 906. [CrossRef]
43. Diniz, F.H.; Kok, K.; Hoogstra-Klein, M.A.; Arts, B. Mapping future changes in livelihood security and environmental sustainability based on perceptions of small farmers in the Brazilian Amazon. *Ecol. Soc.* **2015**, *20*, 26. [CrossRef]
44. Zhao, X. Research on sustainable livelihoods from the perspective of geography: Status, Problems and fields. *Geogr. Res.* **2017**, *10*, 1859–1872.
45. Guo, X.; Zhou, L.; Chen, Y.; Zhao, M. Research on the relationship between farmers' livelihood capital and livelihood strategy under the effect of ecological policy—A case study of Hangjin Banner, Inner Mongolia Autonomous Region. *Chin. J. Agric. Resour. Reg. Plan.* **2018**, *39*, 34–41.
46. Novotny, I.P.; Fuentes-Ponce, M.H.; Lopez-Ridaura, S.; Tittonell, P.; Rossing, W.A.H. Longitudinal analysis of household types and livelihood trajectories in Oaxaca, Mexico. *J. Rural. Stud.* **2021**, *81*, 170–181. [CrossRef]
47. Aliabadi, V.; Ataei, P.; Gholamrezai, S. Identification of the relationships among the indicators of sustainable entrepreneurial ecosystems in agricultural startups. *J. Innov. Knowl.* **2022**, *7*, 100245. [CrossRef]
48. Asante, F.; Guodaar, L.; Arimiyaw, S. Climate change and variability awareness and livelihood adaptive strategies among smallholder farmers in semi-arid northern Ghana. *Environ. Dev.* **2021**, *39*, 100629. [CrossRef]
49. Wang, C.; Yang, Y. Research on the spatial pattern of urban agglomeration in central China. *Prog. Geogr.* **2012**, *31*, 792–798.

agriculture

MDPI

Article

Application of a Quality-Specific Environmental Risk Index for the Location of Hives in Areas with Different Pollution Impacts

Daniel Signorelli [1,2], Luigi Jacopo D'Auria [1,2,*], Antonio Di Stasio [1,2], Alfonso Gallo [1,2], Augusto Siciliano [1,2], Mauro Esposito [1,2], Alessandra De Felice [1] and Giuseppe Rofrano [1,2]

1 Istituto Zooprofilattico Sperimentale del Mezzogiorno, Via Salute 2, 80055 Portici, Italy; daniel.signorelli@izsmportici.it (D.S.)
2 Centro di Referenza Nazionale per l'Analisi e Studio di Correlazione tra Ambiente, Animale e Uomo, IZS Mezzogiorno, Via Salute 2, 80055 Portici, Italy
* Correspondence: jacopoluigi.dauria@izsmportici.it

Abstract: Honeybees and hive products have long been recognized as excellent bio-indicators, as they provide valuable information on the state of the environments concerned. As yet, however, no tool is capable of contextually assessing the level of pollution of different environmental matrices in order to create maps of areas in which to place hives. In this paper, a possible method of creating a regional map to support the implementation of targeted monitoring plans on beekeeping is described. We obtained and identified related-activity pressure factors, which were subsequently combined by means of a multi-criteria approach through the analytic hierarchy process method (AHP). The different levels used were drawn up by first attributing qualitative values; these were converted into quantitative values through scoring elaborations and pair comparisons and then elaborated and standardized by means of different techniques in order to create an index with a spatial distribution of five risk classes throughout the region. To verify the correct execution of the procedure, a consistency ratio method was implemented on this index and validated the reliability of the application as the main source of information for sampling activities on beekeeping products. Creation of the specific environmental risk index enabled us to construct a map displaying the areas of greatest impact on beekeeping activities and a representation of the cumulative effects generated by the different pollutants in the air, water, soil and subsoil compartments. This index may, therefore, constitute an essential tool to support beekeepers in choosing sites for their apiaries.

Keywords: AHP; MCDA; GIS; remote sensing; beekeeping; risk index; environmental pollution

1. Introduction

As economic well-being increases, global pollution levels are rising, to the detriment of the quality of human and animal health [1,2]. The development of population centers over the years in the Campania region has led to a spatial continuity between agricultural areas, urban areas and industrial areas. In the vicinity of large population centers, past urban planning has not provided for the proper division of space like in other parts of the world, resulting in the loss of natural areas and increased pollution levels. Some studies have shown that different levels of exposure to various pollutants cause localized damage to specific organs of the human body [3,4]. Severe environmental pollution also impacts ecosystems and the climate, leading to extreme climate variations and disastrous weather phenomena [5,6]. One of the possible methods of determining pollution by metals, organic pollutants and microplastics is to use bioindicators as tracers [7–9]. The use of hymenoptera, an order of insects encompassing more than 120,000 species worldwide, as bioindicators [10,11], or of beekeeping products [12–14] is well established. Indeed, bee products provide indications of the state of the environment and the chemical contamination present [15]. In particular, the honeybee (*Apis mellifera* Linnaeus, 1758) is a ubiquitous, easy-to breed organism that displays great mobility. Moreover, as the bee's body is covered in hairs, it picks up materials and particulates encountered in the environment. Bees can, therefore, bioaccumulate large amounts of materials from soil, vegetation, air and water. These characteristics make the honeybee a valuable bioindicator and an ideal agent for the easy monitoring of large areas, even in regions where infrastructure is scarce [16]. A study was carried out to calculate an index that would enable us to identify the most appropriate locations for hives in order to minimize the effect of pollution on bees and their products. Determining the most suitable areas for the placement of hives, and the mapping of these areas, is a very important task, as it can reduce colony losses [17] and, at the same time, ensure the necessary pollination services. Indeed, beekeeping contributes to the sustainability and conservation of natural resources through the pollination of a large number of native and cultivated plants [18]. In this context, the data present in the literature on the levels of pollutants in environmental matrices (water, air, soil and plants) in the Campania region of Southern Italy were processed [19]. In this way, a synthetic and specific environmental risk index with regard to beekeeping was produced in order to classify the environment of the Campania region. This index was created by using multi-criteria spatial-indicator construction techniques. These techniques are usually used to determine the choice of a specific site for the realization of industrial activities, as in the case of a photovoltaic field in Ecuador [20] or the determination of an optimal site for a waste-storage facility in Iraq [21]. The problem of understanding how multiple matrices (or levels) influence a certain activity has been analyzed in several studies using multi-criteria decision analysis (MCDA) techniques [22]. MCDA techniques are also used to evaluate spatially distributed data [23] and, especially in studies with a strong spatial component, in combination with GIS systems [24–26]. These approaches focus on a particular type of MCDA technique, the analytic hierarchy process (AHP) in a GIS environment. The synergy between the two techniques facilitates the creation of maps for specific activities, such as planning new coffee plantations in view of climate change [27], or as a support tool for soil-erosion risk assessments [28]. In the field of beekeeping, however, only one such study has been carried out, in the Calabria region of Southern Italy; this focused on the creation of suitability maps through the "fuzzy overlays" technique [29]. The present study aimed to draw up a practical approach to the location of hives in areas with different pollution impacts by using the AHP method in conjunction with spatial analysis techniques in a GIS environment. The novelties and innovations presented in this work are in the use of satellite imagery to assess air pollution levels, producing maps of pollutant concentrations in the lower troposphere and the use of modelist tools to create a regional map to support the implementation of targeted monitoring plans concerning beekeeping.

2. Materials and Methods

The index constructed was derived from the combination of four levels, chosen because they represent the main compartments where different types of pollution occur (Figure 1).

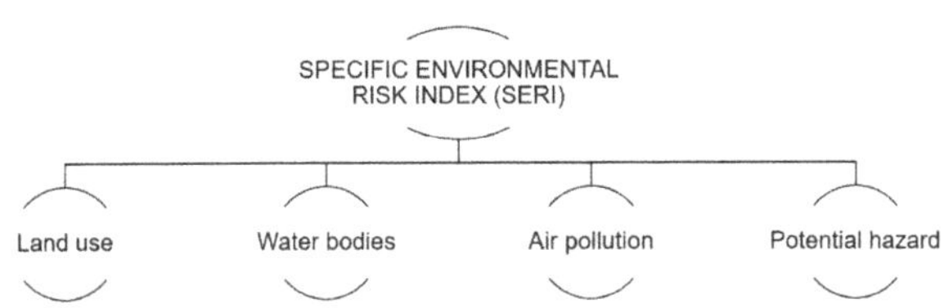

Figure 1. Conceptual model of the specific environmental risk index.

For this purpose, layers of different components were used, combined through an analytic hierarchy process (AHP), a type of multi-criteria decision analysis (MCDA), in a GIS environment. The AHP first decomposes the problem into a hierarchy of sub-problems. The decision-maker then evaluates the relative importance of the various elements by means of pairwise comparisons. The AHP converts these evaluations to numerical values (weights or priorities), which are used to calculate a score for each alternative. A consistency index measures the extent to which the decision-maker has been consistent in her/his responses. The use of decision-support techniques in conjunction with GIS systems is a topic that has been consolidated in recent years. Indeed, in the geospatial field, multi-criteria analysis techniques are increasingly used in synergy with GIS software to generate synthetic indices (Figure 2). In this study, the different layers were elaborated by using QGIS and ESRI® ArcGIS® 10.7 software with plug-ins for performing multi-criteria analyses.

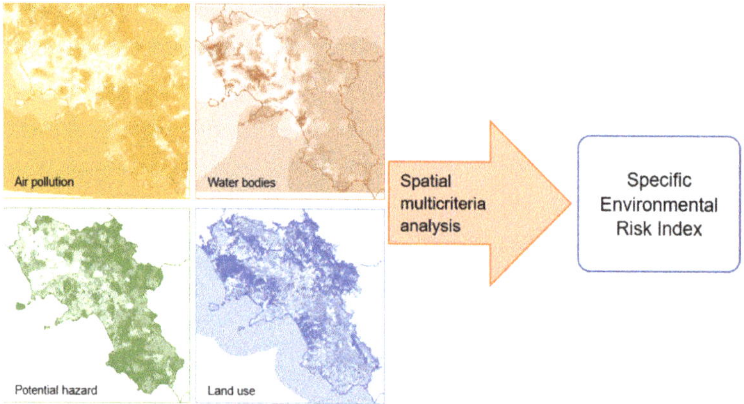

Figure 2. Graphical scheme of index construction.

In the first step, it is necessary to identify the possible areas of greatest impact on beekeeping activities. All potential sources of toxic substances that can interfere with beekeeping products were considered. Thus, four qualitative/quantitative variables, uniformly distributed throughout the territory of the Campania region, were identified: one level for the representation of land uses, one level for the representation of surface water bodies, an information level regarding air quality and an information level associated with the presence of metals in the soil. The different levels were processed in stages using a sequential methodology in order to obtain uniformity in the various workflows of each level used. For simplicity, the general scheme was divided into two stages: a first stage was used to process, correct and standardize the data to be used in the model, and a second stage was used to apply AHP analysis to the standardized data (Figure 3).

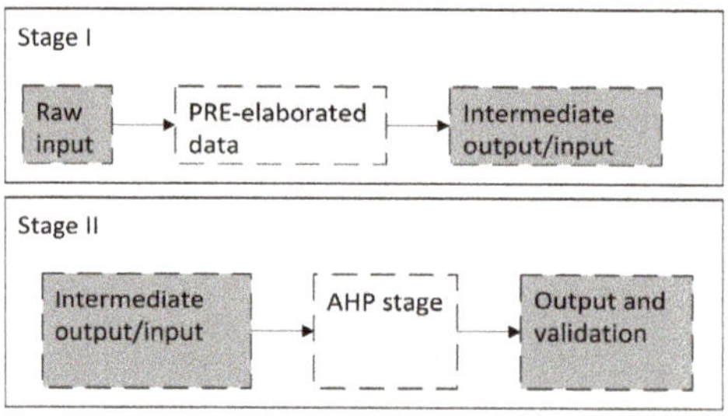

Figure 3. Workflow of the two stages of level elaboration.

In the first stage, the geographical levels to be used in the model were identified, and they were processed and standardized by transforming them into four raster files (intermediate output) using various spatial geostatistics tools, such as inverse distance weighting (IDW) interpolation and statistical classifications using the "natural-breaks" method (Figure 4).

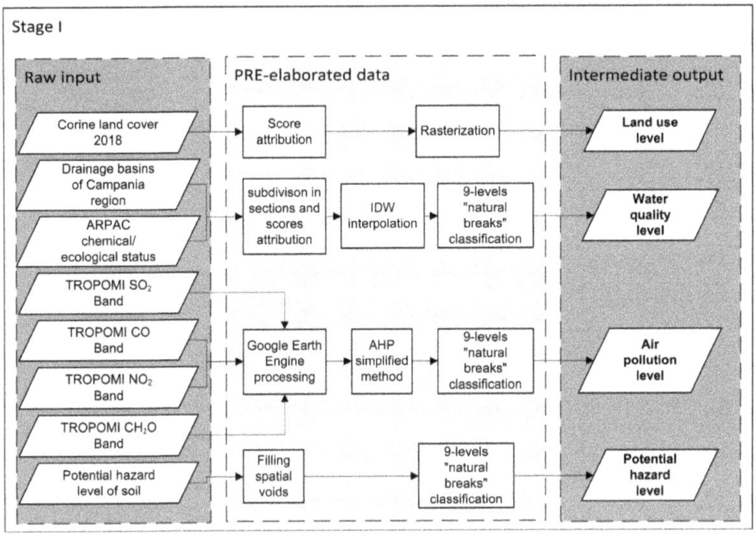

Figure 4. Workflow of first stage of level elaboration.

In the second stage, the actual AHP analysis took place, in which the intermediate levels obtained (output of the first stage and input of the second) were processed by applying weights and scores (Figure 5).

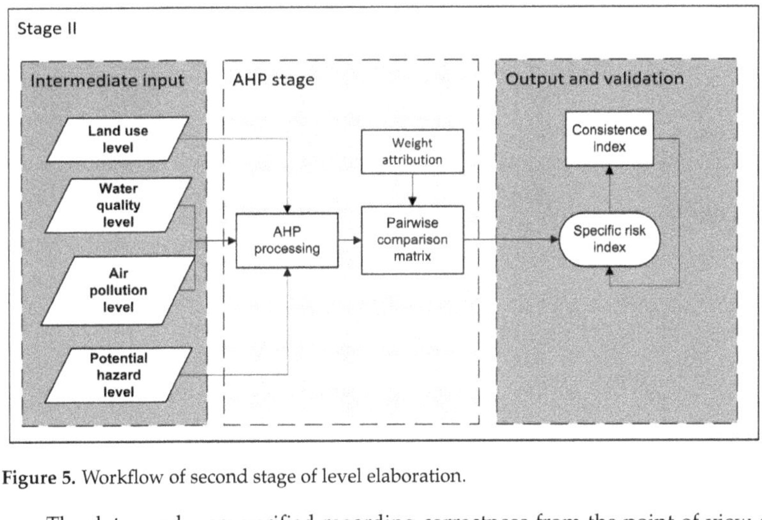

Figure 5. Workflow of second stage of level elaboration.

The data used were verified regarding correctness from the point of view of both numerical and spatial information. In the case of a qualitative variable, a method of scoring by means of a panel [30] was used to obtain a quantitative variable. All layers were processed in order to obtain a representation of coverage classes by using 9 numeric classes and the natural-breaks method [31]. This procedure is mandatory when the analytic hierarchy process (AHP) method is used in a GIS environment [32]. To assign weights to the individual levels, it is necessary to use the pairwise comparison method; this method is used in cross-comparisons between different pairwise-compared levels, in which numerical values of relative importance are assigned (Table 1).

Table 1. Priority scale utilized in pairwise comparisons.

Value	Definition	Explanation
1	Equal importance	Two criteria contribute equally to an objective.
3	Moderate importance	The experience or judgment of a specialist slightly favors one variable over the other.
5	Strong importance	The experience or judgment of a specialist strongly favors one variable over the other.
7	Very strong or proven importance	One criterion is greatly favored over the other; the effect is demonstrable.
9	Extremely important	The evidence favors a criterion as highly as possible.
2, 4, 6, 8	Intermediate values between scale values (when a compromise is required).	

Considering the scarcity of specific information in the literature, a method was devised to assign a weight value to each variable, thus, defining a priority scale among the identified levels. The database of apiaries (Banca Dati Apistica Regionale (BDA-R)) present in the regional territory was used to determine possible spatial interferences between levels (Figure 6).

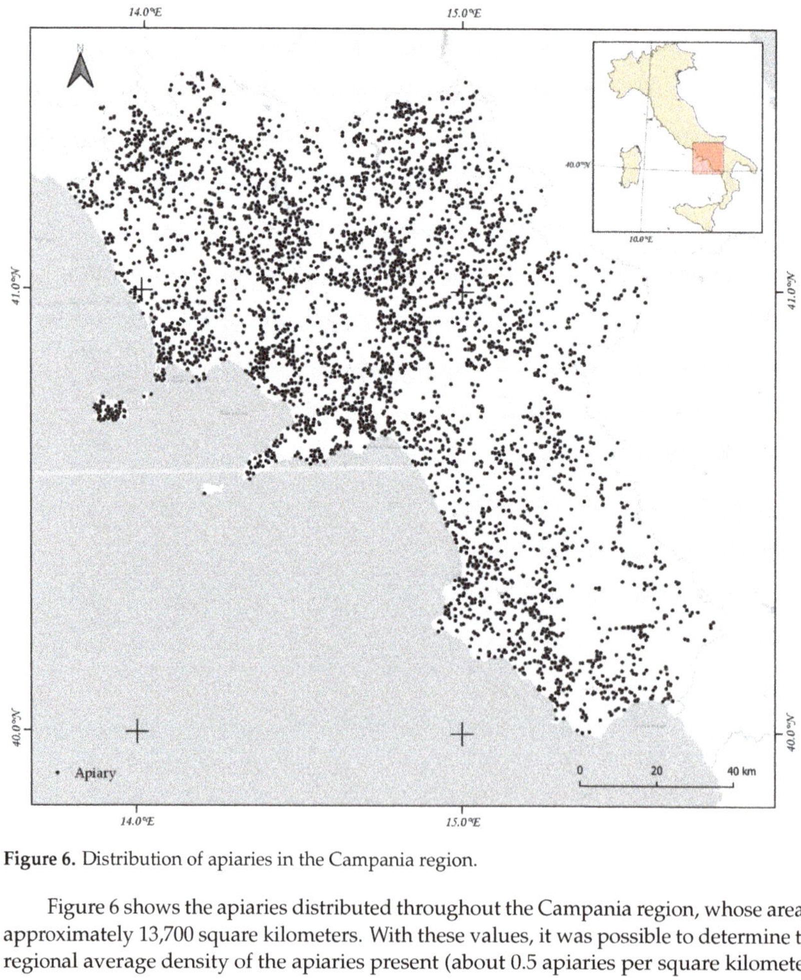

Figure 6. Distribution of apiaries in the Campania region.

Figure 6 shows the apiaries distributed throughout the Campania region, whose area is approximately 13,700 square kilometers. With these values, it was possible to determine the regional average density of the apiaries present (about 0.5 apiaries per square kilometer), which was relatively uniformly distributed. The priority scale among the levels was managed according to the following scheme:

land use > water bodies > air pollution > soil hazard.

The land use level was assigned a higher weight because it has a greater impact in terms of spatial interference, representing the territory in which the apiaries are located. The water bodies level was assigned a lower weight than the land use level; it represents the surface hydrographic network, which may or may not be located near the apiaries, contributing to a moderate interference compared to the apiary points. The air quality level was assigned a lower weight than the first two levels. This variable is represented as a column throughout the troposphere. Considering that the interference is limited to a few meters, a small portion of the column was assigned a moderate interference. The soil hazard level is considered to have a lower weight than all the other levels because the interference associated with this level comes solely from the analytical values of trace elements present in the soil and subsoil. To carry out the pairwise comparisons, a double-entry table (Table 2) was used; the values in the upper triangular matrix were entered in accordance with the priority scale; the values on the main diagonal were 1 for all cells; and the values in the lower triangular matrix were obtained from the reciprocal of each element in the upper triangular matrix.

Table 2. Weight attribution by pairwise comparisons.

	Land Use	Water Bodies	Soil Hazard	Air
Land Use	1	4	4	3
Water Bodies	1/4	1	3	4
Soil Hazard	1/4	1/3	1	1
Air	1/3	1/4	1	1

Table 2 was used to obtain a "priority vector", whereby a percentage weight was attributed to each individual variable by software output in order to determine how much each one contributed to obtaining the index (Figure 7).

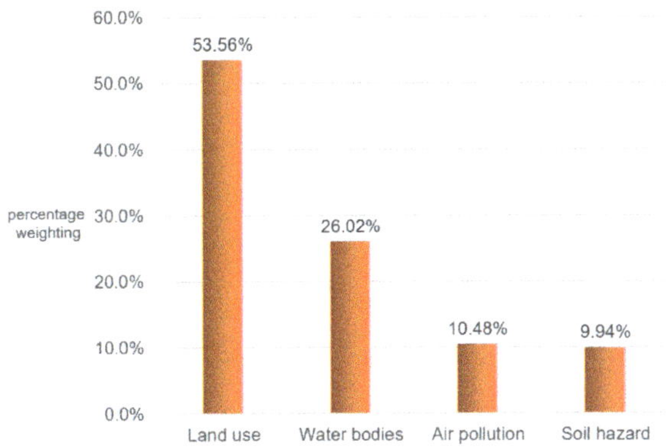

Figure 7. Priority vector of AHP method.

The different levels were processed in the GIS working environment by means of QGIS software and ArcGIS desktop with plug-ins for performing multi-criteria analysis (extAhp 2.0—developed by Oswald Marinoni). In the following sections, the creation of the levels of the different variables is described in detail.

2.1. Quality Level of Surface Water Bodies

To assess the quality of surface water bodies, the analyses presented in the monitoring plan for surface waters [33], which were carried out by the ARPAC (Regional Agency for Environmental Protection in Campania region) in the years 2013–2020, were used. These data are the result of the processing of environmental surveys which qualitatively described the chemical and ecological status of the water at selected points along the watercourses. The classification of the chemical status of surface water bodies is based on the presence of chemicals defined as priority and hazardous substances, such as heavy metals, pesticides, industrial pollutants and endocrine disruptors, among others. These chemical substances are categorized into three groups based on their level of hazard, and distinct environmental quality standards (EQS) are established for each of them for the analysis matrices (water, sediments, biota) where they may be present or accumulate. Non-compliance with the established EQS for each of these substances implies an assessment of "not reaching good chemical status" for the water body. On the other hand, compliance with EQS leads to a "good chemical status" evaluation. The regulation in this regard has been established by Directive 2008/105/EC, updated later by Directive 2013/39/EU, and is implemented in Italy by legislative decree as of 13 October 2015, no. 172. The evaluation of the chemical status of surface water bodies is an important measure for the protection of environment and public health, as contaminated water can pose serious risks to human health and the ecosystem as a whole. Therefore, it is essential to maintain high environmental quality standards to ensure the protection of public health and the sustainability of the environment. The ecological status of inland surface waters, according to Italian legislation, is an indicator that describes the quality of the structure and functioning of aquatic ecosystems. The objective of ecological quality, established by Directive 2000/60/CE, refers to the capacity of the water body to support well-structured and balanced animal and plant communities, which are fundamental biological tools to support the self-purification processes of the water. The classification indicators are represented by biological quality elements, physical-chemical elements in support of biological elements, chemical elements in support of biological elements and hydro morphological elements. The classification is defined by calculating the average value for each parameter analyzed in each year of monitoring and using the worst status obtained in the reference period. The ecological status is defined by the element that is in the worst class according to the general principle, called "one-out, all-out", of Directive 2000/60/CE. To support this level, it was necessary to use the database of surface water bodies reported on the Campania region website (https://www.arpacampania.it/web/guest/acque-superficiali, accessed on 9 January 2023). Processing of the levels was carried out in a GIS environment by means of the QGIS software. The ARPAC sampling points for the determination of the qualitative chemical and ecological status were used to identify the stretches of water to be evaluated. This involved using a multiplicative criterion and normalizing qualitative values to quantitative scores, which were assigned on a scale from 1 to 10 (Table 3). On multiplying the score value by 100, these scores became the inputs for creating a buffer zone for each stretch of water analyzed. This calculation yielded the radius of the buffer zone in meters in order to create a sort of "interference band" around each stretch of water.

Table 3. Ecological and chemical quality scale.

Ecological	Ecological Score	Chemical	Chemical Score
Dry River/N.A.	1	Dry River/N.A.	1
Excellent	2	Good	5
Good	4	Bad	10
Average	6		
Poor	8		
Very Poor	10		

N.A.: not available.

To determine the quality of the surface watercourse, the overall score was calculated by multiplying the ecological score by the chemical score. Subsequently, an interpolation technique (inverse-distance weighting) was implemented by inserting the previous score as an interpolation attribute. This elaboration yielded a raster file (Figure 8), which was used directly in the general model.

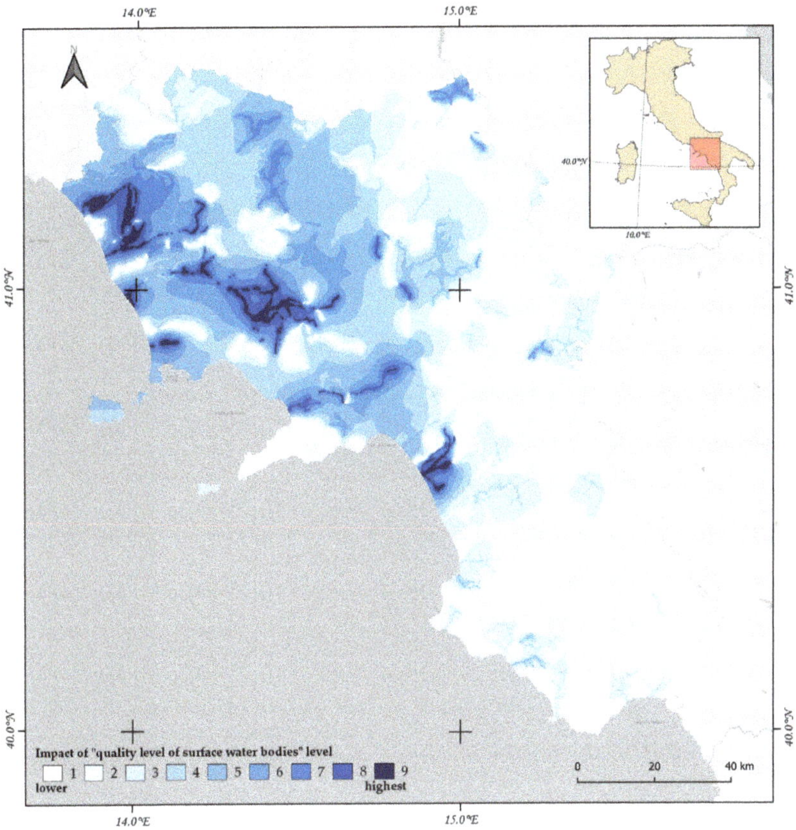

Figure 8. Quality level of surface water bodies in the Campania region.

2.2. Quality Level of Land Use

Data on land cover, land use and the transition between different categories are frequently required for the formulation of sustainable land management and planning strategies. Indeed, they can provide information to support decision-making processes at the EU, national and local levels and enable the effectiveness of environmental policies to be verified. In this study, land-cover data derived from the 2018 edition of the Corine Land Cover (CLC) project (https://land.copernicus.eu/pan-european/corine-land-cover/clc2018, accessed on 19 December 2022) were used; this project was conceived at the European level, specifically for the detection and monitoring of the characteristics of coverage and use of the territory. The CLC outputs are based on the interpretation of satellite photographs by the national teams of the participating states (member states of the European Union and states that cooperate), according to a methodology and a standard nomenclature with the following characteristics: 44 classes at the 3rd hierarchical level of the Corine nomenclature, with some thematic insights at the 4th level regarding Italy. A panel test was distributed to technicians through the Delphi method [30] for the attribution of a quality judgment of the land-cover classes. For each CLC class, these judgments were based on the attribution of a judgment of interference with beekeeping activities on a 9-point qualitative scale, with 1 indicating minimum interference and 9 maximum interference (Table 4).

Table 4. Corine Land Cover quality score attributions.

Impact	Score	Description
Lowest	1	3112—Woods with a prevalence of deciduous oaks (turkey oak and/or downy oak); 3113—Mixed woods mainly with mesophilous and mesothermophilous broad-leaved trees; 3115—Predominantly beech woods; 3122—Forests mainly of montane and oro-Mediterranean pines (black pine and larch, Scots pine, loricate pine); 3124—Woods with a prevalence of larch and/or stone pine; 3125—Woods and plantations predominantly of non-native conifers (Douglasia, insigne pine, white pine); 3131—Mixed woods with a prevalence of deciduous trees; 331—Beaches, dunes, sands; 332—Bare rocks; 333—Areas with sparse vegetation; 334—Areas covered by fires.
Very low	2	3212—Natural grassland with trees and shrubs; 3231—Sclerophyllous vegetation; 3232—Low scrub and garrigue; 324—Transitional woodland shrub; 3241—Young stands after cutting (and/or clear cuts).
Low	3	131—Mineral extraction sites; 3111—Woods with a prevalence of holm oak and/or cork oak; 3114—Predominantly chestnut woods; 3132—Mixed woods with a prevalence of conifers; 3211—Natural grassland prevailingly without trees and shrubs; 323—Sclerophyllous vegetation; 411—Inland marshes; 421—Salt marshes; 511—Water courses; 512—Water bodies; 521—Coastal lagoons; 523—Sea and ocean.
Quite low	4	3117—Woods and plantations mainly of non-native broad-leaved trees (robinia, eucalyptus, ailanthus, etc.); 3121—Woods mainly of Mediterranean pines (stone pine, maritime pine) and cypress trees.
Medium	5	241—Annual crops associated with permanent crops; 242—Complex cultivation patterns; 3116—Broad-leaved forests with a prevalence of hygrophilous species (forests with a prevalence of willows and/or poplars and/or alders, etc.).
High	6	141—Green urban areas; 224—Other permanent crops; 231—Pastures; 243—Land principally occupied by agriculture, with significant areas of natural vegetation.
Quite high	7	112—Discontinuous urban fabric; 123—Port areas; 124—Airports; 133—Construction sites; 142—Sport and leisure facilities; 244—Agro-forestry areas.
Very high	8	111—Continuous urban fabric; 221—Vineyards; 223—Olive groves.
Highest	9	121—Industrial or commercial units; 1211—Industrial areas; 122—Road and rail networks and associated land; 132—Dump sites; 2111—Arable land predominantly without dispersed (line and point) vegetation; 2112—Arable land with scattered (line and point) vegetation; 212—Permanently irrigated land; 222—Fruit trees and berry plantations.

This scoring yielded a qualitative land-cover map (Figure 9), which was then converted to raster-file format by means of GIS software to be used as input for the AHP model.

2.3. *Level of Air Pollution*

Air quality is essential to characterizing the state of the environment and the different pressures on regional territories. In addition to being a secondary source of contamination, air constitutes a possible route of migration of pollutants, owing to the mechanisms of volatilization and dispersion. The pollution of air by numerous pollutants may be of natural or anthropogenic origins (emissions from industrial activities, vehicular traffic, domestic heating and agricultural practices). To obtain a synthetic indicator of air quality, an elaboration of different bands from the Sentinel-5 Precursor satellite was used; Copernicus Sentinel-5 Precursor is a mission dedicated to monitoring our atmosphere and is the result of close collaboration among the European Space Agency, the European Commission, the Netherlands Space Office, industries, data users and scientists. The mission consists of one satellite carrying the TROPOspheric Monitoring Instrument (TROPOMI); this space-borne, nadir-viewing imaging spectrometer covers wavelength bands between ultraviolet and shortwave infrared and can detect daily concentrations of formaldehyde (CH_2O), carbon monoxide (CO), nitrogen dioxide (NO_2) and sulphur dioxide (SO_2) in the tropospheric column (https://s5phub.copernicus.eu/dhus/#/home, accessed on 12 October 2022). Today, few studies use TROPOMI in order to acquire information on air pollutants by using different bands of the on-board spectrometer [34,35]. To download the images and reduce the calculated values, a script was created by means of the Google Earth engine to elaborate all the images from the year 2021 and aggregate these by means of a value-reducer function in order to obtain minimum, median and maximum values at the 95th percentile (https://earthengine.google.com/, accessed on 12 October 2022); this approach yields a single image for each band of air pollutants, which represents the ratio between the moles of substance analyzed and the moles of air (Table 5).

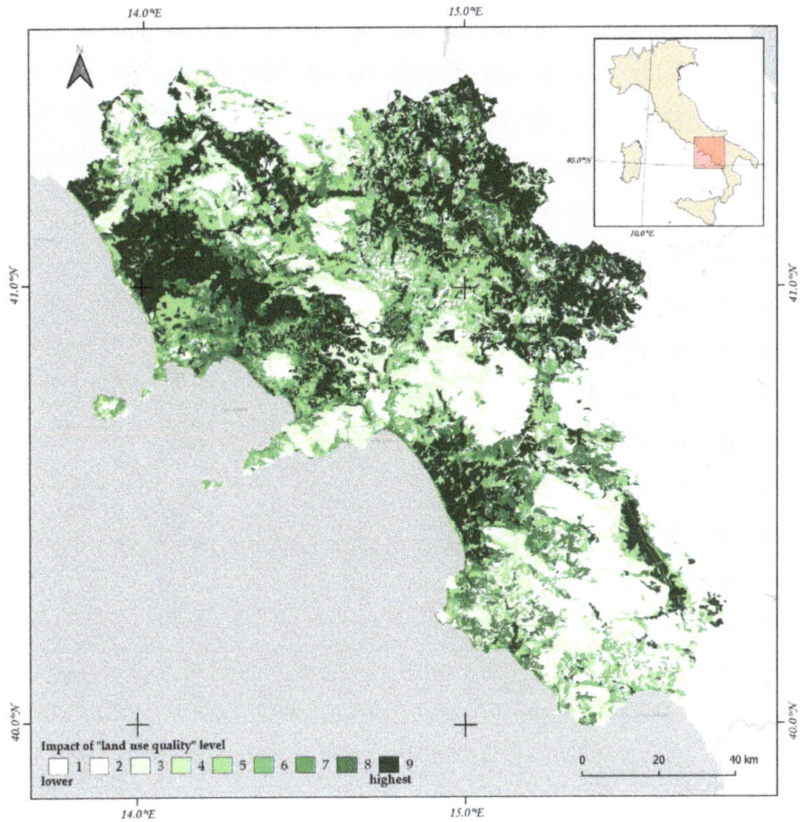

Figure 9. Land use quality levels.

Table 5. The sentinel 5P range of values for each pollutant.

	$\frac{mol\ NO_2}{mol\ air}$	$\frac{mol\ SO_2}{mol\ air}$	$\frac{mol\ CH_2O}{mol\ air}$	$\frac{mol\ CO}{mol\ air}$
minimum	0.0000814	0.0006046	0.0002453	0.0310240
maximum	0.0001271	0.0048747	0.0006101	0.0515890
average	0.0001042	0.0027396	0.0004277	0.0413065

A simplified AHP method, without weighting and pairwise comparisons, was implemented to obtain a single air pollution value; this was discretized by means of the natural-breaks classification, with a value of 1 indicating the lowest concentration and 9 the highest concentration (Figure 10).

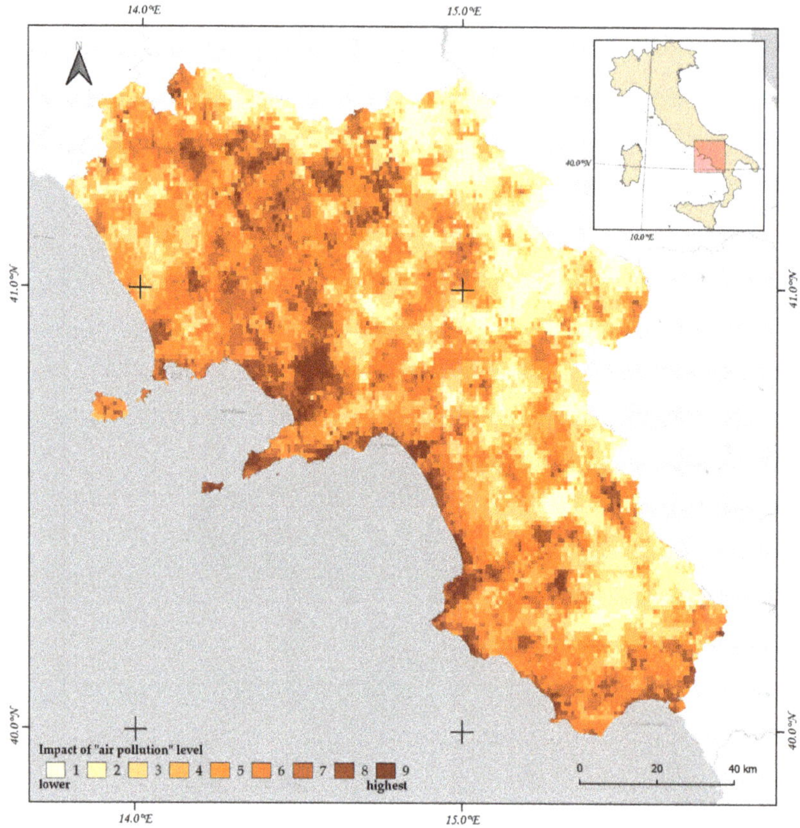

Figure 10. Air pollution levels.

2.4. Potential Hazard Level of Soil

Regarding soil and subsoil, an index modeled on the probability of exceeding the legal limits of pollutants was used. This indicator was constructed within the framework of an agreement between Federico II University and Istituto Zooprofilattico Sperimentale del Mezzogiorno [19]. This index was derived from the analysis of about 4000 soil samples, which were examined for the presence of 53 chemical elements. These data were analyzed by means of spatial statistics models (e.g., inverse distance weighting, kriging, geographically weighted regression) and enabled us to reconstruct continuous areas of concentration that covered the entire regional territory and to estimate the probability of exceeding the legal limits or reference values in areas not covered by sampling [36]. From this analysis, we created a map of the Potential Hazard, which was representative of the number of areas exceeding the CSC (contamination threshold concentrations). This enabled us to identify those areas of the regional territory where multiple inorganic contaminants were present; these were then assigned a relative score according to the level of contamination and were subdivided into classes by means of the "natural-breaks" method (Figure 11).

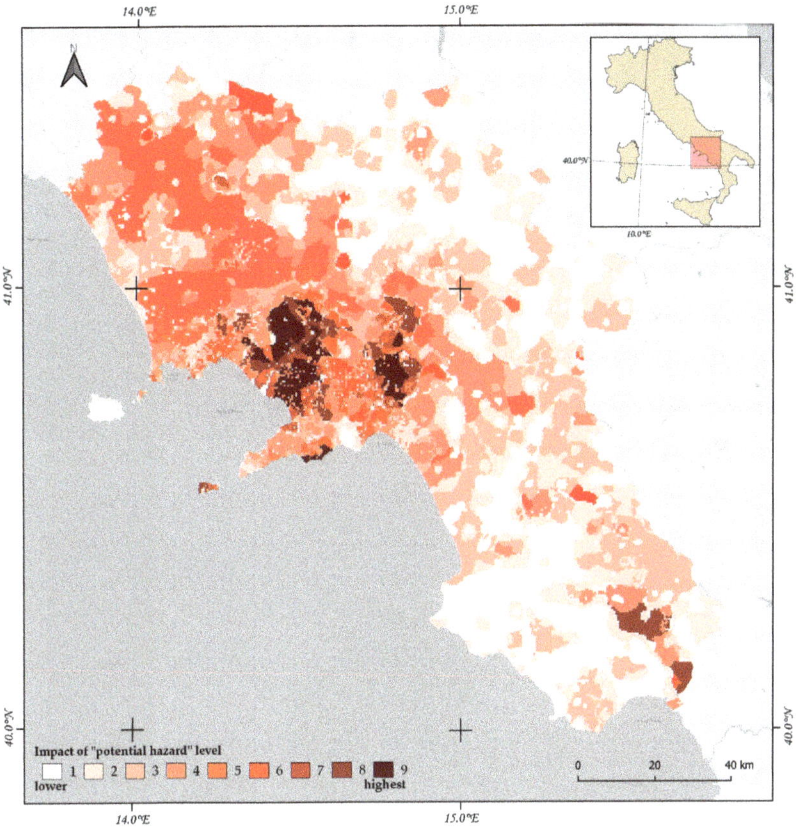

Figure 11. Potential hazard levels of soil.

2.5. *Statistical Verification of Consistency*

To form a statistical evaluation of the correct execution of the AHP method, the consistency index (*CI*) was used [37]. In this index, the value of λ represents the maximum eigenvalue of the matrix and *n* the size of the matrix itself:

$$CI = \frac{(\lambda_{max} - n)}{(n-1)} \quad (1)$$

If the value of the *CI* is zero, then the matrix is consistent; if it deviates from *n*, then the matrix is not perfectly consistent; however, the method used accepts a low degree of inconsistency, as this does not affect the validity of the result obtained. In the first approximation, the maximum eigenvalue of matrix A can be evaluated by referring to the average of the consistencies relative to the individual variables; the result is a maximum eigenvalue of 4.27271, which is close to the dimension *n* of the matrix A. By means of the CI, it is possible to define the *Random Consistency Index* (*RI*), the values of which are available as a function of the size of array A. In this case, for $n = 4$, the RI value is 0.9 (Table 6). At this point, the *Consistency Ratio* (*CR*) of matrix A can be calculated:

$$CR = \frac{CI}{RI} \quad (2)$$

In the present case, the ratio is 0.1, indicating that matrix A is consistent.

Table 6. The Random Consistency Index.

N	3	4	5	6	7	8
RI	0.58	0.9	1.12	1.24	1.32	1.41

3. Results

Elaboration of the specific environmental risk index by means of the AHP method yielded a map of the points of greatest impact on beekeeping activities. The representation of the raster map was split into five risk classes (using the natural-breaks subdivision method) in order to obtain the best separation between the classes (Figure 12). The five risk classes represent different territorial realities. Classes five and flour, which represent "very-high risk" and "high risk", respectively, correspond to areas of high population density, major industrial centers and widespread pollution of water and air. Risk class three, corresponding to "average risk", represents buffer zones, corresponding to areas with high population density but few heavy industries and mostly agricultural enterprises. Risk classes two and one represent "low risk" and "very low risk" values, respectively; these are areas with limited industrial or agricultural activity, low population density and very low or zero air and water pollution levels.

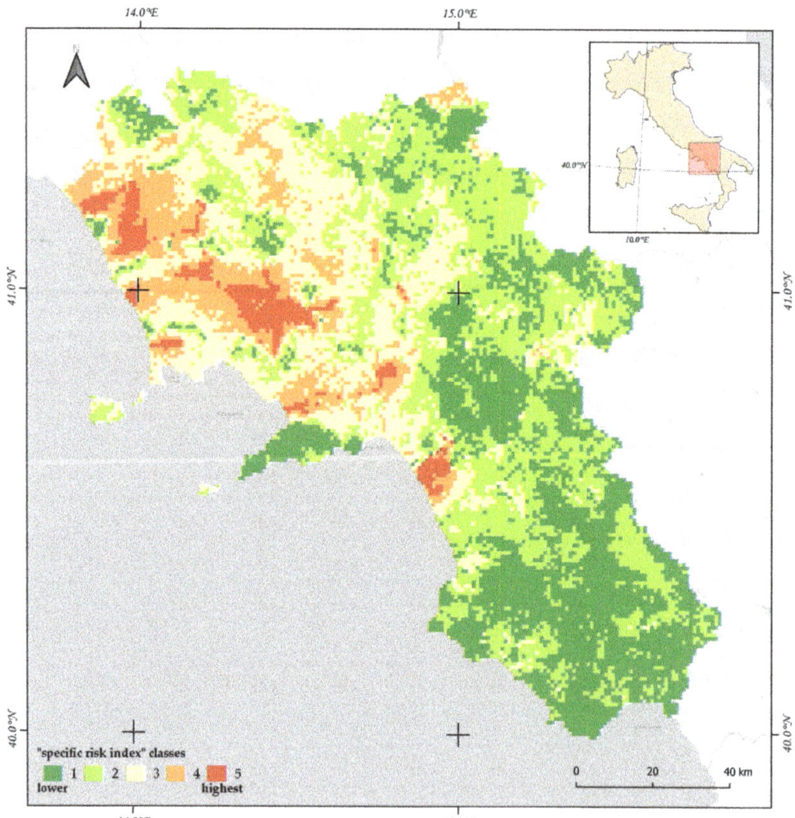

Figure 12. Representation of the specific environmental risk index of the Campania region.

Figure 12 reveals a geographical pattern; in the southern and eastern areas of the Campania region, there are more class one and class two areas, while in the northwest and center there are more class four and class five zones. Overall, however, class one and class two zones account for most of the area of the Campania region, and classes four and five together constitute less than 19% (Figure 13).

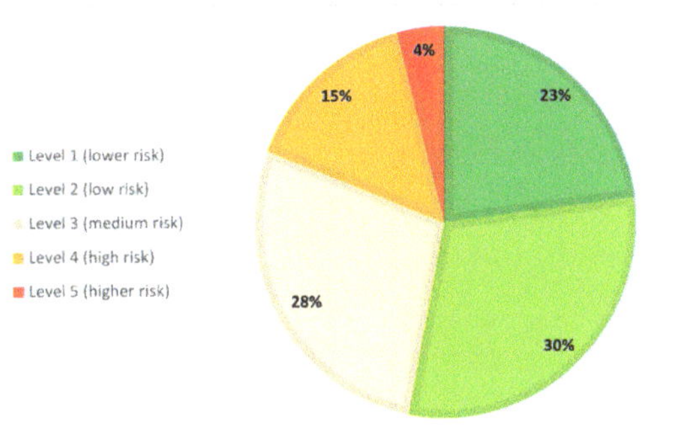

Figure 13. Percentage distribution of the specific environmental risk index in the Campania region.

By making a map overlay between the calculated index and the database of apiaries, it was possible to observe a uniform spatial distribution of all the apiaries surveyed in the regional territory (Figure 14).

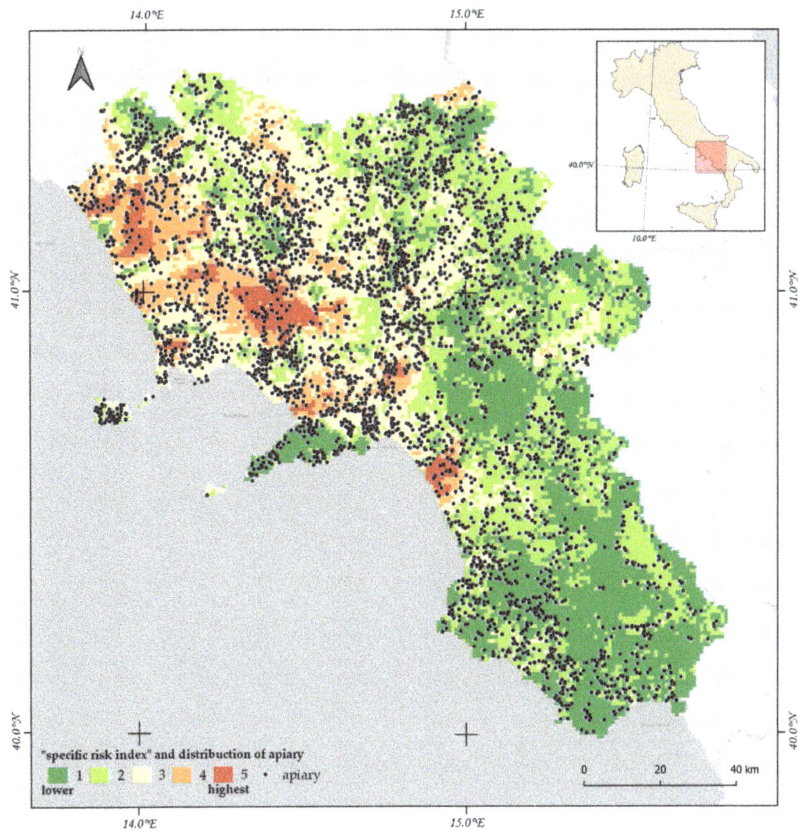

Figure 14. Distribution of apiaries in overlay with the specific risk index.

On those points, a territorial statistic was made by counting the points falling into the different levels of risk, and the relative percentage was calculated (Table 7).

Table 7. Distribution of apiaries in the different risk levels in the Campania region.

Risk Level	Number of Apiaries
Lower risk	18.5%
Low risk	36.8%
Medium risk	26.8%
High risk	14.0%
Higher risk	3.9%

By overlapping the layers, it can be observed that the number of beehives in high-impact areas is significantly lower compared to other areas. These data could be explained because highly polluting anthropogenic activities could coexist in high-impact zones, resulting in concentrations of pollutants to which bees are particularly sensitive. The index helps us evaluate the cumulative effect of pollutants from different environmental compartments. An example of a cumulative effect that may arise is as follows: in the land-cover level compartment, there may be the presence of industrial areas, roads and railways, intensive agricultural areas and a lack of vegetation that is functional to bee nutrition in the interference area corresponding to 3 km^2. In the water quality compartment, there may be the presence of potentially toxic elements whose legal limits are exceeded. In the soil compartment, there may be the presence of inorganic pollutants in the soil and subsoil. In the air quality compartment, there may be the highest concentrations of gaseous pollutants, such as SO_2 and NO_2, due to the geomorphological features of the territory that can lead to photochemical smog effects in particular climatic conditions [38]. This complex condition that arises could create imbalances with significant repercussions on the fitness of honeybees. To validate this hypothesis, studies will be necessary to evaluate the fitness of bees using precision sensors.

4. Discussion

Different papers cited in the introduction part use multi-criteria techniques in applications for geographic information. These techniques use multiple factors in the same analysis with the goal of generating a single indication to support decision-making processes. The difference between multi-criteria techniques is in the way that the criteria are chosen and processed. The aim of this study was to construct an environmental risk index through which to map the Campania region in order to provide a tool to support the implementation of beekeeping-related environmental monitoring plans. Among the papers with the greatest similarities to the topic of beekeeping is Zoccali et al. In fact, the similarity to our work is to generate sensitivity maps for beekeeping activity, whose objective is to evaluate the most suitable areas to carry out beekeeping activity. The application of geostatistical methods aimed at creating beekeeping-specific maps is becoming an increasingly used technique; previous studies have attempted to identify the most suitable areas in which to undertake beekeeping activities. Zoccali et al., for example, argue that honeybees are of fundamental importance to the environment and the economy; therefore, they developed a rapid method of identifying areas suitable for beekeeping, with a view to maximizing productivity and reducing the risk of colony loss. To achieve this goal, they superimposed different levels, such as road networks, temperature, hydrographic networks and CLC, using a fuzzy overlay analysis that did not use the assignment of weights on the different factors used. In this study, data were acquired on all environmental matrices involved in bee production (water, air and soil) in order to gain an overall view of the environment in which bees operate, with a major focus on pollution, taking into account the cumulative effect generated by the different pollutants in the air, water, soil and subsoil compartments. One limitation of the study stems from the regional nature of some of the data used. Indeed, the soil hazard level was derived from work carried out only in Campania, and the quality of surface water was assessed on the basis of surveys carried out by the Campania region's Environmental Protection Agency (ARPA CAMPANIA). To minimize this local approach, the CLC and Sentinel 5P were taken into account. In the future, a standard technique of territorial mapping should be drawn up using nationwide data that are easily retrievable. Further applications should try to use as many global coverage tools as possible so that the method can be easily replicated.

5. Conclusions

The specific risk index can be a very valid tool, both for the construction of bee maps from a production point of view and for the identification of those areas where environmental contamination can be a serious problem for bees and beekeeping products. Environmental pollution is one of the leading causes of mortality in the beekeeping sector and is, therefore, one of the main problems that must be addressed by the scientific community and policymakers. In recent decades, many agricultural practices have been designed to improve agricultural production; however, these have resulted in long-term pressures on the natural environment, negatively affecting native pollinators and reducing pollination services. In addition to developing general governmental agricultural policies, local and regional authorities could take measures to develop beekeeping on a local scale. Such environmental planning approaches are currently gaining increasing attention; there is a need for collaboration between stakeholders and researchers to create incentives so that decisions taken by individuals, communities, companies and governments can promote widely shared values that are compatible with ecosystem services. With this in mind, the environmental index can be a suitable predictive model for assessing environmental impacts and, thus, an essential tool to support beekeepers in choosing sites for their apiaries. In bio-monitoring work, this index tool facilitates the exact identification of points and locations to be considered according to the level of risk. Additionally, it could also provide indications regarding the wellbeing of bees related to pollution issues by predictively monitoring the fitness of honeybees.

Author Contributions: Conceptualization, G.R. and D.S.; methodology, D.S. and L.J.D.; software, D.S.; validation, G.R., L.J.D. and A.D.S.; formal analysis, D.S.; investigation, G.R., L.J.D. and A.D.S.; data curation, D.S. and L.J.D.; writing—original draft preparation, D.S., A.D.S., A.G., A.S., A.D.F. and L.J.D.; writing—review and editing, D.S. and G.R.; supervision, M.E.; project administration, G.R. and M.E.; funding acquisition, G.R. All authors have read and agreed to the published version of the manuscript.

Funding: This research was funded by the Italian Ministry of Health, grant number IZS ME 08-20 RC and IZS ME 08-21 RC.

Institutional Review Board Statement: Not applicable.

Data Availability Statement: The data presented in this study are available on request from the corresponding author.

Conflicts of Interest: The authors declare no conflict of interest.

References

1. Wang, Y.; Sun, K.; Li, L.; Lei, Y.; Wu, S.; Jiang, Y.; Mi, Y. The Impacts of Economic Level and Air Pollution on Public Health at the Micro and Macro Level. *J. Clean. Prod.* **2022**, *366*, 132932. [CrossRef]
2. Wang, X.; Wang, L.; Zhang, Q.; Liang, T.; Li, J.; Bruun Hansen, H.C.; Shaheen, S.M.; Antoniadis, V.; Bolan, N.; Rinklebe, J. Integrated Assessment of the Impact of Land Use Types on Soil Pollution by Potentially Toxic Elements and the Associated Ecological and Human Health Risk. *Environ. Pollut.* **2022**, *299*, 118911. [CrossRef] [PubMed]
3. Kilburn, K.H. Effects of Diesel Exhaust on Neurobehavioral and Pulmonary Functions. *Arch. Environ. Health* **2000**, *55*, 11–17. [CrossRef] [PubMed]
4. Babadjouni, R.M.; Hodis, D.M.; Radwanski, R.; Durazo, R.; Patel, A.; Liu, Q.; Mack, W.J. Clinical Effects of Air Pollution on the Central Nervous System; a Review. *J. Clin. Neurosci.* **2017**, *43*, 16–24. [CrossRef] [PubMed]
5. Delgado, R.C.; de Santana, R.O.; Gelsleichter, Y.A.; Pereira, M.G. Degradation of South American Biomes: What to Expect for the Future? *Environ. Impact Assess. Rev.* **2022**, *96*, 106815. [CrossRef]

6. Abram, N.J.; Henley, B.J.; Gupta, A.S.; Lippmann, T.J.R.; Clarke, H.; Dowdy, A.J.; Sharples, J.J.; Nolan, R.H.; Zhang, T.; Wooster, M.J.; et al. Connections of Climate Change and Variability to Large and Extreme Forest Fires in Southeast Australia. *Commun. Earth Environ.* **2021**, 2, 8. [CrossRef]
7. Yuan, Q.; Wang, P.; Wang, X.; Hu, B.; Liu, S.; Ma, J. Abundant Microbial Communities Act as More Sensitive Bio-Indicators for Ecological Evaluation of Copper Mine Contamination than Rare Taxa in River Sediments. *Environ. Pollut.* **2022**, *305*, 119310. [CrossRef]
8. Saad, D.; Chauke, P.; Cukrowska, E.; Richards, H.; Nikiema, J.; Chimuka, L.; Tutu, H. First Biomonitoring of Microplastic Pollution in the Vaal River Using Carp Fish (Cyprinus Carpio) "as a Bio-Indicator". *Sci. Total Environ.* **2022**, *836*, 155623. [CrossRef]
9. Warner, N.A.; Sagerup, K.; Kristoffersen, S.; Herzke, D.; Gabrielsen, G.W.; Jenssen, B.M. Snow Buntings (Plectrophenax Nivealis) as Bio-Indicators for Exposure Differences to Legacy and Emerging Persistent Organic Pollutants from the Arctic Terrestrial Environment on Svalbard. *Sci. Total Environ.* **2019**, *667*, 638–647. [CrossRef]
10. Widhiono, I.; Pandhani, R.D.; Darsono; Riwidiharso, E.; Santoso, S.; Prayoga, L. Ant (Hymenoptera: Formicidae) Diversity as Bioindicator of Agroecosystem Health in Northern Slope of Mount Slamet, Central Java, Indonesia. *Biodiversitas* **2017**, *18*, 1475–1480. [CrossRef]
11. Reyes-Novelo, E.; Melendez Ramirez, V.; Delfin Gonzalez, H.; Ayala, R. Wild bees (hymenoptera: Apoidea) as bioindicators in the neotropics. *Trop. Subtrop. Agroecosyst.* **2009**, *10*, 1–13.
12. Aldgini, H.M.M.; Abdullah Al-Abbadi, A.; Abu-Nameh, E.S.M.; Alghazeer, R.O. Determination of Metals as Bio Indicators in Some Selected Bee Pollen Samples from Jordan. *Saudi J. Biol. Sci.* **2019**, *26*, 1418–1422. [CrossRef] [PubMed]
13. Kalbande, D.M.; Dhadse, S.N.; Chaudhari, P.R.; Wate, S.R. Biomonitoring of Heavy Metals by Pollen in Urban Environment. *Environ. Monit. Assess.* **2008**, *138*, 233–238. [CrossRef] [PubMed]
14. Bogdanov, S. Contaminants of Bee Products. *Apidologie* **2006**, *37*, 1–18. [CrossRef]
15. Girotti, S.; Ghini, S.; Maiolini, E.; Bolelli, L.; Ferri, E.N. Trace Analysis of Pollutants by Use of Honeybees, Immunoassays, and Chemiluminescence Detection. *Anal. Bioanal. Chem.* **2013**, *405*, 555–571. [CrossRef]
16. Girotti, S.; Ghini, S.; Ferri, E.; Bolelli, L.; Colombo, R.; Serra, G.; Porrini, C.; Sangiorgi, S. Bioindicators and Biomonitoring: Honeybees and Hive Products as Pollution Impact Assessment Tools for the Mediterranean Area. *Euro-Mediterr. J. Environ. Integr.* **2020**, *5*, 62. [CrossRef]
17. Abou-Shaara, H.F.; Al-Ghamdi, A.A.; Mohamed, A.A. A Suitability Map for Keeping Honey Bees under Harsh Environmental Conditions Using Geographical Information System. *World Appl. Sci. J.* **2013**, *22*, 1099–1105. [CrossRef]
18. Marnasidis, S.; Kantartzis, A.; Malesios, C.; Hatjina, F.; Arabatzis, G.; Verikouki, E. Mapping Priority Areas for Apiculture Development with the Use of Geographical Information Systems. *Agriculture* **2021**, *11*, 182. [CrossRef]
19. De Vivo, B. *Monitoraggio Geochimico-Ambientale Dei Suoli Della Regione Campania: Progetto Campania Trasparente*; Aracne: Roma, Italy, 2021; ISBN 9788825540369.
20. Villacreses, G.; Martínez-Gómez, J.; Jijón, D.; Cordovez, M. Geolocation of Photovoltaic Farms Using Geographic Information Systems (GIS) with Multiple-Criteria Decision-Making (MCDM) Methods: Case of the Ecuadorian Energy Regulation. *Energy Rep.* **2022**, *8*, 3526–3548. [CrossRef]
21. Alkaradaghi, K.; Ali, S.S.; Al-Ansari, N.; Laue, J. Combining GIS Applications and Analytic Hierarchy Process Method for Landfill Siting in Sulaimaniyah, Iraq. In *Recent Advances in Environmental Science from the Euro-Mediterranean and Surrounding Regions (2nd Edition) Proceedings of 2nd Euro-Mediterranean Conference for Environmental Integration (EMCEI-2), Tunisia 2019*; Springer International Publishing: Berlin/Heidelberg, Germany, 2021; pp. 1811–1815.
22. Gongora-Salazar, P.; Obadha, M.; Rocks, S.; Fahr, P.; Rivero-Arias, O.; Tsiachristas, A. The Use of Multi-Criteria Decision Analysis (MCDA) to Support Decision-Making in Healthcare: An Updated Systematic Literature Review. *Value Health* **2022**, *in press*. [CrossRef]
23. Ganji, K.; Gharechelou, S.; Ahmadi, A.; Johnson, B.A. Riverine Flood Vulnerability Assessment and Zoning Using Geospatial Data and MCDA Method in Aq'Qala. *Int. J. Disaster Risk Reduct.* **2022**, *82*, 103345. [CrossRef]
24. Fernandes, A.C.P.; Terêncio, D.P.S.; Pacheco, F.A.L.; Fernandes, L.F.S. A Combined GIS-MCDA Approach to Prioritize Stream Water Quality Interventions, Based on the Contamination Risk and Intervention Complexity. *Sci. Total Environ.* **2021**, *798*, 149322. [CrossRef] [PubMed]
25. Ustaoglu, E.; Sisman, S.; Aydınoglu, A.C. Determining Agricultural Suitable Land in Peri-Urban Geography Using GIS and Multi Criteria Decision Analysis (MCDA) Techniques. *Ecol. Modell.* **2021**, *455*, 109610. [CrossRef]
26. Khazaee Fadafan, F.; Soffianian, A.; Pourmanafi, S.; Morgan, M. Assessing Ecotourism in a Mountainous Landscape Using GIS—MCDA Approaches. *Appl. Geogr.* **2022**, *147*, 102743. [CrossRef]
27. Zhang, S.; Liu, X.; Li, R.; Wang, X.; Cheng, J.; Yang, Q.; Kong, H. AHP-GIS and MaxEnt for Delineation of Potential Distribution of Arabica Coffee Plantation under Future Climate in Yunnan, China. *Ecol. Indic.* **2021**, *132*, 108339. [CrossRef]
28. Kucuker, D.M.; Cedano Giraldo, D. Assessment of Soil Erosion Risk Using an Integrated Approach of GIS and Analytic Hierarchy Process (AHP) in Erzurum, Turkiye. *Ecol. Inform.* **2022**, *71*, 101788. [CrossRef]
29. Zoccali, P.; Malacrinò, A.; Campolo, O.; Laudani, F.; Algeri, G.M.; Giunti, G.; Strano, C.P.; Benelli, G.; Palmeri, V. A Novel GIS-Based Approach to Assess Beekeeping Suitability of Mediterranean Lands. *Saudi J. Biol. Sci.* **2017**, *24*, 1045–1050. [CrossRef]
30. Hasson, F.; Keeney, S.; McKenna, H. Research Guidelines for the Delphi Survey Technique. *J. Adv. Nurs.* **2000**, *32*, 1008–1015. [CrossRef]

31. Chen, J.; Yang, S.; Li, H.; Zhang, B.; Lv, J. Research on Geographical Environment Unit Division Based on the Method of Natural Breaks (Jenks). In Proceedings of the International Archives of the Photogrammetry, Remote Sensing and Spatial Information Sciences—ISPRS Archives, Beijing, China, 5–6 December 2013; Volume 40, pp. 47–50.
32. Saaty, T.L. Decision Making with the Analytic Hierarchy Process. *Sci. Iran.* **2002**, *9*, 215–229. [CrossRef]
33. ARPAC—Direzione Tecnica UOC Reti di Monitoraggio e CEMEC Piano di Monitoraggio dei Fiumi Della Campania. 2017. Available online: https://www.arpacampania.it/documents/20182/bcfd4d25-db88-4781-a8b3-4d7d33130024?download=true&_jsfBridgeRedirect=true (accessed on 9 January 2023).
34. Savenets, M.; Dvoretska, I.; Nadtochii, L.; Zhemera, N. Comparison of TROPOMI NO_2, CO, HCHO, and SO_2 Data against Ground-level Measurements in Close Proximity to Large Anthropogenic Emission Sources in the Example of Ukraine. *Meteorol. Appl.* **2022**, *29*, e2108. [CrossRef]
35. Bodah, B.W.; Neckel, A.; Stolfo Maculan, L.; Milanes, C.B.; Korcelski, C.; Ramírez, O.; Mendez-Espinosa, J.F.; Bodah, E.T.; Oliveira, M.L.S. Sentinel-5P TROPOMI Satellite Application for NO2 and CO Studies Aiming at Environmental Valuation. *J. Clean. Prod.* **2022**, *357*, 131960. [CrossRef]
36. Goovaerts, P. *Geostatistics for Natural Reources Evaluation*; Oxford University Press on Demand: Oxford, UK, 1997; ISBN 0195115384.
37. Liu, F.; Zou, S.C.; Li, Q. Deriving Priorities from Pairwise Comparison Matrices with a Novel Consistency Index. *Appl. Math. Comput.* **2020**, *374*, 125059. [CrossRef]
38. Brusseau, M.L.; Matthias, A.D.; Comrie, A.C.; Musil, S.A. Atmospheric Pollution. In *Environmental and Pollution Science*, 3rd ed.; Academic Press: Cambridge, MA, USA, 2019; pp. 293–309. [CrossRef]

agriculture

MDPI

Article

Developing a Portable Spectrometer to Detect Chemical Contaminants in Irrigation Water

Siti Nadhirah Zainurin [1], Wan Zakiah Wan Ismail [1,*], Wan Aina Nadhirah Wan Azlan [1,2], Khairul Nabilah Zainul Ariffin [1] and Wan Maryam Wan Ahmad Kamil [3]

1 Advanced Devices and System, Faculty of Engineering and Built Environment, Universiti Sains Islam Malaysia, Nilai 71800, Negeri Sembilan, Malaysia
2 MECIP Global Engineers Sdn Bhd, Persiaran Guntong, Kerteh 24300, Terengganu, Malaysia
3 School of Physics, Universiti Sains Malaysia, Minden 11800, Pulau Pinang, Malaysia
* Correspondence: drwanzakiah@usim.edu.my

Abstract: Water pollution is a critical issue since it can severely affect health and the environment. The purpose of the study is to develop a portable spectrometer (ESP32-based spectrometer) to detect chemical contaminants in irrigation water by observing the light absorbance of contaminants. ESP32 and a light sensor (photodiode) were respectively, used as the main controller and detector of the portable spectrometer. It was developed based on optical dispersion and Beer–Lambert law theory. The light absorbance of different types of contaminants was displayed in a Blynk application for real-time monitoring. The samples were also tested using a lab-based spectroscopy method, ultraviolet-visible (UV-Vis) spectrometer. The spectral range of the measurement is from 350 nm to 700 nm and the standard error of the ESP32-based spectrometer is from 0.01 to 0.05. Five water samples were tested, consisting of ammonium nitrate, organic pesticide, zinc oxide and two different reservoirs used for irrigation. The absorption peaks of the ammonium nitrate and organic pesticide are 363 nm and 361 nm, respectively. Zinc oxide shows the absorbance peak at 405 nm, whereas both reservoirs show absorbance peaks lie in the region from 300 nm to 370 nm. Therefore, this study shows that different types of contaminants can absorb light only at specific wavelength regions by considering the concentration of samples. The developed ESP32-based spectrometer can be applied for on-site water quality monitoring as it is portable, light, simple and can be monitored in real time using multiple devices.

Keywords: light intensity sensor; photodiode; ESP32-based spectrometer; heavy metal and organic contaminants; pesticide; zinc oxide and agriculture

1. Introduction

Water contamination increases in parallel with technological advances. Pesticides and metal contribute the most to water contamination in agriculture [1,2]. The concentration of metal in soil and water has continuously increased due to industrialization and electronics wastes, which affect human, animals and plants [3]. Heavy metals are chemical elements that have a relatively high density and are poisonous even at low concentrations. Therefore, they are commonly associated with contamination, potential toxicity or ecotoxicity. Metal contaminants that are consumed by plants may cause various health risks when they enter the human body. Phytotoxicity by lead (Pb) causes disorders in normal plant physiological activities, killing cells when organisms are exposed to high concentrations [4]. Untreated wastewater carries a substantial amount of heavy metals and causes potential ecological risks to the environment, food quality, soil health and sustainable agriculture [5]. Metals are toxic at concentrations above ambient conditions, non-biodegradable in the environment and accumulate in animal and plant tissue [6]. The absorption spectrum range of heavy metal ion is from 400 to 500 nm and the absorbance peak for copper ions is around 498 nm [7].

Meanwhile, organic contaminants refer to any chemicals based on carbon such as pesticides, organic solvents and petroleum-based waste. The contaminant is usually produced by the organic nature that has a covalent bond with another compound. These types of contaminants are toxic or carcinogenic not only to humans but also to animals and plants [8]. The organic toxic may affect roots and leaves. Organic pollutants may have toxic effects on plant and root cell structure, biosynthesis, membrane stability and DNA [9]. The absorbance spectrum range for organic matter lies between 280 nm and 380 nm [7]. The intensive use of inorganic fertilizers and pesticides can cause the loss of biodiversity in soils and health problems in plants. Thus, the fertilization leads to water, soil and air pollution [10].

Good water quality monitoring and treatment can produce high quality crops as crop stress can hinder photosynthesis, thereby reducing the growth, fruit weight and yield. Recent studies on water quality include the assessment of water quality using a cyber physical system [11], a filtration system [12] and the estimation of groundwater quality using the integration of a water quality index [13]. Optical sensing is a recent technique that has potential use in water irrigation systems since the method is non-invasive and has a high reliability. Optical techniques can consist of laser application [14], optical sensors [15] and spectroscopy [7,16–19]. Spectroscopy is a reliable physical study used in physics, materials science, chemistry, biochemistry and biomolecular studies since it is suitable for environmental protection and has high detection efficiency [7].

Most spectroscopy studies [17–19] had been conducted using lab-based methods in which a spectrometer or detector was used to measure the optical properties of the samples, and data were displayed in a computer in the lab. These methods are accurate and reliable, but they are complicated and time consuming because samples need to be transported to the lab. It becomes an issue for any consumers who require fast results for quick action. The development of miniaturized analytical devices is important since the devices not only can give fast and reliable outcomes but also can bypass logistic issues via on-site testing or remote confirmation of detection [16,20].

Here, we developed a portable ESP32-based spectrometer to detect the light absorbance of chemical contaminants in irrigation water. The testing was also conducted using a lab-based method (Ocean Optics Spectrometer) to observe the reliability of the developed spectrometer. Optical dispersion and the Beer–Lambert law theory were referred to in this study, and we focus on contamination in irrigation water. We use samples from organic and metal contaminants as they are common issues faced in water irrigation systems in Malaysia. The system was developed using an ESP32 microcontroller to control the spectrometer and a light sensor to sense and measure the light. The output was displayed in a Blynk online application for real-time monitoring using a mobile phone and a laptop. We tested ammonium nitrate, organic pesticides, zinc oxide and water sources from the irrigation system. The sample solutions were prepared by dissolving the chemical compound in the deionized water solution. The presence of contaminants can be detected by their optical properties because they have molecules that can absorb the light. Thus, this study can further assist farmers to monitor irrigation systems that need high quality water for plants. This is because excessive organic contaminants and heavy metals in water can lead to the abnormal growth of plants.

2. Theoretical Study

The spectrometer was developed based on optical dispersion and the Beer–Lambert Law.

2.1. Optical Dispersion

Optical dispersion occurs when light is separated into different individual colors of the spectrum and exposed to diffraction grating [21]. The rainbow color spectrum is produced when the white light is separated into components of different wavelengths or colors. When a light ray passes through one transparent substance and into another, its path bends or refracts. Different wavelengths are separated by modest angles because the refraction index of optical glasses varies by only a few percent across the visible spectrum [21].

2.2. Light Absorbance and Transmittance Based on Beer–Lambert Law

Light absorbance and transmittance in the sample can be found using the Beer–Lambert law. The Beer–Lambert law is a linear relationship between the light absorbance and the concentration of solution. The Beer-Lambert law states that the concentration of a solution is directly proportional to the light absorbance, as shown in Equation (1) [22,23], where A refers to the light absorbance at a given wavelength of light, ε is the molar absorptivity, b is the path length referring to the distance of light passing through the solution and c is the concentration of the compound in the solution.

$$Absorbance(A) = \varepsilon bc = -log\,(T) \tag{1}$$

The light absorbance (A) has an algorithmic relationship to the light transmittance (T) [9]. The light transmittance (T) of the solution is defined as the ratio of the transmitted intensity (I_t) over the incident intensity (I_0). The light transmittance can be defined as in Equation (2) [22–24], where T refers to the light transmittance of a sample, I_t is the transmitted light intensity and I_0 is the incident light intensity.

$$Transmittance(T) = \frac{I_t}{I_0} \tag{2}$$

The incident and transmitted light intensities can be calculated according to Equation (3) [24].

$$Absorbance(A) = -log(T) = -log\frac{I_t}{I_0} \tag{3}$$

3. Materials and Methods

3.1. Sample Preparation

Zinc, ammonium nitrate and organic pesticides were tested because they were among the top environmental toxicants endangering the relevant environment [25]. Water from two reservoirs, used for irrigation was also tested.

3.1.1. Ammonium Nitrate (NH_4NO_3)

Ammonium nitrate can contaminate water when there is a spill from fertilizer. Ammonium nitrate results in small root and small leaf area, which may contribute to a low carbon gain and inhibit crop growth. Ammonium nitrate was prepared by mixing 8.2 g NH_4NO_3 that had around a 0.052–10 µm particle size with 60 mL of deionized water, which produced 0.14 g/mL of sample concentration.

3.1.2. Organic Pesticides

Organic pesticides are chemical compounds or mixtures of chemical compounds that are used to prevent or reduce insect populations. Organic pesticides are made from natural ingredients such as minerals, plants and microorganisms. The elements in pesticides that inhibit plant growth are usually coming from organochlorine DDT, endosulfan, endrin and heptachlor [25]. The larger the organic matter content, the greater the absorption of pesticides.

3.1.3. Zinc Oxide

Zinc oxide is a metal oxide material that has bandgap tailoring, doping with various types of materials and is able to form many structures [26]. Metal contaminations can lead to the decline and deterioration of soil fertility and environment quality [27]. Zinc exists in irrigation water due to chemical spills, which can affect plants. Zinc is a heavy metal compound that has a strong metallic bond. A high concentration of zinc can cause toxicity in plants, which itself can cause the stunting of shoots, the curling and rolling of young leaves and the death of leaf tips and chlorosis. The particle size of zinc oxide used in the system and dissolved in the water was less than 100 nm. A 0.125 g/mL solution was prepared by dissolving 7.5 g of zinc oxide powder in 60 mL of deionized water. The solution was stirred frequently during the testing process to ensure the solution was well mixed.

3.1.4. Water Sources from Irrigation System

The irrigation water sample was taken from different water sources in Nilai. The sample was tested, and types of contaminants could be investigated by observing the spectrum of light absorption. Reservoir I is located near the Faculty of Science and Technology, Universiti Sains Islam Malaysia (USIM). It is used to irrigate the Nursery Landscape, Department of Development and Facilities Management (JPPF), USIM. According to the caretaker of the area, the water may have organic material since it is near the area's sewage. The water source is used twice a day to irrigate the agriculture area. Reservoir II is also located in USIM, but it is rarely used for the irrigation system. The water source may indicate the presence of organic material based on its color. We observe that Reservoir II contains many living things, such as fish, which also can contribute to organic contaminants in the water sample.

3.2. *Spectroscopy Techniques*

Spectroscopy is a non-destructive approach that uses the interaction effect of electromagnetic radiation to identify the energy level and structure of an atomic or molecular substance [28]. Ultraviolet-visible (UV-Vis) spectroscopy methods are used to quantify the transition of ions and highly conjugated electronic compounds by measuring an electron transition corresponding to the excitation of the outer electrons [29]. When light is absorbed by a material, valence (outer) electrons are excited from their normal (ground) states to higher energy (excited) states. Broad absorption peak can be seen instead of sharp absorption peak. Vibrational energy levels are available at each electronic energy level, and transitions can occur to and from the different vibrational levels and, thus, result in peak broadening [29].

We developed a ESP32-based spectrometer and set up the experiment using a lab-based spectrometer, ultraviolet-visible (UV) Ocean Optics spectrometer (Ocean Insight). Ocean Optics spectrometer is commercially used for finding light transmittance, absorbance and fluorescence of any material. The measurements from both spectroscopy techniques were averaged from 10 measurements.

3.2.1. Measuring Light Absorbance Using a Portable Spectroscopy Technique (ESP32-Based Spectrometer)

ESP32 microcontroller was used to read the data waveform generated from the system. Servo motor was used in the system to move a compact-disk (CD) player to provide different light spectrum ranges. Light source (XM-L2 LED Light source, spectrum ranging from ~300 nm to 780 nm) shined the CD player that showed the characteristics of an optical grating. The system was placed in a closed black box to prevent outside light from affecting the samples and to reduce the reflectivity of the internal surfaces. A light sensor (silicon photodiode) was located near the sample to measure the amount of transmitted light. The sensor has high photo sensitivity and radiant sensitivity to measure light from 300 nm to 1200 nm and transform it into electricity. Figure 1 shows the spectrometer that has been developed. Figure 1a,b show the front view and side view of the design to justify the portability of the spectrometer, whereas Figure 1c depicts the position of light source, ESP32 microcontroller and light sensor inside the box. When the light source shined on the CD player, light spectrum was diffracted and the light passed through the sample via the small slit. The enclosure was closed during measurement to reduce effects from surrounding light.

Flow chart of the testing method using ESP32-based spectrometer is depicted in Figure 2. When ESP32 microcontroller was connected to a power source, the servo started to rotate, and various colors of light were exposed to the sample. Light sensor measured the light transmission in the system. After obtaining the light transmission, the light absorbance can be calculated using Beer–Lambert law. If the light absorbance had the maximum peak in the UV-Vis range, the data were analyzed to determine whether the contaminants were organic or heavy metals. Otherwise, if the maximum light absorbance was not in the range of UV-Vis, the system was terminated and another sample was tested. The sample was tested ten times to ensure accuracy of the data.

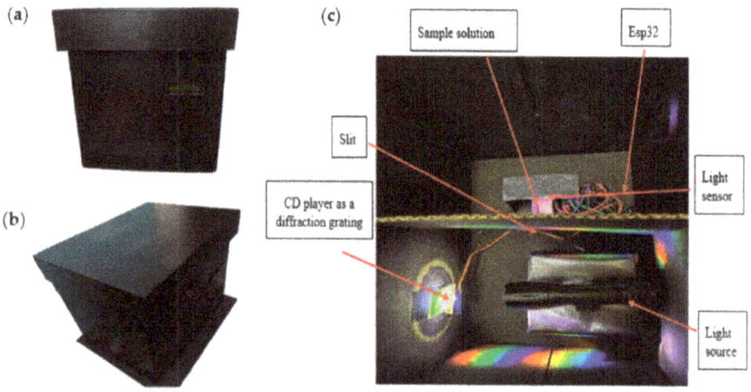

Figure 1. ESP32 spectrometer. (**a**) Front view of the system; (**b**) side view of the system; (**c**) enclosure view of the system.

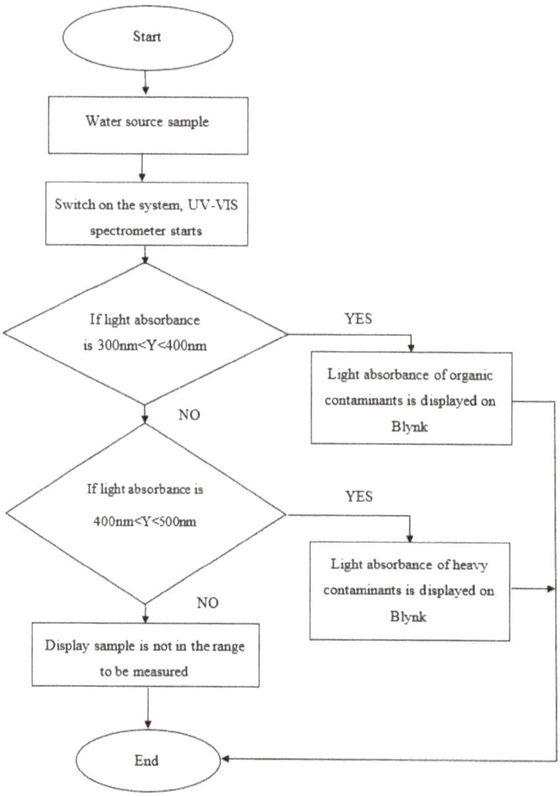

Figure 2. Flow chart of the testing method using ESP32-based spectrometer.

3.2.2. Measuring Light Absorbance Using a Lab-Based Spectroscopy Technique (Ocean Optics Spectrometer)

The measurement of light absorbance using the Ocean Optics Spectrometer was conducted according to a flow chart in Figure 3. Sample was placed in the cuvette and absorbance measurement in OceanView software was applied.

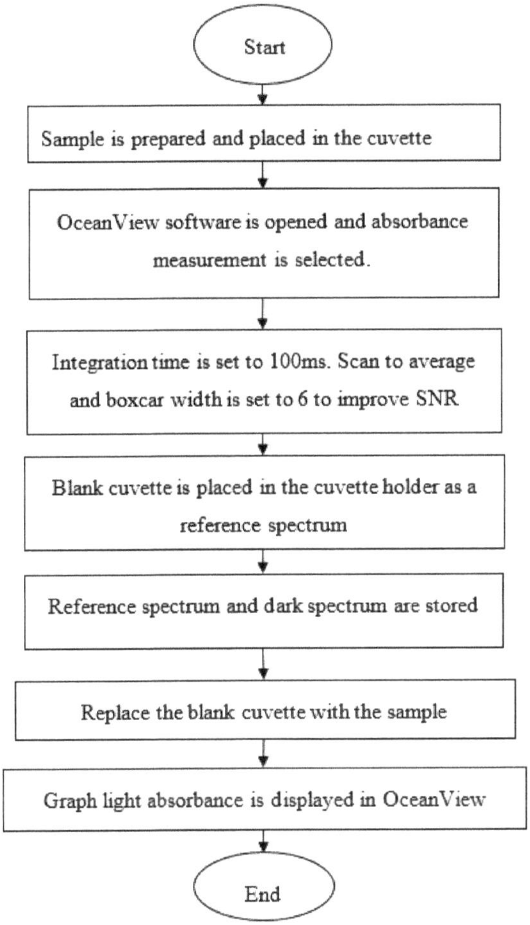

Figure 3. Flow chart of the testing method using the Ocean Optics spectrometer.

Figure 4 shows the experimental setup using the Ocean Optics spectrometer. A light source produced light in a range from 350 to 1000 nm that passed through the plastic cuvette that had samples. The transmitted light was measured by the Ocean Optics spectrometer, which was connected to a computer. The OceanView software was installed in the computer to record the data that were measured by the spectrometer and to display the spectral response. The amount of light transmitted through the sample decreased because some light may have been absorbed by the molecules or particles in the samples, based on the Beer–Lambert law [22,23]. To calibrate the spectrometer, an empty cuvette was shined by the light source, and the light absorbance was stored as a reference spectrum.

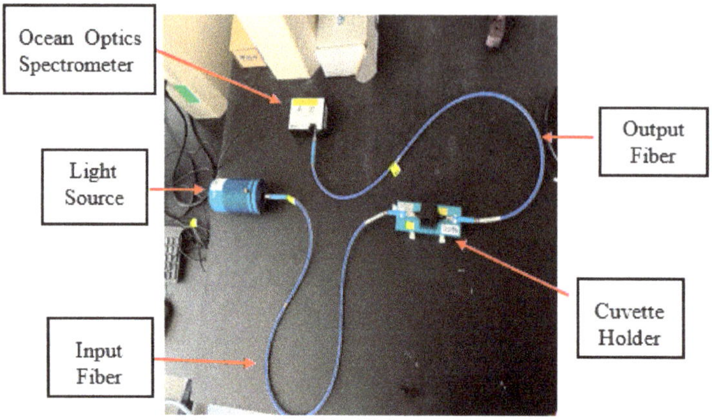

Figure 4. Experimental setup using Ocean Optics spectrometer.

4. Results

4.1. Real-Time Monitoring Using Blynk Application

The sample was tested, and the output was displayed using a Blynk application for real-time monitoring. The data slightly fluctuated due to the sensitivity of the light sensor. Thus, the measurement was repeated ten times. The output was averaged, and the data were plotted using Origin software to observe the absorbance peak. Figure 5 shows the real-time monitoring and sensing method using a mobile phone. The light transmittance, I_t, and the light absorbance, A, can be observed with the Blynk application.

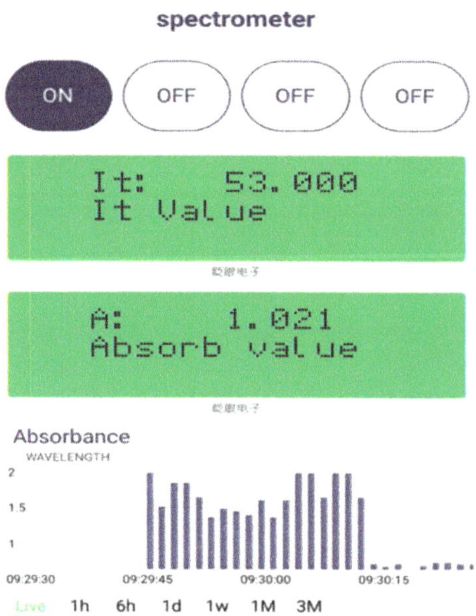

Figure 5. Real-time monitoring and sensing by a mobile phone. *It* value refers to light transmittance and *Absorb* value refers to light absorbance.

In Figure 5, the top display shows four buttons (ON and OFF) to control the movement of the servo motor to vary the spectrum wavelength. The value of the transmitted light, I_t, from the sample is shown on the Blynk display. The light sensor detects light and converts it to an electrical signal output. The transmitted light is then converted to the absorbed light using the Beer–Lambert law. The value of the light sensor is displayed together with the value of the light absorbance at the specific range of the spectrum. Figure 6 depicts the real-time monitoring and sensing by a computer. The servo motor can be turned on or off to change the region of the spectrum wavelength. The output of the absorbance value versus time is displayed by the graph.

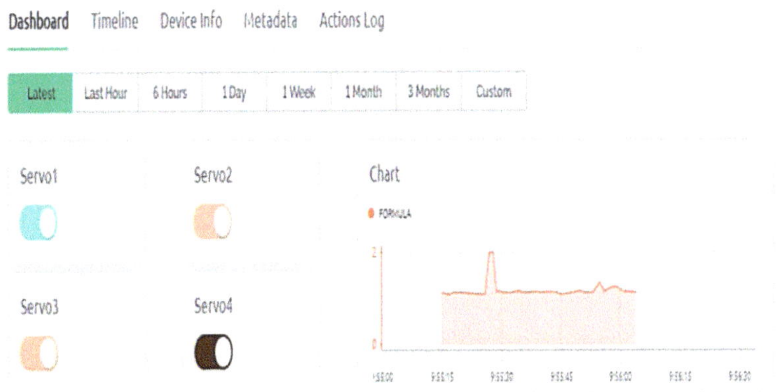

Figure 6. Real-time monitoring and sensing by a computer. Servo 1 to Servo 4 refers to servo motor that can be turned on or off to change the region of the spectrum wavelength.

4.2. *Light Absorbance from ESP32-Based Spectrometer*

The light absorbance using the ESP32-based spectrometer and the Ocean Optics spectrometer was plotted in Origin. Figure 7a–e show the output of the light absorbance (wavelength ranging from 300 nm to 700 nm) from the ESP32-based spectrometer. Figure 7a shows the light absorbance from the pesticide sample with the highest absorbance peak lying in the wavelength ranging from 300 nm to 400 nm. The absorbance peak of ammonium nitrate is also within the wavelength range between 300 nm and 400 nm (Figure 7b), and the absorbance peak of zinc oxide using ESP32-based spectrometer lies between 400 nm and 500 nm with the absorbance peak at 400 nm, as shown in Figure 7c. Meanwhile, Figure 7d shows the light absorption spectrum of Reservoir I and the absorbance peak lies in the region between 300 nm and 400 nm, which indicates that the water may consist of organic contaminants. Reservoir II shows the wavelength range of the absorbance spectrum to be between 300 nm and 400 nm, which also shows that the water sample may consist of organic contaminants (Figure 7e).

4.3. *Light Absorbance from Ocean Optics Spectrometer*

The sample was also tested with a lab-based spectroscopy method. The light absorbance output from the Ocean Optics spectrometer for each water sample is depicted in Figure 8a–e. Figure 8a shows the absorption spectrum of the pesticide sample (from 300 nm to 400 nm) with the absorbance peak at ~361 nm [16]. Meanwhile, Figure 8b shows the absorbance spectrum of ammonium nitrate and the absorption peak lies at ~363. Furthermore, the absorbance peak of zinc oxide (Figure 8c) is ~405 nm and the absorption peaks of Reservoir I (Figure 8d) and Reservoir II (Figure 8e) are ~346 nm and ~361 nm, respectively.

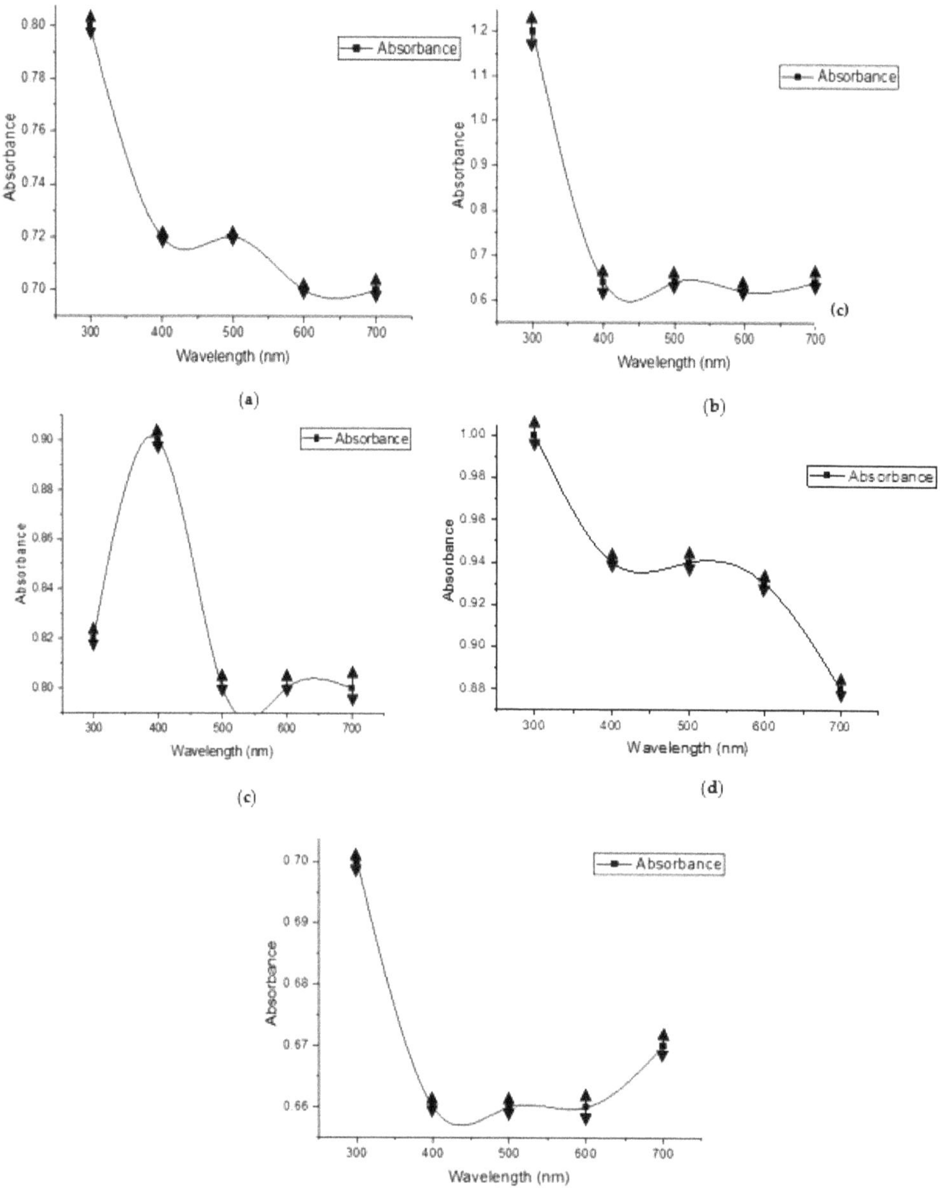

Figure 7. Light absorbance output (wavelength range from ~300 nm to 700 nm) from ESP32-based spectrometer for (**a**) pesticide, (**b**) ammonium nitrate, (**c**) zinc oxide, (**d**) Reservoir I and (**e**) Reservoir II. The value is averaged from 10 measurements, and there is a slight variation (standard error) in the measurement. The standard error is shown by the black arrow.

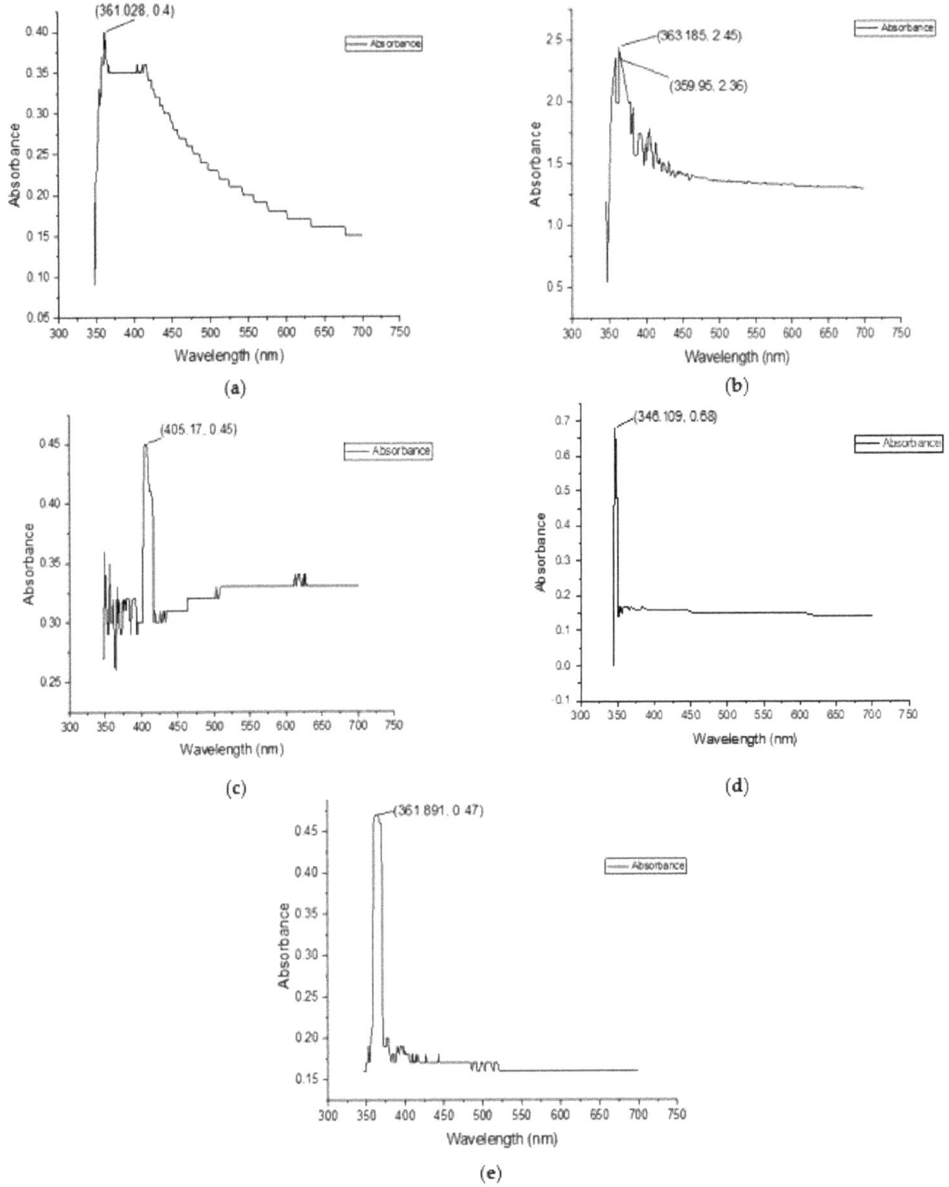

Figure 8. Light absorbance output (wavelength range from 300 nm to 700 nm) from Ocean Optic spectrometer for (**a**) pesticide, (**b**) ammonium nitrate, (**c**) zinc oxide, (**d**) Reservoir I and (**e**) Reservoir II. The value is averaged from 10 measurements.

5. Discussion

Detecting heavy metal and organic contaminants in water is crucial in agriculture. In this research, three contaminants (zinc oxide, ammonium nitrate and organic pesticide), commonly found in irrigation water, and two irrigation water sources (Reservoir I and Reservoir II) were tested. Other chemical elements can also be tested as long as they can absorb light in the ultraviolet and visible spectral ranges. The light source needs to be changed to infrared to detect elements with wavelengths over 800 nm.

The developed spectrometer can display outputs on multiple devices such as mobile phones (Figure 5) and computers (Figure 6). Figure 7a depicts that the highest absorbance of the pesticide sample is in the range from 300 nm to 400 nm. We attribute that to the organic compounds in pesticides, which may comprise of rotenone, neem, pyrethrum and minerals, such as cryolite, boric acid and diatomaceous earth. The chemical compounds also can contribute to the color of the pesticide solution. A previous study conducted by Rohit et al. [30] also showed a quite similar absorbance range of pesticides, from 360 nm to 400 nm.

Meanwhile, the absorption peak of the nitrate sample also lies from 300 nm to 400 nm (Figure 7b). The nitrate ions can be classified as ions that have a strong absorption range in ultraviolet spectrum region, from 280 nm to 320 nm [31]. Most types of UV sources emit in the spectrum region and can interact with nitrate in water to generate hydroxyl radicals [31,32]. When the electromagnetic energy, such as UV light, propagates in aqueous samples, a fraction of energy can be transferred to the ions via the transition of electrons between different energy levels [23,33]. Thus, it shows that nitrate ion achieves the excitation in the region of 300 nm to 400 nm.

Then, the absorption spectrum of zinc oxide lies from 400–500 nm (Figure 7c). The result is similar to the previous studies from Kolli et al. [34] and Darroudi et al. [35], in which the absorbance spectrum of zinc ranged from 350 nm to 500 nm with absorption peaks ~383 nm and at ~357 nm, respectively. This is because the metal bond has strong electron excitation in the respective wavelength region. The electron transition in zinc oxide occurs due to the transition from the valence band to the conduction band. Metal sources in the environment are both anthropogenic and natural. The sources include effluents from wastewater treatment plants, industrial and agricultural wastes, power stations, mining, boating activities, transportation vehicles, urban runoff and geological processes [6].

Reservoir I (Figure 7d) and Reservoir II (Figure 7e) show that the absorbance peaks lie within the wavelength range of 300 nm to 400 nm. This is because the organic material consists of carbon and a hydrogen bond and is matched with the absorption energy from the molecules in the sample solution. The light absorbance of Reservoir II is slightly low compared to Reservoir I since the water is not as turbid as Reservoir I. Reservoir II is not near any sewage area, but the reservoir has many fish, which may also contribute to organic compounds.

In comparison to the ESP32-based spectroscopy method, the lab-based spectrometer (Figure 8) can give higher accuracy due to the continuous reading or light scanning of each sample. The absorption peak of zinc oxide (Figure 8c) is at ~405 nm due to electron transition in the spectrum region. It can be assigned to the intrinsic band-gap absorption of ZnO due to the electron transitions from the valence band to the conduction band (O→Zn) [35]. Then, Ocean Optics spectrometer depicts the absorbance peak at ~346 nm for Reservoir I (Figure 8d) because the organic particles achieve the excitation energy within this region [36]. The Ocean Optics spectrometer also depicts the absorbance peak at ~361 nm for Reservoir II (Figure 8e), which lies from 300 nm to 400 nm. This shows that Reservoirs I and II contain organic contaminants.

Both types of spectrometers have proved the ability to measure the light absorbance to further detect organic and heavy contaminants in water samples. The comparison of system performance between lab-based spectrometer and portable ESP32-based spectrometer is shown in Table 1. These two techniques have advantages and disadvantages, respectively. ESP32-based spectrometer is portable, low-cost and can be monitored in real-time via multiple devices, whereas Ocean Optic spectrometer is non-portable (must be set up with a light source and a computer), requires lab-based monitoring and is high-cost. However, the Ocean Optic spectrometer can give higher accuracy. Thus, it depends on farmers or consumers to choose their preferred method.

Table 1. System comparison between ESP32-based spectrometer and Ocean Optic spectrometer.

ESP32-Based Spectrometer	Ocean Optic Spectrometer
Portable	Non-portable
On-site real-time monitoring	Lab based monitoring
Low cost	High cost
Less accurate	More accurate
Monitor and display result via mobile phone and computer	Monitor and display via computer
Non-invasive	Non-invasive

6. Conclusions

In conclusion, we have developed a portable spectrometer that can measure light absorbance to detect contaminants in water. The spectrometer was developed based on optical dispersion and the Beer–Lambert law. Different types of chemical compounds can produce different absorbance peaks based on their chemical structure. The study investigates the relationship between the types of compounds and absorption spectrum to obtain the light transmittance and absorbance of the sample. The study can be used further to sense any presence of organic and metal contaminants in irrigation water. The spectrometer consists of the ESP32 microcontroller, a servo motor to control spectra movement, a light sensor to measure light transmittance and Blynk software to display the output of the spectrometer. An experiment to measure light absorbance using a lab-based spectrometer, Ocean Optics spectrometer, was also set up. The organic contaminants have an absorbance peak that lies between 300 nm and 400 nm, whereas the heavy metal contaminants show a peak that lies between 400 nm and 500 nm. The irrigation water from two different water sources (Reservoir I and Reservoir II) was tested using the ESP32 and Ocean Optics spectrometers and both sources show the presence of organic contaminants. Thus, we believe that the ESP32-based spectrometer can reliably be used in irrigation systems since the device is portable and can be monitored in real-time. The other advantages of the ESP32-based spectrometer are that it is low-cost, non-invasive and can differentiate types of contaminants based on optical properties. It is also a simple and fast method for quantitative estimation.

In the future, a larger battery capacity will be used to ensure the spectrometer can last longer. We will perform simultaneous detection of elements by combining a few contaminants into one sample. We will test more heavy metal elements such as copper and iron. Rivers and lakes, especially polluted rivers will be tested as well. The ESP32-based spectrometer can be further upgraded to find concentrations or the molar absorptivity of biological macromolecules and organic molecules.

Author Contributions: Conceptualization, supervision, project administration, funding acquisition, W.Z.W.I.; writing—original draft preparation, W.A.N.W.A.; experiment, writing—review and editing, S.N.Z.; funding acquisition and review, K.N.Z.A.; reviewing and proof reading, W.M.W.A.K. All authors have read and agreed to the published version of the manuscript.

Funding: This research was funded by the Ministry of Higher Education, Malaysia under the Fundamental of Research Grant Scheme, grant number FRGS/1/2021/WAB02/USIM/02/1 and USIM-MMU Matching grant (USIM/MG/MMU-PPKMT-ZDA/FKAB/SEPADAN-S/70822). The APC was funded by Universiti Sains Islam Malaysia (USIM).

Institutional Review Board Statement: Not applicable.

Data Availability Statement: Not applicable.

Acknowledgments: We would like to acknowledge our funder, the Ministry of Higher Education, Malaysia (MOHE), all members of Faculty of Engineering and Built Environment (FKAB), USIM and our families.

Conflicts of Interest: We declare no conflict of interest that can influence the representation or interpretation of the reported research results. The funders had no role in the design of the study; in the collection, analyses, or interpretation of data; in the writing of the manuscript; or in the decision to publish the results.

References

1. Food and Agriculture Organization of the United Nations. Water for Sustainable Food and Agriculture: A report produced for the G20 Presidency of Germany. 2017. Available online: https://www.fao.org/3/i7959e/i7959e.pdf (accessed on 10 April 2023).
2. Food and Agriculture Organization of the United Nations. The State of the World's Land and Water Resources for Food and Agriculture 2021—Systems at Breaking Point. 2022. Available online: https://www.fao.org/land-water/solaw2021/en/ (accessed on 10 April 2023).
3. Ciaramella, B.R.; Corinzia, S.A.; Cosentino, S.L.; Testa, G. Phytoremediation of Heavy Metal Contaminated Soils Using Safflower. *Agronomy* **2022**, *12*, 2302. [CrossRef]
4. Hernández-Pitalúa, D.; Hernández-Orduña, M.G.; Martínez-Escalante, G.A.; Lagunes-Gómez, I. Influence of Lead (Pb) and Its Relationship with the pH of Water on the Growth of Creole Maize (*Zea mays* L.). *Agriculture* **2022**, *12*, 749. [CrossRef]
5. Hashem, I.A.; Abbas, A.Y.; Abd El, A.E.-N.H.; Salem, H.M.; El-Hosseiny, O.E.; Abdel-Salam, M.A.; Saleem, M.H.; Zhou, W.; Hu, R. Potential of rice straw biochar, sulfur and ryegrass (*Lolium perenne* L.) in remediating soil contaminated with nickel through irrigation with untreated wastewater. *PeerJ* **2020**, *6*, 2020. [CrossRef]
6. Hall, L.W.; Anderson, R.D. A Comparison of Sediment Metal Concentrations as Potential Stressors to Resident Benthic Communities in an Agricultural Waterbody. *Agriculture* **2022**, *12*, 1029. [CrossRef]
7. Guo, Y.; Liu, C.; Ye, R.; Duan, Q. Advances on water quality detection by uv-vis spectroscopy. *Appl. Sci.* **2020**, *10*, 6874. [CrossRef]
8. Hashimi, M.H.; Hashimi, R.; Ryan, Q. Toxic Effects of Pesticides on Humans, Plants, Animals, Pollinators and Beneficial Organisms. *Asian Plant Res. J.* **2020**, *5*, 37–47. [CrossRef]
9. Zhang, C.; Feng, Y.; Liu, Y.; Chang, H.; Li, Z.; Xue, J. Uptake and translocation of organic pollutants in plants: A review. *J. Integr. Agric.* **2017**, *16*, 1659–1668. [CrossRef]
10. Chandini, R.; Kumar, R.; Prakash, O. The Impact of Chemical Fertilizers on our Environment and Ecosystem. *Res. Trends Environ. Sci.* **2019**, *4*, 69–86. Available online: https://www.researchgate.net/publication/331132826 (accessed on 5 April 2023).
11. Zainurin, S.N.; Ismail, W.Z.W.; Mahamud, S.N.I.; Ismail, I.; Jamaludin, J.; Ab. Aziz, N.A. Integration of Sensing Framework with A Decision Support System for Monitoring Water Quality in Agriculture. *Agriculture* **2023**, *13*, 1000. [CrossRef]
12. Razman, N.A.; Wan Ismail, W.Z.; Razak, M.H.A.; Ismail, I.; Jamaludin, J. Design and analysis of water quality monitoring and filtration system for different types of water in Malaysia. *Int. J. Environ. Sci. Technol.* **2023**, *20*, 3789–3800. [CrossRef]
13. Sevda, T. Estimation of groundwater quality using an integration of water quality index, artificial intelligence methods and GIS: Case study, Central Mediterranean Region of Turkey. *Appl. Water Sci.* **2023**, *13*, 15.
14. Wan Ismail, W.Z.; Goldys, E.M.; Dawes, J.M. Extended emission wavelength of random dye lasers by exploiting radiative and non-radiative energy transfer. *App. Phys. B* **2016**, *122*, 40. [CrossRef]
15. Peteinatos, G.G.; Korsaeth, A.; Berge, T.W.; Gerhards, R. Using optical sensors to identify water deprivation, nitrogen shortage, weed presence and fungal infection in wheat. *Agriculture* **2016**, *6*, 24. [CrossRef]
16. Laganovska, K.; Zolotarjovs, A.; Vázquez, M.; Mc Donnell, K.; Liepins, J.; Ben-Yoav, H.; Karitans, V.; Smits, K. Portable low-cost open-source wireless spectrophotometer for fast and reliable measurements. *HardwareX* **2020**, *7*, e00108. [CrossRef]
17. Zhang, H.; Zhang, L.; Wang, S.; Zhang, L.S. Online water quality monitoring based on UV–Vis spectrometry and artificial neural networks in a river confluence near Sherfield-on-Loddon. *Environ. Monit. Assess.* **2022**, *194*, 9. [CrossRef]
18. Kamil, N.A.I.M.; Nor'aini, Z.S.; Wan Ismail, W.Z.; Balakrishnan, S.R.; Juliza Jamaludin, J.; Ismail, I. Investigating the Quality of Milk using Spectrometry Technique and Scattering Theory. *Eng. Technol. Appl. Sci. Res.* **2021**, *11*, 7111–7117. [CrossRef]
19. Kamil, N.A.I.M.; Ismail, W.Z.; Wan Ismail, I.; Jamaludin, J.; Hanasil, N.S.; Ibrahim, R.K.R. Analysis of Milk from Different Sources Based on Light Propagation and Random Laser Properties. *Photonics* **2021**, *8*, 486. [CrossRef]
20. Rateni, G.; Dario, P.; Cavallo, F. Smartphone-Based Food Diagnostic Technologies: A Review. *Sensors* **2017**, *17*, 1453. [CrossRef]
21. Transportation, L.F.S. Chapter 12 Linear Fiber-Optic Signal Transportation. In *Modern Cable Television Technology*, 2nd ed.; Elsevier: Amsterdam, The Netherlands, 2004; pp. 509–557.
22. Listiaji, P.; Suparta, G.B. Low-cost imaging spectrophotometer system for absorbance measurement. *J. Phys. Conf. Ser.* **2020**, *1567*, 042093. [CrossRef]
23. Wilson, K.; Walker, J. *Principles and Techniques of Biochemistry and Molecular Biology*, 7th ed.; John Wiley & Sons, Ltd.: Hoboken, NJ, USA, 2011; Volume 39.
24. Agilent Technologies. The Basics of UV-Vis Spectrophotometry A Primer. 2021. Available online: https://www.agilent.com/cs/library/primers/public/primer-uv-vis-basics-5980-1397en-agilent.pdf (accessed on 10 April 2023).
25. Tzanetou, E.N.; Karasali, H. A Comprehensive Review of Organochlorine Pesticide Monitoring in Agricultural Soils: The Silent Threat of a Conventional Agricultural Past. *Agriculture* **2022**, *12*, 728. [CrossRef]
26. Hamid, N.; Suhaimi, S.; Othman, M.Z.; Wan Ismail, W.Z. A Review on Thermal Evaporation Method to Synthesis Zinc Oxide as Photocatalytic Material. *Nano Hybrids Compos.* **2021**, *31*, 55–63. [CrossRef]

27. Mlangeni, A.T.; Raab, A.; Kumambala, P.; Monjerezi, M.; Matumba, L.; Feldmann, J. Evaluation of Metal (loids) Concentrations in Soils of Selected Rice Paddy Fields in Malawi. *Agronomy* **2022**, *12*, 2349. [CrossRef]
28. Cooper, M.; Michael, K. Chapter 2: Spectroscopy: How we know what we know about the structure of matter in OCLUE. In *Organic Chemistry, Life, the Universe & Everything*; Michigan State University Libraries: East Lansing, MI, USA, 2020; ISBN 978-1-62610-102-9.
29. The Royal Society of Chemistry. Ultraviolet/visible spectroscopy. In *Modern Chemical Techniques*; The Royal Society of Chemistry: London, UK, 2015; pp. 92–115.
30. Rohit, J.V.; Kailasa, S.K. Cyclen dithiocarbamate-functionalized silver nanoparticles as a probe for colorimetric sensing of thiram and paraquat pesticides via host–guest chemistry. *J. Nanoparticle Res.* **2014**, *16*, 2585. [CrossRef]
31. Sohbatzadeh, F.; Bagheri, H.; Safari, R. Effect of electrical discharge in water on concentration of nitrate solution. *Chin. Phys. B* **2017**, *26*, 025101. [CrossRef]
32. Keen, O.S.; Love, N.G.; Linden, K.G. Nitrate Photochemistry in the Context of Water Reclamation. In *Water Reclamation and Sustainability*; Elsevier: Amsterdam, The Netherlands, 2014; pp. 229–246. [CrossRef]
33. Yeshno, E.; Arnon, S.; Dahan, O. Real-time monitoring of nitrate in soils as a key for optimization of agricultural productivity and prevention of groundwater pollution. *Hydrol. Earth Syst. Sci.* **2019**, *23*, 3997–4010. [CrossRef]
34. Manyasree, D.; Kiranmayi, P.; Kolli, V.R. Characterization and antibacterial activity of zno nanoparticles synthesized by co precipitation method. *Int. J. Appl. Pharm.* **2018**, *10*, 224–228. [CrossRef]
35. Zak, A.K.; Razali, R.; Majid, W.H.A.; Darroudi, M. Synthesis and characterization of a narrow size distribution of zinc oxide nanoparticles. *Int. J. Nanomed.* **2011**, *6*, 1399–1403. [CrossRef]
36. Schaller, C. *Structure & Reactivity in Organic, Biological and Inorganic Chemistry*; College Saint Benedict/Saint John's University: Queens, NY, USA, 2014.

agriculture

MDPI

Article

Will Social Network Relationship Significantly Enhance Farmers' Participation in the Supply of Small Water-Saving Irrigation and Water Conservancy Facilities in China?

Lei Wang [1,2] and Yiwen Zhao [1,*]

[1] Business School, Hohai University, No.200, Jinlingbei Road, Changzhou 213022, China
[2] Center for Systems Engineering and Management Innovation, Changzhou 213022, China
* Correspondence: zhaoyiwen1129@126.com; Tel.: +86-18912488757

Abstract: The supply level of irrigation and water conservancy is related to the utilization efficiency of water resources, the production level of farmers, and the supply quality of agricultural products, especially relating to national food security and stability. Based on 1169 pieces of data collected from field surveys in three provinces of China in 2019, an evaluation system of social network relationships was constructed from five aspects: network scale, network tightness, trust and commitment, social atmosphere and sense of belonging, and social participation. These five aspects are the channels for farmers to obtain information. A binary logistic model was used to analyze the impact of the social network relationships on farmers' participation in small water-saving irrigation and water conservancy facilities supply, and the key factors affecting farmers' behavior were selected. The purpose of this study is to further improve the evaluation system of social network relationships and the study of the effect of social network relationships on farmers' water-saving behaviors, enrich relevant theories and provide a feasible path for the implementation of water-saving irrigation from the macro initiative level. The results show that higher network closeness and policy satisfaction, water management experience, agricultural insurance, strong family decision-making power, etc., will reduce the likelihood of farmers participating in the supply of small water-saving irrigation and water conservancy facilities; increasing network compactness will increase the possibility of farmers' participation in the supply; trust and commitment, social ethos and sense of belonging, social participation, and other factors have no significant influence on farmer participation behavior. It can be seen that network tightness and network scale play an important role in the behavior of farmers' participation in public affairs. In conclusion, social network relationships will affect farmers' participation in the supply of irrigation and water conservancy facilities, but different dimensions have different influences on it.

Keywords: social network relationship; small water-saving irrigation; water conservancy facilities; farmers' participation; binary logistic model

1. Introduction

China is one of the 13 countries with the poorest per capita water resources in the world. Agriculture is the industry that consumes the most water in China. According to China Water Resources Bulletin 2021, agricultural water consumption accounts for 61.5% of China's total water consumption, while irrigation water accounts for 90% to 95% of agricultural water consumption. Therefore, the mode of agricultural irrigation has a crucial impact on the contradiction between the supply and demand of water resources. However, the utilization rate of farmland irrigation water in China is low, and the effective utilization coefficient is only 0.568, while it is as high as 0.7–0.8 in developed countries [1]. It can be seen that China's agricultural water is still in the stage of inefficient and extensive development. The Ministry of Water Resources, the Chinese Academy of Engineering, and other departments predict that only if agricultural water use maintains zero or negative growth can China ensure water security and ecological security and then ease the contradiction between the supply and demand of water resources. Developing water-saving irrigation is an important way to improve agricultural water efficiency and reduce agricultural water consumption. At present, small water-saving irrigation facilities are mainly promoted by the government, and how to guide farmers to actively participate in the construction of farmland water conservancy is an urgent problem that needs to be solved in the present and the future. Farmers, as micro-decision-making bodies engaged in agricultural production, have the freedom to choose to participate in the construction of farmland water conservancy or not [2]. Zuo, Z.Y. used a conditional value method to analyze farmers' willingness to pay for water-saving irrigation facilities in North China groundwater over-extraction areas and believed that farmers only had demand for modern water-saving irrigation technology, but the low comparative benefits of agriculture and the low risk tolerance of farmers directly limited farmers' ability to pay for modern water-saving irrigation facilities, and the investment in the adoption of new technology should be jointly committed to by farmers and the government [3].

In addition, according to the FAO report (2019, the State of Food Security and Nutrition in the World), more than 820 million people in the world did not have enough food in 2018, world hunger increased for the third consecutive year, and food security remained uncertain. Mabry (1996) believes that one-third of the world's grain harvest is due to the construction and management of small water-saving irrigation and water conservancy projects, especially in developing countries, which play a very important role [4,5]. At present, nearly half of China's cultivated land is still "fields on hill tops which depend on rains for water", and the main reason is that the shortage of small water-saving irrigation and water conservancy facilities has not been effectively solved [6]. During the period of the people's commune (1958–1978), wherein all of the means of production and the public property of the villages were owned, and unified management was exercised over the farmers and the affairs of the villages, the land was owned by the people's commune, and the irrigation and water conservancy facilities were mainly supplied by the commune with a high degree of organization. Therefore, the shortage of irrigation and water conservancy facilities has greatly improved since the founding of the People's Republic of China (1949). However, after the implementation of the household contract responsibility system (1978-), the land was distributed to farmers, which broke the original situation of the collective supply of irrigation and water conservancy facilities, and the degree of organization decreased [7]. As a result, the construction of irrigation and water conservancy facilities has been slow or even stagnated for a long time [8]. From the 1990s, the construction of irrigation and water conservancy facilities began to recover. However, with the diversification of the crop planting structures, the contradiction between the public supply of small water-saving irrigation and water conservancy facilities, which can only be used to irrigate food crops, and the individual demands of the household management of farmers, which require more timely and convenient irrigation, has become increasingly fierce. Therefore in the current household contract responsibility system based on finely cultivated land, the participation of farmers is an effective way to solve this contradiction [9].

Farmers' participation in supply is based on a certain region, and many individual farmers participate in the process of collective cooperation. Theoretically speaking, the goals of an individual and the collective are not necessarily identical, but people want prestige, respect, and friendship, as well as other social and psychological goals, and when there is no economic incentive for individuals to contribute to the interests of the group, social incentives may drive people to do so [10]. The essence of farmers' participation in supply is to achieve compatibility between the goals of an individual and the collective among the relevant stakeholders in a certain region. As a special organization group, farmers learn from the social network relationship formed on the basis of geography, blood relationships, and kinship to form similar ideas and to influence the decision-making behavior of farmers [11], which can push the combination of micro individual behavior with macro collective behavior and promote the realization of collective action. For example, in a clan village, the 'clan leader' of the clan has high prestige and is also the leader of the villagers' group. The members of the clan support the leadership of the 'clan leader', and the social network formed by the 'clan leader' as the center guides the villagers to form norms with certain functions of encouragement and restraint, which can influence the behaviors of the members of the clan [12]. Therefore, it is of great theoretical significance and feasibility to introduce the concept of social network relationships into the study of farmers' participation in the supply of small-scale irrigation and water conservancy facilities. As a result, it is of great significance to study social network relationships and promote rural public governance to introduce education for farmers based on social network relationships into the supply problem of farmers participating in small-scale irrigation and water conservancy facilities.

The concept of the social network was first proposed by R. Brown. Later, many scholars interpreted the connotation of the social network, and the description of the single dimension has changed to multi-dimensional [13]. Economic behavior is restricted by the social network, and the appropriate embedding of social network relationships can improve the quality of organizational resource acquisition [14–19]; effective network relationship embedding enables people to obtain information content through effective information channels [20]. It also provides the basis of social trust for people and provides decision-making-related information for people, which is more accurate and effective in the process of communication [14]. In addition, it improves social learning and the cooperation willingness and efficiency of participants through information sharing [21]. From the perspective of institutional economics, an effective social network makes it easier to realize collective action by reducing the cost of collective action, and the success of cooperation makes social networks expand [22,23]. At the same time, effective social network relationship also plays an important role in risk sharing [24,25]. In the field of the supply of public goods, good social network relations contribute to enhancing trust and expanding social exchange, thus improving the supply of public goods. The larger the stock of social capital between the farmers and villages, the higher the level of trust in the village, and the more actively the farmers participate in the supply of public goods [13,22,26,27]. However, because farmers may have different concerns within social network relationships, different dimensions of social capital have different impacts on Farmers' participation behavior [28]. Although most scholars believe that social networks can promote the realization of collective action and improve management efficiency, some scholars put forward the opposite view. The constraints of network relationships will reduce people's choices and then affect people's behavior, and people have free-riding behavior [29]. It leads to the lack of farmers' ability to participate in governance and the dilemma of cooperation, and the impact on Farmers' willingness to participate is relatively complex [30]. Affected by the social network, some people may have a herd mentality, and a few people's non-participation behaviors will lead to the majority of people not participating in the behavior, which may lead to the failure of cooperation [31,32]. Therefore, the negative influence of social network relationships on the spontaneous supply of rural public goods cannot be ignored, which may make the spontaneous supply of rural public goods face the threat of failure.

In summary, scholars have undertaken relevant research on the index evaluation system where social network relations, relationship networks, and trust and norms are regarded as the three dimensions of social capital [33]. However, it is still relatively one-sided and fails to cover such factors as the strength of the network relationship and the interpersonal environment of the village where the farmers live. Although scholars have recognized the importance of social network relationships to members' participation, technology expansion, and resource acquisition, few scholars take the subdivision dimension of social network relationship as the influencing factors of farmers' participation in the supply of farmland and water conservancy facilities. In addition, most scholars study rural household social network systems from the perspective of social capital. For example, Li, J.P.

analyzed the influence of social capital on farmers' willingness to participate and pay in the cooperative supply of water-saving irrigation facilities by selecting the dimensions of social capital, such as social networks, trust, reputation, participation, norms, and reciprocity. Xu, L. estimated farmers' social capital from three dimensions of network, trust, and norms and analyzed the influence of social capital on farmers' willingness to participate in irrigation management reform. Cai, Q.H and Zhu, Y.C empirically analyzed the mechanisms that influence farmers' participation in village collective action from the dual perspectives of social capital and income gap [2,34,35]. Therefore, starting from the social network relationship, this article attempts to analyze the channels through which these dimensions affect human behavior using the structure of the relationship network. To be brief, the article pays more attention to the channels and paths that social network relationships affect farmers' individual behaviors. This article draws on the previous research results and combines the field research in Shaanxi, Ningxia, and Henan provinces in 2019 to build a social network relationship evaluation system and then combines the Logistic model to focus on whether and how the social network relationship will affect farmers' participation in the supply of irrigation and water conservancy facilities so as to further improve the evaluation system for social network relationships and the study of the effect of social network relationships on farmers' water-saving behaviors, enrich relevant theories, and provide a feasible path for the implementation of water-saving irrigation from the macro initiative level.

2. Materials and Methods

2.1. Model Construction

2.1.1. Coefficient of Variation Method

In order to calculate the value of each dimension of social network relationships, it is necessary to select a method to determine the weight of the lowest index in the index system. Since the objective weighting method is not based on the subjective judgment of human beings, it reduces the biased performance of expert weighting and objectively reflects the relative importance of indicators. This paper selects the objective weighting method—the coefficient of variation method to determine the weight of each index [36]. The basic idea of the variation coefficient method is: in the multi-index comprehensive evaluation, if a certain index has a large degree of variation in the observed values of all of the evaluated objects, it can clearly distinguish the level of each evaluated object in this aspect, then the index should be given a larger weight; conversely, it should be given a smaller weight. The coefficient of variation and the weight of each index were calculated as in Formulas (1)~(3).

$$V_i = \frac{\sigma_i}{\overline{x_i}} \quad (1)$$

$$W_i = \frac{V_i}{\sum_{i=1}^{n} V_i} \quad (2)$$

$$F = \sum_{i=1}^{n} (W_i \times I_i) \quad (3)$$

In Formula (1), σ_i is the standard deviation of index i and $\overline{x_i}$ is the average value of index i. In Formula (2), V_i is the coefficient of variation of the index i, and W_i is the weight of each index obtained after normalizing V_i. In Formula (3), F is the comprehensive evaluation index, and Ii is the single evaluation score of the index i after standardization.

2.1.2. Logistic Regression Model

The linear regression model is suitable for regression analysis with continuous dependent variables but not for regression analysis with discrete dependent variables. In general, discrete data are used to measure peoples' behavior selection, so when discrete data are used as a dependent variable, the linear regression model is no longer applicable. The discrete selection model originated from Fechner's conditional binary reflex research, and then Warner applied it to the field of economic research. The binary logistic regression model is a kind of discrete choice model that refers to a multivariate analysis method of the relationship between the binary observation results and some influential factors, which can make up for the deficiency of the simple regression model when dealing with discontinuous variables [37,38]. The model can clearly explain the correlation between the dependent variables and independent variables and reveal the quantitative change law between the dependent variables and the independent variables. However, this model also has some limitations. The model theoretically requires the sample size to be greater than 50 for parameter estimation and cannot solve the problem of "multicollinearity" [39,40]. Therefore, it is necessary to adopt a large sample when applying this model and try to choose independent variables.

The model is shown in Equation (4). In Formula (4), Y is the dependent variable, which obeys a binomial distribution, and the values are 0 and 1, respectively, the independent variable is $X_1, X_2 \ldots X_n$. It is the same as the linear regression model. β_0 is a constant term and β_i is the partial regression coefficient corresponding to X_i $(i = 1, 2 \ldots, m)$.

$$LogP_{it}(Y = 1) = \beta_0 + \beta_1 X_1 + \beta_2 X_2 + \ldots + \beta_n X_n \quad (4)$$

2.2. Farmers' Social Network Relationship System

2.2.1. The Data Source

Shaanxi, Ningxia, and Henan provinces have large agricultural output and good agricultural development. In terms of geographical location and ecological environment, Henan Province is mainly plains, Shaanxi Province is mainly hilly, and Ningxia is short of water. In terms of planting structure, Shaanxi Province mainly produces grain crops; Henan Province mainly produces food crops and cash crops; and Ningxia mainly produces cash crops. These three provinces are representative in terms of geographical location, ecological environment, and planting structure.

Therefore, the study conducted in 2019 took Shaanxi, Ningxia, and Henan provinces as the research objects and selected representative provinces and cities for the questionnaire distribution and statistical analysis. The questionnaire can be divided into four aspects: the basic information of the farmers (collecting the basic information of the farmers, such as the head of household, age, gender, etc.), the supply of small water-saving irrigation and water conservancy facilities, and the overall supply of small water-saving irrigation and water conservancy facilities, and the basic situation of the social capital of the farmers.

The basic information of the farmers was collected to explore whether their own characteristics will affect their participation behavior. The status of farmers' participation in the supply of small water-saving irrigation and water conservancy facilities is to understand the participation behavior of the farmers and obtain the dependent variable data of the binary logistic regression model; the purpose of studying the overall supply of small water-saving irrigation and water conservancy facilities is to explore the supporting effect of the government on farmers' behavior from the other hand by understanding the current construction and management of small water-saving irrigation and water conservancy facilities. The basic situation of farmers' social capital involves each dimension of farmers' social network relationships, and the social network relationship of each dimension is taken as the independent variable of the binary Logistic regression model to explore the influence of the social network relationships of different dimensions on farmers' participation behavior. In the process of issuing the questionnaires, 3 cities were selected from each province, 1–2 counties were selected from each city, 4 townships were selected from each county, and 5 villages were selected from each township. Overall, 80 samples were randomly selected from each village for questionnaire distribution and statistical analysis. The random sampling method was adopted to sample the counties, townships, villages, and farmers. A total of 1440 samples were collected, of which 1169 were valid samples, and the effective rate of samples was 81.12%. See Appendix Table A1 for details. The research areas are shown in Figure 1.

Figure 1. The main survey areas (Shanxi, Henan, Ningxia) in China.

2.2.2. Index System Construction

A social network is an overall network of formal and informal relationships among specific groups of people, including the direct relationships between people and the indirect relationships formed through the nodes of material, environmental, and cultural sharing, which can also be called relationships [41,42]. Social capital refers to the social resources people use through social networks, which are different social networks [43]. However, most scholars do not distinguish between the two when constructing the index system, and the social capital they mentioned is actually the social network. Referring to the achievements of predecessors on the construction of social network relationship index systems, this study constructs an index system of social network relationships using five dimensions: network scale, network compactness, trust and commitment, social atmosphere and sense of belonging, and social participation.

Network scale refers to the scope and breadth of social network coverage; it reflects the range of resources that people can obtain information from social networks [44]. Network compactness refers to the frequency and quality of the mutual connection between individuals, and it reflects the dynamic degree of sharing material resources and knowledge resources [45]. Trust and commitment can be expressed as the degree of trust and trust in neighbors, relatives, the residents of the whole village, government departments, and civil servants, and it reflects the informal values and norms shared among the members [46]. It is worth noting that farmers only make general judgments about different relationships. Social ethos and a sense of belonging are the background of the village on which social network relations depend. Social participation refers to the members' investment in public affairs, reflecting the degree of members' participation in farmland affairs [47,48]. Social learning is influenced by these four aspects and could change an individual's behavior [49].

Previous studies generally divided the social network relationship variables into two target levels and added one of the indicators of social ethos and sense of belonging on the basis of which, in order to reflect people's emotional strength to the collective and confidence in the development of the village [39–41], as shown in Table 1.

Table 1. Evaluation System of farmers' social network relationship.

Level 1 Sub Target Layer	Level 2 Sub Target Layer
Network size	Number of general friends (A1); number of close friends (A2); number of relatives and friends with their own business (A3); number of relatives and friends working in government agencies (A4); number of relatives and friends working in public institutions (A5); and the number of relatives and friends living in cities (A6)
Network compactness	The number of recent visits with neighbors (B1); recent return visits with neighbors (B2); recent contact with relatives and friends (B3); number of relatives and friends visited on New Year's Day (B4); the number of red packets sent out and recovered on the annual festival (B5;, and the voluntary help between relatives and friends (B6)
Trust and commitment	Trust in neighbors (C1); trust in relatives (C2); trust in the whole village residents (C3); trust in government departments and civil servants (C4); trust in neighbors to keep keys (C5); and trust neighbors to take care of children (C6)
Social ethos and sense of belonging	The sense of security for village life (D1); the sense of honor for village life (D2); the identity of doing well without leaving a name (D3); the sense of responsibility to support natural disasters and man-made disasters (D4); whether the village has a common vision for a better life (D5); and whether the neighbors have mutual help (D6)
Social participation	Whether or not to participate in arts and entertainment organizations (E1); whether to join the party and League organizations (E2); whether to participate in the village committee (E3); whether to participate in rural credit cooperatives or various cooperatives and trade associations (E4); whether the villagers will collectively help the villagers in sudden difficulties (E5); and whether the village plans collective public affairs according to their actual needs (E6)

3. Results

3.1. Weight Determination and Evaluation

In this study, the coefficient of variation method is used to calculate the weight of the variable (index). According to each group of data obtained from the questionnaire survey, the evaluation value of the five variables in the first-level target layer is calculated from the weight of each variable in the secondary target layer and the actual survey data. The calculated weights are shown in Figure 2. According to the survey data and the calculated weight, the index value of each sample is calculated. There are 1169 samples, each sample has four calculated values, and there will be 4676 pieces of data to be presented. Because the output data will take up a lot of space, it will not be displayed here.

3.2. Logistic Regression Model

3.2.1. Variable Selection

Only the model was introduced in 2.2, and the dependent and independent variables should be identified now. In this paper, whether the farmers participate in the supply of irrigation and water conservancy facilities is selected as the dependent variable. The independent variables include the social network relationship variables and the control variables. Among them, the social network relationship variables include network scale, network compactness, trust and commitment, social ethos and sense of belonging, and social participation; the control variables are agricultural insurance (whether to obtain agricultural insurance), policy effect satisfaction (the government's support for farmers' participation), the number of sons, participation in farmers' cooperatives, water conservancy management experience, family decision-making power, and technology. Skills, health status, the number of the labor force, and cultivated land area. See Table 2 for details.

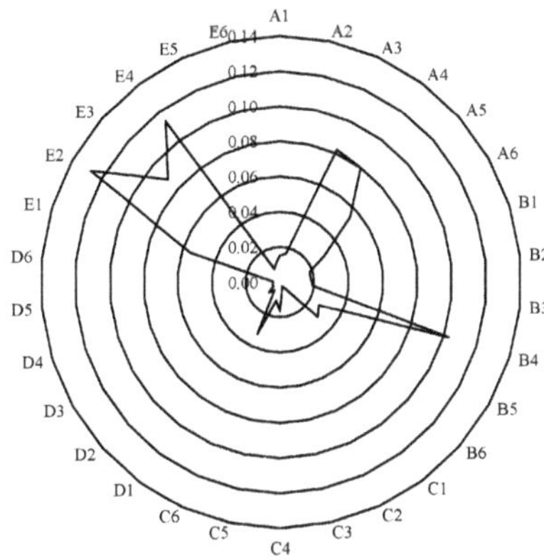

Figure 2. Index weight distribution of level 2 sub target layer.

Table 2. The related variables of driving factors of social network relationship.

			Variable Name	Variable Definition	Mean Value	Standard Deviation
Dependent Variable			**Participation or Not**	**Yes = 0, No = 1**	**0.27**	**0.45**
Independent variables	Social network		network size	The score is calculated by the secondary target level index	0.04	0.04
			network compactness	The score is calculated by the secondary target level index	0.07	0.04
			trust and commitment	The score is calculated by the secondary target level index	0.30	0.10
			social ethos and sense of belonging	The score is calculated by the secondary target level index	0.72	0.12
			social participation	The score is calculated by the secondary target level index	0.05	0.09
	control variable	Influence of insurance and policy	agricultural insurance	Purchase or not, yes = 0, no = 1	0.88	0.30
			satisfaction of policy effect	Is the policy of "one case, one discussion" effective yes = 1, general = 2, no = 3	2.90	0.36
		Influence of production conditions	farmers' cooperative	participation or not, yes = 0, no = 1	0.95	0.23
			water management experience	have water management experience or not, yes = 0, no = 1	0.99	0.12
			technical skills	use technical skills or, yes = 0, no = 1	0.83	0.38
			number of labor	population with working capacity (person)	1.98	0.78
			cultivated land area	total family cultivated land area (MU)	8.13	4.99
		other impacts	number of sons	number of sons owned (person)	1.34	0.72
			family decision making power	head of household, yes = 0, no = 1	0.45	0.50
			health condition	very poor = 1, poor = 2, general = 3, relatively healthy = 4, very healthy = 5	4.24	0.84

Note: household head, skills, health status, water conservancy manager, farmer cooperatives, and agricultural insurance all indicate the individual characteristics of the interviewees, the same below.

3.2.2. Regression Results

In this paper, the binary logistic model was analyzed using SPSS26.0 software(IBM, New York, USA). The regression results are shown in Table 3.

Table 3. The regression results of driving factors of social network relationship.

		Variable Name	B	SE	Wals	Sig	Exp(B)
Social network		network size X1	2.569 *	1.508	2.901	0.089	13.055
		network compactness X2	−15.954 ***	2.521	40.053	0.000	0.000
		trust and commitment X3	0.938	0.807	1.351	0.245	2.555
		social ethos and sense of belonging X4	−0.456	0.596	0.585	0.444	0.634
		social participation X5	−0.258	0.708	0.133	0.715	0.772
Control variable	Influence of insurance and policy	satisfaction of policy effect X6	−0.678 ***	0.154	19.453	0.000	0.508
		agricultural insurance X7	−0.510 **	0.199	6.540	0.011	0.601
	Influence of production conditions	farmers' cooperative X8	0.290	0.388	0.557	0.455	1.336
		water management experience X9	−1.147 **	0.499	5.283	0.022	0.318
		technical skills X10	−0.655 ***	0.172	14.529	0.000	0.519
		number of labor X11	0.339 ***	0.092	13.590	0.000	1.404
		cultivated land area X12	0.000	0.003	0.000	1.000	1.000
	Other impacts	number of sons X13	0.180 *	0.109	2.719	0.099	1.197
		health condition X14	−0.068	0.083	0.665	0.415	0.934
		family decision-making power X15	−0.303 *	0.158	3.649	0.056	0.739
		constant β_0	5.116 ***	1.494	11.730	0.001	166.680

Note: *, **, ***, respectively, represent the significance level of 10%, 5% and 1%.

4. Discussion

The scores of the five indicators of the social network relationships from high to low are social atmosphere and sense of belonging, trust and commitment, network compactness, social participation, and network scale. The average score for the social ethos and sense of belonging is the highest, which indicates that farmers have a relatively high sense of identity with the local culture due to the relatively fixed and consistent public moral concepts generated by Chinese civilization within the organization. Farmers are constrained by the basic rules of "acquaintance society" and follow the behavioral logic of "one's own people" [50], so they have high trust in themselves. Rural areas are relatively closed, and the communication between farmers and the outside world is relatively less. Due to the influence of the market economy concept on farmers, farmers regard the development of the family economy as the top priority, which may reduce the frequency of communication with others and the social participation activities that will not bring direct benefits to farmers.

It can be seen from Table 2 that most of the farmers are non-households because the farmers were selected randomly during the fieldwork; there is only one head for each family in China. Most of them have no water conservancy management experience and lack technical skills. The grass-roots water conservancy managers are mainly responsible for the management and maintenance of basic water conservancy facilities as well as the scheduling of water resources. In China, each village generally has 2–4 grass-roots water conservancy managers, and the farmers who are engaged in agricultural production are older, and most of them have no technical skills nowadays. The survey found that the health of farmers is generally good, each household has about two workers, and each family has about one son. The average total arable land of the family is 8.13 mu, which means that each labor force is responsible for farming and irrigating an area of about 4 mu. However, 97% of the farmers are not water managers, their active concern and participation in irrigation and water conservancy facilities supply will be relatively low, and according to the feedback of the peasant household questionnaire, the farmers are not satisfied with the effect of the "one case, one discussion" policy. In this way farmers cannot be encouraged to participate in the supply of irrigation and water conservancy facilities from the policy level. The policy of "one case, one discussion" means that the village must hold a villagers' assembly for a public matter of production, with the participation of more than two-thirds of the village's farmer representatives and the approval of more than half of the farmers, and all of the participants have the right to supervise. It is to further standardize the rural collective production public welfare enterprise financing management system. However, due to poor operability, financial opacity, and other reasons, the mean of Table 2 shows that the system did not give full play to its institutional advantages, resulting in villager dissatisfaction. Farmer cooperatives are conducive to establishing cooperative relations among farmers and coordinating their limited supply capacity of labor, irrigation and water conservancy facilities so as to achieve better supply effects. However, 94.4% of the farmers stated that they did not participate in farmers' cooperatives, which shows that the degree of agricultural production organization in China is still low. Agricultural cooperatives are a kind of spontaneous organization, lacking standardized organization and supervision systems. The investigation found that farmers did not receive the expected income when they participated in agricultural cooperatives. At the same time, 79.9% of the farmers did not participate in the agricultural insurance, which would increase the risk faced by farmers relating to agricultural production.

4.1. The Impact of Social Network on Farmers' Participation in Small Water-Saving Irrigation and Water Conservancy Supply

From the regression results in Table 3, it can be seen that the variable coefficient of the network scale is positive and passes the test at a significance level of 10%. Therefore, the larger the network scale, the weaker the willingness of farmers to participate in the supply of small water-saving irrigation and water conservancy facilities. Coleman also believed that network size would affect members' participation behavior, but he believed that the impact was positive [51]. With the expansion of the network scale, the scope and breadth of the social network coverage will also increase. The expansion of the social network has realized valuable communication and cooperation, which has great social value. In addition, people will receive more experience and knowledge from social learning, which affects their decision-making behavior [52]. For example, with the continuous development of communication equipment, farmers have access to more and more information sources. Their decision-making is not only based on their own experience but is also affected by other people's opinions, government policies, and other factors. Information communication enables farmers to obtain more alternative strategies. As rational-economic individuals, farmers are bound to make decisions that can maximize their own interests. Compared with the other incomes of farmers, agricultural investment returns are lower than other industries, and it may not be an optimal choice to participate in the construction of farmland facilities for farmers.

The influence of network compactness on the dependent variable is significant, and the coefficient was negative, indicating that the higher the network compactness, the more willing farmers are to participate in the construction of irrigation and water conservancy facilities. This is consistent with the conclusion of Hu Jinyan and Zhang Bo, who also believe that network compactness can influence people's behavior more than network size [53]. The network compactness is mainly reflected in the frequency of the connection between the farmers. Network compactness has an important impact on the diffusion of information, which in turn affects the decision-making behavior of the farmers [54]. When the network density is high, the interaction between the farmers and the surrounding villagers is more frequent. The effective transmission of information between the farmers is to enhance the social learning of farmers, which promotes the resonance of the same demand for rural public facilities and strengthens the demand. In this way, farmers are more likely to take cooperative actions and reach a consensus.

The factors of trust and commitment have no significant impact on farmers' participation in small water-saving irrigation and water conservancy supply, which is inconsistent with Ostrom's view [55]. It is also inconsistent with Dong Li's research. He believes that the higher the level of interpersonal trust of farmers, the more likely they are to form emotional and value identities with other farmers, and they are more willing to credit or act on the advice of other farmers and the possibility of reaching a cooperation contract between farmers is higher [56]. From the perspective of "the Pattern of Difference Sequence" in China, the trust between farmers is mainly based on blood kinship, and the scope of trust follows the "principle of proximity" in the process of expansion [57]. This pattern of differential order forms a differential contract relationship; that is, the closer the relationship, the stronger the contract. According to this theory, the stronger the contractual relationship is, the easier it is to reduce the transaction cost, which is conducive to the realization of collective action. However, irrigation and water conservancy facilities belong to quasi-public goods [58,59], which are non-exclusive and have a long payback period. From the perspective of input–output analysis, it is difficult for farmers to accurately evaluate their input and income because agricultural income has both market and natural risks. Therefore, even if the level of trust and commitment is high, farmers may not be able to determine exactly whether they are involved in the supply of water conservancy facilities.

Social ethos and sense of belonging had no significant impact on farmers' participation in small water-saving irrigation and water conservancy supply. Social ethos and sense of belonging belong to the category of social identity, which can affect the emotion and cohesion between the individuals in a group. However, the study finds that social identity does not have a significant impact on individual behavior, which may be due to group norms [60–62]. In the villages with good social ethos, the degree of organization is higher, and the degree of homogenization of the farmers is also higher. Most of these villages are located in the main grain-producing areas, and over long-term agricultural production, they have formed a system of irrigation and water conservancy facilities suitable for local irrigation. Perhaps farmers are reluctant to break away from this situation. In addition, due to the relatively closed rural environment, resulting in the farmers' cognitive limitations, the enthusiasm for the reform of the construction and management of irrigation and water conservancy facilities is insufficient.

Social participation has no significant impact on farmers' willingness to participate in small water-saving irrigation and water conservancy construction. This conclusion is inconsistent with the conclusion of Yan Tingwu. Yan Tingwu believes that farmers who have participated in the public affairs of the government or other organizations have a broader vision, thus showing a higher willingness to participate [63]. Social participation is the concern, understanding, and behavioral input of farmers to the current situation and activities of all aspects of social life [28]. According to the theory of social identity, social participation can make farmers obtain the identity of being a member of a group [64], and identity will increase the supply of more social resources [65]. Social participation in rural areas mainly includes election, training, entertainment activities, village affairs, etc.; these activities mainly reflect the political rights of farmers, the improvement of skills, or the inheritance of culture. In other words, the use of social rights by farmers does not affect their participation in the supply of irrigation and water conservancy facilities, which may be because most of the participation behaviors of farmers are formed by the spontaneous organization of farmers and have nothing to do with social rights. Although China implements the system of "one case, one discussion", the construction of irrigation and water conservancy facilities involves many subjects, and the interests of each subject are difficult to balance, the effect of the system is not obvious. This can be reflected in the mean value of the satisfaction with the policy effect (Table 2).

4.2. The Impact of Other Factors on Farmers' Participation in Small Water-Saving Irrigation and Water Conservancy Supply Behavior

The variable coefficient of agricultural insurance is negative, and a 5% significance level shows that the farmers without agricultural insurance are more willing to participate in the supply of farmland and water conservancy facilities. Farmers' cooperatives fail to pass the significance test, which has little impact on farmers' participation behavior, probably because local cooperatives fail to provide technical guidance to farmers, resulting in their low degree of participation and the low degree of farmer organization. The coefficient of the water management experience variable is negative, passing a significance test of 10%, which means that the farmers without water management experience are more willing to participate in the supply. This is consistent with the conclusion of Wang, B. and Wang, H., who concluded that water conservancy management experience would affect farmers' participation behaviors [66]. The variable coefficient of technical skills is also negative and passes the significance test; that is, when farmers have relevant skills, their willingness to supply water conservancy facilities will decrease. The number of the labor force passed the significance test at the level of 10%, which means that with an increased labor force, farmers' willingness to participate tends towards "non-participate". The cultivated land area has no significant impact on farmers' participation in the supply of irrigation and water conservancy facilities, while Wang, B. and Zhu, Y.C. believe that the participation behavior of farmers is affected by the area of irrigated farmland [67]. The cultivated land area includes irrigated areas and non-irrigated areas. If there is a large difference in the composition of the cultivated land area, it may be the reason for this phenomenon. The variable coefficient of the independent variable number of sons is positive, which means that with an increased number of sons, the willingness of farmers to participate is weakened. In China, the more sons there are in a family, the heavier the burden on the family will be. Farmers often take the income of migrant workers as the main income of the family, and they will not pay more attention to agricultural production, which weakens the enthusiasm for the supply of agricultural production facilities. The head of the household (through a 10% significance test) has a significant negative effect on the dependent variables. The head of the household plays an important role in household decision-making and has a broader perspective than the other family members, and the more likely they are to believe that non-agricultural production is likely to produce more benefits, the easier it is for them to "not participate". The possible reason is that with the popularization of machine technology, the manual labor of farmers is greatly reduced, and the health status of farmers is no longer the main obstacle for them to engage in agricultural activities, so it has a weak impact on farmer participation in the supply of water conservancy facilities. This is not consistent with the research of Yang, X.Y and Zhu, Y.C. [68]. The survey areas they chose were underdeveloped, with a low degree of mechanization, and agricultural production was strongly dependent on the health status of the farmers.

5. Conclusions and Suggestions

5.1. Conclusions

In this paper, 1169 pieces of data from a field survey are used to analyze the influence of social network relationships on farmer participation in the supply of water conservancy facilities. The results show that:

Firstly, social network relationships will affect the farmers' behavior. This is consistent with the research conclusions of Cui, B.Y., Zhang, Z.G., and other scholars [56,69]. However, different scholars have different conclusions about the impact of each dimension variable on farmer behavior under the social network. The possible reason is that different scholars choose farmers who provide different public goods or participate in different public affairs, such as environmental protection investment behavior, irrigation system management, and rural public goods [50,67,69]. In the social network relationship, network size and network compactness have a significant impact on farmer participation in farmland and water conservancy facilities, while trust and commitment, social ethos and sense of belonging, and social participation have no significant impact on farmer participation in irrigation and water conservancy facilities.

Secondly, the expansion of the network scale will decrease the possibility of farmer participation in the supply of irrigation and water conservancy facilities.

Thirdly, the higher the network density of the farmers is, the more likely they are to participate in the supply of water conservancy facilities.

Fourthly, good policy effects, having agricultural insurance, water conservancy management experience, strong family decision-making power, technical skills, increase in the labor force, and the number of sons will inhibit the positive participation of farmers in the supply of water conservancy facilities. However, farmer cooperatives, health status, and cultivated land area have no significant influence on the dependent variables.

In general, social network variables affect farmer participation in the supply of irrigation and water conservancy facilities, but different dimensions have different impacts on the individual behaviors of the farmers.

5.2. Suggestions

According to the above conclusions, this paper puts forward the following suggestions:

Firstly, government departments should strengthen publicity and carry out more collective activities and, at the same time, build a good platform to strengthen the communication among farmers and the contact between villagers and village cadres. In this way, farmers' demand for rural public goods can be further manifested and provide guarantees for realizing the collective supply of the public goods needed by farmers.

Secondly, the government should encourage the dissemination of positive information to farmers and avoid the distortion caused by negative information or false information relating to farmers' behavior, and provide a good humanistic foundation for rural governance.

Thirdly, in public governance, it is necessary to conduct regional governance. The relevant incentive policies of the state should be more in line with the needs of farmers to improve their living standards and pay attention to the timely feedback of policy effects..

Fourthly, building industrial chains for processing and selling agricultural products can provide channels for the sale of agricultural products, improve the efficiency of agricultural production, increase farmers' agricultural income, enhance farmers' confidence in agricultural production, and finally, provide good conditions for agricultural development. In addition, due to the fragility of agricultural production, the government should support agricultural insurance to provide farmers with security in relation to agricultural production.

6. Research Prospect

Due to objective reasons such as length and subjective factors such as the author's research ability, there are still some valuable research issues that have not been discussed in depth in this article, including the following:

First, the data used in this study are only limited to northwest China, mainly studying arid and semi-arid areas with water shortage, which has certain limitations in terms of regional characteristics. However, it is necessary to study the behaviors of farmers' participation in the supply of farmland water conservancy facilities in regions with different regional characteristics and agricultural development conditions. Therefore, surveys in other parts of China can be conducted in the future to obtain national data for further analysis.

Second, the marginal effect of regression analysis is not taken into account in the empirical link, so the explanation is not straightforward enough. The marginal effect can further analyze the influence of the independent variables on the dependent variables. It can show how much more likely farmers are to implement water-saving behavior when the variable changes by one unit.

Third, this paper does not discuss the mediating effect of "social learning" on social network relationships and the individual behaviors of farmers. In the future, we can further measure the social relationships with individual farmers as the core and measure the quality of the social network relationships.

Author Contributions: Conceptualization, L.W.; methodology, L.W.; software, L.W.; validation, L.W.; formal analysis, L.W.; investigation, L.W.; resources, L.W.; data curation, L.W.; writing—original draft preparation, Y.Z.; writing—review and editing, Y.Z.; visualization, L.W.; supervision, L.W.; project administration, L.W.; funding acquisition, L.W. All authors have read and agreed to the published version of the manuscript.

Funding: This research was funded by National Natural Science Foundation of China (41501126); Changzhou Soft Science Project (522012712); Changzhou Soft Science Project (520045112).

Institutional Review Board Statement: Not applicable.

Informed Consent Statement: Informed consent was obtained from all subjects involved in the study. Informed consent was obtained from all subjects involved in the study.

Data Availability Statement: The data of this paper were obtained through the author's field research.

Acknowledgments: All individuals included in this section have consented to the acknowledgement.

Conflicts of Interest: The authors declare no conflict of interest.

Appendix A

Table A1. Statistical range and sample size of the questionnaire.

<table>
<tr><th colspan="2" rowspan="2">Statistical Area</th><th colspan="3">Number of Counties, Towns and Villages in The Statistical Area</th><th colspan="2">Number of Statistical Samples/Household</th></tr>
<tr><th>County</th><th>Township</th><th>Village</th><th>By Prefecture Level City</th><th>By Region</th></tr>
<tr><td rowspan="3">Hetao Area(Ningxia)</td><td>Yinchuan City</td><td>1</td><td>4</td><td>20</td><td>160</td><td rowspan="3">480</td></tr>
<tr><td>Shizuishan City</td><td>1</td><td>4</td><td>20</td><td>160</td></tr>
<tr><td>Wuzhong City</td><td>1</td><td>4</td><td>20</td><td>160</td></tr>
<tr><td rowspan="3">Guanzhong Area (Shaanxi Province)</td><td>Baoji City</td><td>1</td><td>4</td><td>20</td><td>160</td><td rowspan="3">480</td></tr>
<tr><td>Xianyang City</td><td>1</td><td>4</td><td>20</td><td>160</td></tr>
<tr><td>Weinan City</td><td>1</td><td>4</td><td>20</td><td>160</td></tr>
<tr><td rowspan="2">the Central Plains (Henan Province)</td><td>Kaifeng City</td><td>2</td><td>8</td><td>40</td><td>320</td><td rowspan="2">480</td></tr>
<tr><td>Xinxiang City</td><td>1</td><td>4</td><td>20</td><td>160</td></tr>
<tr><td colspan="2">total</td><td>9</td><td>36</td><td>180</td><td colspan="2">1440</td></tr>
</table>

Note: Kaifeng city is mainly divided into grain-crop region and cash-crop region, with obvious regional distribution. In order to understand the situation of different regions, two different types of counties were selected as samples.

References

1. China Water Resources Bulletin 2021. *Water Resour. Dev. Manag.* **2022**, *8*, 85.
2. Li, J.P. An Empirical Study of the Willingness to Collective Supply of Water-saving Irrigation Facilities: Base on Social Capital Theory. *J. Lanzhou Commer. Coll.* **2014**, *30*, 118–126.
3. Zuo, Z.Y. Farmers' willingness to pay for modern water-saving irrigation technology in groundwater overextraction area of North China: Based on the conditional value investigation of Ningjin County, Dezhou City, Shandong Province. *J. Agrotech. Econ.* **2016**, *6*, 32–46.
4. Mabry, J. Cananls and Communities. In *Small water-saving irrigation Systems;* University of Arizona Press: Tucson, AZ, USA, 1996.
5. Sinyolo, S.; Mudhara, M.; Wale, E. The Impact of Smallholder Irrigation on Household Welfare: The Case of Tugela Ferry Irrigation Scheme in KwaZulu—Natal, South Africa. *Water SA* **2014**, *40*, 145–156. [CrossRef]
6. Cai, Q.; Zhu, Y.; Chen, Q. Can Social Networks Increase Households' Contribution to Public-Good Provision in Rural China? The Case of Small Hydraulic Facilities Construction. *China Agric. Econ. Rev.* **2016**, *20*, 148–169. [CrossRef]
7. Lam, W.F.; Ostrom, E. Analyzing the Dynamic Complexity of Development Interventions: Lessons from an Irrigation Experiment in Nepal. *Policy Sci.* **2010**, *43*, 1–25. [CrossRef]

8. Wang, L.; Du, D. Analysis on the Development Process and Current Situation of Irrigation and Water Conservancy Facilities Supply in China. *China Water Conserv.* **2017**, *11*, 14–17.
9. Hin, L.M.; Han, Y.Q. Social Capital, Peasant Household Differentiation and Village Collective Action. *Resour. Dev. Mark.* **2019**, *4*, 527–532.
10. Olson, M. *The Logic of Collective Action: Public Goods and the Theory of Groups*; Harvard University Press: Cambridge, MA, USA, 1971.
11. Qiu, L. The Mechanism of Clan Network in Rural Public Goods Supply—Based on the Investigation of Shawan Village in Jiangxi Province. *J. Hunan Agric. Univ. (Soc. Sci.)* **2020**, *21*, 44–48.
12. Qi, X.L.; Wu, J.Q. A Study on the Governance of Large Villages under the Background of Clans—A Case study of Huangcun village in West Guangdong. *Agric. Technol. Econ.* **2019**, *2*, 70–77.
13. Sabucedo, J.M.; Dono, M.; Seoane, G. The Importance of Protesters' Moral: Moral Obligation as a Key Variable to Understand Collective Action. *Front. Psychol.* **2018**, *3*, 418. [CrossRef]
14. Granovetter, M. Economic Action and Social Structure: The Problem of Embeddedness. *Am. J. Sociol.* **1985**, *91*, 481–510. [CrossRef]
15. Buchanan, J. *On Democratic Finance*; Commercial Press: Beijing, China, 1993.
16. Paustian, K. Agriculture, Farmers and GHG Mitigation: A New Social Network? *Carbon Manag.* **2012**, *3*, 253–257. [CrossRef]
17. Uzzi, B. Social Structure and Competition in Interfirm Networks: The Paradox of Embeddedness. *Adm. Sci. Q.* **1997**, *42*, 35–67. [CrossRef]
18. Morse, E.A.; Fowler, S.W.; Lawrence, T.B. The Impact of Virtual Embeddedness on New Venture Survival: Overcoming the Liabilities of Newness. *Entrep. Theory Pract.* **2007**, *6*, 139–159. [CrossRef]
19. Wu, L.Y.; Wang, C.J.; Chen, C.P.; Pan, L.Y. Internal Resources. External Network and Competitiveness during the Growth Stage: A Study Taiwanese High-tech Ventures. *Entrep. Theory Pract.* **2008**, *5*, 529–549. [CrossRef]
20. Granovetter, M. The Strength of Weak Ties. *Am. J. Sociol.* **1973**, *78*, 1360–1380. [CrossRef]
21. Krishna, A.; Uphoff, N. *Mapping and Measuring Social Capital through Assessment of Collective Action to Conserve and Develop Watersheds in Rajasthan. India*; Cambridge University Press: Cambridge, UK, 2003.
22. Isham, J.; Kähkönen, S. How do Participation and Social Capital Affect Community-based Water Projects? Evidence from Central Java, Indonesia. In *The Role of Social Capital in Development: An Empirical Assessment*; Cambridge University Press: Cambridge, UK, 2002.
23. Durlauf, S.; Fafchamps, M. Social Capital. In *Handbook of Economic Growth*; Aghion, P., Durlauf, S., Eds.; Elsevier: Amsterdam, The Netherlands, 2004.
24. Munshi, K.; Rosenzweig, M. *Why is Mobility in India So Low? Social Insurance, Inequality and Growth*; NBER Working Paper; Brown University: Providence, RI, USA, 2009.
25. Manuela, A.; Giacomo, D.G.; Marcos, A.R.; Imran, R. *Insurance in the Extended Family*; Working Paper; University of Arizona: Tucson, AZ, USA, 2008.
26. Fafchamps, M. Development and social capital. *J. Dev. Stud.* **2006**, *42*, 1180–1198. [CrossRef]
27. Ostrom, E. Collective action and the Evolution of Social Norms. *J. Econ. Perspect.* **2000**, *14*, 137–158. [CrossRef]
28. Miao, S.S. Research on Farmers' Cooperative Participation Behavior of Small Water Conservancy Facilities from the Perspective of Multi-dimensional Heterogeneity of Social Capital. *China Popul. Resour. Environ.* **2014**, *24*, 46–54.
29. Coleman, J. *Rational Action Theory*; Social Sciences Academic Press: Beijing, China, 1990.
30. Alejandro, P. Social Capital: Its Origin and Application in Modern Sociology. In *Yang Xuedong's Translation of Social Capital and Social Development*; Social Sciences Academic Press: Beijing, China, 1995.
31. Mao, S.L. The Way of Public Affairs Managenent. *J. Jiangsu Adm. Institute* **2010**, *1*, 100–105.
32. Wang, L. 'Hollowing out', Population, Rural Social Capital and Rural Public Goods Supply: A Litenature Review. *J. Spec. Zone Econ.* **2013**, *11*, 132–133.
33. Putnam, R.D. Making Democracy Work. In *Civic Traditions in Modern Italy*; Princeton University Press: Princeton, NJ, USA, 1993.
34. Xu, L.; Luo, D.L.; Liu, A.J. Effects of social capital on farmers' willingness to participate in irrigation management reform. *Resour. Sci.* **2015**, *37*, 1287–1294.
35. Cai, Q.H.; Zhu, Y.C. Influences of Social Capital and Income Inequality on Village Collective Action—Based on Farmers' Participation in the Maintenance of Small Irrigation Systems in Three Provinces. *J. Public Manag.* **2016**, *13*, 89–100+157.
36. Liu, Q.; Zhu, Y.C. Research on the Impact of Social Capital on Farmers' Participation in Small water-saving irrigation and Water Conservancy Supply Behavior. *Agric. Technol. Econ.* **2015**, *12*, 32–41.
37. Wang, J.C.; Guo, Z.G. *Logistic Regression Model: Method and Application*; Higher Education Press: Beijing, China, 2001.
38. Zhang, W.T.; Dong, W. *Advanced Course of SPSS Statistical Analysis*; Higher Education Press: Beijing, China, 2013.
39. Lubin, J.H. An empirical evaluation of the use of condition data. *Biometrika* **1981**, *68*. [CrossRef]
40. Shen, H.B.; Xu, Y. Applicable Conditions and Limitations of Logistic Regression Model. *Chin. Public Health* **1991**, *3*, 127–129.
41. Li, Z. Research on the Influence of Social Network Embedding on entrepreneurial Performance of College Students. *Educ. Dev. Res.* **2017**, *21*, 24–30.
42. Mitchell, J. *The Concept and Use of Social Networks*; Manchester University Press: Manchester, UK, 1969.
43. Lin, N. *Social Capital*; Shanghai People's Publishing House: Shang Hai, China, 2004.

44. Ahuja, G. Collaboration Networks, Structural Holes, and Innovation: A Longitudinal Study. *Adm. Sci. Q.* **2000**, *45*, 425–455. [CrossRef]
45. Xi, L.; Peng, C.; Yang, H. The Impact of Resource Patching on the Synergy of Dual Innovation: The Regulatory Role of Environmental Dynamics. *Tech. Econ.* **2017**, *36*, 1–5.
46. Fukuyama, F. *Trust: The Scocial Vitues and Creation of Prosperity*; Free Press: New York, NY, USA, 1995.
47. Herman, A. *Transfer Participation: Individual Interests and Collective Action*; Shanghai People Press: Shanghai, China, 2008.
48. Gorton, M.; Sauer, J.; Peshevski, M. The Dimensions of Social Capital and Rural Development: Evidence from Water Communities in the Republic of Macedonia. *Eur. Assoc. Agric. Econ.* **2010**. [CrossRef]
49. Argyris, C.; Schon, D.A. Organizational Learning: A Theory of Action Perspective. *Addison* **1978**, *12*, 103–108. [CrossRef]
50. Chen, B.F. Acquaintance Society: An Ideal Study of village Order Mechanism. *Society* **2011**, *1*, 223–241.
51. Coleman, J.S. Social Capital in the Creation of Human Capital. *Am. J. Sociol.* **1998**, *94*, 95–120. [CrossRef]
52. Song, X.F.; Jiang, W.; Liu, X.H.; Lu, H.; Tian, Z.H.; Du, X.J. A Survey of Game Theory as Applied to Social Networks. *Tsinghua Sci. Technol.* **2020**, *25*, 734–742. [CrossRef]
53. Hu, J.Y.; Zhang, B. Social Network, Private Financing and Family Entrepreneurship: An Empirical Analysis Based on the Differences between Urban and Rural areas in China. *Financ. Res.* **2014**, *10*, 148–163.
54. Jiang, C.X.; Chen, Y.; Liu, K.J.R. Evolutionary Social Information Diffusion Analysis. In Proceedings of the Global Communications Conference, Austin, TX, USA, 8–12 December 2014; pp. 2911–2916.
55. Ostrom, E. *The Meaning of Social Capital and Its Link to Collective Action*; Social Science Electronic Publishing: London, UK, 2008.
56. Dong, L.; Li, H.; Yu, Y.L.; Han, Y.X. Study on the Influence of Multi-Dimensional Social Capital on the Management of Domestic Waste Provided by Farmers in Poverty-Stricken Areas. *Ecol. Econ.* **2020**, *9*, 198–203.
57. Fei, X.T. *Fertility System in Rural China*; Peking University Press: Beijing, China, 1998.
58. Buchanan, J.M. An Economic Theory of Clubs. *Economic* **1965**, *32*, 1–14. [CrossRef]
59. Samuelson, P.A. A Note on the Pure Theory of Consumer's Behaviour. *Economic* **1938**, *17*, 61–71. [CrossRef]
60. Bruner, M.W.; Boardley, I.D.; Coté, J. Social Identity and Prosocial and Antisocial Behavior in Youth Sport. *Psychol. Sport Exerc.* **2014**, *15*, 56–64. [CrossRef]
61. Bruner, M.W.; Boardley, I.D.; Allan, V.; Forrest, C.; Root, Z.; Côté, J. Understanding Social Identity and Intrateam Moral Behavior in Competitive Youth Ice Hockey: A Narrative Perspective. *Sport Psychol.* **2017**, *31*, 173–186. [CrossRef]
62. Bruner, M.W.; Boardley, I.D.; Benson, A.J.; Wilson, K.S.; Root, Z.; Turnnidge, J.; Sutcliffe, J.; Côté, J. Disentangling the Relations between Social Identity and Prosocial and Antisocial Behavior in Competitive Youth Sport. *J. Youth Adolesc.* **2018**, *47*, 1113–1127. [CrossRef] [PubMed]
63. Yan, T.W.; He, K.; Zhang, J.B. Analysis of the Impact of Social Capital on Farmers' Willingness to Invest in Environmental Protection—An Empirical Study on the Recycling of Agricultural Waste in Hubei Province. *China Popul. Resour. Environ.* **2016**, *1*, 158–164.
64. Haslam, S.A.; Jetten, J.; Postmes, T.; Haslam, C. Social Identity, Health and Well-Being: An Emerging Agenda for Applied Psychology. *Appl. Psychol.* **2009**, *58*, 1–23. [CrossRef]
65. Jetten, J.; Haslam, C.; Haslam, S.A.; Dingle, G.; Jones, J.M. How Groups Affect Our Health and Well-Being: The Path from Theory to Policy. *Soc. Issues Policy Rev.* **2014**, *8*, 103–130. [CrossRef]
66. Wang, B.; Wang, H.; Zhu, Y.C. The Impact of Institutional Capacity on the Governance Performance of Farmland Irrigation System: Based on the Survey Data of Six Provinces in the Yellow River Irrigation Area. *China Popul. Resour. Environ.* **2019**, *8*, 122–129.
67. Wang, B.; Zhu, Y.C. Labor Outflow and Farmers' Participation in Village Collective Action Selection—Taking Farmers' Participation in the Supply of Small Irrigation and Water Conservancy Facilities as an Example. *Resour. Environ. Arid. Areas* **2018**, *12*, 49–54.
68. Yang, X.Y.; Zhu, Y.C. Impact of Grass-roots Organizations and Social Networks on Water Management and Protection of Small-scale Farmers. *China Rural. Water Conserv. Hydropower* **2020**, *4*, 97–101.
69. Cui, B.Y.; Zhang, Z.G. Analysis on the Influencing Factors of Rural Public Goods Supply Behavior—Based on the Theoretical Analysis Framework of Embedded Social Structure. *J. Nanjing Agric. Univ. (Soc. Sci. Ed.)* **2009**, *1*, 25–31.

agriculture

MDPI

Article

Risk Assessment of Heavy Metals Contamination in Soil and Two Rice (*Oryza sativa* L.) Varieties Irrigated with Paper Mill Effluent

Arwa A. AL-Huqail [1], Pankaj Kumar [2,*], Ebrahem M. Eid [3,4], Bashir Adelodun [5,6], Sami Abou Fayssal [7,8], Jogendra Singh [2], Ashish Kumar Arya [9], Madhumita Goala [10], Vinod Kumar [2] and Ivan Širić [11,*]

1 Department of Biology, College of Science, Princess Nourah bint Abdulrahman University, P.O. Box 84428, Riyadh 11671, Saudi Arabia
2 Agro-ecology and Pollution Research Laboratory, Department of Zoology and Environmental Science, Gurukula Kangri (Deemed to Be University), Haridwar 249404, Uttarakhand, India
3 Biology Department, College of Science, King Khalid University, Abha 61321, Saudi Arabia
4 Botany Department, Faculty of Science, Kafrelsheikh University, Kafr El-Sheikh 33516, Egypt
5 Department of Agricultural and Biosystems Engineering, University of Ilorin, Ilorin PMB 1515, Nigeria
6 Department of Agricultural Civil Engineering, Kyungpook National University, Daegu 41566, Korea
7 Department of Agronomy, Faculty of Agronomy, University of Forestry, 10 Kliment Ohridski Blvd, 1797 Sofia, Bulgaria
8 Department of Plant Production, Faculty of Agriculture, Lebanese University, Beirut 1302, Lebanon
9 Department of Environmental Science, Graphic Era (Deemed to be University), Dehradun 248002, Uttarakhand, India
10 Nehru College, Pailapool, Affiliated Assam University, Silchar 788098, Assam, India
11 University of Zagreb, Faculty of Agriculture, Svetosimunska 25, 10000 Zagreb, Croatia
* Correspondence: rs.pankajkumar@gkv.ac.in (P.K.); isiric@agr.hr (I.Š.)

Abstract: Heavy metal pollution from industrial wastewaters has become an issue of global concern. These wastewaters are frequently used for inland irrigation which possess a serious risk of heavy metal contamination of both soil and cultivated crops. The problem is more common in developing countries like India where industrial wastewaters are often discharged without appropriate treatments. Therefore, this study aimed at assessing the impact of paper mill effluent for irrigation on the growth, productivity, and heavy metal accumulation potential of two rice (*Oryza sativa* L.) varieties (PB-1121 and PR-121). Water, soil, and rice crop samples were obtained from the vicinity of Saharanpur city, Uttar Pradesh, India, and subsequently analyzed for selected physicochemical and heavy metal parameters. Results showed that paper mill effluent and nearby Kali River water had significant ($p < 0.05$) loads of pollutants that impacted the soil properties. Moreover, the maximum plant height (123.48 ± 4.86 and 98.83 ± 2.02 cm); total chlorophyll (6.70 ± 0.25 and 6.64 ± 0.17 mg/g); leaf carotenoids (0.67 ± 0.08 and 0.63 ± 0.05 mg/g); starch content (71.08 ± 2.05 and 72.60 ± 1.63%); amylose content (25.10 ± 1.32 and 20.28 ± 1.24%); crop yield (4270.20 ± 75.12 and 5830.58 ± 63.10 kg/ha); and straw yield (5472.05 ± 93.90 and 6683.76 ± 61.26 kg/ha) of PB-1121 and PR-121, respectively, were observed using Kali River water irrigation source as compared to paper mill effluent and borewell water. The order of heavy metal accumulation in rice straw and grain followed: Fe > Mn > Zn > Cu > Cr > Cd. Similarly, the bioaccumulation factors for heavy metals in rice straw and grain were observed as >1 and <1, respectively. Moreover, the target hazard quotient (THQ) showed that there was no potential health risk of heavy metal in rice irrigated with contaminated water supply except for the PR-121 variety irrigated with Kali River water where the health risk index (HRI) value exceeded the limit (>1). As a result, the findings of this study provide useful information about the current state of heavy metal pollution and the health risks associated with rice crops irrigated with contaminated water sources.

Keywords: crop contamination; health risks; heavy metals; industrial pollution; wastewater irrigation

1. Introduction

Water insecurity and pollution of limited available water resources are of the major challenges of the twenty-first century. According to the report of the United Nations [1], around 80% of wastewater is discharged into the environment without any pre-treatment. Among these wastewaters, industrial ones contain dangerous heavy metals, nitrates, nitrites, organic carbon, and organic and inorganic phosphorus of high concentration levels threatening humans and aquatic lives [2]. A direct result of such wastewaters' discharge into water sources is the eutrophication and the constitution of a toxin-producing waterborne pathogens' platform [3]. The impact study of such wastewaters' reuse in crop irrigation is usually investigated but lacks reliability in most cases [4]. It was reported that the type of wastewater and the sensitivity of plant species regulate the negative impact of industrial wastewater on cultivated crops [5]. Although these authors outlined that industrial wastewater did not affect seed germination of lettuce (*Lactuca sativa*) cucumber (*Cucumis sativus*), and muskmelon (*Cucumis melo*), it decreased the root and shoot length of *C. sativus*. However, the application of industrial wastewaters in common wheat (*Triticum aestivum*) resulted in reductions in seed germination, seedling fresh and dry weights, vigor and tolerance indices, plant height, number of leaves, root fresh and dry weights, shoot fresh and dry weights [6]. The pesticide residues, oils as well as heavy metals found in industrial wastewater are considered threatening contaminants that impart disorders in soil microbiota functioning and deleterious effects on crop productivity, crop composition, and thus pose as potential human health risks [7].

On the other hand, soil chemical composition may contribute to heavy metals' leaching and their increased mobility when soil pH decreases [8]. In this context, positive correlations were found between soil pH and amounts of manganese (Mn), cadmium (Cd), and lead (Pb) with Cd and Pb being the most highly mobile heavy metals in soil [9]. Besides their increased mobility and leaching potential in low pH soil, heavy metals are high competitors to adsorption and subsequent uptake by plants [10], which inhibits other essential metals' absorption by the latter. Regarding soil microbiota, it was reported that chromium (Cr) significantly limits soil bacteria and *Actinomycetes* populations [11]. These microbiotas have shown a decrease in their biological activity in acidic soils, translated by the inhibition of free nitrogen fixation and therefore a decrease in arable land fertility [12]. The strong correlation between soil organic matter and available nitrogen on one hand and richness and diversity in soil microbiota on the other hand further explains how heavy metals directly affect the latter [13]. Furthermore, two Gram-positive isolates found naturally in soil (*Paenibacillus* sp. and *Bacillus thuringeinsis*) have shown some tolerance for increased Cd and zinc (Zn) concentrations for the former and copper (Cu) concentration for the latter [14]. Although these findings are promising, most soil microbiotas including the two isolates still suffer from increased heavy metal concentrations as being highly sensitive, pushing them to adopt defensive mechanisms to survive [15]. Moreover, a human intervention consisting of the raise of soil pH can contribute to the suppression of Cd and Cu accumulation in wheat grains [16]. However, once soil microbiotas are affected, crop yield, composition, and quality are consequently altered. Lead-contaminated wheat fields showed reduced chlorophyll *a* and *b* and carotenoid contents besides increased human health risk. Potassium, phosphorus, and protein contents were also decreased by 55, 60, and 81%, respectively. The number of seeds per plant, seed weight per plant, and harvest index were reduced by 90, 88, and 61%, respectively [6]. Another study mentioned that high Cd concentrations in *L. sativa*, *Pichia* spp. and *Azotobacter chroococcum* soils hindered roots' growth and development, reduced yield, and inhibited seeds germination [17].

Rice (*Oryza sativa* L.) crop is considered the staple food for over half of the global population. According to a recent report, around 510 million metric tons of rice are produced annually in the world [18]. China and India are the leading producers and consumers (around 50% of the world's production) and their traditional cuisine relies on this nutritious crop. By the end of 2022, around 515 million metric tons of rice are expected to be produced with 25.04% in India alone (https://worldpopulationreview.com/

country-rankings/rice-production-by-country (accessed on 20 September 2022)). Rice is a rich source of antioxidants, carbohydrates, minerals, proteins, and vitamins essential for the human diet and health [19]. However, these outstanding properties risk being seriously affected if rice fields are subjected to heavy metal contamination. Rice grains tend to actively bioaccumulate heavy metals originating from contaminated soils; therefore, soil properties are considered the most influencing factor [16]. Rice grown in the vicinity of Turag Riverbank (Bangladesh), in which large disposal of industrial effluents occur, was reported to be highly infested leading to low fertility and yield. High Cd, Pb, and nickel (Ni) concentrations were observed in both water and soil being above the safe limits. In brief, these industrial effluents resulted in dangerous soil degradation (decrease in soil quality by 100%), making these fields inappropriate for subsequent cultivation [20]. Moreover, dermal diseases were detected among nearby locals which poses a serious human health risk.

Star Paper Mill is a paper manufacturing mill located in Saharanpur District, Uttar Pradesh. Its production is estimated at 2500 tons per day [21]. The factory disposes of its effluent through a long canal into the Kali River. These effluents are subsequently and directly used by farmers to irrigate PB-1121 and PR-121 hybrid rice varieties. These varieties are the major and most prestigious species cultivated in India owing to their high productivity and superior quality attributes. Thus, the current study aims to investigate the risk associated with potentially toxic heavy metals accumulation in soil and two rice varieties (PB-1121 and PR-121) irrigated with paper mill effluent.

2. Materials and Methods

2.1. Description of Study Area

The present study was carried out in the suburban vicinity of Saharanpur city, Uttar Pradesh, India. The study site was in north-west of the Saharanpur city with several residential and agricultural establishments (29°56′20.20″ N and 77°34′17.10″ E). Herein, rice (*Oryza sativa* L.), wheat (*Triticum aestivum* L.), and sugarcane (*Saccharum officinarum* L.) are the most grown crops. In this region, Star Paper Mill Ltd. is the biggest industrial unit of the district with estimated paper production of more than 2500 tons per day [21]. This paper mill generates a huge load of effluent, which is released into the nearby river, namely Kali. This mill discharges its water in a 2 m wide and 3 km long drain, which passes through agricultural fields and connects to the Kali River. Several farmers of nearby fields directly utilize the effluent from the drain to irrigate their lands. Moreover, several farmers also exploit the water from the Kali River to irrigate their crops. Figures 1 and 2a,b show the location of Star Paper Mill Ltd. and selected sampling sites. In this study, a total of nine sampling sites were taken, including three borewell irrigated sites as control, three sites irrigated with absolute paper mill effluent, and three sites irrigated with Kali River water (after the disposal of paper mill effluent).

2.2. Description of Studied Rice Varieties

The two selected rice varieties viz., Pusa Basmati (PB) 1121 and PR-121 are commonly grown in North India due to their specific marketable and profitable features. PB-1121 is a popular rice variety developed with help of Basmati 370 and Type 3 varieties. With exceptional elongation, expansion, and aromatic features, this variety best suits the soils of the Saharanpur district thereby making it highly marketable and preferred by farmers. PB-1121 achieves a plant height of 120–130 cm, matures in 130–150 days, and gives an average yield of 5500 kg/ha [22]. Additionally, this variety is highly preferred for creating more resistant and beneficial crop seeds. On the other hand, PR-121 is a semi-dwarf and high-yielding rice variety developed by Punjab Agricultural University, Ludhiana, India. PR-121 achieves an average plant height of 98 cm, has long slender translucent grains, and gives a yield of 7601 kg/ha in 140 days [23]. PR-121 is resistant to nearly 10 pathogenic diseases including bacterial blight. These distinct features make this variety popular in the Saharanpur district. Therefore, the current study was carried out to assess the impact of paper mill effluent for irrigation on soil and selected rice varieties (PB-1121 and PR-121) grown in the semi-urban vicinity of Saharanpur city.

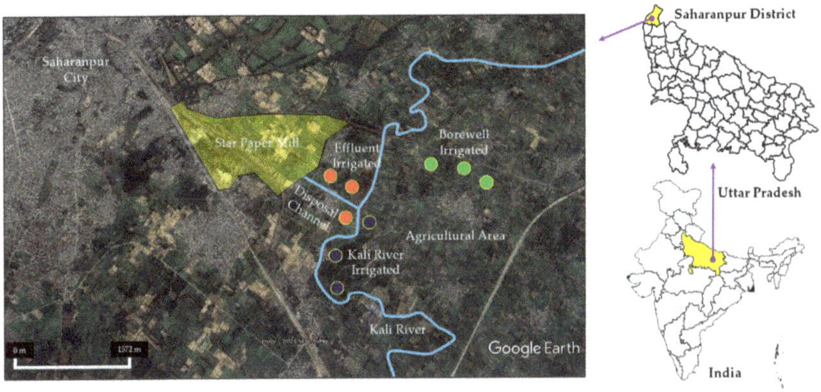

Figure 1. Map view of the Star Paper Mill situated in Saharanpur district of Uttar Pradesh, India, and its adjoining agricultural areas (Source: Google Earth Pro).

Figure 2. (**a**) Disposal channel carrying effluent of Star Paper Mill and (**b**) disposal point connecting to Kali River with a view of rice fields (Photos: Jogendra Singh).

2.3. Sample Collection and Chemical Analysis

For the current study, water, soil, and rice plant samples were collected from July to November 2021. The water samples (borewell, paper mill effluent, and Kali River water) were collected in 5 L capacity plastic cans while soil and rice samples were collected in 2 kg capacity zip-locked polyethylene bags. The soil and water samples were collected in three phases, i.e., vegetative growth stage (0–10 days after transplanting), reproductive stage (60–70 days), and grain filling (100–140 days). However, the straw and grain samples were collected after the final harvesting and drying stage. Purposely, the water and soil samples were analyzed for selected physicochemical and heavy metal parameters following the standard protocols described by APHA [24] and AOAC [25]. In this, pH, electrical conductivity, and total dissolved solids were determined by using a microprocessor-based multi-meter (1611, ESICO, Parwanoo, India). Biochemical and chemical oxygen demands (BOD_5 and COD) were estimated by using a digital dissolved oxygen meter (1801, ESICO, Parwanoo, India) and an open-reflux digester, respectively. Nitrogen contents were determined following Kjeldahl's method [26]. Organic matter of soil was analyzed following Walkley and Black method [27]. Similarly, the contents of potassium (K) were determined by using a flame-photometer (1382, ESICO, Parwanoo, India) while phosphorus was determined by using a UV-Vis spectroscopy (Cary 60, Agilent Technologies, Santa Clara, CA, USA) as previously described by Kumar et al. [28]. Finally, the contents of six heavy metals (Cd, Cr, Cu, Fe, Mn, and Zn) were determined by using atomic absorption spectroscopy (Analyst 800, PerkinElmer, Waltham, Massachusetts, USA) [29]. The detection limit of AAS for selected heavy metals was (Cd: 3 μg/L; Cr: 4 μg/L; Cu: 4 μg/L; Fe: 5 μg/L; Mn: 3 μg/L; and Zn: 3 μg/L). The certified reagent materials and analytical conditions are described earlier in our previous study [28]. Selective hollow cathode lamps were cast-off at optimum current and operated in accordance with the manufacturer's standard operating procedures (SOPs). The slit width was set to 0.5 nm, and the AAS was run with an air/acetylene gas mixture. Calibration curves were created using standard heavy metal solutions (0 as the control, 0.5, 1, 5, 10, 50, and 100 mg/L). The maximum recovery percentage (>98%) was used to perform quality assurance of the heavy metal analysis results. All analyses were carried out in triplicate.

2.4. Data Analysis

Heavy metals are known to accumulate in plant tissue, but the amount accumulated by each plant varies. The bioaccumulation factor (BF) is a measure of how much a plant can absorb in comparison to the surrounding soil. This indicator is very important because it can help researchers understand how plants react to contamination and subsequent amelioration of the contaminated sites [6]. In this study, BF was used to assess the heavy metal accumulation potential efficiency of two rice varieties irrigated with different water sources. The form of the index is given in Equation (1):

$$\text{Bioaccumulation factor (BF)} = \text{HMR}/\text{HMS} \quad (1)$$

where, HMR and HMS represent the heavy metal concentration in rice (grain) and soil, respectively. Also, the health risk of heavy metal accumulation in two rice varieties was investigated using the target hazard quotient (THQ) tool [30]. The form of the index is given in Equation (2):

$$\text{THQ} = 10^{-3} \times \frac{Ef \times Ed \times Ir \times HMc}{Bw \times Cp \times Rd} \quad (2)$$

where, *Ef* is exposure frequency, *Ed* is exposure duration (365 days), *Ir* is the rice ingestion rate, *HMc* is the heavy metal concentration in rice sample (mg/kg), *Bw* is the average body weight (70 Kg for adult and 16 Kg for child group), *Cp* is the consumption period (25,550 days for adult and 5840 days for child groups), *Rd* is the reference doses (Cd: 5.0×10^{-4}; Cr: 3.0×10^{-3}; Cu: 4.2×10^{-2}; Fe: 7.0×10^{-1}; Mn: 1.4×10^{-2}, and Zn: 3.0×10^{-1}), and 10^{-3} is the conversion factor, respectively. The combined toxicity of heavy metal intake from rice varieties was calculated by using the following Equation (3):

$$\text{HRI} = \sum \text{THQ}_{\text{Cd} + \text{Cr} + \text{Cu} + \text{Fe} + \text{Mn} + \text{Zn}} \quad (3)$$

Moreover, the data obtained in this study were tested using a one-way analysis of variance (ANOVA) to derive the significant difference between water source, irrigated soils, and rice plant growth/yield parameters. The level of statistical significance was adjusted as the probability (p) < 0.05.

2.5. Software and Tools

In this study, Microsoft Excel 2019 (Microsoft, Redmond, WA, USA) and OriginPro 2022b (OriginLab, Northampton, MA, USA) software packages were used for the data analysis and visualization.

3. Results and Discussion

3.1. Characteristics of Effluent and Agricultural Soils

The chemical analysis results of the studied irrigation water sources are shown in Table 1. Results revealed that all water sources had a pH value within the safe limits for surface disposal and inland irrigation set by the Central Pollution Control Board of India (CPCB) and Bureau of Indian Standards (BIS), respectively, with the following decreasing order: paper mill effluent > borewell > Kali River water. Paper mill effluent and Kali River water both had significantly higher EC values compared to borewell water. All water sources' TDS were below the safe limit for surface disposal, and their values in paper mill effluent > Kali River water surpassed the borewell's TDS value. A TDS level lower than 700 mg/L is considered adequate for crop irrigation; whereas, a range of 700–2000 mg/L may be slight to moderately restricted for use [31]. A similar trend was observed regarding BOD where paper mill effluent and Kali River water had 200.0-fold and 69.2-fold higher contents, respectively, compared to borewell. The former surpassed the safe limit for both surface disposal and inland irrigation. However, COD levels were in the following decreasing order: paper mill effluent > Kali River water > borewell with the first two sources having higher values than the latter. High BOD and COD levels in irrigation water led to lower oxygen availability for crops as a result of decreased soil permeability and therefore a direct effect on plant growth [32]. Similarly, paper mill effluent and Kali River water had significantly higher TKN values compared to borewells. The higher TKN level in Kali River water compared to paper mill effluent simulates possible discharge of sewage or leached manure already applied to nearby crop fields. Total K and total P contents were higher in paper mill effluent and Kali River water compared to the borewell. It was reported that high K levels can decrease water infiltration in irrigated soils [33], whereas the high P levels in water sources (probably originating from extensive use of fertilizers) can accelerate eutrophication and lead to ravaging algal blooms [34]; thus, causing devastating impacts on rice fields.

Table 1. Average characteristics of irrigation water sources (borewell, paper mill effluent, and Kali River water) of the study area.

Parameters	Irrigation Water Sources			Limit for Surface Disposal [a]	Limit for Inland Irrigation [b]
	Borewell	Paper Mill Effluent	Kali River Water		
pH	7.56 ± 0.11 b	8.14 ± 0.24 c	6.09 ± 0.37 a	5.50–9.00	5.50–9.00
Electricidal Conductivity (EC: dS/m)	0.61 ± 0.35 a	5.30 ± 0.09 c	4.07 ± 0.18 b	*Na*	*Na*
Total Dissolved Solids (TDS: mg/L)	110.36 ± 9.12 a	1409.62 ± 15.02 c	850.04 ± 29.05 b	1900.00	*Na*
Biochemical Oxygen Demand (BOD: mg/L)	4.05 ± 0.35 a	810.09 ± 20.38 c	280.26 ± 11.47 b	100.00	30.00
Chemical Oxygen Demand (COD: mg/L)	8.20 ± 0.26 a	1610.50 ± 18.01 c	709.45 ± 17.60 b	250.00	250.00
Total Kjeldahl's Nitrogen (TKN: mg/L)	8.12 ± 0.54 a	110.43 ± 3.70 c	140.06 ± 10.31 b	100.00	100.00
Total Potassium (K: mg/L)	6.41 ± 2.29 a	106.93 ± 4.58 c	82.24 ± 7.40 b	*Na*	*Na*
Total Phosphorus (*p*: mg/L)	4.46 ± 0.18 a	98.20 ± 5.07 c	31.03 ± 3.17 b	*Na*	5.00
Cadmium (Cd: mg/L)	*Bdl*	0.69 ± 0.08 a	3.70 ± 0.46 b	2.00	2.00
Chromium (Cr: mg/L)	*Bdl*	5.43 ± 0.12 b	2.50 ± 0.23 a	2.00	*Na*
Copper (Cu: mg/L	0.03 ± 0.01 a	1.90 ± 0.06 b	4.85 ± 0.39 c	3.00	3.00
Iron (Fe: mg/L)	0.62 ± 0.03 a	5.62 ± 0.17 b	11.07 ± 0.74 c	1.00	3.00
Manganese (Mn: mg/L)	0.38 ± 0.04 a	3.88 ± 0.21 b	6.28 ± 0.29 c	1.00	2.00
Zinc (Zn: mg/L)	0.26 ± 0.02 a	2.54 ± 0.19 b	9.64 ± 0.41 c	15.00	5.00

Different letters (a–c) indicate a significant difference ($p < 0.05$) between irrigation water sources; *Bdl*: below detectable limits; *Na*: not available; [a]: Central Pollution Control Board of India (CPCB); [b]: Bureau of Indian Standards (BIS).

In the case of heavy metals, Cd and Cr levels were very low in the borewell; therefore, they were undetectable. Kali River water had both Cd and Cr levels higher than the safe limit for surface disposal and inland irrigation. High Cd levels in irrigation water can lead to the disruption of plant mechanisms, nutrient uptake, and translocation in water and plants besides soil contamination [35], whilst high Cr levels in irrigation water can lead to oxidative stress, thus inhibiting some crop growth parameters [36] and subsequent bioaccumulation affecting human health [37]. Similarly, Cu, Fe, Mn, and Zn levels were higher in paper mill effluent and Kali River water than in borewell water. Specifically, Cu and Zn levels in paper mill effluent were below the safe limits for both surface disposal and inland irrigation, whereas Fe and Mn levels within the same irrigation water source surpassed these limits. Although the Zn level in Kali River water was below the safe limit for surface disposal, it was higher than that for inland irrigation. High Cu levels in irrigation water were associated with severe phytotoxicity symptoms [38]. Moreover, the Fe-enrichment of irrigation water was previously reported to adversely affect rice productivity [39], whereas high Mn levels in irrigation water were associated with potential toxicity stress imposed on cultivated crops [40]. Furthermore, reduced crop growth was earlier reported due to high zinc levels in water irrigation; this impact is accentuated if salt stress occurs [41].

Results presented in Table 2 depicted that although there were significant differences between irrigated soils in terms of pH, all of them were neutral. The rice crop is to some extent tolerant to acidity. It was reported that a pH of 6 is suitable for this crop's cultivation and pH values ranging between 5 and 8 are acceptable [42]. In this study, paper mill effluent- and Kali River water-irrigated soils had a higher EC compared to borewell-irrigated soils. Moreover, soils irrigated with paper mill effluent and Kali River water had a higher organic matter content compared to borewell-irrigated soils. This may improve soil aeration and reduce soil compaction around roots [43]. Also, TKN increased in soils irrigated with paper mill effluent and Kali River water. Such increases simulate increased ammonia in soils which were reported to halt the growth of crops, i.e., alfalfa [44]. Similarly, paper mill effluent and Kali River water-irrigated soils were richer in K than those irrigated using borewells. Although K is an essential component for crop yield improvement and soil enrichment, extremely high amounts can inhibit Mg uptake by crops [45]. For the heavy metals (Cd, Cr, Cu, Fe, Mn, and Zn), their contents increased maximally in paper mill effluent irrigated soils followed by Kali River water. Increased heavy metals in soils were associated with reduced microbial populations, their diversity, and activity; cytotoxic effect on soils, and phytotoxicity in plants leading to reduced plant growth and chlorosis [46].

Table 2. Average characteristics of soils irrigated with borewell, paper mill effluent, and Kali River water in the study area.

Parameters ^	Irrigated Soils		
	Borewell	**Paper Mill Effluent**	**Kali River Water**
pH	7.24 ± 0.05 b	7.69 ± 0.07 c	6.80 ± 0.20 a
Electricidal Conductivity (EC: dS/m)	2.27 ± 0.16 a	4.80 ± 0.22 c	3.10 ± 0.19 b
Organic Matter (OM: %)	2.86 ± 0.12 a	5.10 ± 0.42 bc	4.74 ± 0.34 b
Total Kjeldahl's Nitrogen (TKN: mg/kg)	124.16 ± 3.09 a	261.20 ± 10.08 c	210.42 ± 7.25 b
Potassium (K: mg/kg)	56.72 ± 8.26 a	220.89 ± 6.07 c	190.65 ± 10.81 b
Total Phosphorus (TP: mg/kg)	34.61 ± 5.32 a	128.48 ± 3.31 c	110.29 ± 8.03 b
Cadmium (Cd: mg/kg)	0.21 ± 0.05 a	5.10 ± 0.12 b	6.16 ± 1.15 bc
Chromium (Cr: mg/kg)	1.07 ± 0.09 a	9.89 ± 0.21 c	4.32 ± 0.38 b
Copper (Cu: mg/kg)	2.90 ± 0.14 a	16.43 ± 0.10 b	29.60 ± 2.42 c
Iron (Fe: mg/kg)	16.07 ± 0.26 a	47.21 ± 1.03 b	59.07 ± 3.95 c
Manganese (Mn: mg/kg)	1.50 ± 0.09 a	8.43 ± 1.04 b	11.51 ± 0.46 c
Zinc (Zn: mg/kg)	2.43 ± 0.12 a	12.08 ± 0.67 b	18.02 ± 1.28 c

^: The unit of all parameters is mg/kg except for pH, EC, and OM; Different letters (a–c) indicate a significant difference ($p < 0.05$) between irrigated soil parameter values.

3.2. Growth and Productivity of Selected Rice Varieties

Rice yield can be significantly influenced by water quality. Low germination rates, reduced plant growth, and increased susceptibility to disease can all be caused by poor water quality. In extreme cases, it can even cause crop death. However, when water contains sufficient levels of bioavailable nutrients, rice yield is impacted positively. In general, however, effluent irrigation can have a positive impact on rice yields, especially in areas where water is scarce. Additionally, using effluent to irrigate crops can help to reduce water pollution and the spread of disease. At the same time, if effluent contains noxious chemicals, it may be dangerous for both soil health and crop plants [47]. Table 3 shows the impact of irrigation water sources on the growth and productivity of selected rice varieties. Results showed that the average height of PB-1121 did not vary significantly ($p < 0.05$) among all irrigated soils. However, PR-121 outlined a significant increase in plant height in Kali River-irrigated soils compared to borewell-irrigated as control. This increase may assume an overexpression of *OsMPH1* which regulates rice crop height and subsequently grain yield [48]. In addition, total chlorophyll and carotenoid contents increased significantly in PB-1121 and PR-121 varieties when irrigated with paper mill effluent and Kali River water as compared to borewell water. This result simulates an interaction between *NRL3* and *NAL9/VYL* regulating the leaf morphology of rice and inducing the accumulation of chlorophyll [49]. Such an increase was previously reported to be the consequence of *OsDXS2* and *OsGLK1* overexpression [50]. In another study, increased total chlorophyll and carotenoid contents per leaf area were outlined as responses to increased salinity in wheat [51]. Our findings also confirm this simulation as high salinity levels were found in irrigation water sources used and in soils of grown fields. In the case of harvested rice grain properties, the starch and amylose content increased significantly ($p < 0.05$) in PB-1121 and PR-121 varieties when irrigated with paper mill effluent and Kali River water. Increased starch content in rice was associated with better consumer health pronounced by improved digestibility [52]. Increased rice quality was mainly attributed to an increased amylose percentage within rice grains. Moreover, the yield of both rice varieties increased when irrigated with paper mill effluent and the Kali River. As aforementioned, increased grain yields can be attributed to the overexpression of *OsMPH1* [48]. Besides that, the increased leaf area surface may have promoted the photosynthesis mechanism and subsequently total chlorophyll and grain yield. Similarly, the straw yield of both varieties also improved significantly.

Table 3. Impact of different irrigation water sources on growth and productivity performance of two rice varieties.

Parameters	Irrigated Soils and Rice Varieties					
	Borewell		Paper Mill Effluent		Kali River	
	PB-1121	PR-121	PB-1121	PR-121	PB-1121	PR-121
Plant Height (cm)	113.47 ± 5.02 c	93.82 ± 3.68 a	118.38 ± 7.15 c	97.64 ± 2.40 ab	123.48 ± 4.86 cd	98.83 ± 2.02 b
Total Chlorophyll (mg/g)	5.09 ± 0.13 a	5.12 ± 0.10 a	5.86 ± 0.22 bc	5.45 ± 0.17 b	6.70 ± 0.25 d	6.64 ± 0.17 d
Leaf Carotenoids (mg/g)	0.51 ± 0.05 a	0.53 ± 0.09 a	0.60 ± 0.07 ab	0.57 ± 0.04 a	0.67 ± 0.08 b	0.63 ± 0.05 b
Starch Content (%)	60.70 ± 3.70 a	62.76 ± 2.10 a	67.39 ± 1.82 b	69.98 ± 2.28 b	71.08 ± 2.05 bc	72.60 ± 1.63 c
Amylose Content (%)	21.01 ± 1.20 a	19.49 ± 0.97 a	23.35 ± 1.70 ab	20.09 ± 0.69 b	25.10 ± 1.32 bc	20.28 ± 1.24 bc
Crop Yield (kg/ha)	3710.50 ± 150.40 a	5244.02 ± 69.10 d	4050.92 ± 81.53 b	5566.12 ± 131.09 e	4270.20 ± 75.12 c	5830.58 ± 63.10 f
Straw Yield (kg/ha)	5110.17 ± 86.53 a	6035.20 ± 80.18 c	5318.17 ± 30.85 b	6481.30 ± 45.73 d	5472.05 ± 93.90 b	6683.76 ± 61.26 e

Different letters (a–f) indicate a significant difference ($p < 0.05$) between irrigated rice parameter values.

3.3. Heavy Metal Accumulation in Selected Rice Varieties

The results of heavy metal concentrations in straw and grain of grown rice varieties in different irrigated soils are reported in Table 4. Specifically, Cd, Cr, Cu, Fe, Mn, and Zn accumulated in the straw of PB-1121 and PR-121 rice varieties were in the following ranges: 0.19–0.27 mg/kg and 0.20–0.31 mg/kg; 0.82–1.40 mg/kg and 0.92–1.51 mg/kg; 13.52–21.67 mg/kg and 14.10–23.04 mg/kg; 178.76–290.53 mg/kg and 212.65–310.98 mg/kg; 78.54–112.67 mg/kg and 62.40–105.43 mg/kg; 19.26–28.66 and 21.58–30.32 mg/kg. In this study, both varieties' grains enclosed acceptable ranges of Cd (0.03–010 mg/kg), Cr (0.15–0.41 mg/kg), Cu (1.30–2.10 mg/kg), Fe (14.09–24.29 mg/kg), Mn (1.48–2.81 mg/kg) and Zn (1.87–3.21 mg/kg) being lower than the safe limits (0.1–0.2 mg/kg, 1.0 mg/kg,

20.0 mg/kg, 425.5 mg/kg, 500.0 mg/kg, and 100.0 mg/kg, respectively) set for cereals and legumes [53–55]. Moreover, the bioaccumulation factor (BF) of heavy metals in the straw of both rice varieties followed a decreasing order of Fe > Mn > Zn > Cu > Cr > Cd. However, higher bioaccumulation of heavy metals was encountered in paper mill effluent irrigated rice followed by Kali River and borewell. The bioaccumulation factors in rice straw of PB-1121 and PR-121 are depicted in Figure 3. In fact, all bioaccumulation factors in rice grains were acceptable (<1.0) [56], except for Mn in PB-1121 and PR-121 varieties irrigated with borewell. Although high heavy metal accumulation was observed in both soil and water, they did not have serious impacts on the final product. This was also observed in the case of other crops grown on industrial media with high heavy metal accumulation [28]. This may be attributed to a high Mn adsorption and/or high Mn transport across the plasma membrane into root cell walls [57]. Moreover, increased Mn concentrations in food can lead to neurotoxicity expressed by neuroinflammation, oxidative stress, and cognitive dysfunction possibly impairing Parkinson's disease [58]. Rice straw is generally burned in fields resulting in increased greenhouse gas emissions and further pollutant enrichment of rice soils and crops in subsequent cycles [59].

Table 4. Impact of different irrigation water sources on heavy metal accumulation (mg/kg) in grains of two rice varieties.

Rice Parts	Heavy Metals	Irrigated Soils and Rice Varieties					
		Borewell		Paper Mill Effluent		Kali River	
		PB-1121	PR-121	PB-1121	PR-121	PB-1121	PR-121
Straw	Cd	0.19 ± 0.02 a	0.20 ± 0.03 a	0.23 ± 0.02 a	0.25 ± 0.03 ab	0.27 ± 0.04 b	0.31 ± 0.02 b
	Cr	0.82 ± 0.05 a	0.92 ± 0.08 ab	1.40 ± 0.13 c	1.51 ± 0.09 c	1.10 ± 0.15 b	1.22 ± 0.07 b
	Cu	13.52 ± 0.28 a	14.10 ± 0.51 a	18.16 ± 1.02 b	20.45 ± 0.92 b	21.67 ± 1.36 bc	23.04 ± 0.65 c
	Fe	178.76 ± 6.91 a	212.65 ± 10.75 b	254.02 ± 12.08 c	272.18 ± 20.78 c	290.53 ± 12.45 d	310.98 ± 21.04 d
	Mn	78.54 ± 3.35 b	62.40 ± 6.02 a	110.78 ± 11.21 cd	90.65 ± 9.42 c	112.67 ± 8.22 d	105.43 ± 7.60 cd
	Zn	19.26 ± 0.70 a	21.58 ± 0.56 a	25.02 ± 1.40 b	27.02 ± 1.01 b	28.66 ± 0.95 bc	30.32 ± 1.24 c
Grain	Cd	0.03 ± 0.01 a	0.04 ± 0.02 a	0.06 ± 0.02 ab	0.07 ± 0.01 b	0.09 ± 0.02 bc	0.10 ± 0.02 c
	Cr	0.15 ± 0.02 a	0.16 ± 0.03 a	0.39 ± 0.06 b	0.41 ± 0.04 b	0.31 ± 0.02 b	0.36 ± 0.05 c
	Cu	1.60 ± 0.05 a	1.68 ± 0.04 a	1.73 ± 0.04 ab	1.82 ± 0.12 b	2.02 ± 0.09 c	2.10 ± 0.11 c
	Fe	14.09 ± 0.36 a	15.04 ± 0.82 a	20.11 ± 1.27 b	21.18 ± 0.96 b	22.30 ± 1.55 bc	24.29 ± 0.83 c
	Mn	1.72 ± 0.20 ab	1.48 ± 0.13 a	2.34 ± 0.32 c	2.05 ± 0.07 c	2.81 ± 0.16 d	2.57 ± 0.29 c
	Zn	1.87 ± 0.03 a	1.92 ± 0.09 a	2.52 ± 0.14 b	2.69 ± 0.08 b	3.10 ± 0.25 c	3.21 ± 0.08 c

Same letters (a–d) indicate no significant difference ($p < 0.05$) between irrigated rice parameter values.

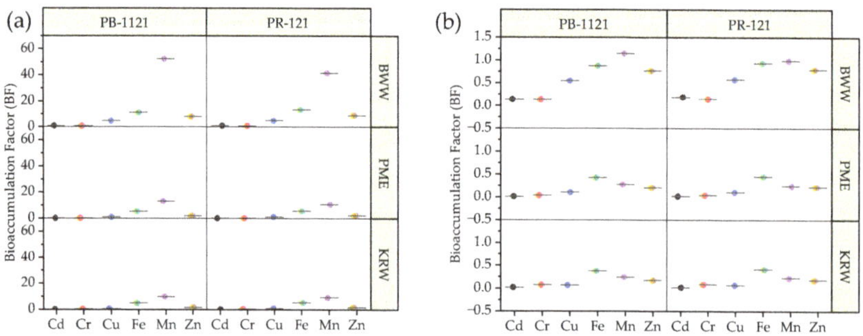

Figure 3. Bioaccumulation factor (BF) of heavy metals in (**a**) straw and (**b**) grain of PB-1121 and PR-121 rice varieties irrigated with different water sources in Saharanpur district of Uttar Pradesh, India (BWW: borewell, PME: paper mill effluent, and KRW: Kali River water; the same color dots correspond to values of same heavy metal).

3.4. Health Risk Assessment of Heavy Metals Accumulation

Heavy metals are abundant in the environment and can be absorbed by crops. Rice is an important food source for Indians, so it is critical to assess the health risk of heavy metals in rice. The study of THQ and HRI gives a broader spectrum of possible health risks of food consumption by humans [30]. An HRI value of 1.0 or less indicates that the heavy metal is unlikely to be harmful to human health. However, HRI values greater than 1.0 indicates that the heavy metal may be hazardous to one's health. Table 5 summarizes the obtained THQs and HRIs found in both rice varieties grown in different irrigated soils. Surprisingly, child THQs were always higher than adult ones despite the rice variety and irrigated soil. Heavy metal levels in Indian varieties reported in that study were much higher than found in the current one simulating different soil physicochemical characteristics and water quality used in rice growing and production besides the variability in used crop varieties. Child and adult HRIs followed the following decreasing order: PR-121 Kali River > PB-1121 Kali River > PR-121 Paper Mill Effluent > PB-1121 Paper Mill Effluent > PR-121 Borewell > PB-1121 Borewell. It is noteworthy that the PR-121 variety had higher HRIs compared to PB-1121 in all irrigated soils which simulates a possible effect of consumed varieties on human health. All HRIs except child HRI of PR-121 variety in Kali River-irrigated soil were below 1; therefore, they are considered safe. The alarming increase in child HRI may be attributed to Cd content which can lead to heart, kidney, and lung diseases as well as nervous tissue disorders [60]. Thus, special attention should be given to the PR-121 rice variety originating from soils irrigated with Kali River water to children. Previously, Lien et al. [60] found in the Yilan area (Taiwan) an endangering health risk of rice on children pronounced by high HRIs and attributed that to Cd levels being concomitant with our findings. A similar observation was acknowledged by Guo et al. [61] regarding the direct influence of Cd levels on rice HRI in Chinese polluted paddy soils simulating possible carcinogenic impacts. Therefore, the THQ and HRI approaches successfully detected minor and possible risk(s) associated with grown rice varieties in different irrigated soils of Saharanpur District.

Table 5. Target hazard quotient (THQ)-based health risk assessment of heavy metals accumulation in two rice varieties irrigated with different water sources.

Heavy Metals	Health Risk Variables	Irrigated Soils					
		Borewell		Paper Mill Effluent		Kali River	
		PB-1121	PR-121	PB-1121	PR-121	PB-1121	PR-121
Cadmium (Cd)	THQ (Child)	0.156	0.175	0.212	0.231	0.298	0.387
	THQ (Adult)	0.124	0.132	0.148	0.156	0.172	0.280
Chromium (Cr)	THQ (Child)	0.147	0.150	0.221	0.228	0.197	0.292
	THQ (Adult)	0.120	0.121	0.152	0.155	0.141	0.148
Copper (Cu)	THQ (Child)	0.136	0.137	0.138	0.140	0.145	0.147
	THQ (Adult)	0.115	0.116	0.116	0.117	0.119	0.120
Iron (Fe)	THQ (Child)	0.019	0.120	0.027	0.028	0.019	0.032
	THQ (Adult)	0.018	0.019	0.011	0.012	0.013	0.024
Manganese (Mn)	THQ (Child)	0.115	0.099	0.156	0.137	0.187	0.171
	THQ (Adult)	0.049	0.042	0.067	0.059	0.080	0.073
Zinc (Zn)	THQ (Child)	0.116	0.016	0.018	0.018	0.019	0.012
	THQ (Adult)	0.012	0.013	0.013	0.014	0.014	0.014
HRI	Child	0.689	0.697	0.772	0.782	0.865	1.041
	Adult	0.438	0.443	0.507	0.513	0.539	0.659

4. Conclusions

The findings of this study concluded that effluent release from Star Paper Mill Ltd., Saharanpur, India had significant untreated pollutants that exceeded the permissible limits of both inland irrigation and surface discharge. Effluent disposal showed a significant ($p < 0.05$) increase in the water quality of Kali River thereby exceeding its pollution level. In addition, the soils irrigated with paper mill effluent and Kali River water showed a comparatively high level of physicochemical and heavy metal parameters suggesting potential contamination. The two rice varieties (PB-1121 and PR-121) irrigated with paper mill effluent and Kali River water showed a significant ($p < 0.05$) increase in growth, yield, and productivity. However, the PR-121 variety showed relatively higher growth, yield, and heavy metal accumulation in straw and grains as compared to PB-1121. The health risk investigation showed potential health risks of heavy metal exposure through consumption of rice irrigated with paper mill effluent and Kali River water. Further studies on biomonitoring of other toxic heavy metals in paper mill effluent and Kali River is highly recommended while local farmers are advised not to use the noxious paper mill effluent or contaminated Kali River water for their field irrigation.

Author Contributions: Conceptualization, P.K.; data curation, J.S.; formal analysis, J.S.; funding acquisition, A.A.A.-H., E.M.E. and I.Š.; investigation, P.K. and M.G.; methodology, P.K.; project administration, A.A.A.-H.; resources, P.K. and V.K.; software, P.K. and M.G.; supervision, P.K.; validation, A.A.A.-H., E.M.E., B.A., S.A.F., J.S., A.K.A., M.G., V.K. and I.Š.; writing—original draft, P.K. and S.A.F.; writing—review & editing, A.A.A.-H., E.M.E., B.A., A.K.A., M.G., V.K. and I.Š. All authors have read and agreed to the published version of the manuscript.

Funding: This research was funded by Princess Nourah bint Abdulrahman University Researchers Supporting Project number (PNURSP2022R93), Princess Nourah bint Abdulrahman University, Riyadh, Saudi Arabia.

Institutional Review Board Statement: Not applicable.

Informed Consent Statement: Not applicable.

Data Availability Statement: Not applicable.

Acknowledgments: This is joint work from the members of the Sustainable Agro-Environment International Research Group (SAEIRG). The authors express their gratitude to Princess Nourah bint Abdulrahman University Researchers Supporting Project number (PNURSP2022R93), Princess Nourah bint Abdulrahman University, Riyadh, Saudi Arabia. All individuals included in this section have consented to the acknowledgment.

Conflicts of Interest: The authors declare no conflict of interest.

References

1. UN WWAP. *The United Nations World Water Development Report 3*; UN WWAP: Geneva, Switzerland, 2012.
2. Roy, S.; Garg, A.; Garg, S.; Tran, T.A. *Advanced Industrial Wastewater Treatment and Reclamation of Water*; Springer: Cham, Switzerland, 2022; ISBN 978-3-030-83810-2.
3. Ilyas, M.; Ahmad, W.; Khan, H.; Yousaf, S.; Yasir, M.; Khan, A. Environmental and Health Impacts of Industrial Wastewater Effluents in Pakistan: A Review. *Rev. Environ. Health* **2019**, *34*, 171–186. [CrossRef] [PubMed]
4. Adelodun, B.; Kumar, P.; Odey, G.; Ajibade, F.O.; Ibrahim, R.G.; Alamri, S.A.M.; Alrumman, S.A.; Eid, E.M.; Kumar, V.; Adeyemi, K.A.; et al. A Safe Haven of SARS-CoV-2 in the Environment: Prevalence and Potential Transmission Risks in the Effluent, Sludge, and Biosolids. *Geosci. Front.* **2022**, *13*, 101373. [CrossRef]
5. Uzma, S.; Azizullah, A.; Bibi, R.; Nabeela, F.; Muhammad, U.; Ali, I.; Rehman, Z.U.; Häder, D.P. Effects of Industrial Wastewater on Growth and Biomass Production in Commonly Grown Vegetables. *Environ. Monit. Assess.* **2016**, *188*, 328. [CrossRef]
6. Kanwal, A.; Farhan, M.; Sharif, F.; Hayyat, M.U.; Shahzad, L.; Ghafoor, G.Z. Effect of Industrial Wastewater on Wheat Germination, Growth, Yield, Nutrients and Bioaccumulation of Lead. *Sci. Rep.* **2020**, *10*, 11361. [CrossRef] [PubMed]
7. Azimi, A.; Azari, A.; Rezakazemi, M.; Ansarpour, M. Removal of Heavy Metals from Industrial Wastewaters: A Review. *ChemBioEng Rev.* **2017**, *4*, 37–59. [CrossRef]
8. Król, A.; Mizerna, K.; Bożym, M. An Assessment of PH-Dependent Release and Mobility of Heavy Metals from Metallurgical Slag. *J. Hazard. Mater.* **2020**, *384*, 121502. [CrossRef]
9. Sintorini, M.M.; Widyatmoko, H.; Sinaga, E.; Aliyah, N. Effect of PH on Metal Mobility in the Soil. *IOP Conf. Ser. Earth Environ. Sci.* **2021**, *737*, 012071. [CrossRef]
10. Campillo-Cora, C.; Conde-Cid, M.; Arias-Estévez, M.; Fernández-Calviño, D.; Alonso-Vega, F. Specific Adsorption of Heavy Metals in Soils: Individual and Competitive Experiments. *Agronomy* **2020**, *10*, 1113. [CrossRef]
11. Lenart, A.; Wolny-Koładka, K. The Effect of Heavy Metal Concentration and Soil Ph on the Abundance of Selected Microbial Groups within Arcelormittal Poland Steelworks in Cracow. *Bull. Environ. Contam. Toxicol.* **2013**, *90*, 85–90. [CrossRef]

12. Bravo, S.; Amorós, J.A.; Pérez-de-los-Reyes, C.; García, F.J.; Moreno, M.M.; Sánchez-Ormeño, M.; Higueras, P. Influence of the Soil PH in the Uptake and Bioaccumulation of Heavy Metals (Fe, Zn, Cu, Pb and Mn) and Other Elements (Ca, K, Al, Sr and Ba) in Vine Leaves, Castilla-La Mancha (Spain). *J. Geochem. Explor.* **2017**, *174*, 79–83. [CrossRef]
13. Hu, X.; Wang, J.; Lv, Y.; Liu, X.; Zhong, J.; Cui, X.; Zhang, M.; Ma, D.; Yan, X.; Zhu, X. Effects of Heavy Metals/Metalloids and Soil Properties on Microbial Communities in Farmland in the Vicinity of a Metals Smelter. *Front. Microbiol.* **2021**, *12*, 2347. [CrossRef] [PubMed]
14. Rathnayake, I.V.N.; Megharaj, M.; Bolan, N.; Naidu, R. Tolerance of Heavy Metals by Gram Positive Soil Bacteria. *Int. J. Environ. Eng.* **2010**, *2*, 191–195.
15. Chmielowska-Bąk, J.; Gzyl, J.; Rucińska-Sobkowiak, R.; Arasimowicz-Jelonek, M.; Deckert, J. The New Insights into Cadmium Sensing. *Front. Plant Sci.* **2014**, *5*, 245. [CrossRef] [PubMed]
16. Xu, D.; Shen, Z.; Dou, C.; Dou, Z.; Li, Y.; Gao, Y.; Sun, Q. Effects of Soil Properties on Heavy Metal Bioavailability and Accumulation in Crop Grains under Different Farmland Use Patterns. *Sci. Rep.* **2022**, *12*, 9211. [CrossRef]
17. Vardhan, K.H.; Kumar, P.S.; Panda, R.C. A Review on Heavy Metal Pollution, Toxicity and Remedial Measures: Current Trends and Future Perspectives. *J. Mol. Liq.* **2019**, *290*, 111197. [CrossRef]
18. Shahbandeh, M. Rice—Statistics & Facts. Available online: https://worldpopulationreview.com/country-rankings/rice-production-by-country (accessed on 31 August 2022).
19. Vici, G.; Perinelli, D.R.; Camilletti, D.; Carotenuto, F.; Belli, L.; Polzonetti, V. Nutritional Properties of Rice Varieties Commonly Consumed in Italy and Applicability in Gluten Free Diet. *Foods* **2021**, *10*, 1375. [CrossRef]
20. Afrad, M.S.I.; Monir, M.B.; Haque, M.E.; Barau, A.A.; Haque, M.M. Impact of Industrial Effluent on Water, Soil and Rice Production in Bangladesh: A Case of Turag River Bank. *J. Environ. Health Sci. Eng.* **2020**, *18*, 825–834. [CrossRef]
21. Singh, J.; Kumar, V.; Kumar, P.; Kumar, P.; Yadav, K.K.; Cabral-Pinto, M.M.S.; Kamyab, H.; Chelliapan, S. An Experimental Investigation on Phytoremediation Performance of Water Lettuce (*Pistia stratiotes* L.) for Pollutants Removal from Paper Mill Effluent. *Water Env. Res.* **2021**, *93*, 1543–1553. [CrossRef]
22. Singh, V.; Singh, A.K.; Mohapatra, T.; Ellur, R.K. Pusa Basmati 1121–A Rice Variety with Exceptional Kernel Elongation and Volume Expansion after Cooking. *Rice* **2018**, *11*, 19. [CrossRef]
23. Bharaj, T.S.; Mangat, G.S.; Kaur, R.; Singh, K.; Singh, N. PR 121: A New Semi-Dwarf High Yielding Variety of Rice (*Oryza sativa* L.). *J. Res. (Punjab Agric. Univ.)* **2014**, *51*, 202–203.
24. Apha, A. *Standard Methods for the Examination of Water and Wastewater*, 22nd ed.; American Public Health Association: Washington, DC, USA, 2012.
25. AOAC International. *AOAC Official Methods of Analysis*, 21st ed.; AOAC International: Rockville, MD, USA, 2019.
26. Chromý, V.; Vinklárková, B.; Šprongl, L.; Bittová, M. The Kjeldahl Method as a Primary Reference Procedure for Total Protein in Certified Reference Materials Used in Clinical Chemistry. I. A Review of Kjeldahl Methods Adopted by Laboratory Medicine. *Crit. Rev. Anal. Chem.* **2015**, *45*, 106–111. [CrossRef] [PubMed]
27. Walkley, A.; Black, I.A. An Examination of the Degtjareff Method for Determining Soil Organic Matter, and a Proposed Modification of the Chromic Acid Titration Method. *Soil Sci.* **1934**, *37*, 29–38. [CrossRef]
28. Kumar, V.; Thakur, R.K.; Kumar, P. Assessment of Heavy Metals Uptake by Cauliflower (*Brassica oleracea* Var. Botrytis) Grown in Integrated Industrial Effluent Irrigated Soils: A Prediction Modeling Study. *Sci. Hortic.* **2019**, *257*, 108682. [CrossRef]
29. Kumar, P.; Eid, E.M.; Al-Huqail, A.A.; Širić, I.; Adelodun, B.; Abou Fayssal, S.; Valadez-Blanco, R.; Goala, M.; Ajibade, F.O.; Choi, K.S.; et al. Kinetic Studies on Delignification and Heavy Metals Uptake by Shiitake (*Lentinula edodes*) Mushroom Cultivated on Agro-Industrial Wastes. *Horticulturae* **2022**, *8*, 316. [CrossRef]
30. Kumar, P.; Kumar, V.; Eid, E.M.; Al-Huqail, A.A.; Adelodun, B.; Abou Fayssal, S.; Goala, M.; Arya, A.K.; Bachheti, A.; Andabaka, Ž.; et al. Spatial Assessment of Potentially Toxic Elements (PTE) Concentration in Agaricus Bisporus Mushroom Collected from Local Vegetable Markets of Uttarakhand State, India. *J. Fungi* **2022**, *8*, 452. [CrossRef] [PubMed]
31. Ayers, R.S.; Westcot, D.W. *Water Quality for Agriculture—FAO Irrigation and Drainage*; FAO: Roma, Italy, 1985; Volume NO. 29.
32. Balkhair, K.S.; Ashraf, M.A. Field Accumulation Risks of Heavy Metals in Soil and Vegetable Crop Irrigated with Sewage Water in Western Region of Saudi Arabia. *Saudi J. Biol. Sci.* **2016**, *23*, S32–S44. [CrossRef]
33. Rengasamy, P.; Marchuk, A. Cation Ratio of Soil Structural Stability (CROSS). *Soil Res.* **2011**, *49*, 280–285. [CrossRef]
34. Riemersma, S.; Little, J.; Ontkean, G.; Moskal-Hebert, T. Phosphorus Sources and Sinks in Watersheds: A Review. *Alta. Soil Phosphorus Limits Proj.* **2006**, *5*, 82.
35. Haider, F.U.; Liqun, C.; Coulter, J.A.; Cheema, S.A.; Wu, J.; Zhang, R.; Wenjun, M.; Farooq, M. Cadmium Toxicity in Plants: Impacts and Remediation Strategies. *Ecotoxicol. Environ. Saf.* **2021**, *211*, 111887. [CrossRef]
36. Singh, D.; Sharma, N.L.; Singh, C.K.; Kumar Sarkar, S.; Singh, I.; Lal Dotaniya, M. Effect of Chromium (VI) Toxicity on Morpho-Physiological Characteristics, Yield, and Yield Components of Two Chickpea (*Cicer arietinum* L.) Varieties. *PLoS ONE* **2020**, *15*, e0243032. [CrossRef]
37. Malakar, A.; Snow, D.D.; Ray, C. Irrigation Water Quality—A Contemporary Perspective. *Water* **2019**, *11*, 1482. [CrossRef]
38. Shiyab, S. Phytoaccumulation of Copper from Irrigation Water and Its Effect on the Internal Structure of Lettuce. *Agriculture* **2018**, *8*, 29. [CrossRef]
39. Murphy, T.; Phan, K.; Yumvihoze, E.; Irvine, K.; Wilson, K.; Lean, D.; Poulain, A.; Laird, B.; Chan, L.H.M. Effects of Arsenic, Iron and Fertilizers in Soil on Rice in Cambodia. *J. Health Pollut.* **2018**, *8*, 180910. [CrossRef] [PubMed]

40. Chen, W.; Wang, Z.; Jin, M.; Ferré, T.P.A.; Wang, J.; Huang, J.; Wang, X. Effect of Sodium Chloride and Manganese in Irrigation Water on Cotton Growth. *Agron. J.* **2018**, *110*, 900–909. [CrossRef]
41. Redondo-Gómez, S.; Andrades-Moreno, L.; Mateos-Naranjo, E.; Parra, R.; Valera-Burgos, J.; Aroca, R. Synergic Effect of Salinity and Zinc Stress on Growth and Photosynthetic Responses of the Cordgrass, Spartina Densiflora. *J. Exp. Bot.* **2011**, *62*, 5521–5530. [CrossRef]
42. Azman, E.A.; Jusop, S.; Ishak, C.F.; Ismail, R. Increasing Rice Production Using Different Lime Sources on an Acid Sulphate Soil in Merbok, Malaysia. *Pertanika J. Trop. Agric. Sci.* **2014**, *37*, 223–247.
43. King, A.E.; Ali, G.A.; Gillespie, A.W.; Wagner-Riddle, C. Soil Organic Matter as Catalyst of Crop Resource Capture. *Front. Environ. Sci.* **2020**, *8*, 50. [CrossRef]
44. Duan, R.; Fedler, C.B. Preliminary Field Study of Soil TKN in a Wastewater Land Application System. *Ecol. Eng.* **2015**, *83*, 1–4. [CrossRef]
45. Xu, X.; Du, X.; Wang, F.; Sha, J.; Chen, Q.; Tian, G.; Zhu, Z.; Ge, S.; Jiang, Y. Effects of Potassium Levels on Plant Growth, Accumulation and Distribution of Carbon, and Nitrate Metabolism in Apple Dwarf Rootstock Seedlings. *Front. Plant Sci.* **2020**, *11*, 904. [CrossRef]
46. Okereafor, U.; Makhatha, M.; Mekuto, L.; Uche-Okereafor, N.; Sebola, T.; Mavumengwana, V. Toxic Metal Implications on Agricultural Soils, Plants, Animals, Aquatic Life and Human Health. *Int. J. Environ. Res. Public Health* **2020**, *17*, 2204. [CrossRef]
47. Jung, K.; Jang, T.; Jeong, H.; Park, S. Assessment of Growth and Yield Components of Rice Irrigated with Reclaimed Wastewater. *Agric. Water Manag.* **2014**, *138*, 17–25. [CrossRef]
48. Zhang, Y.; Yu, C.; Lin, J.; Liu, J.; Liu, B.; Wang, J.; Huang, A.; Li, H.; Zhao, T. OsMPH1 Regulates Plant Height and Improves Grain Yield in Rice. *PLoS ONE* **2017**, *12*, e0180825. [CrossRef] [PubMed]
49. Chen, W.; Sheng, Z.; Cai, Y.; Li, Q.; Wei, X.; Xie, L.; Jiao, G.; Shao, G.; Tang, S.; Wang, J.; et al. Rice Morphogenesis and Chlorophyll Accumulation Is Regulated by the Protein Encoded by Nrl3 and Its Interaction with Nal9. *Front. Plant Sci.* **2019**, *10*, 175. [CrossRef] [PubMed]
50. Li, Z.; Gao, J.; Wang, B.; Xu, J.; Fu, X.; Han, H.; Wang, L.; Zhang, W.; Deng, Y.; Wang, Y.; et al. Rice Carotenoid Biofortification and Yield Improvement Conferred by Endosperm-Specific Overexpression of OsGLK1. *Front. Plant Sci.* **2022**, *13*, 951605. [CrossRef] [PubMed]
51. Shah, S.H.; Houborg, R.; McCabe, M.F. Response of Chlorophyll, Carotenoid and SPAD-502 Measurement to Salinity and Nutrient Stress in Wheat (*Triticum aestivum* L.). *Agronomy* **2017**, *7*, 61. [CrossRef]
52. Ordonio, R.L.; Matsuoka, M. Increasing Resistant Starch Content in Rice for Better Consumer Health. *Proc. Natl. Acad. Sci. USA* **2016**, *113*, 12616–12618. [CrossRef]
53. FAO/WHO. *Report of the Thirty Eight Session of the Codex Committee on Food Hygiene*; FAO: Rome, Italy, 2007.
54. FAO/WHO. *Report of the Twenty Eighth Session*; ALINORM 05/28/12; FAO: Rome, Italy, 2005.
55. USDA. *National Food Safety Standard of Maximum Levels of Contaminants in Foods*; USDA: Washington, DC, USA, 2018.
56. Khan, Z.I.; Mansha, A.; Saleem, M.H.; Tariq, F.; Ahmad, K.; Ahmad, T.; Awan, M.U.F.; Abualreesh, M.H.; Alatawi, A.; Ali, S. Trace Metal Accumulation in Rice Variety Kainat Irrigated with Canal Water. *Sustainability* **2021**, *13*, 13739. [CrossRef]
57. Hu, A.Y.; Zheng, M.M.; Sun, L.M.; Zhao, X.Q.; Shen, R.F. Ammonium Alleviates Manganese Toxicity and Accumulation in Rice by Down-Regulating the Transporter Gene OsNramp5 through Rhizosphere Acidification. *Front. Plant Sci.* **2019**, *10*, 1194. [CrossRef]
58. Peres, T.V.; Schettinger, M.R.C.; Chen, P.; Carvalho, F.; Avila, D.S.; Bowman, A.B.; Aschner, M. Manganese-Induced Neurotoxicity: A Review of Its Behavioral Consequences and Neuroprotective Strategies. *BMC Pharmacol. Toxicol.* **2016**, *17*, 57. [CrossRef]
59. Ahmad, R.; Hadi, F.; Jan, A.U.; Ditta, A. Straw Incorporation in Contaminated Soil Enhances Drought Tolerance but Simultaneously Increases the Accumulation of Heavy Metals in Rice. *Sustainability* **2022**, *14*, 10578. [CrossRef]
60. Lien, K.W.; Pan, M.H.; Ling, M.P. Levels of Heavy Metal Cadmium in Rice (*Oryza sativa* L.) Produced in Taiwan and Probabilistic Risk Assessment for the Taiwanese Population. *Environ. Sci. Pollut. Res.* **2021**, *28*, 28381–28390. [CrossRef] [PubMed]
61. Guo, B.; Hong, C.; Tong, W.; Xu, M.; Huang, C.; Yin, H.; Lin, Y.; Fu, Q. Health Risk Assessment of Heavy Metal Pollution in a Soil-Rice System: A Case Study in the Jin-Qu Basin of China. *Sci. Rep.* **2020**, *10*, 11490. [CrossRef] [PubMed]

agriculture

MDPI

Article

Use of Bio-Preparations to Reduce Ammonia Emissions from Cattle Farming: Effects of Manure Storage Time and Ventilation Intensity

Reda Mažeikienė * and Rolandas Bleizgys

Agriculture Academy, Vytautas Magnus University, 44248 Kaunas, Lithuania
* Correspondence: redos.laiskai@gmail.com

Abstract: Ammonia emissions are both an environmental and health issue. Biological preparations (also known as biopreparations) have attracted interests as a tool to reduce ammonia emissions from livestock. However, little is known about their effectiveness in modulating evaporating NH_3 from manure. The aim of this study was to evaluate the effectiveness of one biopreparation product (MycroZyme© Micron Bio-Systems Ltd., Bridgwater, UK) by studying the dynamics of NH_3 gas evolution, taking into account factors such as manure storage time (fresh or stored) and the application method (sprayed or mixed). Experimental studies were performed with control manure (manure without biopreparation) and manure treated with the biopreparation. Data showed that the use of biopreparation slowed down the ammonia emission from manure, so the ammonia concentration in the wind tunnel decreased on average from 21.69 ± 0.9 to 14.43 ± 1.5 ppm. The effect of the biopreparation on ammonia evaporation was higher in the presence of more intensive ammonia emission conditions; i.e., when the manure was fresh, there was no crust on the surface, and the airflow over the manure was intense, and there was a large gradient of ammonia concentration on the manure surface. Based on the obtained results, recommendations were made to reduce ammonia emissions in litter-less cowsheds, where liquid manure is stored.

Keywords: cattle manure; biopreparation; ammonia emissions; cowsheds

1. Introduction

Dairy cattle manure is one of the most polluting wastes in the agricultural sector. Intensive dairy farming produces large amounts of manure with high concentrations of organic matter, including nitrogen [1–3]. Indeed, the livestock farming sector is the main source of atmospheric forms of nitrogen, such as ammonia, worldwide [4]. Livestock emits 83–91% of total ammonia emissions to the environment (i.e., about 8–9 million tonnes per year) [4–7]. It is therefore important to have a good understanding of the key factors influencing such emissions in order to develop effective strategies to reduce air pollution.

Ammonia is considered an atmospheric pollutant that reacts in the atmosphere due to its high solubility in water to form ammonium (NH_4^+) compounds [6]. These compounds can have significant adverse effects on the biodiversity of natural and semi-natural ecosystems due to acidity and eutrophication in both terrestrial and aquatic environments. Ammonia can also be dangerous to human health [4,6–8]. In addition, during the evaporation of NH_3 from livestock manure, farmers lose valuable nutrient nitrogen resources because the loss of ammonia-N reduces the quality of manure fertilizers, and agricultural systems often require additional nitrogen supplementation using mineral nitrogen fertilizers [6,7]. Therefore, animal manure needs to be properly treated and disposed of in order to minimise the risk to the environment and at the same time to recover nutrients [9].

Several technological [4,6], chemical [6,8], and biological [4,7–13] preparations are known to reduce the release of NH_3 from manure. The technological and chemical methods help in reducing the release of NH_3 from manure, but they have several shortcomings, such as the high costs for equipment, difficulties in implementing, and potential cause creation of by-products. Air pollution can be reduced by a variety of preventive (aeration, alkali, temperature sanitation, etc.), technological (scrubbers, biofilters, ozonation, irradiation), and chemical (synthetic essential oils, terpenes, glycols, ethylene oxide, etc.) methods. Chemicals are camouflage and therefore only change the threshold of discomfort but do not eliminate the causes of unpleasant odours. The harmfulness of deodorizing compounds has not been established [11]. Other chemicals that are designed to disinfect a contaminated object are toxic and destroy both pathogenic and probiotic microflora, and their residual products can remain toxic for long periods of time [12]. Therefore, innovative farming needs to find the most natural ways to reduce harmful emissions to the environment. Recently, there has been an increasing focus on biological preparations, also called biopreparations, that are based on bacterial cultures or enzymes being environmentally friendly and not requiring special equipment for their use [10–13]. According to the known mode of the action, biopreparations are divided into several groups that include agents that improve the biodegradability of manure and consist of strains of microorganisms and/or enzymes, acidifying additives that lower the pH of manure, absorbent additives, and urease inhibitors [4]. Various biopreparations consisting of microorganisms and enzymes of different compositions are most widely used on farms [9–13]. However, little is known about the effectiveness of these additives in evaporating NH_3 from manure. Therefore, the aim of this research was to determine the influence of one commercially available biopreparation product on the evaporation of ammonia from manure, focusing mainly on the effects of application method, manure storage time, and ventilation intensity. Based on the findings of the study, the overarching goal was to develop recommendations for the use of biological preparations in cattle housing. We hypothesized that the use of the biopreparation can reduce ammonia emissions from manure, and this effect will depend on the duration of application as well as on the application method.

2. Materials and Methods

2.1. Experimental Conditions

The research was carried out at the Institute of Mechanics, Energy and Biotechnology Engineering, the Thermal Energy Processes and Emissions Laboratory at Vytautas Magnus University, Lithuania. Liquid cattle manure used for the research was collected from the site near the milking room in a semi-deep cowshed, using the technique described in detail by our team earlier [10]. Briefly, manure samples were collected in 40 L chambers in the cowshed. To ensure the homogeneity of the manure in the chambers, the manure was thoroughly mixed in the barn with a manual mixer and then loaded into the chambers.

The biopreparation product used in this experiment was a probiotic composition (MycroZyme©) purchased from Micron Bio-Systems Ltd., Bridgwater, UK, and containing strains of *Bacillus subtilis*, *Bacillus licheniformis*, and *Bacillus amyloliquefaciens* in 7 billion bacterial colony forming units (CFU) per mL). Before use, the biopreparation was diluted (5 mL in 100 mL water) with water free of active chlorine based on the protocol of the company.

2.2. Methodology and Equipment for Measuring Ammonia Emissions from Manure

We performed eight trials to determine ammonia emissions from the stored manure without or with treatment using the biopreparation. In each trial, the first storage tank of liquid cattle manure with a capacity of 25 L was sprayed with the 100 mL diluted solution of the biopreparation. The second storage tank (chamber No. 2) with the same capacity (25 L) of liquid cattle manure was not sprayed and served as control. Ammonia emissions trials from the manure treated with biopreparations or not were performed in a wind tunnel bench, using the method described in detail earlier [10]. The manure was stored in both chambers under the same aerobic conditions at a constant temperature of 18.7 ± 0.2 °C. Periodically, every 10 days, each manure chamber was placed in a wind tunnel, and ammonia emissions from the manure were determined. The manure was mixed with a wooden spatula before each test. Each emission test was performed for up to

1.5 h. The test was performed at a constant air flow above the manure and at a constant ambient temperature. The air velocity in the 100 mm diameter duct was determined before each test at 0.1 ± 0.02 ms^{-1}. To analyse the influence of air ventilation velocity on ammonia emissions, studies were performed at air velocities of 0.1 and 0.7 m s^{-1} as described earlier [10]. The pH of the manure and the dry matter content of the manure were determined before the tests. Laboratory tests were performed for 3 months.

An analyser GME700 (laser spectroscopy) was used to measure the ammonia (NH_3) gas concentration (Supplementary Figure S1). The measuring range of NH_3 was from 0 to 2000 ppm. Air temperature and humidity were measured by temperature and humidity sensors connected to the "Almemo 2590-9" system. Measuring range was for temperature from 30 to 60 °C, relative humidity from 5 to 98%, and the accuracy of the device was ±0.1%.

2.3. Statitical Analysis

Data were tested for normality (Bonferroni test), and descriptive statistics was performed, including the means and the indicators of the variance for all variables measured. The reliability of statistical data was assessed according to the Student's *t*-test. The standard error and the cut-off for the minimum confidence interval were calculated at the statistical significance level of $p < 0.05$.

The dependency of the obtained variables over time was analysed using polynomial regression analyses. The goodness of the fit was evaluated by the determination coefficient (R^2).

3. Results and Discussion

3.1. Ammonia Gas Emissions from Cattle Manure Mixed or Sprayed with Biopreparation with Various Storage Times

Data of the changes in ammonia emissions from fresh manure either mixed and homogenized with biopreparation or without being mixed with biopreparation (control) are shown in Figure 1. In the control manure, there was an increase of the ammonia concentration in the wind tunnel from initially 10–15 ppm to around 25 ppm after 33 min, the time when a plateau was reached, as supported by the quadratic regression. However, when the biopreparation were mixed, the polynomial regression revealed a cubic term, indicating in the first 10 min a greater emission but decreasing trend of ammonia release up to 20 ppm after 33 min, likely suggesting an effect of the biopreparation starting after around 33 min of the activity.

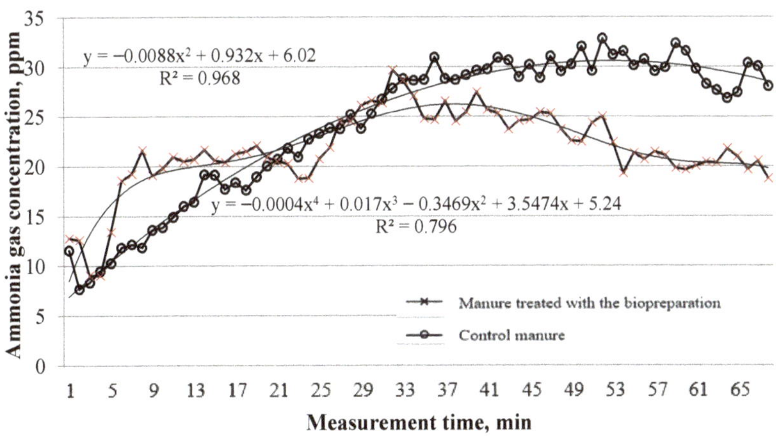

Figure 1. Ammonia concentration when fresh liquid cattle manure was mixed and homogenized with biopreparation or not (control) as measured in the wind tunnel up to 67 min.

When biopreparation was sprayed onto the fresh manure, there was a different curve of ammonia emission, which is shown in Figure 2. Accordingly, after spraying the manure on the surface with the biopreparation, a significant difference again with a cubic term in the values of ammonia concentration in the wind tunnel was found. As expected, the control manure performed the same as in the first case of mixed biopreparation shown in Figure 1.

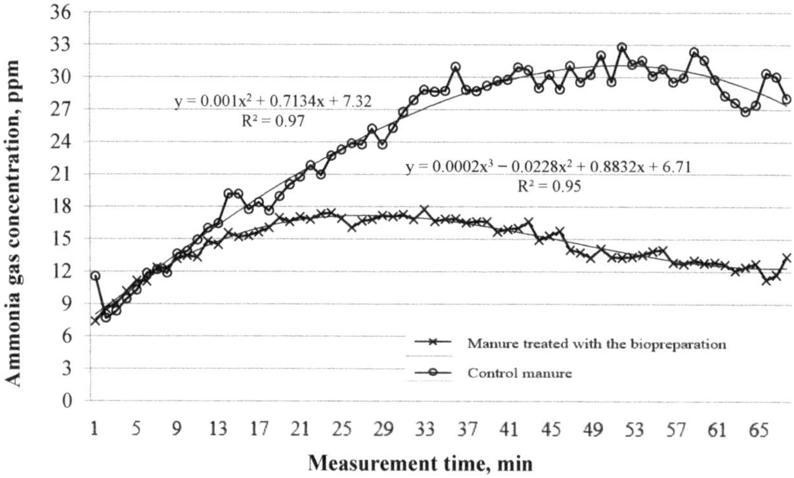

Figure 2. Change in the ammonia concentration in the wind tunnel with time after addition of manure sprayed with biopreparation (biologically treated manure) or the manure without additives (control).

Our data showed that there were significant (1.5-fold) reductions in ammonia emissions from the manure sprayed with the biopreparation on the fresh manure. As a whole, the ammonia concentrations ranged from 7 to 33 ppm during the testing. At the beginning of the testing, an increase in ammonia emissions was observed from both manures, which persisted for most of the study period, and then, the emissions from the control manure (manure without additives) stabilised. After 21–25 min, the ammonia emission from the manure sprayed with the biopreparation decreased polynomially, and after one hour in the wind tunnel, the concentration was less than 12 ppm. Based on the results of this measurement, it can be stated that spraying manure with the biopreparation reduced ammonia emissions, and therefore, the concentration in the wind tunnel decreased on average from 21.69 ± 0.9 to 14.43 ± 1.5 ppm. During the tests (within 70 min), the average ammonia concentration in the wind tunnel after the loading of the control manure was 24.3 ppm and after the loading of the manure treated with the biopreparation was 14.5 ppm, i.e., 40% lower.

The difference in ammonia emissions from the control manure and the manure treated with the biopreparation after certain periods confirms the results obtained previously in the manure of broilers [13] and laying hens [14] and shows the effectiveness of the biopreparation in influencing the targeted processes in manure and in reducing ammonia gas release into the environment. Rational use of the selected measures [15] can effectively reduce odours and ammonia gas emissions from the cattle manure. The ammonia reshuffle into simple compounds occurs in two stages. The first stage is nitrification, in which ammonia is oxidized to nitrite (nitritation), and then, nitrite is oxidized to nitrate (nitratation), whereas the second step is denitrification, in which nitrate is decomposed to nitrogen gas [16]. An explanation of the stronger effect of spraying biopreparation on the ammonia emission than of mixing can be the direct effect of the bacterial strains in the biopreparation on the pH in the surface of the manure. Bacterial probiotics are known to decrease in pH by producing lactic acid [17]. A lower pH can inhibit the urease activity [18] and hence degradation of manure ammonium nitrogen into ammonia gas.

After 10 days, another trial of ammonia emissions was performed without mixing the manure in both chambers, maintaining a constant air velocity of 0.1 ms^{-1} in the air duct. The results are presented in Figure 3.

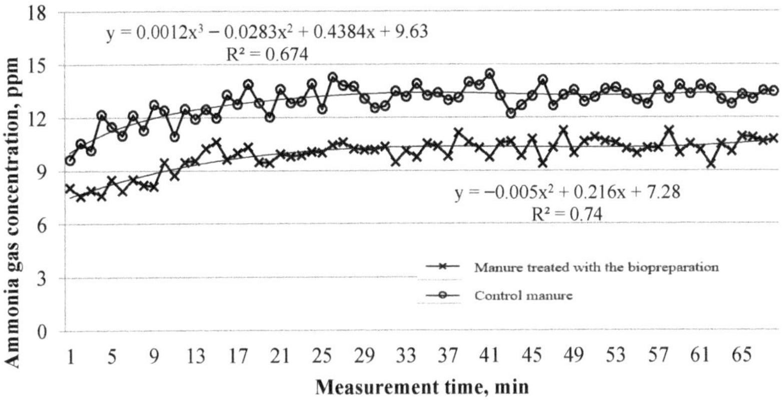

Figure 3. Change in ammonia concentration in the wind tunnel with the addition of the manure treated with the biopreparation and the control manure without additives (10 days after the addition of the biopreparation without mixing the manure).

Taking into account the changes in the manure treated with the biopreparation and the control manure presented in Figure 3, a uniform and insignificant change in gas concentration was found in both test manures when evaluated in parallel during the evaluation of the intensity of ammonia gas dispersion after 10 days. The ammonia emissions from both types of manure changed insignificantly: in the wind tunnel, the concentration varied from 9 to 15 ppm in the case of the control manure and from 7 to 11 ppm in the case of the manure treated with the biopreparation; i.e., the ammonia emissions were 23.2% lower. After 10 days of application of the biological preparation to the cattle manure, the ammonia emissions from the biologically treated manure decreased on average by two times, and regarding the manure without additives, the ammonia emissions were reduced by 1.89 times. Thus, no significant difference was observed between the determined ammonia emission values.

Manure was mixed in both chambers, and comparative studies of ammonia emissions were performed. The results are shown in Figure 4. After mixing the manure, the intensity of the ammonia emissions differed less from the two types of manure. In the wind tunnel, the average ammonia concentration over 70 min was 18.1 ppm for the control manure and 15.3 ppm for the manure treated with the biopreparation; i.e., the intensity of the ammonia emissions differed by 15.4%.

Forty-four days after the first spraying of the biopreparation on the manure surface, the manure was re-sprayed with the biopreparation, and 10 days after the second spraying, comparative ammonia emissions studies were performed by testing the control manure and the manure treated with the biopreparation. Studies were performed by increasing the ventilation intensity of the chamber so that the air velocity in the air duct was increased to 0.7 m/s (Figure 5).

The manure was mixed in both chambers prior to the ammonia emissions studies. At the beginning of the study, the ammonia emission from the manure in both chambers increased, and after 13–15 min, the emission process stabilised and continued to change slightly: in the wind tunnel, the average ammonia concentration was 8.2 ppm with the addition of the control manure and 7.3 ppm with the addition of the manure treated with the biopreparation, a difference of 11.0%.

Rather similar results were obtained after 66 days of the storage of the manure, which are shown in Figure 6. The values of ammonia emissions overlapped each other, and after 10 min, the ammonia gas concentrations in the wind tunnel are stabilized at a level around 11–12 ppm.

The fact that the concentration of ammonia is lower, and the effect of biopreparation on ammonia release is less can be explained with the building of the crust on the surface of the manure with time. As the manure dries, a crust forms on its outer layer, which acts as a protection against ammonia diffusion [19].

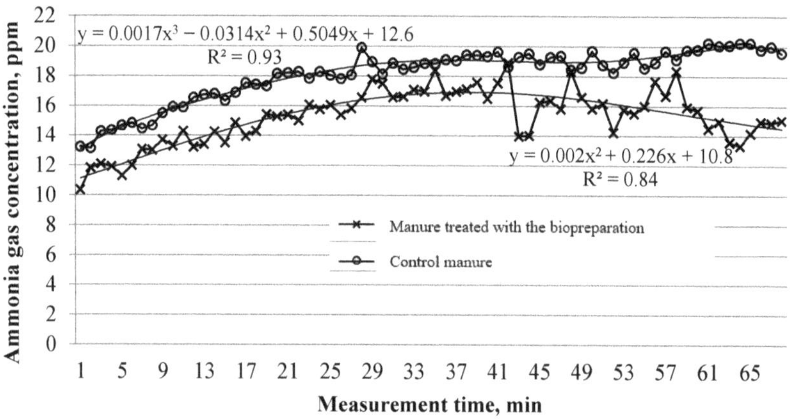

Figure 4. Change in ammonia concentration in the wind tunnel with the addition of the manure treated with the biopreparation and the control manure without additives (10 days after the addition of the biopreparation by mixing the manure).

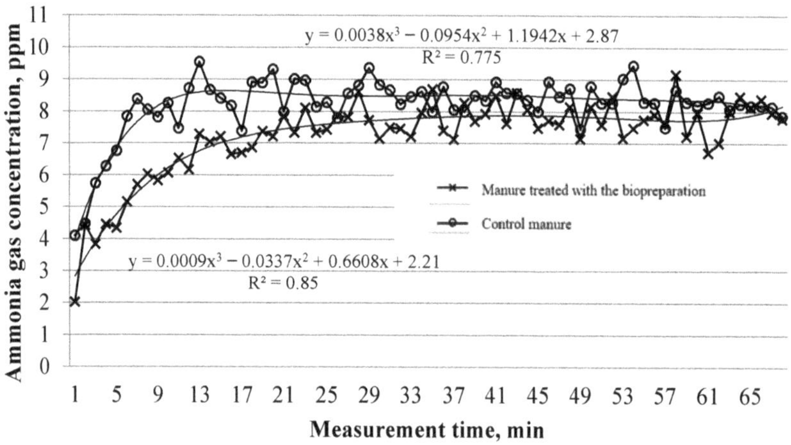

Figure 5. Change in ammonia concentration in the wind tunnel with the addition of the manure treated with the biopreparation and the control manure without additives (55 days after the addition of the biopreparation, the manure is mixed).

A summary of the changes in the ammonia in ppm and % relative to the control for various storage times is shown in Figure 7. According to the summarised indicators of ammonia emissions from the manure treated with the biopreparation and the control manure, differences were found only in certain periods. Significant reduction of the ammonia emissions from the manure treated with the biopreparation is achieved immediately after spraying of the biopreparation and by increasing air velocity after 55 days of manure storage, and a reduction in ammonia emissions from manure without additives was found after 10 days and also by increasing the chamber ventilation intensity. Comparing the ammonia emissions from the manure without additives and the manure treated with the biopreparation, the largest difference was found only after spraying the biopreparation and also without mixing the manure. The smallest differences in ammonia concentration were observed after 40 days of manure storage, and this could be explained by the decrease in the efficacy of the biopreparation.

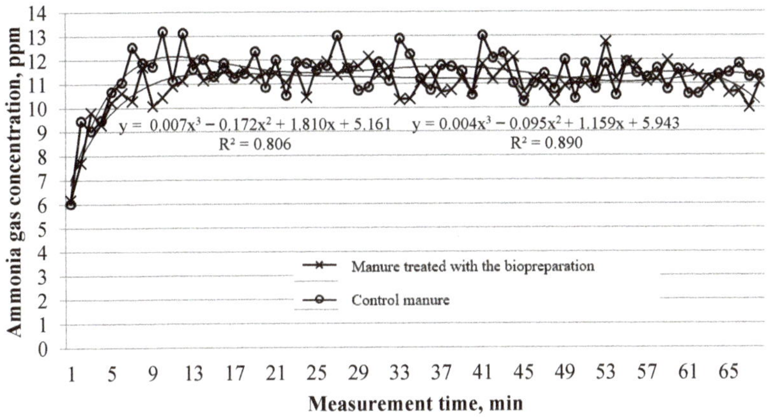

Figure 6. Change in ammonia concentration in the wind tunnel with the addition of the manure treated with the biopreparation and the control manure without additives (66 days after the addition of the biopreparation).

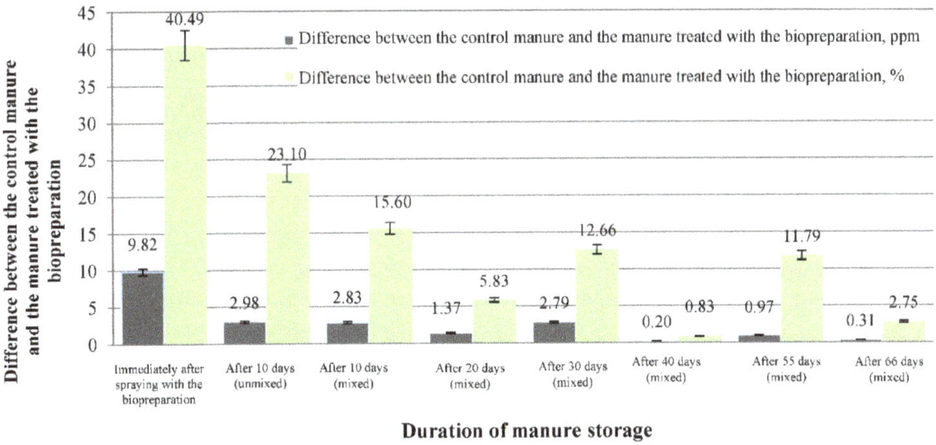

Figure 7. Difference between the average ammonia concentration values in the wind tunnel (ppm and %) after the addition of the control manure without additives and the manure treated with the biopreparation in the tunnel, depending on manure storage time 35.

3.2. *Ammonia Gas Emissions in an Environment of Variable Ventilation Intensity*

The diffusion of ammonia into the environment is influenced by air velocity. During intensive ventilation, the concentration of ammonia in the air above the manure is low, which promotes more intensive evaporation of ammonia [18]. However, with a strong air flow, the top layer of manure can dry out more intensively, which in turn reduces the evaporation of ammonia [19–21]. Figure 8 shows changes in the ammonia emissions from the manure treated with the biopreparation and the manure without additives with two different air velocities. Throughout the study, the concentration of ammonia in the biologically treated manure chamber increased and reached higher values than in the previous study after 55 days of manure storage (Figure 5). The increasing concentration of ammonia gas may have been influenced by other factors such as intensified activity of urobacteria for urease production or changes in chemical composition of the manure [22,23].

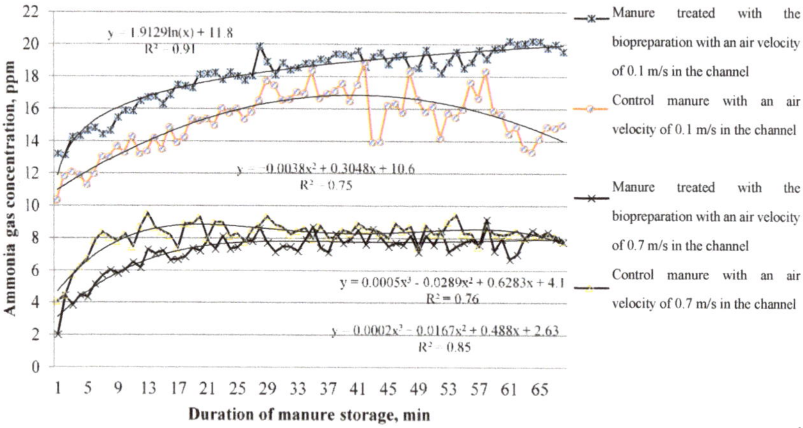

Figure 8. Change in ammonia concentration in the wind tunnel with the addition of the manure treated with the biopreparation and the control manure without additive at two different ventilation velocities.

In the present study, the values of ammonia gas concentration in the wind tunnel with the manure treated with the biopreparation and the control manure without additives varied 80% in parallel throughout the whole study period. However, the ammonia gas concentration values were on average 16±1 ppm for low (v = 0.1 m/s) and 7 ± 2 for the high (v = 0.7 m/s) air velocity. Theoretical studies have shown that the process of evaporation of ammonia from manure is determined by many variable and interrelated factors, one of the most important of which is air velocity. As the air velocity varies within the recommended limits in cowsheds (up to 2 m s^{-1} in summer), ammonia emissions change even more significantly. As the air velocity increases from 0.1 to 2.0 m s^{-1}, the ammonia emissions increase by as much as 10 times. As air velocity increases, its effect on emissions decreases. In the air velocity range 0–0.5 m s^{-1}, as the air velocity increases by 0.1 m s^{-1}, the ammonia emissions increase on average by 39.4%, and in the air velocity range 1.5–2.0 m s^{-1}, the emissions increase, respectively, by only 4.7%.

The obtained results confirm the statements of other researchers [24–28]: as the ventilation intensity increases, the ammonia concentration in the room decreases, and the emission increases disproportionately. Therefore, according to previous scientific studies, a 4.5-fold increase in ventilation intensity decreases the NH_3 concentration in the chamber by 1.9–2.4-fold, and the emission increases by 1.7–2.5-fold ($p < 0.05$). However, in order to accurately compare the actual emission values with the results of the studies by other authors, all factors (air flow, the content of dry matter in manure, the chemical composition of manure, the protein content of manure, temperature, manure storage time, and the drying of manure surface) must be considered, on which the effect of ventilation intensity on gaseous emissions from manure depends.

As the ventilation intensity increases, the ammonia concentration in the chamber decreases, and the emission increases rapidly [29]. As the ventilation intensity increases from 18.5 m^3 (h $m^2)^{-1}$ to 217.0 m^3(h $m^2)^{-1}$, the ammonia concentration decreases from 31 ppm to 7 ppm, and the emission increases from 406.4 mg $m^{-2}{\cdot}h^{-1}$ to 1124.2 mg $m^{-2}{\cdot}h^{-1}$.

The researchers found a reliable correlation between the intensity of ammonia emission from manure and the concentration of ammonia above manure, which is expressed as a function of a second-order polynomial (R^2 = 0.96) (Figure 9).

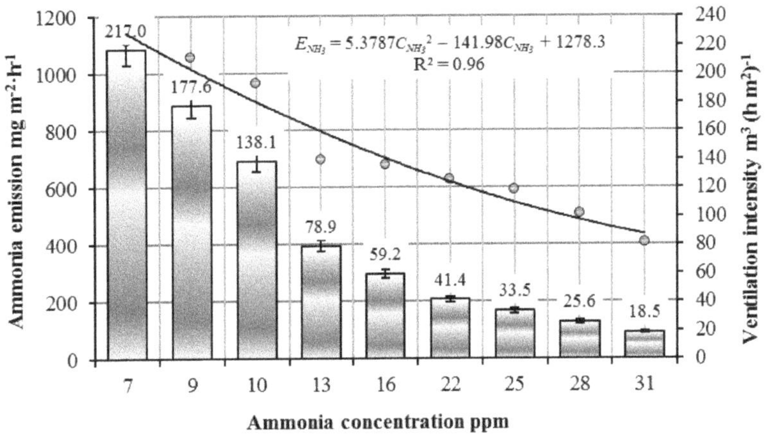

Figure 9. Dependence of ammonia emission (line) from the liquid cattle manure on ammonia concentration at different ventilation intensities (column).

As the ventilation intensity increases, the ammonia concentration in the chamber decreases, and the emission increases rapidly. Studies with manure stored for various periods, of which the surface had unevenly dried before crushing and formed a crust, showed different ammonia emissions from the manure as well as different effects of ventilation intensity on ammonia emissions.

4. Conclusions and Recommendations

In conclusion, the study showed that spraying manure with the biopreparation product reduced ammonia emissions by as much as 40%. The effect of the biopreparation on ammonia evaporation was higher in the presence of more intensive ammonia emission; i.e., when the manure is fresh, there is no crust on the surface, the air flow over the manure is intense, and there is a large gradient of ammonia concentration on the manure surface. Nevertheless, the process of ammonia evaporating from the manure is very complex. It depends on many factors: temperature, air velocity, ventilation intensity, pH, farming system (pastures, livestock housing, waste management), nitrogen content in feed and manure, carbon-to-nitrogen ratio in manure, and manure surface area. In order to make the most efficient use of the biopreparation, more detailed, experimental studies are needed to assess the influence of various different parameters on the interaction processes between the manure and the biopreparations.

Based on the findings of this research, the following recommendations can be made to reduce ammonia emissions in litter-less cowsheds, where liquid manure is stored:

1. The smallest differences in ammonia emission from both types of manure were obtained after 40 days, and an almost identical trend of ammonia emission from both chambers was found. This indicates the longest possible duration of efficacy of the biopreparation solution.
2. To reduce ammonia emissions and air pollution in barns, it is recommended to spray manure with a biopreparation. The effect of the biopreparation will be higher during the warm period, especially at temperatures above 20 °C.
3. After estimating the intensity of ammonia emissions from manure of different humidity, it is recommended to use the biopreparation in more liquid manure, from which gas emissions are significantly higher. It is recommended to use the biopreparation to

reduce ammonia emissions in litter-less cowsheds where liquid manure accumulates. Its use is in line with the innovative tendencies of modernisation of cowsheds to implement technologies designed for liquid manure in litter-less cowsheds.

4. By controlling the microclimatic factors in the manure environment, it is possible to influence the processes that take place on the manure surface and slow down the ammonia evaporation process from the manure, intensifying the crust formation and reducing the ammonia concentration gradient in the vertical above the manure. A greater effect of the biopreparation on the evaporation of ammonia can be obtained when the manure is not mixed.

Supplementary Materials: The following supporting information can be downloaded at: https://www.mdpi.com/article/10.3390/agriculture12101626/s1, Figure S1: Ammonia gas concentration measurement system with laser analyser GME700.

Author Contributions: Conceptualization, R.B. and R.M.; methodology, R.B. and R.M.; software, R.M.; validation, R.M.; investigation, R.M.; data curation, R.M.; writing—original draft preparation, R.M.; writing—review and editing, R.M.; visualization, R.M. All authors have read and agreed to the published version of the manuscript.

Funding: This research received no funding.

Institutional Review Board Statement: Not applicable.

Informed Consent Statement: Not applicable.

Data Availability Statement: The data presented in this study are available on request from the corresponding author. The data are not publicly available due to internal laboratory restrictions.

Conflicts of Interest: The authors declare no conflict of interest.

References

1. Rico, C.; García, H.; Rico, J.L. Physical–anaerobic–chemical process for treatment of dairy cattle manure. *Bioresour. Technol.* **2011**, *102*, 2143–2150. [PubMed]
2. Haselmann, A.; Zehetgruber, K.; Fuerst-Waltl, B.; Zollitsch, W.; Knaus, W.; Zebeli, Q. Feeding forages with reduced particle size in a total mixed ration improves feed intake, total-tract digestibility, and performance of organic dairy cows. *J. Dairy Sci.* **2019**, *102*, 8839–8849. [PubMed]
3. Khiaosa-ard, R.; Zebeli, Q. Cattle's variation in rumen ecology and metabolism and its contributions to feed efficiency. *Livest. Sci.* **2014**, *162*, 66–75.
4. Van der Stelt, B.; Temminghoff, E.J.M.; Van Vliet, P.C.J.; Van Riemsdijk, W.H. Volatilization of ammonia from manure as affected by manure additives, temperature and mixing. *Bioresour. Technol.* **2007**, *98*, 3449–3455. [PubMed]
5. Sanz, A.; Misselbrook, T.H.; Sanz, M.J.; Vallejo, A. Use of an inverse dispersion technique for estimating ammonia emission from surface-applied slurry. *Atmos. Environ.* **2010**, *44*, 999–1002.
6. McIlroy, J.P.; McGeough, K.L.; Laughlin, R.J.; Carolan, R. 2019. Abatement of ammonia emissions from dairy cow house concrete floor surfaces through additive application. *Biosyst. Eng.* **2019**, *188*, 320–330.
7. Martínez-Lagos, J.; Salazar, F.; Alfaro, M.; Misselbrook, T. Ammonia volatilization following dairy slurry application to a permanent grassland on a volcanic soil. *Atmos. Environ.* **2013**, *80*, 226–231. [CrossRef]
8. Hagenkamp-Korth, F.; Haeussermann, A.; Hartung, E.; Reinhardt-Hanisch, A. Reduction of ammonia emissions from dairy manure using novel urease inhibitor formulations under laboratory conditions. *Biosyst. Eng.* **2015**, *130*, 43–51.
9. Zubair, M.; Wang, S.; Zhang, P.; Ye, J.; Liang, J.; Nabi, M.; Zhou, Z.; Tao, X.; Chen, N.; Sun, K.; et al. Biological nutrient removal and recovery from solid and liquid livestock manure: Recent advance and perspective. *Bioresour. Technol.* **2020**, *301*, 122823.
10. Borowski, S.; Gutarowska, B.; Durka, K.; Korczyński, M.; Opaliński, S.; Kołacz, R. Biological deodorization of organic fertilizers. *Przemysł Chem.* **2010**, *89*, 318–323.
11. Gutarowska, B.; Matusiak, K.; Borowski, S.; Rajkowska, A.; Brycki, B. Removal of odorous compounds from poultry manure by microorganisms on perlite—Bentonite carrier. *J. Environ. Manag.* **2014**, *141*, 70–76. [CrossRef]
12. Naujokienė, V.; Bagdonienė, I.; Bleizgys, R.; Rubežius, M. Biotreatment Effect on Dynamics of Cattle Manure Composition and Reduction of Ammonia Emissions from Agriculture. *Agriculture* **2021**, *11*, 303. [CrossRef]
13. Eglite, S.; Ilgaza, A.; Butka, M. Reduction of ammonia emissions by applying probiotics on litter in a commercial breeding poultry house. *J. Artic. Agron. Res.* **2021**, *19* (Suppl. 2), 1015–1022.
14. Ali, M.; Zubair, M.; Rosyidi, A.; Amin, M. Screening of ammonia-degrading bacteria to reduce ammonia content in the manure of laying hens. *Earth Environ. Sci.* **2020**, *441*, 012138. [CrossRef]

15. Zhua, X.; Burgerb, M.; Doaneb, T.A.; Horwathb, W.R. Ammonia oxidation pathways and nitrifier denitrification are significant sources of N2O and NO under low oxygen availability. PNAS. *Agric. Sci.* **2013**, *110*, 6328–6333.
16. Tang, R.Y.; Wu, Z.L.; Wang, G.Z.; Liu, W.C. The effect of Bacillus amyloliquefaciens on productive performance of laying hens. Italian. *J. Anim. Sci.* **2018**, *17*, 436–441. [CrossRef]
17. Emerson, K.; Russo, R.C.; Lund, R.E.; Thurston, R.V. Aqueous ammonia equilibrium calculations: Effect of pH and temperature. *J. Fish. Res. Bd. Can.* **1975**, *32*, 2379–2383. [CrossRef]
18. Ahring, B.K. Perspectives for anaerobic digestion. *Adv. Biochem. Eng./Biotechnol.* **2003**, *81*, 1–30.
19. Brose, G. Emission von klimarelevanten gasen, ammoniak und geruch aus einem milchviehstall mit schwerkraftlüftung. Ph.D. Thesis, Universität Hohenheim, Stuttgart, Germany, 2000; p. 136.
20. Zhang, G.; Bjerg, B.; Strøm, J.S.; Morsing, S.; Kai, P.; Tong, G.; Ravn, P. Emission effects of three different ventilation control strategies-A scale model study. *Biosyst. Eng.* **2008**, *100*, 96–104.
21. Demmers, T.G.M.; Burgess, L.R.; Short, J.L.; Phillips, V.R.; Clark, J.A.; Wathes, C.M. First experiences with methods to measure ammonia emissions from naturally ventilated cattle buildings in the UK. *Atmos. Environ.* **1998**, *32*, 285–293. [CrossRef]
22. Aguerre, M.J.; Wattiaux, M.A.; Powell, J.M. Emissions of ammonia, nitrous oxide, methane, and carbon dioxide during storage of dairy cow manure as affected by dietary forage–to–concentrate ratio and crust formation. *J. Dairy Sci.* **2012**, *95*, 7409–7416. [CrossRef] [PubMed]
23. Zhang, N.; Bai, Z.; Winiwarter, W.; Ledgard, S.; Luo, J.; Liu, J.; Guo, Y.; Ma, L. Reducing Ammonia Emissions from Dairy Cattle Production via Cost-Effective Manure Management Techniques in China. *Environ. Sci. Technol.* **2019**, *53*, 11840–11848. [CrossRef]
24. Philippe, F.; Cabaraux, J.; Nicks, B. Ammonia emissions from pig houses: Influencing factors and mitigation techniques. *Ecosyst. Environ.* **2011**, *141*, 245–260. [CrossRef]
25. Samer, M.; Ammon, C.; Loebsin, C.; Fiedler, M.; Berg, W.; Sanftleben, P.; Brunsch, R. Moisture balance and tracer gas technique for ventilation rates measurement and greenhouse gases and ammonia emissions quantification in naturally ventilated buildings. *Build. Environ.* **2012**, *50*, 10–20. [CrossRef]
26. Blanes–Vidal, V.; Hansen, M.N.; Pedersen, S.; Rom, H.B. Emissions of ammonia, methane and nitrous oxide from pig houses and slurry: Effects of rooting material, animal activity and ventilation flow. *Agric. Ecosyst. Environ.* **2008**, *124*, 237–244. [CrossRef]
27. Kang, J.H.; Lee, S.J. Improvement of natural ventilation in a large factory building using a louver ventilator. *Build. Environ.* **2008**, *43*, 2132–2141. [CrossRef]
28. Bagdonienė, I. Ammonia emissions from cattle manure in the environment of changing microclimate factors. Ph.D. Thesis, Alexandras Stulginskis University Academy, Kaunas, Lithuania, 2013; p. 91.
29. Bagdonienė, I.; Bleizgys, R. Ammonia emissions from dairy cattle manure under variable ventilation rates. *Ann. Anim. Sci.* **2014**, *14*, 141–151. [CrossRef]

agriculture

MDPI

Article

Spatial Distribution, Pollution Characteristics, and Health Risk Assessment of Heavy Metals in Soils from a Typical Agricultural County, East China

Linsong Yu [1,2,*], Hongbo Liu [1], Weidong Liu [3], Pinrui Qin [3], Jian Yu [3], Bing Zhou [2], Fugui Zhang [4], Ziwan Chen [2], Yuyan Zhao [5] and Zeming Shi [2]

1 Shandong Institute of Geophysical and Geochemical Exploration, Jinan 250013, China
2 School of Earth Sciences, Chengdu University of Technology, Chengdu 610059, China
3 Shandong Provincial Bureau of Geology & Mineral Resources, Jinan 250013, China
4 Institute of Geophysical & Geochemical Exploration, Chinese Academy of Geological Sciences, Langfang 065000, China
5 College of Geo-Exploration Science and Technology, Jilin University, Changchun 130012, China
* Correspondence: sean_yls@163.com

Abstract: This study collected 1135 topsoil samples of cultivated land in Laiyang County, eastern China, to analyze the distribution, pollution and health risks of heavy metals (As, Cd, Cu, Cr, Hg, Ni, Pb and Zn). The results show that: (1) the levels of eight heavy metals at some sites were higher than the risk screening values in China, and there was heavy metal pollution. (2) An improved geoaccumulation index was used to evaluate soil pollution. The average value of I_{geo} before and after improvement was 0.32 (I_{ml}, no pollution to medium pollution) and −0.04 (I_{ol}, no pollution), respectively. (3) Hg and Cd were identified as the main contributors to ecological risk in this study, with a cumulative ecological risk contribution percentage > 65%. The results of the potential ecological risk index (*PERI*) show that 9.3% of the sampling sites were considered to have moderate ecological risk. (4) As, Pb, Ni and Cu made a contribution of >95% in terms of non-carcinogenic risk to adults and children through different exposure routes, and different soil intake routes posed no non-carcinogenic risk to adults; there was a sampling site with $HQ_{\text{Children}} > 1$, which was consistent with the non-carcinogenic risk site, indicating that children in this location and surrounding areas are more likely to face dual health risks. Therefore, it is necessary to promote the risk management of heavy metals in the study area in order to safely use soil resources.

Keywords: cultivated soil; heavy metal; spatial analysis; ecological risk assessment; health risk assessment

1. Introduction

As the main component of cultivated land, the quality of soil can directly affect crop growth, food safety, and even biological health [1–5]. According to the *National Survey Bulletin on Soil Pollution* (2014) [6], 19.4% of the cultivated land in China exceeded the designated standards for heavy metals, among which cadmium (Cd), nickel (Ni), copper (Cu), arsenic (As), mercury (Hg), and lead (Pb) were identified as the main pollutants. The *Global Assessment of Soil Pollution: Report* (2021) [7] pointed out that the growing demand for agricultural products and industrial systems, as well as the growing global population, has led to the serious pollution of cultivated soil. The accumulation of heavy metals in cultivated soil leads to declines in soil fertility and soil function, which affect the yield and quality of crops [3]. Moreover, these heavy metals can enter the human body via the food chain, thereby posing a considerable threat to human health [8]. Consequently, the Joint Environmental Protection Agency has listed the eight heavy metals Pb, Cd, chromium (Cr), Hg, As, Cu, zinc (Zn), and Ni as priority targets for pollutant control [9,10].

Cultivated soil is one of the prominent sinks of heavy metals. A combination of high geological background, secondary enrichment during pedogenesis, and human activities can contribute varying degrees of heavy metal pollution, which often exceed the respective standards for cultivated soil [11,12]. Heavy metals are characterized by high toxicity, non-biodegradability, and bioaccumulation in cultivated soils [13–17], and can enter the human body directly via skin contact, inhalation, or ingestion, thereby posing a potential risk to health [18,19], examples of which include *Minamata disease* caused by Hg pollution [20,21] and *Itai-itai disease* attributable to the long-term consumption of Cd-contaminated rice [22,23]. Consequently, it is necessary to study the enrichment, spatial distribution, and sources of heavy metals in soil, particularly from the perspective of pollution and health risk assessments. In this regard, the most effective strategy for dealing with the pollution of cultivated land is to conduct comprehensive soil environmental quality and pollution assessments, examine the potential ecological and health risks of heavy metals, and take proactive measures to avoid the high labor and capital costs of future pollution treatment and landscape restoration [24,25]. Among the analytical approaches currently adopted to determine the correlations and sources of heavy metals in soil, principal component analysis (PCA), cluster analysis, and the positive matrix factorization (PMF) model are routinely applied, whereas the ecological risk index and health risk assessments are mainly used to evaluate the ecological impact of heavy metals and determine the risk of exposure to soil-borne heavy metals [26–31]. Moreover, some scholars [13,25] are of the opinion that in order to effectively and objectively evaluate the pollution status and risks posed by soil-borne heavy metals in a specific area, it is necessary to concurrently evaluate the spatial distribution and pollution characteristics of these metals, as well as conducting risk assessments.

The area selected for investigation in the present study is located in Laiyang County in the east of China, which from a geomorphological perspective is a self-contained small watershed system, with cultivated land distributed on either side of the main river. The watershed area is small and the water system is simple, and as such, is less affected by external interference factors. By studying such small watersheds, it is possible to more accurately identify local heavy metal distribution, pollution, and sources, and conduct health risk characteristic assessments. Focusing on this hilly agricultural area of Laiyang County, we set out with the following research objectives: (1) to analyze the content and distribution characteristics of heavy metals in cultivated soil, based on heavy metal reference values, correlations, and surface geochemical characteristics; (2) to analyze the pollution characteristics of heavy metals in cultivated land using an improved geoaccumulation index method; and (3) to assess the potential ecological risks of heavy metals in cultivated soil and the health risks associated with human exposure. The results of this study will contribute to furthering our understanding of the spatial distribution and risk characteristics of soil-borne heavy metals in cultivated land lying within small watersheds, as well as the accurate classification management and safe utilization of soils in typical small watersheds with high geological backgrounds.

2. Materials and Methods

2.1. Study Area

Laiyang County, located in the east of Shandong Province (Figure 1), is the main area in China for the cultivation of Laiyang pears, a renowned high-quality agricultural product. The geographical coordinates of the study area, which covers approximately 1091 km^2, are 120°31′–121°00′ E, 36°48′–37°09′ N. The main land-use type is arable land, which accounts for 83% of the total. The study area lies within the north temperate monsoon zone. There are four distinct seasons, a year-round southeasterly prevailing wind, and sufficient sunlight. Annual averages for precipitation, temperature, relative humidity, sunshine hours, wind speed, and frost-free periods are 800 mm, 11.2°C, 73%, 2996 h, 2.7 m/s, and 173 days, respectively. The study area is typical of the hilly landforms in eastern Shandong Province, with the terrain sloping from north to south (Figure 1a). Mountains, hills, and plains account for 21.5%, 47.06%, and 31.43% of the total area, respectively. The area is located along the northeastern margin of the Jiaolai Basin of the North China Plate, where northeasterly trending fault structures have developed. Rock outcrops include Archean, Proterozoic basement metamorphic rock series, Mesozoic Cretaceous, Cenozoic Paleogene, Neogene, and magmatic rocks, which are mainly distributed in the northern part of the study area. On the basis of the relationship between the rocks and the soil-forming parent materials, the outcrops are mainly classified as Quaternary sediments, sandstones, volcanic rocks, metamorphic rocks, granites, diorites, and basalt parent materials (Figure 1b). The soil types in the study area mainly include brown, riparian fluvo-aquic, coarse bone, and cinnamon soils. The study area is dominated by agriculture, and the non-metallic minerals are mainly quarried for construction material [32–34].

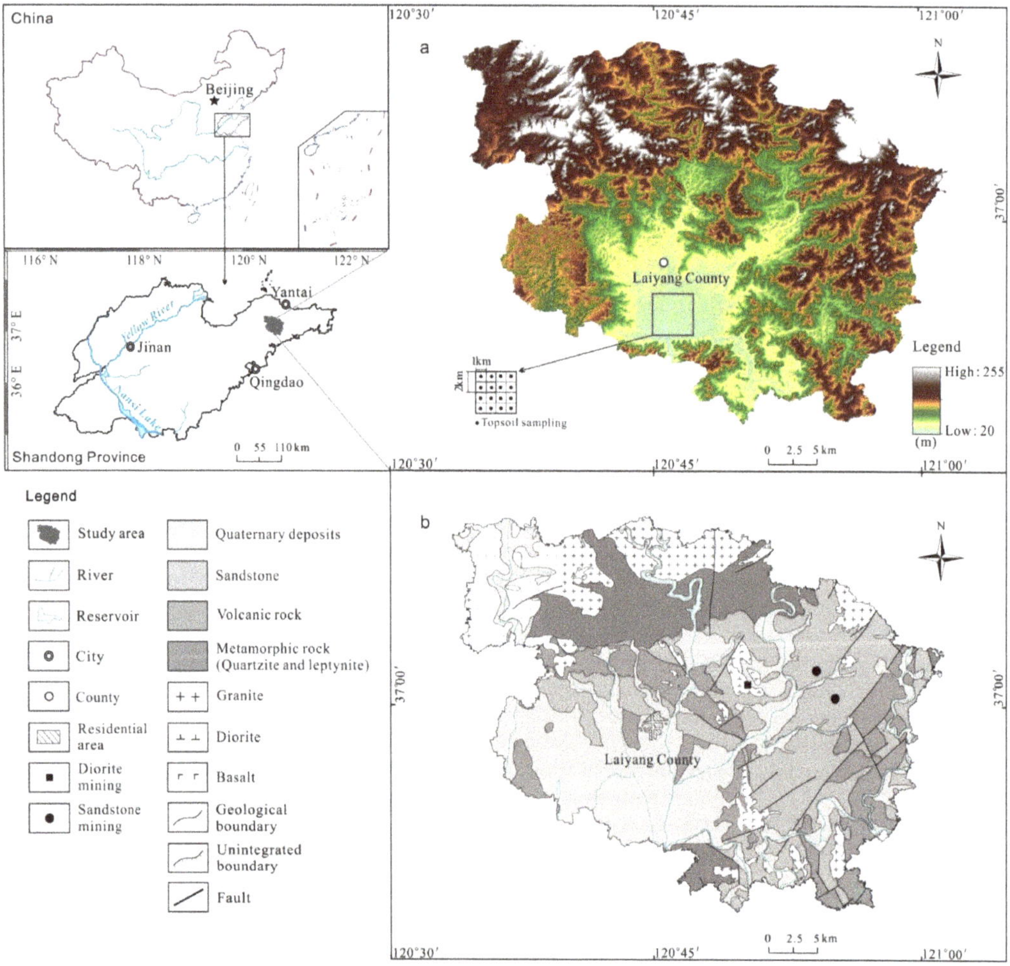

Figure 1. (**a**) Location and (**b**) geological map of the study area.

2.2. Sampling and Analysis

Topsoil samples were collected from 1135 sampling sites in cultivated land in the study area at a sampling density of 1 point km^{-2} (Figure 1a). The topsoil samples were extracted to a depth of 20 cm, and at each sample site we collected of five sub-samples, which were mixed to give a single composite sample, the fresh mass of which was at least 1500 g.

The soil samples were placed in a ventilated area to dry naturally. Then, they were passed through a 10-mesh (pore size 2 mm) nylon sieve. For each sample, 100g was obtained via an aliquot device for sample preparation, ground, and passed through a 200-mesh (pore size 75 μm) nylon sieve for analysis. The techniques applied for heavy metal analyses were as follows. (1) Pb, Ni, Cr, Zn, and Cu were determined by X-ray fluorescence spectrometry (Axiosmax; PANalytical B.V, Almelo, The Netherlands), for which 4.00 g aliquots of sampled soils were initially introduced evenly into a low-pressure polyethylene plastic ring, which was placed on a press, and subjected to powder pressing prior to analysis. (2) Cd was determined by graphite furnace atomic absorption spectrometry (GF-AAS, PE600; Thermo Elemental, Waltham, MA, USA). Aliquots (0.2500 g) of powdered sample were placed in a polytetrafluoroethylene crucible, and digested by the addition of a mixed acid solution ($HF–HClO_4–HNO_3$). The samples were then shaken to obtain a constant volume for testing. (3) As and Hg were determined by hydride generation–atomic fluorescence spectrometry (AFS9750; Beijing Haiguang Instrument, Beijing, China). Soil samples (0.5000 g) were placed in colorimetric tubes, and dissolved by the addition of aqua regia and incubation in a water bath. Following incubation, the suspensions were made up to volume by the addition of 10% hydrochloric acid and shaken well prior to determination. The pH of samples was determined using a potentiometric method (PHS-3C; Shanghai Precision Scientific Instrument Co., Ltd., Shanghai, China), the values of which are displayed to two decimal points. The detection limits set for As, Cd, Cr, Cu, Hg, Ni, Pb, and Zn were 0.2, 0.02, 2, 1, 0.0003, 2, 1, and 2 mg kg^{-1}, respectively, which are all below the minimum requirements of the *Land Quality Geochemical Evaluation Specifications* (DZ/T0295-2016). Prior to use, all glassware used in analyses was soaked in 10% nitric acid solution for 24 h, all reagents were of analytical grade, and the water used for analyses was ultrapure. For analytical purposes, we determined values for 50 duplicate samples and four national standard soil materials (GBW-07403) for quality control. The sample repeatability inspection pass rate and the abnormal point repeated inspection pass rate were established to be 92.1–97.5%, and 95.3–99.6%, respectively, and the values obtained for the standard samples were all within the accepted error range.

2.3. Statistical Analysis

SPSS 19.0 (IBM Inc., Armonk, NY, USA) and Origin 2022 (Origin Lab Corporation., Northampton, MA, USA) were used for data processing and statistical analysis. The number of decimal places of the reported values for raw data and various statistical parameters were determined by comprehensively considering the national standard detection limit and the reliable value of the actual detection limit in the laboratory. The suitability of the sample data was assessed using the Kaiser–Meyer–Olkin (KMO) and Bartlett spherical tests. Maps were produced using ArcGIS 10.2 (Esri, Redlands, CA, USA), and graphics were generated using CorelDraw X8 (Corel, Ottawa, ON, Canada).

2.4. Evaluation Method

2.4.1. The Geoaccumulation Index Method and Its Improvement

The geoaccumulation index [35] is a quantitative index that is used to assess the degree of heavy metal pollution in soils and sediments [36,37]. The index takes into account the extant natural geological processes, and can directly reflect the level of heavy metal pollution through changes in the measured values of heavy metal elements relative to the environmental background value [14]. It is calculated using the following formula:

$$I_{geo} = \log_2 \frac{C_i}{K \times B_i}, \quad (1)$$

where I_{geo} is the geoaccumulation index (abbreviated as I_{ol} herein); C_i is the concentration of heavy metal i in the surface soil of the study area (mg kg^{-1}); Bi is the geochemical background value of heavy metal i in the soil (mg kg^{-1}); and K is the change conversion coefficient (the general value of which is 1.5). A change in geoaccumulation index values can reflect the pollution characteristics of soil. The corresponding grading standards are as follows [14,35,37]: unpolluted ($I_{geo} < 0$), unpolluted to moderately polluted ($0 \leq I_{geo} < 1$), moderately polluted ($1 \leq I_{geo} < 2$), moderately to severely polluted ($2 \leq I_{geo} < 3$), severely polluted ($3 \leq I_{geo} < 4$), severely to extremely polluted ($4 \leq I_{geo} < 5$), and extremely polluted ($I_{geo} \geq 5$).

However, the uncertainty in the selection of evaluation parameters and the size of samples used in the traditional evaluation of the cumulative index results in certain differences in the evaluation of soil accumulation, and the large span of the degree of pollution classification standards also contributes to inaccuracies in the evaluation results [29,38]. To overcome these problems, in the present study, we adopted an enhanced version of the geoaccumulation index method [38]. Calculations performed using the Nemerrow index method [15,39] are based on the sub-index results of a single factor, taking the maximum sub-index and the mean value of the sub-index, which can not only reflect the influence of a single factor but also deals with the comprehensive situation of multiple factors. Using this method, the average values of heavy metal concentrations in soil at sampling sites in the study area and the square mean of the maximum value of heavy metal concentrations were used to replace the single heavy metal concentration in formula (1). Accordingly, this approach not only comprehensively takes into consideration the pollution of heavy metals at each sampling point, but also highlights the severity of pollution. It can thus effectively evaluate the degree of specific heavy metal pollution in the watershed [38]. The formulae used to derive values for the improved geoaccumulation index are as follows:

$$P_{avel} = \frac{1}{n}\sum\nolimits_{i=1}^{n} \frac{C_n}{K \times B} \quad (2)$$

$$P_{maxl} = \frac{C_{max}}{K \times B} \quad (3)$$

$$I_{ml} = \log_2 \sqrt{\left(P_{avel}^2 + P_{maxl}^2\right)/2}, \quad (4)$$

where P_{ave1} is the average value of the heavy metal concentration variation index of n sampling points in the study area; C_n is the heavy metal concentration of the nth sampling site (mg kg^{-1}); and B is the environmental background value of a specific heavy metal, which in this study is the geochemical background value of topsoil in Shandong Province (mg kg^{-1}). P_{max1} is the variation index of heavy metal concentrations at the sampling point with the highest concentration; C_{max} is the maximum concentration of heavy metals (mg kg^{-1}); and I_{m1} is the improved geoaccumulation index. The value of the coefficient K remains 1.5, and we used the grading standard of the in situ cumulative index method.

2.4.2. Potential Ecological Risk Index

The potential ecological risk index (*PERI*) was first proposed by the Swedish scholar Hakanson [40] and combines toxicological, environmental, chemical, and ecological effects to express the potential ecological risks of heavy metals via intuitive and interpretable quantitative values [30,41–43]. It is calculated using the following formulae:

$$E_r^i = T_r^i \times \frac{C^i}{C_n^i} \quad (5)$$

$$PERI = \sum\nolimits_{i=1}^{m} E_r^i, \quad (6)$$

where C_i is the topsoil content of heavy metal element i (mg kg^{-1}); ${C_n}^i$ is the geochemical background value of heavy metal element i in the topsoil of Shandong Province (mg kg^{-1}); ${E_r}^i$ is the potential ecological hazard index of heavy metal element i; and ${T_r}^i$ is the toxicity response coefficient of soil heavy metal i, which refers to the heavy metal toxicity level sequence proposed by Hakanson [40]: Hg (=40) > Cd (=30) > As (=10) > Pb = Cu = Ni(=5) > Cr (=2) > Zn (=1). *PERI* serves as a comprehensive ecological risk index for a specific environment, which is the sum of all risk factors for each heavy metal in the cultivated soil environment. The single element potential ecological hazard index ($E_r{}^i$) is divided into the following five grades from low to high [31,44]: $E_r{}^i < 40$ represent a slight ecological hazard; $40 \leq E_r{}^i < 80$, a moderate ecological hazard; $80 \leq E_r{}^i < 160$, a considerable ecological hazard; $160 \leq E_r{}^i < 320$, a strong ecological hazard; and $E_r{}^i \geq 320$, a significantly strong ecological hazard. *PERI* is divided into the following four grades from low to high [31]: $PERI < 150$ represents a slight ecological risk; $150 \leq PERI < 300$, a moderate ecological risk; $300 \leq PERI < 600$, a strong ecological risk; and $PERI \geq 600$, a significantly strong ecological risk.

2.4.3. Health Risk Assessment

The health risk associated with heavy metals is dependent on two factors, one of which is the level of environmental pollution, including the concentration, form, and toxic effects of heavy metals [26,45], and the second is human exposure behavior, including the behavior and characteristics of humans exposed to environmental heavy metals [18,27,46]. Heavy metals in soils can enter the human body through hand–oral ingestion, skin contact, and inhalation via attachment to dust particles, thereby posing a potential risk to health [14,47–50]. In this study, we applied three exposure methods to calculate the daily average exposure to topsoil in the study area. The evaluation models included non-carcinogenic risk and carcinogenic risk models, the values for which were obtained using the following formulae:

$$ADD_{ing} = C \times \frac{IngR \times EF \times ED \times CF}{BW \times AT}, \tag{7}$$

$$ADD_{inh} = C \times \frac{InhR \times EF \times ED \times CF}{PEF \times BW \times AT}, \tag{8}$$

$$ADD_{derm} = C \times \frac{SA \times AF \times ABF \times EF \times ED \times CF}{BW \times AT}, \tag{9}$$

$$HI = \sum_{i=1}^{n} HQ = \sum_{i=1}^{n} \frac{ADD(ingestion,\ inhalation,\ dermal)_i}{RfD_i}, \tag{10}$$

$$TCR = \sum_{i=1}^{m} CR_i = \sum_{i=1}^{m} ADD_i(ingestion,\ inhalation,\ dermal) \times SF_i, \tag{11}$$

where *C* is the topsoil heavy metal content (mg kg^{-1}); *ADD* is the average daily exposure (mg kg^{-1}); *RfD* is the non-carcinogenic reference dose of the exposure route; *HQ* is the non-carcinogenic index of heavy metal *i*; and *HI* is the total non-carcinogenic hazard index of heavy metal at the sampling point. If $HI < 1$, the health risk is small or negligible, whereas when $HI \geq 1$, there is a non-carcinogenic risk. SF_i is the carcinogenic risk coefficient of heavy metal *i*; *CR* is the carcinogenic risk of heavy metal *i*; and *TCR* is the total carcinogenic risk of heavy metals As and Cd. A *TCR* value $< 1 \times 10^{-6}$ is indicative of no risk, *TCR* values from 10^{-6} to 10^{-4} indicate an acceptable risk, and a *TCR* value $> 1 \times 10^{-4}$ denotes a high risk [14,26,47–51]. The parameters of the three exposure models are shown in Table 1, and the risk slope coefficient and reference dose of each heavy metal health risk assessment are shown in Table 2.

Table 1. Parameters for exposure models.

Factors	Description		Unit	Values		Source
				Adult	Child	
C	The topsoil heavy metal content		mg kg^{-1}			This study
IngR	Ingestion rate		mg d^{-1}	100	200	[9,26]
InhR	Inhalation rate		m^3 d^{-1}	7.5	15	[9,26]
EF	Exposure frequency		d a^{-1}	350	350	[49,51]
ED	Exposure duration		a	24	6	[9,26]
CF	Conversion factor		kg mg^{-1}	1×10^{-6}	1×10^{-6}	[14,26]
SA	Exposed skin area		cm^2	5700	2800	[9,26]
AF	Skin adherence factor		mg cm^{-2} d^{-1}	0.07	0.2	[9,26]
ABS	Dermal absorption factor		unitless	0.01	0.01	[14]
PEF	Particle emission factor		m^3 kg^{-1}	1.36×10^{-9}	1.36×10^{-9}	[9,26]
BW	Average body weight		kg	62.5	16	[26,49]
AT	Average time	non-carcinogens	d	ED × 365		[47,48]
		carcinogens		70 × 365		

Table 2. Reference doses (*RfDs*) for non-carcinogenic metals and slope factors (*SFs*) for carcinogenic metals (mg kg^{-1}).

	RfD			Source	*SF*			Source
	Ingestion	Dermal	Inhalation		Ingestion	Dermal	Inhalation	
As	3.00×10^{-4}	1.23×10^{-4}	3.00×10^{-4}	[26,50]	1.50	1.50	4.30×10^{-3}	[46]
Cd	1.00×10^{-4}	1.00×10^{-5}	1.00×10^{-4}	[14]	6.10	6.10	1.80×10^{-3}	[46]
Cu	4.00×10^{-2}	1.20×10^{-2}	4.02×10^{-2}	[14]	–	–	–	
Cr	3.00×10^{-3}	6.00×10^{-5}	2.86×10^{-5}	[26,50]	–	–	42.00	[46]
Hg	3.00×10^{-4}	2.10×10^{-5}	3.00×10^{-4}	[14,26]	–	–	–	
Ni	2.00×10^{-2}	5.40×10^{-3}	2.06×10^{-2}	[26]	–	–	8.40×10^{-1}	[46]
Pb	3.50×10^{-3}	5.25×10^{-4}	3.25×10^{-3}	[14]	–	–	–	
Zn	3.00×10^{-1}	6.00×10^{-2}	3.00×10^{-1}	[26]	–	–	–	

3. Results and Discussion

3.1. Heavy Metal Contents

The descriptive statistics (Table 3) indicate that the pH of the topsoils in the study area generally followed a normal distribution, with recorded values of between 4.66 and 7.82. The average value of pH was 6.23, and the standard deviation was 0.63. There were few data deviating from the mean. The contents of heavy metals ω(As), ω(Cd), ω(Cu), ω(Cr), ω(Hg), ω(Ni), ω(Pb), and ω(Zn) ranged from 1.9 to 66.6, 0.026 to 1.836, 8.4 to 251.7, 11.4 to 611.1, 0.006 to 0.79, 3.4 to 222.1, 6.3 to 327.2, and 263.4 to 345.3 mg kg^{-1}, respectively. Compared with the geochemical background values of topsoils in Shandong Province [52], the average values of Cu, Cr, Hg, Ni, Pb, and Zn in the soil were significantly higher, and the proportions of sites at which levels exceeded the corresponding background values were 66.6%, 76.6%, 42.2%, 57.6%, 54.8% and 52.9%, respectively. Compared with the average values of the topsoil in China [11], the average values of Cu, Cr, and Ni in the soil were higher, and the proportion of sites at which levels exceeded the average values were 65.5%, 79.2%, and 57.8%, respectively. This indicated that the study area had a background of high contents of Cu, Cr, and Ni. Compared with the median values of the world topsoil [11], the average contents of As, Cu, and Cr were relatively high, and compared with the *Agricultural Land Soil Pollution Risk Control Standards* (GB 15618-2018), the number of sampling sites at which topsoil levels of As, Cd, Cu, Cr, Hg, Ni, Pb, and Zn exceeded the respective risk screening values were 5, 6, 24, 44, 1, 1, 3, and 1. In addition, it can be seen from Figure 2 that when the pH value was between 6.0 and 7.5, the contents of As, Cu, Cr, Pb, and Ni were the highest. Statistical analyses revealed the heavy metals As, Cd, Hg, and Pb to be strongly positively skewed, thereby indicating that there were a few samples with large values, which contributed to an extension of the right-hand side of the curve.

Table 3. Heavy metal concentrations in surface soils (N = 1135) (mg kg^{-1}).

		pH	As	Cd	Cu	Cr	Hg	Ni	Pb	Zn
Minimum		4.66	1.9	0.026	8.4	11.4	0.006	3.4	6.3	23.4
Maximum		7.82	66.6	1.836	251.7	611.1	0.79	222.1	327.2	345.3
Mean		6.23	8.01	0.118	33.33	89.26	0.037	33.63	25.37	69.46
Median		6.2	7.3	0.103	26.8	73.1	0.029	28.9	24.2	64.8
S.E.		0.02	0.13	0.003	0.65	1.8	0.001	0.63	0.43	0.76
S.D.		0.63	4.24	0.1	21.9	60.66	0.04	21.11	14.36	25.75
CV (%)		10.1	52.9	82.3	65.7	68	98.5	62.8	56.6	37.1
Skewness		0.11	4.2	9.11	3.15	3.59	10.31	3.32	12.14	2.71
Kurtosis		−0.7	41.71	129.3	15.94	17.01	180.06	14.89	212.2	18.04
Shandong topsoil (1)		7.32	8.6	0.132	22.6	62	0.031	27.1	23.6	63.3
China topsoil (2)		—	11.2	0.097	23	61	0.065	27	26	74
World topsoil (3)		—	6	0.35	30	70	0.06	50	35	90
SEQRCS (4)	pH < 6.5	—	30	0.3	50	150	0.5	70	90	200
	6.5 < pH < 7.5	—	25	0.3	100	200	0.6	100	120	250
	pH > 7.5	—	20	0.6	100	250	1	190	170	300

(1) Background values of soil geochemistry in Shandong Province, East China [52]. (2) China National Environmental Monitoring Centre (1990) and Wei F S (1991), N = 4095 [11]. (3) Median values of soil element content in different countries, F. Bowen (1979) [11]. (4) Soil environmental quality risk control standard for soil contamination of agricultural land, China. (GB15618-2018) (MEEC, 2018).

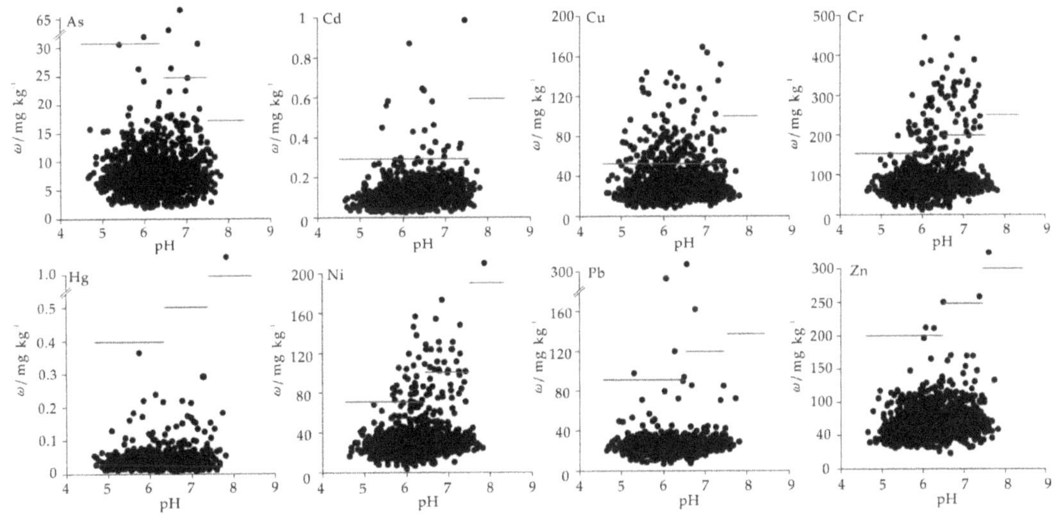

Figure 2. Relationships between soil heavy metal contents and pH. Note: The red line represents the risk screening value corresponding to different pH values.

The coefficient of variation (CV) is a normalized measure of the degree of dispersion of a probability distribution, and generally, the greater the degree of dispersion, the greater the CV value. On the basis of the grading of the CV described previously [53–55], the CV of pH in this study was 10.1%, which indicated a low degree of variation, and the data dispersion was not large. With respect to the assessed heavy metals, we obtained CV values of 52.9%, 65.7%, 68%, 62.8%, 56.6%, and 37.1% for As, Cu, Cr, Ni, Pb, and Zn, respectively, indicating medium to high variation, whereas the CV values of 82.3% and 98.5% obtained for Cd and Hg, respectively, indicated high to extremely high variation.

In general, the eight heavy metals were characterized by moderate and high variation, indicating that their data were discrete and that there were certain differences with respect to spatial distribution.

3.2. Correlation and Spatial Distribution of Heavy Metals

The results obtained from the statistical analysis of the 1135 samples revealed a sampling moderation value of 0.539 for the Kaiser–Meyer–Olkin (KMO) measure of sampling accuracy, indicating a partial correlation between variables [56]. Concurrently, we obtained a significant Bartlett spherical test value of 3376.389 ($p < 0.001$), indicating that the result of the PCA of variable sample data could reflect the relationships among chemical elements [57–59]. Based on the data validity test, the principal component analysis of eight heavy metals was carried out. According to previous research [60], it was proposed that the closer the load variable line, the stronger the correlation in the PCA component diagram. Therefore, in Figure 3, heavy metal elements in the study area are divided into two groups, namely, Ni and Cr comprising one group and Zn, Pb, Cr, Hg, As, and Cu clustered in the other. In addition, the correlation plot (Figure 4) in the study area was shown to be nearly consistent with the principal component plot (Figure 3). However, the cluster analysis diagram in Figure 4 shows some more detailed information. In addition to the strong correlation between Cr and Ni, there was also a certain correlation between Cu and Zn, while the correlation of the other four elements was much weaker. This could reflect the spatial distribution characteristics of the medium and high variation in element content in the study area.

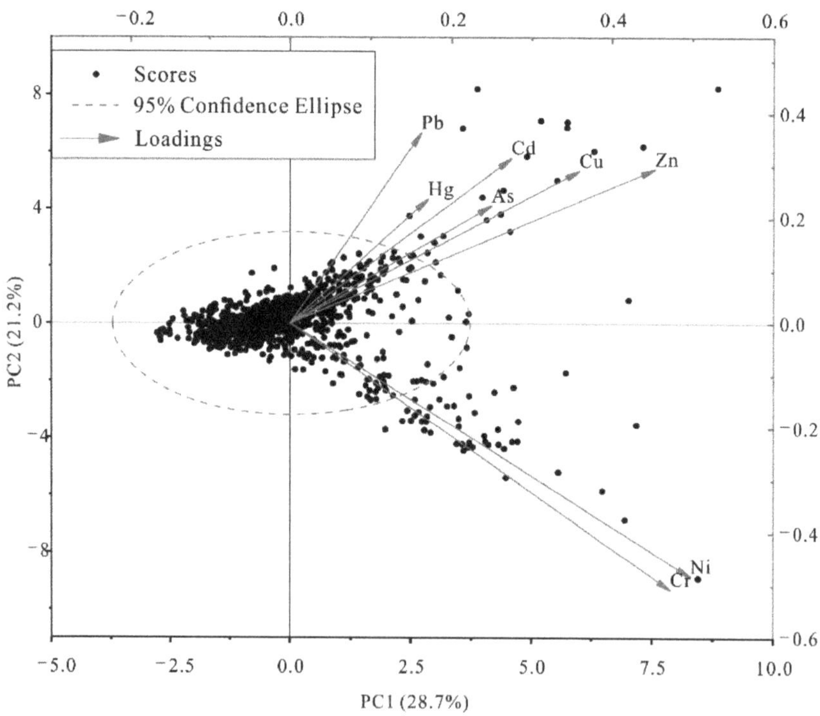

Figure 3. Principal component analysis plot of eight heavy metals in the study area.

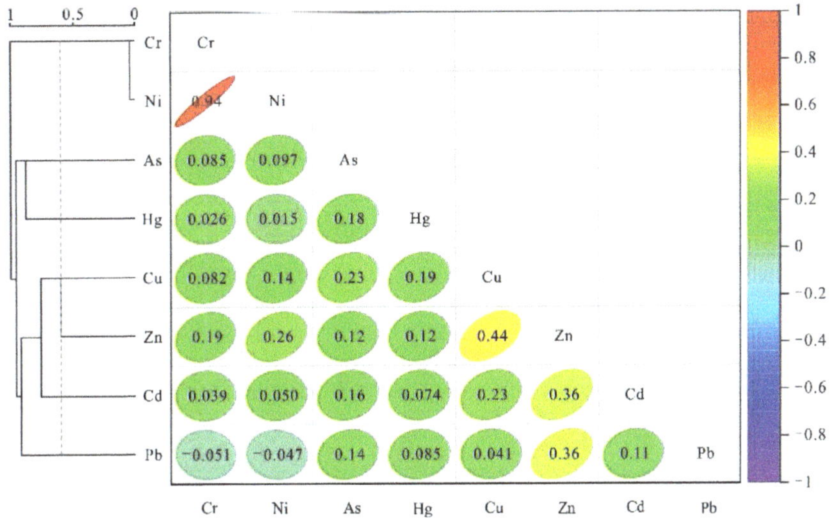

Figure 4. Correlation plot of eight heavy metals in the study area.

pH is often regarded as the most important variable of soil [61,62], which reflects the acid–base environment of heavy metal elements and can contribute to determining the distribution of soil chemical elements [63]. On the basis of our analysis of the spatial distribution of pH in the study area, we established that soils to the north of the study area are acidic to weakly acidic (Figure 5), whereas those in the central area (in the vicinity of the building stone quarrying area) and southern (Laiyang County and its south) area tend to be neutral. To a large extent, the distribution characteristic of pH reflects the influence of the soil-forming parent material and human activities, whereas the spatial distribution characteristics of heavy metals in the study area (Figure 5) were found to reflect the distribution of soil pH. The study area is characterized by high natural background levels of Cu, Cr, and Ni. As shown in Figure 3, Cu and Cr and Ni show quite distinct patterns of spatial distribution, with some authors proposing that the strong correlation between Cr and Ni is associated with the parent material or processes of soil formation [12,64]. In contrast, the distributions of Hg, Cd, Pb, Cu, and Zn are generally considered to be more reflective of human activities [14,47–50,53–55]. Whereas we detected no obvious anomaly (any significant high value) in the distribution of Cu near urban areas, the distributions of Hg, Cd, and Pb have clearly been influenced to varying extents by the human activities in urban areas and the quarrying of building stone. This view is consistent with the conclusion of Khan N et al. [65] on heavy metals in soils along the Swat River Basin, Khyber Pakhtunkhwa, Pakistan. Specifically, in the granite-rich region to the north of the study area, we obtained notably low values for As and Pb, whereas Cd, Cr, Hg, and Ni showed mixed distribution patterns of low and medium-high abundance. In the region underlain by metamorphic rock, Cr, Cu, Ni, and Zn are characterized by a belt of middle to high values distributed in a northeasterly direction, whereas in soils formed above basic volcanic rocks in the east, there is a zone characterized by high Cr and Ni values oriented in a north–south direction. In the sandstone area in the east, we recorded mainly medium to high values for Cd, Cu, Pb, and Zn, whereas high values were obtained for As, Hg, Cu, and Pb in soil collected from the north–northeast continuous high-value points, which tends to be indicative of the geological structure of the study area. In the area underlain by Quaternary rocks, Cd, Hg, Pb, and pH showed high value distributions in Laiyang County and the area to the south.

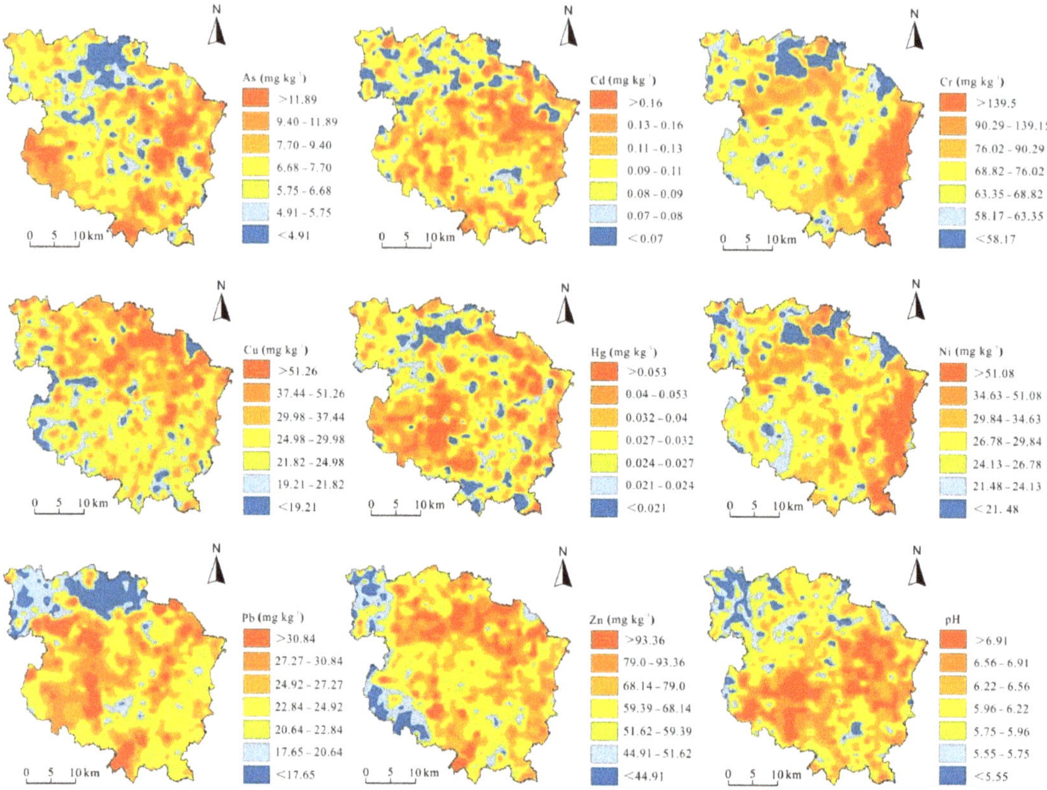

Figure 5. Spatial distributions of pH and eight heavy metals in soils of the study area.

3.3. Levels of Heavy Metal Contamination

Taking the geochemical background values of surface soil in Shandong Province as reference values, we statistically analyzed the land accumulation indices of heavy metals in cultivated soil within the study area (Table 4). The Igeo of As, Cd, Cu, Cr, Hg, Ni, Pb, and Zn ranged from −2.76 to 2.37, −2.96 to 3.21, −2.01 to 2.89, −3.03 to 2.72, −2.91 to 4.09, −3.58 to 2.45, −2.49 to 3.21, and −2.02 to 1.86, respectively. The assessment results show that the pollution sites ($I_{geo} \geq 0$) in the study area were primarily unpolluted to moderately polluted and moderately to severely polluted. Among them, Cd, Hg, and Pb each severely pollute one site. It can be seen from Table 4 that the average of I_{geo} was ordered as: Cu (−0.23) > Cr (−0.25) > Ni (−0.45) > Zn (−0.53) > Pb (−0.57) > Hg (−0.60) > As (−0.83) > Cd (−0.96). The order of the number of sites exceeding the background values was: Cu (359) > Cr (255) > Hg (189) > Ni (177) > Zn (142) > As (99) > Cd (84) > Pb (58). This finding thus indicated that cultivated soil in the study area was polluted to varying degrees by the eight heavy metals, among which, more than 10% of the monitored sites were found to be characterized by Cu, Cr, Hg, Ni, and Zn pollution. By comparing values obtained for the geoaccumulation index before and after performance improvement (Table 4 and Figure 6), we obtained average pre- (I_{ol}) and post- (I_{ml}) improvement values of 0.32 and −0.04, respectively, equating to unpolluted to moderately polluted and unpolluted sites, respectively. The corresponding $I_{geo} > 0$ sites prior to and after improvement were 677 and 421, respectively. Using the improved evaluation procedure, we found that there had been a significant reduction in the number of sites initially classified as polluted based on I_{ml} indices, and those sites initially placed in the severely polluted were downgraded to moderately to severely polluted. As shown in Figure 7, polluted soils as determined based on I_{ol} indices were adjudged to be widely distributed, and most sites surveyed in the study area were placed in the unpolluted to moderately polluted category. The sites characterized by moderate to severe pollution, both before and after index improvement, were found to be primarily concentrated around the Laiyang urban area, the sites of sandstone quarrying, and the southern region of the study area, which broadly correspond to the distribution of the sandstone parent material. From a comparison of Table 4 and Figure 7, it can be seen that the distribution of improved I_{ml} data and pollution were more centralized and reasonable, and are likely to provide a more accurate reflection of the levels of heavy metal pollution in soils of the study area. This tends to be consistent with the findings of previous studies that have demonstrated that the scientific and reasonable assessment of heavy metal pollution requires not only an assessment of a combination of multiple indicators [13,22] but also certain algorithm corrections [38].

Table 4. Results obtained for evaluations using a heavy metal geoaccumulation index.

		As	Cd	Cu	Cr	Hg	Ni	Pb	Zn	I_{ol}	I_{ml}
Minimum		−2.76	−2.96	−2.01	−3.03	−2.91	−3.58	−2.49	−2.02	−1.03	−1.31
Maximum		2.37	3.21	2.89	2.72	4.09	2.45	3.21	1.86	4.09	3.60
Mean		−0.83	−0.96	−0.23	−0.25	−0.60	−0.45	−0.57	−0.53	0.32	−0.04
Unpolluted	$I_{geo} < 0$	1036	1051	776	880	946	958	1077	993	458	714
Degree of pollution	$0 \leq I_{geo} < 1$	92	71	285	178	148	124	45	136	474	314
	$1 \leq I_{geo} < 2$	6	9	65	69	33	51	10	6	171	98
	$2 \leq I_{geo} < 3$	1	3	9	8	7	2	2	0	29	8
	$3 \leq I_{geo} < 4$	0	1	0	0	0	0	1	0	2	1
	$4 \leq I_{geo} < 5$	0	0	0	0	1	0	0	0	1	0
	$I_{geo} \geq 5$	0	0	0	0	0	0	0	0	0	0
Proportion of polluted sites/%		8.7	7.4	31.6	22.4	16.7	15.6	5.1	12.5	59.6	37.1

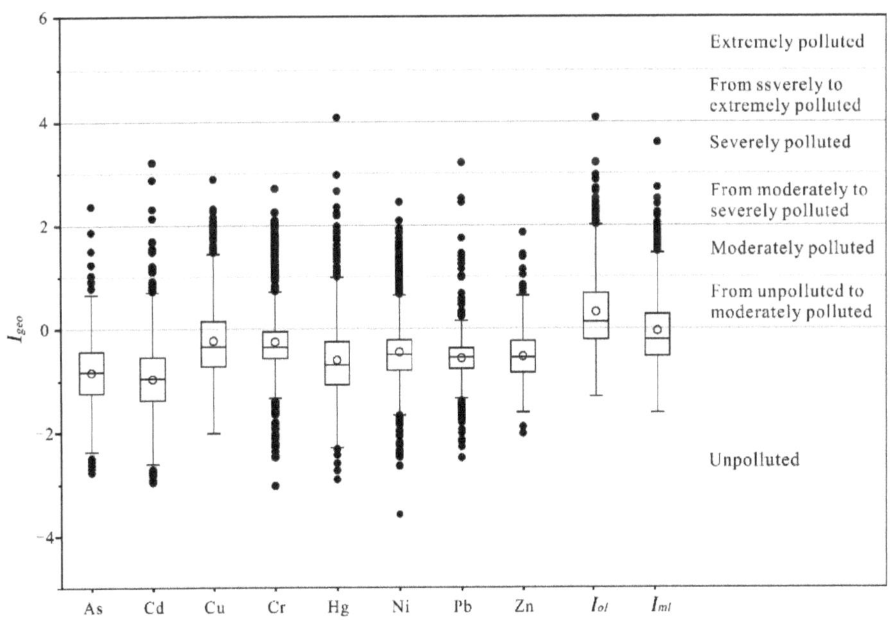

Figure 6. A boxplot of geoaccumulation index (I_{geo}) distribution in the study area.

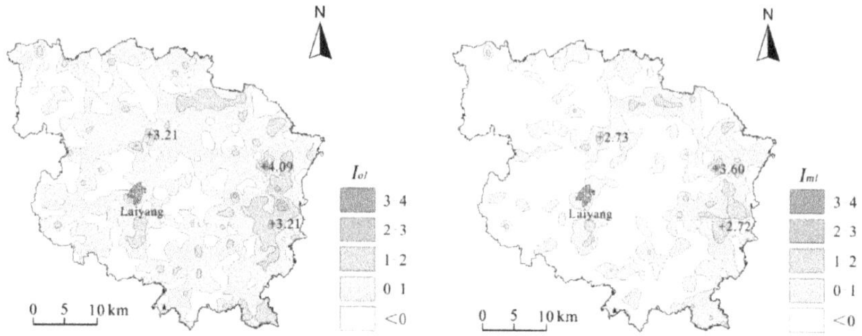

Figure 7. Spatial distribution of geoaccumulation index (I_{geo}) values. Note: + denotes the maximum of I_{geo}.

3.4. *Risk Assessment of Heavy Metals*

3.4.1. Potential Ecological Risk Assessment

Taking the geochemical background values of surface soil in Shandong Province as reference values, we statistically analyzed the potential ecological risk index of heavy metal elements in cultivated soil in the study area (Table 3), on the basis of which, we determined the proportional ecological risk of heavy metal elements in the study area (Figure 8). According to the calculation results (Table 5), the average E_r values of the assessed metals could be arranged in the following descending order: Hg (47.2) > Cd (26.8) > As (9.3) > Cu (7.4) > Ni (6.2) > Pb (5.4) > Cr (2.9) > Zn (1.1). In general, Cr and Zn presented only a slight ecological hazard, whereas As, Cd, Cu, Hg, Ni, and Pb showed different degrees of ecological hazard, exceeding standard values at 3, 122, 1, 487, 1, and 3 sites, respectively, with corresponding rates of 0.3%, 10.7%, 0.1%, 42.9%, 0.1%, and 0.3%, respectively. The spectrum of potential ecological risks shown in Figure 8 indicates that the level of ecological risk of the assessed heavy metals could be ordered as follows: Hg > Cd > As > Cu > Ni > Pb > Cr > Zn. This is consistent with that reflected by the average E_r values. This indicates that Hg and Cd were the key factors presenting an ecological risk in the study area. Table 5 shows that the percentage contribution rates of Hg and Cd to the potential ecological risk were 6.9–92.7% and 3.4–83.2%, respectively, with corresponding mean and median values of 42.1% and 25.5% and 40.9% and 24.3%, respectively. Moreover, the cumulative ecological risk contributed by Hg and Cd was estimated to be >65%, thereby indicating that Hg and Cd are the main heavy metals posing an ecological risk in the soil of the study area. The CV values obtained for Hg and Cd ecological risk were 29.3% and 38.9%, respectively, indicating a moderate variation and that the distribution of data was relatively consistent. Comparatively, the CV values obtained for Cu, Cr, and Ni, the three elements with high background levels, were greater than 50%, thereby indicating a medium-strong variance, which also implies differences in the ecological risks attributable to differences in geological background levels. *PERI* analysis revealed that soil at 1029 sampling points in the study area posed a slight ecological risk, whereas that at 106 sites, accounting for 9.3% of the total, posed a moderate to strong ecological risk. Patterns of the spatial distribution of *PERI* (Figure 9) revealed a concentration in the urban area and to the east and the north of Laiyang. This pattern corresponds to the spatial distribution and pollution levels of heavy metals, indicating that the content, pollution, and associated risk of heavy metals in the study area are interrelated variables, and also highlights the objectivity of the comprehensive evaluation indicators.

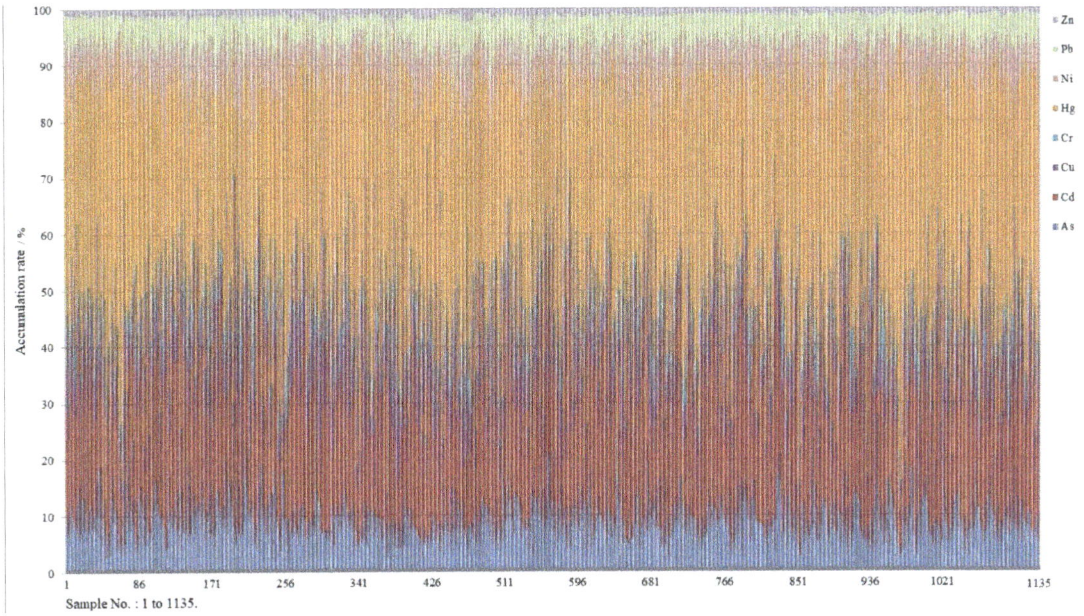

Figure 8. Accumulation rate of the ecological risk (E_r) of heavy metals.

Table 5. Statistical table of the potential ecological risk index of soil heavy metals and the proportion of sites at which values exceeded standard values.

	Er									*PERI*	
		As	Cd	Cu	Cr	Hg	Ni	Pb	Zn		
Minimum		2.2	5.8	1.9	0.4	8.0	0.6	1.3	0.4		30.7
Maximum		77.4	417.3	55.7	19.7	1018.8	41.0	69.3	5.5		1099.1
Mean		9.3	26.8	7.4	2.9	47.2	6.2	5.4	1.1		106.2
S.D.		4.9	22.0	4.8	2.0	46.5	3.9	3.0	0.4		56.9
Exceeds the standard number of points	$E_r < 40$	1132	1014	1134	1135	648	1134	1132	1135	$PERI < 150$	1029
	$40 \leq E_r < 80$	3	108	1	0	390	1	3	0	$150 \leq PERI < 300$	93
	$80 \leq E_r < 160$	0	12	0	0	72	0	0	0	$300 \leq PERI < 600$	12
	$160 \leq E_r < 320$	0	2	0	0	22	0	0	0	$PERI \geq 600$	1
	$Er \geq 320$	0	2	0	0	3	0	0	0		
Contribution rate of E_r/%	Minimum	1.1	3.4	1.0	0.2	6.9	0.7	0.6	0.1		
	Maximum	31.7	83.2	32.1	15.5	92.7	29.5	41.0	3.8		
	Mean	9.3	25.5	7.3	2.9	42.0	6.3	5.5	1.1		
	Median	8.8	24.3	6.3	2.5	40.9	5.4	5.2	1.1		
	CV/%	40.1	38.9	52.6	59.7	29.3	55.6	44.5	39.5		

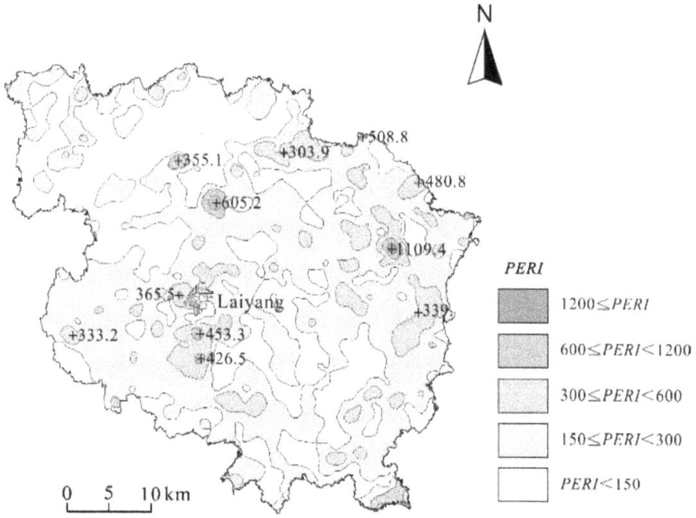

Figure 9. Spatial distributions of the potential ecological risk index (*PERI*). Note: + denotes the maximum PERI values.

3.4.2. Health Risk Assessment

In order to assess the risk posed by heavy metal pollution in cultivated soils, it is necessary to gain an estimate of the level of exposure. Consequently, to estimate the risk of human exposure to contaminated soil via inhalation, ingestion, and skin contact, we used the total combined concentration of heavy metals to calculate the hazard quotient (*HQ*). The results of our non-carcinogenic health risk assessment revealed that the non-carcinogenic health risk index of the same element for the three different routes of exposure could be ordered as follows: $HQ_{ing} > HQ_{dermal} > HQ_{inh}$. Furthermore, we found that for each of the assessed heavy metals, the average *HQ* value for adults and children was <1 (Table 6). The non-carcinogenic risks posed by the eight heavy metals to adults and children could be ordered as As > Pb > Ni > Cu > Cd > Zn > Hg > Cr and As > Pb > Ni > Cu > Zn > Hg > Cd > Cr, respectively. In general, we found that at none of the sampling sites in the study area was there a non-carcinogenic risk posed to human health by soil-borne heavy metals. The non-carcinogenic health risk index of adults (HI_{adults}) was between 1.08×10^{-2} and 2.07×10^{-1}, both of which are <1, indicating that there was no non-carcinogenic health risk to adults. Comparatively, the non-carcinogenic health risk index assessed for children ($HI_{children}$) was between 1.31×10^{-1} and 3.01, among which 10 samples contained a single heavy metal with an *HQ* value > 1 (As in 9 samples and Pb in 1 sample), and 17 samples with an *HI* value > 1, accounting for 1.5% of the total. These findings accordingly indicated that the levels of soil heavy metals in the areas where these samples were collected posed a potential non-carcinogenic health risk to children. Furthermore, we found that the contents of As, Pb, Cr, Ni, and Cu in these sites exceed the corresponding risk screening values (GB15618–2018). We also estimated that the percentage contribution of As, Pb, Ni, and Cu to *HI* in adults and children was >95% (Table 6), and these four heavy metals were the main factors influencing the non-carcinogenic risk. It can be seen from Figure 10 that the sampling sites with *HI* > 1 are mainly distributed in the cultivated land at the downstream of the sandstone mining region in the east and the southeast of Laiyang County. This pattern could be explained by the high background levels associated with an area of volcanic rock in the east. Moreover, children are potentially more readily exposed to contaminated soil than are adults, owing to their particular diet and behavioral habits [66,67], which contribute to higher *HI* values. The increased non-cancer risk found in soil in the eastern part of the study area can be ascribed primarily to the contents of As, Pb, and Cr in natural sources.

We found that the *HQ* and *HI* values obtained in the present study are significantly higher than those previously reported for this area by Zhuo et al. (2020) [14], Liu et al. (2021) [68], and Hua et al. (2022) [69], which we suspect could be attributed to our grid-intensive soil sampling across the entire region, rather than assessments based on sporadic soil sampling.

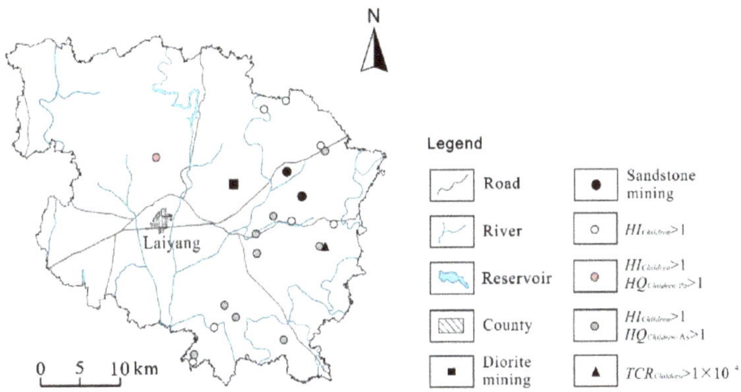

Figure 10. Spatial distribution of *HI*, *HQ*, and total cancer risk (*TCR*) in the study area.

Given that As and Cd have high toxicity coefficients, they are often used to calculate carcinogenic risk [26,70]. The distributions of Cr and Ni tend to be closely correlated and have high background levels in the study area. Consequently, in the present study, we focused on the carcinogenic risks posed by As, Cd, Cr, and Ni. The results of our carcinogenic health risk assessment (presented in Table 7) show that the carcinogenic health risk indices (*CR*) for different routes of As and Cd exposure could be ordered as $CR_{ing} > CR_{dermal} > CR_{inh}$. Furthermore, we established that the carcinogenic risk of the assessed heavy metals to adults and children from high to low was As > Cr > Cd > Ni. It can be seen from Table 7 that the average total cancer risk (*TCR*) for adults and children was 8.08×10^{-6} and 1.44×10^{-5}, respectively, both of which exceed 1×10^{-6}, which is considered an acceptable level [9,47], and in the case of children, one of the *TCR* values was greater than 1×10^{-4}, which is considered an extremely high risk level, [9,26,47]. These findings accordingly tend to indicate that both local adults and children are exposed to a potential risk of cancer. Notably, we found that the sampling site indicating a high carcinogenic risk to children coincides with that identified as posing a non-carcinogenic risk (Figure 8), thereby indicating that children living in the vicinity of this site are potentially exposed to double health risks. However, some scholars [26,27,46,70] have proposed that values of carcinogenic risk between 10^{-6} and 10^{-4} are acceptable, and thus if this standard were to be implemented, the carcinogenic risk in the study area would accordingly be deemed acceptable. However, it should be emphasized that the highest value of 1.44×10^{-5} we obtained for carcinogenic risk to children exceeded 1×10^{-4}, indicating that the local carcinogenic risk in the study area should not be ignored. Although there are many cases of human exposure to heavy metals, which may pose certain health risks, it should be noted that in our assessment, we considered only the intake of these elements via direct contact with soil. In general, humans are mainly exposed to heavy metals via less direct exposure routes, such as the consumption of contaminated food products (e.g., vegetables and meat) and water [14,26,46,51]. Consequently, in order to successfully identify common exposure pathways of potential concern, it is recommended to apply a multipath risk assessment analysis for the relevant environmental media [26,27,46].

Table 6. *HQ* and *HI* assessments of health risks posed to adults and children by soil in the study area.

HQ-non-canc.		As	Cd	Cu	Cr	Hg	Ni	Pb	Zn	*HI*	Proportion of *HQ*
Adults	Maximum	1.34×10^{-1}	1.41×10^{-2}	1.14×10^{-2}	6.78×10^{-4}	6.61×10^{-3}	2.04×10^{-2}	1.89×10^{-1}	2.21×10^{-3}	2.07×10^{-1}	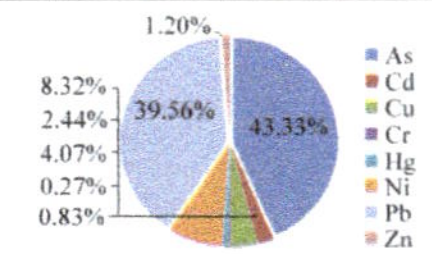
	Minimum	3.81×10^{-3}	1.96×10^{-4}	3.81×10^{-4}	1.26×10^{-5}	5.19×10^{-5}	3.12×10^{-4}	3.65×10^{-3}	1.50×10^{-4}	1.08×10^{-2}	
	Mean + S.E.	1.61×10^{-2} $+7.489 \times 10^{-6}$	9.06×10^{-4} $+ 6.564 \times 10^{-7}$	1.51×10^{-3} $+ 8.106 \times 10^{-7}$	9.90×10^{-5} $+5.930 \times 10^{-8}$	3.06×10^{-4} $+ 2.661 \times 10^{-7}$	3.09×10^{-3} $+ 1.709 \times 10^{-6}$	1.47×10^{-2} $+ 7.322 \times 10^{-6}$	4.44×10^{-4} $+ 1.454 \times 10^{-7}$	3.71×10^{-2} $+ 1.181 \times 10^{-5}$	
	S.D.	8.50×10^{-3}	7.45×10^{-4}	9.92×10^{-4}	6.73×10^{-5}	3.02×10^{-4}	1.94×10^{-3}	8.31×10^{-3}	1.65×10^{-4}	1.34×10^{-2}	
	Amount (>1)	0	0	0	0	0	0	0	0	0	
Children	Maximum	2.84	2.52×10^{-2}	8.60×10^{-2}	3.14×10^{-3}	4.61×10^{-2}	1.53×10^{-1}	1.17	1.64×10^{-2}	3.01	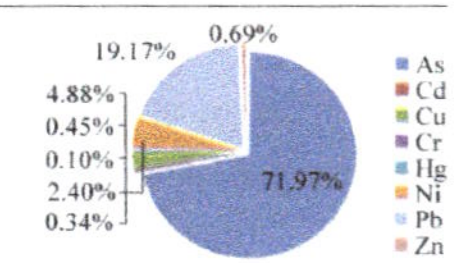
	Minimum	8.11×10^{-2}	3.50×10^{-4}	2.87×10^{-3}	5.86×10^{-5}	3.62×10^{-4}	2.35×10^{-3}	2.30×10^{-2}	1.11×10^{-3}	1.31×10^{-1}	
	Mean + S.E.	3.42×10^{-1} $+1.595 \times 10^{-4}$	1.62×10^{-3} $+1.172 \times 10^{-6}$	1.14×10^{-2} $+ 6.590 \times 10^{-6}$	4.59×10^{-4} $+ 2.749 \times 10^{-7}$	2.13×10^{-3} $+ 1.850 \times 10^{-6}$	2.32×10^{-2} $+ 1.286 \times 10^{-5}$	9.11×10^{-2} $+ 4.511 \times 10^{-5}$	3.30×10^{-3} $+ 1.075 \times 10^{-6}$	3.30×10^{-3} $+ 3.736 \times 10^{-4}$	
	S.D.	1.81×10^{-1}	1.33×10^{-3}	7.48×10^{-3}	3.12×10^{-4}	2.10×10^{-3}	1.46×10^{-2}	5.12×10^{-2}	1.22×10^{-3}	1.99×10^{-1}	
	Amount (>1)	9	0	0	0	0	0	1	0	17	

Table 7. The carcinogenic health risk index (*CR*) of ingestion, dermal contact, and inhalation of soils for adults and children in the study area.

CR-canc.		As			Cd			*CR*				*TCR*
		RfD_{ing}	RfD_{dermal}	RfD_{inh}	RfD_{ing}	RfD_{dermal}	RfD_{inh}	As	Cd	Cr (RfD_{inh})	Ni (RfD_{inh})	
Adults	Maximum	5.48×10^{-5}	2.19×10^{-6}	2.10×10^{-11}	6.14×10^{-6}	2.45×10^{-7}	2.43×10^{-13}	5.70×10^{-5}	6.39×10^{-6}	5.50×10^{-6}	3.99×10^{-8}	5.91×10^{-5}
	Minimum	1.56×10^{-6}	6.24×10^{-8}	6.00×10^{-13}	8.53×10^{-8}	3.41×10^{-9}	3.37×10^{-15}	1.63×10^{-6}	8.87×10^{-8}	1.03×10^{-7}	6.12×10^{-10}	1.82×10^{-6}
	Mean + S.E.	6.59×10^{-6} $+ 3.075 \times 10^{-9}$	2.63×10^{-7} $+ 1.225 \times 10^{-10}$	2.53×10^{-12} $+ 1.181 \times 10^{-15}$	3.95×10^{-7} $+ 2.855 \times 10^{-10}$	1.57×10^{-8} $+ 1.137 \times 10^{-11}$	1.56×10^{-14} $+ 1.128 \times 10^{-17}$	6.86×10^{-6} $+ 3.198 \times 10^{-9}$	4.10×10^{-7} $+ 2.969 \times 10^{-10}$	8.03×10^{-7} $+ 4.802 \times 10^{-10}$	6.05×10^{-9} $+ 3.348 \times 10^{-12}$	8.08×10^{-6} $+ 3.313 \times 10^{-9}$
	S.D.	3.49×10^{-6}	1.39×10^{-7}	1.34×10^{-12}	3.24×10^{-7}	1.29×10^{-8}	1.28×10^{-14}	3.63×10^{-6}	3.37×10^{-7}	5.45×10^{-7}	3.80×10^{-9}	3.76×10^{-6}
	Amount ($>10^{-4}$)	0	0	0	0	0	0	0	0	0	0	0
Children	Maximum	1.03×10^{-4}	2.87×10^{-6}	9.41×10^{-12}	6.14×10^{-6}	2.45×10^{-7}	2.43×10^{-13}	1.06×10^{-4}	1.23×10^{-5}	6.16×10^{-6}	7.46×10^{-8}	1.08×10^{-4}
	Minimum	2.93×10^{-6}	8.20×10^{-8}	2.68×10^{-13}	8.53×10^{-8}	3.41×10^{-9}	3.37×10^{-15}	3.01×10^{-6}	1.71×10^{-7}	1.15×10^{-7}	1.14×10^{-9}	3.30×10^{-6}
	Mean + S.E.	1.24×10^{-5} $+ 5.753 \times 10^{-9}$	3.46×10^{-7} $+ 1.612 \times 10^{-10}$	1.13×10^{-12} $+ 5.278 \times 10^{-16}$	3.95×10^{-7} $+ 2.855 \times 10^{-10}$	1.57×10^{-8} $+ 1.137 \times 10^{-11}$	1.56×10^{-14} $+ 1.128 \times 10^{-17}$	1.27×10^{-5} $+ 5.912 \times 10^{-9}$	7.92×10^{-7} $+ 5.736 \times 10^{-10}$	8.99×10^{-7} $+ 5.383 \times 10^{-10}$	1.13×10^{-8} $+ 6.247 \times 10^{-12}$	1.44×10^{-5} $+ 6.088 \times 10^{-9}$
	S.D.	6.53×10^{-6}	1.83×10^{-7}	5.99×10^{-13}	3.24×10^{-7}	1.29×10^{-8}	1.28×10^{-14}	6.71×10^{-6}	6.51×10^{-7}	6.11×10^{-7}	7.09×10^{-9}	6.91×10^{-6}
	Amount ($>10^{-4}$)	1	0	0	0	0	0	0	0	0	0	1

4. Conclusions

In this study, a new combination of multiple statistical methods and the improved geoaccumulation index method was used to investigate the heavy metal pollution in topsoil in Laiyang County. Specifically, the ecological risk and health risk of heavy metals in topsoil were quantitatively evaluated and spatially analyzed. The results show the following: (1) The proportion of sites with Cu, Cr, and Ni levels exceeding the background value was more than 55%. Combined with their distribution and rock characteristics, it was considered that the study area is a typical geological high background area of the above three elements. The content of heavy metals (As, Cd, Cu, Cr, Hg, Ni, Pb, and Zn) in some sampling sites exceeded the risk screening value (GB15618-2018), indicating that those heavy metals were polluting. (2) The eight heavy metals in the study area had pollution points of different degrees, among which, the proportion of Cu, Cr, Hg, and Zn pollution points was more than 10%, being mainly unpolluted to moderately polluted and moderately polluted to severely polluted. The improved I_{geo} was used to generate statistics on the pollution. It was found that the average value of I_{ol} was 0.32 (unpolluted to moderately polluted), and the average value of I_{ml} was -0.04 (unpolluted). (3) Mercury and cadmium were identified as the main contributors to the ecological risk in the study area, with the cumulative ecological risk contribution percentage of >65%. The results of *PERI* showed that 9.3% of the sampling sites were considered to have moderate ecological risk. (4) The study found that As, Pb, Ni and Cu contributed more than 95% to the non-carcinogenic risk. The sampling sites with $HI_{children} > 1$ were mainly distributed in the east of the study area; however, different routes of soil intake posed no non-carcinogenic risk to adults. Simultaneously, the risk of cancer in adults and children was acceptable. However, the location identified as posing high carcinogenic risk to children was consistent with the non-carcinogenic risk location identified by this study, which indicated that children in this location and the surrounding areas were more likely to face dual health risks.

It is worth noting that the assessment of I_{ml} found that the distribution of contaminated soil is relatively centralized and reasonable, which can accurately reflect the pollution degree of the study area. This new exploration may help to objectively evaluate the spatial distribution of heavy metal pollution and support the risk management of heavy metals. In addition, the analysis of a typical agricultural county will also provide some reference for other cultivated soil investigations.

Author Contributions: Conceptualization, L.Y., W.L. and Z.S.; methodology, L.Y. and P.Q.; writing—original draft preparation, L.Y., W.L., J.Y., B.Z. and Z.C.; writing—review and editing, L.Y. and Z.S.; visualization, L.Y.; supervision, F.Z. and Y.Z.; project administration, W.L.; funding acquisition, H.L. All authors have read and agreed to the published version of the manuscript.

Funding: This research was funded by the Shandong Geological Exploration Project (grant nos. 2019(44) and 2020(58)) and the Science and Technology Innovation Project (grant no. KY202227) of Shandong Provincial Bureau of Geology & Mineral Resources.

Informed Consent Statement: Not applicable.

Data Availability Statement: The data presented in this study are available on request from the corresponding authors. The processed data are not publicly available as the data are also part of an ongoing study.

Acknowledgments: The authors would like to thank the editors and reviewers for their valuable comments and suggestions.

Conflicts of Interest: The authors declare no conflict of interest.

References

1. Alexander, P.D.; Alloway, B.J.; Dourado, A.M. Genotypic variations in the accumulation of Cd, Cu, Pb and Zn exhibited by six commonly grown vegetables. *Environ. Pollut.* **2006**, *144*, 736–745. [CrossRef] [PubMed]
2. Díaz, R.O.; Fonticiella, M.D.; Arado López, J.O.; Borrell Munoz, J.L.; D'Alessandro, R.K.; López, P.N. Spatial distribution and contamination assessment of heavy metals in urban topsoils from Las Tunas City, Cuba. *Bull. Environ. Contam. Toxicol.* **2013**, *91*, 29–35. [CrossRef] [PubMed]
3. Balkhair, K.S.; Ashraf, M.A. Field accumulation risks of heavy metals in soil and vegetable crop irrigated with sewage water in western region of Saudi Arabia. *Saudi J. Biol. Sci.* **2016**, *23*, 32–44. [CrossRef]
4. Zhang, D.; Huang, R.; Gao, X.S. Spatial Characteristics and Potentical Risk Factor of Heavy Metals in Cultivated Land in the Transition Zone of a Mountain Plain. *Environ. Sci.* **2022**, *43*, 946–956. [CrossRef]
5. Ma, Q.H.; Zhang, G.H.; Geng, R. Present condition analysis of sloping farmland soil quality in the eastern water erosion zone of China. *Sci. Soil Water Conserv.* **2017**, *15*, 36–42. [CrossRef]
6. Ministry of Environmental Protection of the PRC; Ministry of Land and Resources of the PRC. *National Survey Bulletin on Soil Pollution*; Ministry of Environmental Protection of the PRC: Beijing, China; Ministry of Land and Resources of the PRC: Beijing, China, 2014.
7. Food and Agriculture Organization of the United Nations (FAO); United Nations Environment Programme (UNEP). *Global Assessment of Soil Pollution: Report*; Food and Agriculture Organization of the United Nations: Rome, Italy, 2021. [CrossRef]
8. Zhao, D.J.; Wang, X.Q. Distribution, sources and potential ecological risk of heavy metals in the floodplain soils of the karst area of Yunnan, Guizhou, Guangxi. *China Environ. Sci.* **2020**, *40*, 1609–1619. [CrossRef]
9. US EPA. *Supplemental Guidance for Developing Soil Screening Levels for Superfund Sites*; Office of Emergency and Remedial Response: Washington, DC, USA, 2002; pp. 4–24.
10. Chai, L.; Wang, X.; Ma, L.; Cheng, Z.X.; Su, L.M.; Wang, Y.H. Sources appointment of heavy metals in cultivated soils of Lanzhou based on PMF models. *China Environ. Sci.* **2020**, *40*, 3919–3929. [CrossRef]
11. Chi, Q.H.; Yan, M.C. *Handbook of Elemental Abundance for Applied Geochemistry*; Geological Press: Beijing, China, 2007.
12. Zhang, F.G.; Peng, M.; Wang, H.Y.; Ma, H.H.; Cheng, H.X. Ecological risk assessment of heavy metals at township scale in the high background of heavy metals, Southwestern China. *Environ. Sci.* **2020**, *41*, 4197–4209. [CrossRef]
13. Jia, Z.; Li, S.; Wang, L. Assessment of soil heavy metals for eco-environment and human health in a rapidly urbanization area of the upper Yangtze Basin. *Sci. Rep.* **2018**, *8*, 3256. [CrossRef]
14. Ahmad, W.; Alharthy, R.D.; Zubair, M.; Ahmed, M.; Rafique, S. Toxic and heavy metals contamination assessment in soil and water to evaluate human health risk. *Sci. Rep.* **2021**, *11*, 17006. [CrossRef]
15. Miao, R.Q.; Yan, Y.F.; Bai, Y.; Deng, A. Assessment of heavy metal contamination of soil in Kunming. *Adv. Mater. Res.* **2015**, *1092*, 774–779. [CrossRef]
16. Chen, Z.; Xu, J.; Duan, R.; Lu, S.; Hou, Z.; Yang, F.; Peng, M.; Zong, Q.; Shi, Z.; Yu, L. Ecological Health Risk Assessment and Source Identification of Heavy Metals in Surface Soil Based on a High Geochemical Background: A Case Study in Southwest China. *Toxics* **2022**, *10*, 282. [CrossRef] [PubMed]
17. Canpolat, Ö.; Varol, M.; Okan, Ö.Ö.; Eriş, K.K.; Çağlar, M. A comparison of trace element concentrations in surface and deep water of the Keban Dam Lake (Turkey) and associated health risk assessment. *Environ. Res.* **2020**, *190*, 110012. [CrossRef] [PubMed]
18. Zhao, H.R.; Xia, B.C.; Fan, C.; Zhao, P.; Shen, S.L. Human health risk from soilheavy metal contamination under different land uses near Dabaoshan Mine, Southern China. *Sci. Total Environ.* **2012**, *417–418*, 45–54. [CrossRef]
19. Rutigliano, F.A.; Marzaioli, R.; De Crescenzo, S.; Trifuoggi, M. Human health risk from consumption of two common crops grown in polluted soils. *Sci. Total Environ.* **2019**, *691*, 195–204. [CrossRef]
20. Ekino, S.; Susa, M.; Ninomiya, T.; Imamura, K.; Kitamura, T. Minamata disease revisited: An update on the acute and chronic manifestations of methyl mercury poisoning. *J. Neurol. Sci.* **2007**, *15*, 131–144. [CrossRef]
21. Hara, N.; Saito, H.; Takahashi, K.; Takeda, M. Lower urinary tract symptoms in patients with Niigata Minamata disease: A case-control study 50 years after methyl mercury pollution. *Int. J. Urol.* **2013**, *20*, 610–615. [CrossRef]
22. Aziz, R.; Rafiq, M.T.; Li, T.; Liu, D.; He, Z.; Stoffella, P.J.; Sun, K.; Xiao, Y. Uptake of cadmium by rice grown on contaminated soils and its bioavailability/toxicity in human cell lines (Caco-2/HL-7702). *J. Agric. Food Chem.* **2015**, *63*, 3599–3608. [CrossRef]
23. Pei, N.; Liang, X.F.; Qin, X.; Zhao, L.J.; Huang, Q.; Xu, Y.; Sun, Y.B. Remediation and persistent stability effects of sepiolite on cadmium-contaminated paddy soil. *J. Agro-Environ. Sci.* **2022**, *41*, 0277–0284. [CrossRef]
24. Hu, Y.; He, K.; Sun, Z.; Chen, G.; Cheng, H. Quantitative source apportionment of heavy metal(loid)s in the agricultural soils of an industrializing region and associated model uncertainty. *J. Hazard. Mater.* **2020**, *391*, 122244. [CrossRef]
25. Varol, M.; Gündüz, K.; Sünbül, M.R. Pollution status, potential sources and health risk assessment of arsenic and trace metals in agricultural soils: A case study in Malatya province, Turkey. *Environ. Res.* **2021**, *202*, 111806. [CrossRef] [PubMed]
26. Liu, Y.; He, Z.H.; Niu, X.K.; Zhang, D.; Pan, B. Health Risk Assessment of Soil Heavy Metals in a Small Watershed of a Mining Area in Yunnan. *Environ. Sci.* **2022**, *43*, 936–945. [CrossRef]
27. Wei, R.P.; Wang, X.; Tang, W.L.; Yang, Y.; Yang, L. Bioaccumulations and potential human health risks assessment of heavy metals in ppk-expressing transgenic rice. *Sci. Total Environ.* **2020**, *710*, 136496. [CrossRef]

28. Yu, L.; Wan, F.; Fan, H.; Kang, G.; Liu, H.; Wang, D.; Xu, J. Spatial Distribution, Source Apportionment, and Ecological Risk Assessment of Soil Heavy Metals in Jianghugongmi Producing Area, Shandong Province. *Environ. Sci.* **2022**, *43*, 307–319. [CrossRef]
29. Ju, T.N.; Lei, M. Geo-accumulation Index Method to Optimize the Evalution Method of Polymetallic Environment Quality: Taking Developed Agricultural Areas as an Example. *Environ. Sci.* **2022**, *43*, 957–964. [CrossRef]
30. Guillén, M.T.; Delgado, J.; Albanese, S.; Nieto, J.M.; Lima, A.; Vivo, B.D. Heavy metals fractionation and multivariate statistical techniques to evaluate the environmental risk in soils of Huelva Township (SW Iberian Peninsula). *J. Geochem. Explor.* **2012**, *119*, 32–43. [CrossRef]
31. Zhu, D.; Wei, Y.; Zhao, Y.; Wang, Q.; Han, J. Heavy metal pollution and ecological risk assessment of the agriculture soil in xunyang mining area, Shaanxi province, northwestern China. *Bull. Environ. Contam. Toxicol.* **2018**, *101*, 178–184. [CrossRef]
32. Zhang, G.S.; Jia, X.M.; Xiao-Hang, M.; Qian, Z.; Shi, C.Y.; Zhang, D.F. The warning research of soil enzyme activities on heavy metal-contamination of brown earth in Shandong. *Plant Nutr. Fertil. Sci.* **2004**, *10*, 272–276. [CrossRef]
33. Liu, P.; Zhao, H.J.; Wang, L.L. Analysis of Heavy Metal Sources for Vegetable Soils from Shandong Province, China. *Agric. Sci. China* **2011**, *10*, 11. [CrossRef]
34. Zhang, F.; Chang, C.P.; Li, J.; Song, S.; Cai, W.J.; Zhao, G.S.; Li, F.D. Health risk assessment of heavy metals in groundwater in rural area of Jiaodong Peninsula:a case study in Laiyang City, Shandong Province. *Environ. Sci. Technol.* **2013**, *36*, 180–186.
35. Muller, G. Index of geoaccumulation in sediments of the Rhine River. *Geojournal* **1969**, *2*, 108–118.
36. Krzysztof, L.; Wiechuła, D.; Korus, I. Metal contamination of farming soils affected by industry. *Environ. Int.* **2004**, *30*, 159–165. [CrossRef]
37. Zhuo, H.; Wang, X.; Liu, H.; Fu, S.; Ren, L. Source analysis and risk assessment of heavy metals in development zones: A case study in rizhao, china. *Environ. Geochem. Health* **2020**, *42*, 135–146. [CrossRef] [PubMed]
38. Liu, Z.H.; Meng, R.H.; Dai, H.X.; Hong, Q.Y.; Yan, Q.H.; Liu, Y.T.; Wang, H.T.; Chen, T. Evaluation of heavy metals pollution in surface sediments using an improved geo-accumulation index method. *J. Agro-Environ. Sci.* **2019**, *38*, 2157–2164. [CrossRef]
39. Nemerow, N. *Scientific Stream Pollution Analysis*; Scripta Book Co.: Washington, DC, USA, 1974.
40. Hakanson, L. An ecological risk index for aquatic pollution control.a sedimentological approach. *Water Res.* **1980**, *14*, 975–1001. [CrossRef]
41. Gong, Q.; Deng, J.; Xiang, Y.; Wang, Q.; Yang, L. Calculating pollution indices by heavy metals in ecological geochemistry assessment and a case study in parks of Beijing. *J. Earth Sci.* **2008**, *19*, 230–241. [CrossRef]
42. Mungai, T.M.; Owino, A.A.; Makokha, V.A.; Gao, Y.; Yan, X.; Wang, J. Occurrences and toxicological risk assessment of eight heavy metals in agricultural soils from Kenya, Eastern Africa. *Environ. Sci. Pollut. Res. Int.* **2016**, *23*, 18533–18541. [CrossRef]
43. Long, Z.J.; Huang, Y.; Zhang, W.; Shi, Z.L.; Yu, D.M.; Chen, Y.; Liu, C.; Wan, R. Effect of different industrial activities on soil heavy metal pollution, ecological risk, and health risk. *Environ. Monit. Assess.* **2021**, *193*, 20. [CrossRef]
44. Zhang, L.; Liu, J. In situ relationships between spatial–temporal variations in potential ecological risk indexes for metals and the short-term effects on periphyton in a macrophyte-dominated lake: A comparison of structural and functional metrics. *Ecotoxicology* **2014**, *23*, 553–566. [CrossRef]
45. Ferreira, B.L.; De Miguel, E. Geochemistry and risk assessment of street dust in Luanda, Angola: A tropical urban environment. *Atmos. Environ.* **2005**, *39*, 4501–4512. [CrossRef]
46. Li, Y.W.; Wang, J.; Ju, T.Z.; Wang, L.N.; Zhang, S.N.; Zha, X.H. Heavy metal pollution characteristics and human health risk assessment in soils from different functional areas of Baiyin, Gansu, China. *Chin. J. Ecol.* **2017**, *36*, 1408–1418. [CrossRef]
47. Ministry of Environmental Protection. *Chinese Population Exposure Parameter Manual. Adult Volume*; China Environmental Science Press: Beijing, China, 2013.
48. US EPA. *Risk Assessment Guidance for Superfund Volume I Human Health Evaluation Manual (Part A)*; Office of Emergency and Remedial Response: Washington DC, USA, 1989.
49. US EPA. *Risk Assessment Guidance for Superfund Volume I: Human Health Evaluation Manual (Part E, Supplemental Guidance for Dermal Risk Assessment) Final*; Office of Emergency and Remedial Response: Washington DC, USA, 2004.
50. US EPA. *Exposure Factors Handbook: 2011 Edition*; Office of Health and Environmental Assessment: Washington, DC, USA, 2011; [EPA/600/R-09/052F].
51. Duan, X.L. *Exposure Factors Handbook of Chinese Population*; China Environmental Science Press: Beijing, China, 2013.
52. Yu, L.; Zhang, F.; Zang, K.; He, L.; Wan, F.; Liu, H.; Zhang, X.; Shi, Z. Potential Ecological Risk Assessment of Heavy Metals in Cultivated Land Based on Soil Geochemical Zoning: Yishui County, North China Case Study. *Water* **2021**, *13*, 3322. [CrossRef]
53. Wilding, L.P. *Spatial Variability: Its Documentation, Accommodation and Implication to Soil Surveys*; PUDOC: Wageningen, The Netherlands, 1985; pp. 166–194.
54. Ju, T.N.; Wu, X.; Shi, H.D.; Gao, F.J.; Li, X.Z.; Wang, Y.R.; Luan, T.; Fan, P. Heavy metal pollution and ecological risk assessment of arable land soil in Haigou small watershed. *J. Environ. Eng. Technol.* **2018**, *8*, 556–562. [CrossRef]
55. Sun, G.J.; Li, W.H.; Zhu, C.G. Spatial variation analysis of topsoil bulk density in the Yili Valley, Xin jiang. *Resour. Sci.* **2016**, *38*, 1222–1228. [CrossRef]
56. Barnett, V. *Environmental Statistics: Methods and Applications*; Wiley: Hoboken, NJ, USA, 2005.
57. Li, Q.; Cao, Y.; He, L.S.; Wang, Y.F.; Gong, C.; He, S.H. Spatial distribution characteristics and source analysis of soil heavy metal at typical smelting industry sites. *Environ. Sci.* **2021**, *42*, 5930–5937. [CrossRef]

58. Qu, M.K.; Li, W.D.; Zhang, C.R.; Wang, S.Q.; Yang, Y.; He, L.Y. Source apportionment of heavy metals in soils using multivariate statistics and geostatistics. *Pedosphere* **2013**, *23*, 437–444. [CrossRef]
59. Ullah, R.; Khan, N.; Ali, K.; Khan, M.E.H.; Jones, D.A. Screening of Xanthium strumarium (IAPS) Growing on Abandoned Habitats in Khyber Pakhtunkhwa, Pakistan: Perspectives for Phytoremediation. *Appl. Sci.* **2021**, *11*, 11704. [CrossRef]
60. Tian, H.Z.; Xue, Z.G.; Cheng, K.; Qu, Y.P.; Chai, F.H.; Hao, J.M. Trend and characteristics of atmospheric emissions of Hg, As and Se from coal combustion in China,1980–2007. *Atmos. Chem. Phys.* **2010**, *10*, 11905–11919. [CrossRef]
61. Hsu, S.L.; Hung, J.; Wallace, A. Soil pH Variation Within a Soil. I. pH Variation in Soil Pores Observed in a Column-Leaching Method. *Commun. Soil Sci. Plant Anal.* **2004**, *35*, 319–329. [CrossRef]
62. Müller, H.S.; Dechow, R.; Flessa, H. Inventory and assessment of pH in cropland and grassland soils in Germany. *J. Plant Nutr. Soil Sci.* **2022**, *185*, 145–158. [CrossRef]
63. Chi, Q.H.; Ma, S.M. Comparison between chemical composition of bedrocks in the upper reaches and that of alluvial plain soils in the lower reaches of a drainage area. *Geol. Bull. China* **2008**, *27*, 188–195. [CrossRef]
64. Nikos, N.; José Antonio, R.M. Multiscale analysis of heavy metal contents in soils: Spatial variability in the Duero River basin (Spain). *Geoderma* **2012**, *189*, 554–562. [CrossRef]
65. Khan, N.; Khan, J.; Ullah, R.; Ali, K.; Jones, D.A.; Khan, M.E.H. Heavy Metals Contaminants in Watercress (Nasturtium officinale R.BR.): Toxicity and Risk Assessment for Humans along the Swat River Basin, Khyber Pakhtunkhwa, Pakistan. *Sustainability* **2022**, *14*, 4690. [CrossRef]
66. Murgueytio, A.M.; Evans, R.G.; Sterling, D.A.; Clardy, S.A.; Shadel, B.N.; Clements, B.W. Relationship between lead miningand blood lead levels in children. *Arch. Environ. Health* **1998**, *53*, 414–423. [CrossRef] [PubMed]
67. Yang, Q.Q.; Li, Z.Y.; Lu, X.N.; Lu, X.M.; Duan, Q.N.; Huang, L.; Bi, J. A review of soil heavy metal pollution from industrial and agricultural regions in China: Pollution and risk assessment. *Sci. Total Environ.* **2018**, *642*, 690–700. [CrossRef]
68. Liu, H.; Zhang, Y.; Yang, J.; Wang, H.; Hu, W. Quantitative source apportionment, risk assessment and distribution of heavy metals in agricultural soils from southern Shandong Peninsula of China. *Sci. Total Environ.* **2021**, *767*, 144879. [CrossRef] [PubMed]
69. Hua, C.; Zhuo, H.M.; Kang, A.L.; Fang, Z.T.; Zhu, M.Y.; Dong, M.X.; Wang, J.C.; Ren, L. Contamination, risk assessment and source apportionment of the heavy metals in the soils of apple orchard in Qixia City, Shandong Province, China. *Stoch. Environ. Res. Risk Assess.* **2022**, *36*, 2581–2595. [CrossRef]
70. Yu, F.; Zhang, Y.W.; Yan, M.S.; Wang, R.; Zhang, F.L.; Zhong, K.Q.; Zhu, H.S.; Luo, K. Heavy metal pollution and human health risks assessment of soil and crops near the mercury ore in Chongqing. *Environ. Chem.* **2022**, *41*, 536–548. [CrossRef]

MDPI AG
Grosspeteranlage 5
4052 Basel
Switzerland
Tel.: +41 61 683 77 34

Agriculture Editorial Office
E-mail: agriculture@mdpi.com
www.mdpi.com/journal/agriculture

www.ingramcontent.com/pod-product-compliance
Lightning Source LLC
Chambersburg PA
CBHW061800210926
51850CB00006B/603